Systemic fungicides

Systemic fungicides

Second edition

Edited by R. W. Marsh, O.B.E.
with the assistance of R. J. W. Byrde
and D. Woodcock

Longman
London and New York

Longman Group Limited London

Associated companies, branches and representatives throughout the world

Published in the United States of America by Longman Inc., New York

First published 1972
Second edition 1977

Library of Congress Cataloging in Publication Data
Marsh, R. W.
Systemic fungicides.

Bibliography: p.
Includes index.
1. Systemic fungicides. I. Byrde, Robert Jocelyn Walter, joint author. II. Woodcock, David, joint author. III. Title.
SB951.3.M36 1977 632′.952 76-49542
ISBN 0-582-44167-6

Printed in Great Britain by Butler & Tanner Ltd, Frome and London

Contents

Preface to the first edition ix
Preface to the second edition xi
Acknowledgements xiii
List of contributors xiv

R. L. Wain and G. A. Carter
1 Nomenclature and definitions 1
Mode of action of systemic compounds

R. L. Wain and G. A. Carter
2 Historical aspects 6
Early investigations Development from mid-1930s The assessment of therapeutic and systemic activity The search for new chemotherapeutants Conclusion

D. Woodcock
3 Structure–activity relationships 32
Organophosphorus compounds Antibiotics Carboxylic acid anilides Heterocyclic compounds Aliphatic and alicyclic compounds Organotin compounds Aromatic compounds

D. Woodcock
4 Toxicological considerations 85
Organophosphorus compounds Antibiotics Carboxylic acid anilides Heterocyclic compounds Aromatic compounds Residues

S. H. Crowdy
5 Translocation 92
Entry into the free space within the tissues Apoplastic movement Symplastic movement Translocation of systemic fungicides Conclusion

A. E. Dimond and S. Rich

6 Effects on physiology of the host and on host/pathogen interactions 115

Introduction Effects of systemic fungicides on the host Production of fungitoxicants in plants Stomatal closure as a physical barrier Systemic compounds that alter growth of the host Treatments that block pathogenic processes Future prospects

A. Kaars Sijpesteijn

7 Effects on fungal pathogens 131

Introduction Interference with respiration Interference with protein synthesis Interference with nucleic acid synthesis and nuclear and cell division Interference with synthesis and function of cell walls and membranes Mode of action still uncertain General remarks on mode of action, selectivity and resistance

J. W. Vonk and A. Kaars Sijpesteijn

8 Metabolism 160

Introduction Benzimidazoles and related compounds Oxathiins and related compounds Organophosphorus esters Chloroneb Pyrimidines Tridemorph Triforine Prothiocarb

J. Dekker

9 Resistance 176

Introduction Occurrence of acquired resistance Origin of resistance Mechanism of resistance Mutation for resistance *in vitro* Emergence of resistance in the field Manipulation of fungicide resistance Conclusions

E. Evans

10 Efficient use of systemic fungicides 198

Economic considerations Biological activity and systemicity Seed treatment Soil treatment Leaf and stem treatments Post-harvest treatments Timing systemic fungicide treatments Future strategy for chemotherapy

D. H. Brooks and N. G. Buckley

11 Results in practice – I. Cereals and grasses 213

Introduction Barley and wheat Rice Maize Grasses and turf

D. M. Spencer

11 Results in practice – II. Glasshouse crops 240

Introduction Disease control in the glasshouse Use of selected systemic fungicides Methods of application in glasshouses Tolerance Non-fungicidal effects of benzimidazole compounds Biological control Present trends: future needs

R. B. Maude

11 Results in practice – III. Vegetable crops 259

Introduction Seed-borne diseases Soil-borne diseases Leaf diseases Post-harvest diseases

R. J. W. Byrde and V. W. L. Jordan

11 Results in practice – IV. Fruit crops 274

Introduction Temperate fruit Sub-tropical fruit Tropical fruit Conclusions

E. B. Smalley

11 Results in practice – V. Forestry 294

Introduction Root diseases Foliage diseases Stem diseases and trunk cankers Vascular wilt diseases (other than Dutch elm disease) Dutch elm disease

References 320

Alphabetical list of systemic fungicides in commercial use 389

Index 391

Preface to the first edition

Critics will doubtless say that this book was out of date before it was published, thus confirming the rapid progress now being made in the study of systemic fungicides. This accelerated progress started around 1968 and it will be noted that of the 1 174 references 591 are papers published since 1967.

Because research in this field is developing so rapidly we felt that this was not the time to produce a definitive account of systemic fungicides. Instead we have aimed at making a relatively brief survey of the position now reached, including an outline of the basic principles of systemic fungicidal action, a summing-up of the present state of knowledge in the various branches of the subject, examples of practical applications and a look at possible future developments.

Experimental work on systemic fungicides is now worldwide, but the major centres of research have been in the Netherlands, England and the United States. We have been specially fortunate in securing the authoritative contributions of Professor Dekker and Dr Kaars Sijpesteijn from the Netherlands, of Professor Wain, Professor Crowdy, Dr Woodcock, Dr Byrde, Dr Brooks, Dr Spencer, Dr Carter, Mr Maude and Dr Evans from this country and of Dr Dimond from USA. Dr Dimond's death on 4 February 1972 was a serious loss to plant pathology: his contribution to *Systemic Fungicides* was his last completed manuscript.

This book is aimed at research workers in the field of fungicides, manufacturers of crop protection products, students of agriculture and horticulture and all concerned practically with fungicides and their use.

R. W. Marsh

Long Ashton Research Station, 1972

Preface to the second edition

In 1972, when the first edition of this book appeared, the commercial use of systemic fungicides still retained an element of novelty. In the ensuing five years many new systemics have come on to the market, the range of crops treated has been much expanded and the distribution of systemic fungicide treatments has become worldwide. These developments are reflected in the changes made in the second edition.

An entirely new section on the use of systemic fungicides in forestry has been added to Chapter 11 by Dr Smalley, the leading US worker in this field. This provides for the first time a detailed account of the application of systemics to trees, in particular for combating Dutch elm disease. In addition, Dr Smalley gives a comprehensive survey of the uses of systemic fungicides on other diseases of mature trees, in forest nurseries and for Christmas-tree production.

The noteworthy additions to Chapter 3 (Dr Woodcock) include an account of the continued interest in substituted benzimidazoles and triazoles in the search for systemic fungicidal activity. Some triazinobenzimidazoles containing long alkyl substituents (from Hoechst) are effective against a range of Phycomycetes – the major group of fungi that has hitherto been unassailable by systemic treatments. Substitution in the 1- position of 1,2,4-triazoles has led to exciting new fungicides (from Bayer), one of which, triadimefon, possesses not only protectant properties but also eradicant and systemic activity against a wide range of plant pathogens, especially powdery mildews.

Dr Woodcock's Chapter 4 gives additional information on mammalian toxicity and includes new material on residues, tolerances and methods of analysis.

In Chapter 5 Professor Crowdy records, among other recent findings, instances of translocation of systemic fungicides downwards from leaves to roots. Chapter 6, on effects on physiology of the host and on host/pathogen interaction, written for the first edition by the late Dr Dimond, has been brought up to date by his colleague, Dr Saul Rich. Dr Kaars Sijpesteijn's Chapter 7 on effects on fungal pathogens now includes considerations of interference with protein synthesis, with nucleic acid synthesis and with cell division. This chapter is supplemented by a new contribution by Dr Vonk and Dr Kaars Sijpesteijn on metabolism of systemic fungicides, detailing their various breakdown processes.

The widespread emergence of pathogen isolates resistant to several systemic fungicides has given much cause for concern. Dr Dekker, in Chapter 9, provides a timely and comprehensive account of the present position in this rapidly-changing field, with special reference to the genetic basis of fungal resistance.

Dr Evans' chapter on commercial methods of application has been expanded to cover economic aspects and a future strategy for chemotherapy. His death a few weeks after the completion of this chapter has deprived crop protection of one of its leading exponents in Britain.

In Chapter 11 (Results in practice) the section on cereals by Dr Brooks now has an additional author, Dr Buckley, and has been extended to include grasses and turf. Dr Spencer's survey of results on glasshouse crops has new material on flowers (carnations, chrysanthemums, roses), pot plants and other ornamentals, and mushrooms. In the section on results on vegetables, Mr Maude includes a reappraisal of methods of application of systemics for the control of root infections, in particular, club root. He also refers to a new development in the treatment of virus diseases – soil applications of carbendazim preventing symptom expression of beet western yellows virus in lettuce and of tobacco mosaic virus in leaves of tobacco. Results on fruit, compiled by Dr Byrde with the help of Dr Jordan, demonstrate the wider coverage of the use of systemics on tropical and sub-tropical crops.

Finally, a summary list of systemic fungicides now in commercial use, giving chemical, common and trade names, with names of manufacturers, has been added immediately following the References, which now number 1818.

The authors thank the reviewers of the earlier edition, who combined generous appreciation with helpful criticism from which the present edition has benefited.

R. W. Marsh
D. Woodcock
R. J. W. Byrde

Long Ashton Research Station
February 1977

Acknowledgements

We are grateful to the following for permission to reproduce copyright material:

Academic Press Inc. for material from *Environmental Toxicology of Pesticides* (Eds. Matsumura *et al.*) by Noguchi (1972) and for material from *Pesticide Biochemistry and Physiology Journal* **5** by White and Thorn (1975); Society of Chemical Industry and authors for Table I and Table II on p. 44 of *Pesticide Science* **2** by Jank & Grossman (1971); British Crop Protection Council for Tables 4 and 5 on pp. 460 and 461 of *Proc. 6th Br. Insectic. Fungic. Conf.* **2** by Ten Haken & Dunn (1971) and a table from *Proc. 8th Br. Insectic. Fungic. Conf.* **2** by Finney *et al.* (1975); Cambridge University Press for Tables 3 and 4 on p. 469 of *Ann. appl. Biol.* **57** by Pluijers & Kaars Sijpesteijn (1966) and Table 3 on p. 479 of *Ann. appl. Biol.* **61** by Kaars Sijpesteijn *et al.* (1968); Gordon and Breach for Tables 4 and 5 of the *Proc. of the 2nd Inter. cong. of Pest. Chem.* (1971) by Clifford *et al.*; Koninklijke Nederlandse Chemische Vereniging and authors for part of Table V on p. 816 of *Rec. Trav. Chim.* **79,** 807–822 (1960) by Van den Bos *et al.*; Nederlandse Planteziektenkundige Vereniging for Tables 3 and 4 on p. 136 and Table 5 on p. 137 of *Neth. J. Pl. Path.* **74** by Tempel *et al.* (1968); Springer Verlag for Table 1 on p. 95 of *Residue Reviews* **25** by Misato (1969), Table V on p. 103 of *Residue Reviews* **25** by Misato (1969), Table 1 in column 8 of *U.S. 3,249,499* by von Schmeling *et al.* (1966) and Table IV in column 10 of *U.S. 3,249,499* by von Schmeling *et al.* (1966); Faculteit Landbouwwetenschappen, University of Ghent, for Tables 1 and 2 on p. 737 of *Meded. Rijksfaculteit Landbouw-weten-schaffen*, Gent **32** by Pommer & Kradel (1967) also Table 3 on p. 738 and Table 2 on p. 1219 of *Meded. Landbhoogesch Opzoek Stns, Gent* **27** by Dekker (1962); the U.S. Government Printing Office for a table from the *E.P.A. Compendium of Registered Pesticides* **II** (1973).

We have been unable to trace the following authors and would appreciate any information that would enable us to do so:

The authors for Table 1 on p. 701 of *Agr. Biol. Chem.* **34** by Maeda *et al.* (1970).

For permission to redraw diagrams we are grateful to Academic Press Inc., to Annual Reviews Inc., and to Blackwell Scientific Publications Ltd.

List of contributors

D. H. Brooks Ph.D. Products Acquisition and Licensing Manager, ICI Ltd, Plant Protection Division, Fernhurst, Haslemere, Surrey

N. G. Buckley Ph.D. Senior Technical Representative, Agricultural Chemicals Division, ICI United States Inc.

R. J. W. Byrde B.Sc., Ph.D. Head of Plant Pathology Section, Long Ashton Research Station; Reader, University of Bristol

G. A. Carter Ph.D. Senior Scientific Officer (Plant Pathologist), Agricultural Research Council Growth Substances and Systemic Fungicides Unit, Wye College (University of London), nr Ashford, Kent

S. H. Crowdy B.Sc., Ph.D., A.R.C.S. Professor of Botany, University of Southampton

A. E. Dimond (the late) Ph.D. Chief, Department of Plant Pathology and Botany, Agricultural Experiment Station, New Haven, Connecticut, USA

J. Dekker Ph.D. Professor; Director of Laboratorium voor Fytopathologie, Wageningen, The Netherlands

E. Evans (the late) Ph.D., D.Sc. Senior Plant Pathologist, Fisons Pest Control Ltd, Chesterford Park Research Station, nr Saffron Walden, Essex

V. W. L. Jordan Ph.D. Research Associate in Plant Pathology, Long Ashton Research Station, University of Bristol

A. Kaars Sijpesteijn Ph.D. Head of Department of Biochemistry and Microbiology, Institute for Organic Chemistry of the Organisation for Applied Scientific Research in the Netherlands, TNO, Utrecht

R. B. Maude B.A., M.Sc. Plant Pathologist, National Vegetable Research Station, Wellesbourne, Warwick

Saul Rich Ph.D. Chief, Department of Plant Pathology and Botany, Agricultural Experiment Station, New Haven, Connecticut, U.S.A

E. B. Smalley Ph.D. Professor of Plant Pathology and Forestry, University of Wisconsin, Madison, Wisconsin, USA

D. M. Spencer Ph.D. Head of Mycology and Bacteriology Department, Glasshouse Crops Research Institute, Rustington, Littlehampton, Sussex

J. W. Vonk Ph.D. Chemist of the Department of Biochemistry and Microbiology, Institute for Organic Chemistry of the Organisation for Applied Scientific Research in the Netherlands, TNO, Utrecht

R. L. Wain C.B.E., D.Sc., F.R.I.C., F.R.S. Professor of Agricultural Chemistry, Wye College, University of London; Honorary Director, Agricultural Research Council Growth Substances and Systemic Fungicides Unit

D. Woodcock M.Sc., Ph.D., D.Sc., F.R.I.C. Head of Organic Chemistry Section, Long Ashton Research Station; Reader, University of Bristol

Nomenclature and definitions

R. L. Wain and G. A. Carter

Types of fungicide

A fungicide may be broadly defined as a compound which kills or inhibits the growth of fungi although compounds which inhibit without killing are more accurately termed fungistats. Those fungicides used to control fungal diseases of plants are frequently divided into three types as follows:

Protective – these provide protection against infection at the site of application.

Eradicant – these cure an established infection at the site of application.

Systemic – these can prevent the development of disease on regions of the plant away from the site of application.

Clearly the three groups are distinguished on the basis of time of application relative to infection and uptake by, and mobility within, plant tissues.

Prevention versus cure

There are two general approaches to the control of infectious plant disease: prevention and cure. **Prevention** or **prophylaxis** seeks to prevent infection by inhibiting the pathogen before it has penetrated the host and become established within its tissues. Compounds capable of achieving this are called **protective fungicides** or **protectants**; most fungicides are of this type. **Cure** or **therapy** seeks to eliminate a pathogen after it is already established within the tissues of its host. Compounds capable of achieving this are called **chemotherapeutants** or **curative fungicides**.

It may not always be easy to determine precisely when a plant has become infected, but penetration of the host cuticle is a useful criterion for practical purposes.

Movement within plant tissues

Apart from the injection of fungicides into the stems of trees, field applications of fungicides are made to plant surfaces as seed dressings, foliar sprays or soil drenches around the roots. The great majority of fungicides when applied in this way do not penetrate the underlying tissues in effective quantities. These compounds cannot cure an established internal infection nor be carried within the plant to untreated areas, although some surface redistribution can occur through volatility or diffusion in water films. In general these fungicides can be used only as protectants although they may act as chemotherapeutants against those fungi, e.g. powdery mildews (*Erysiphaceae*) which, after penetrating their host, develop externally upon the surface of the plant.

A few fungicides can penetrate into underlying tissues in effective quantities and these are potential therapeutants for a much wider range of diseases. If, following penetration, effective amounts are translocated within the plant, defining **translocation** as the movement of a compound within the plant body to tissues remote from the site of application, then they are also **systemic**.

The relationship between mobility within the plant and potential use in disease control is summarised below:

Degree of mobility	*Potential use as fungicide*
Not absorbed	Protectant Therapeutant for surface pathogens
Absorbed – not translocated	*Topical chemotherapeutant or eradicant*
Absorbed – translocated	Systemic

Advantages of systemic fungicides

When non-systemic compounds are applied to a plant, areas missed in the initial application remain unprotected, as does any subsequent growth. Furthermore, the surface deposits are exposed to weathering. Repeated applications are therefore necessary to ensure continued protection against disease. A further limitation of non-systemic compounds is that it is difficult to use them for the protection of inaccessible organs such as roots.

Since they are able to move freely from applications made at readily-available sites to distant, untreated tissues, systemic compounds suffer none of these disadvantages. It may be possible to delay the application of systemics until disease symptoms are apparent, thereby eliminating the need for routine and often wasteful pre-symptom applications. One application of a systemic compound cannot, however, be expected to give continuous protection to the whole plant because translocation may be limited to certain tissues, and the compound will be subjected to dilution and possibly metabolic breakdown as the plant grows (see Chapters 5 and 7). The main disadvantages of systemics have been their narrow spectrum of activity and the rapidity with which fungi develop resistance towards them (see Chapter 9).

Site of action of systemics

Although applications of systemic fungicides are usually described as 'protecting' plant tissues from infection their action is often therapeutic in that it occurs within the host plant after the pathogen has penetrated.

Systemic protective activity *sensu stricto* occurs when, for example, a compound applied to roots is translocated to the leaves and exuded on to the leaf surface there inhibiting the fungus prior to penetration of the host. Both types of systemic activity have been observed.

Thus Bent (1970) found that root applications of dimethirimol and ethirimol affected germ-tube emergence and appressorium formation in powdery mildew conidia dusted upon the leaf surface of treated plants. By contrast, examination of wheat leaves taken from plants treated with three other systemic anti-mildew compounds showed that inhibition of mildew development became apparent only after penetration had commenced (Dekker and van der Hoek-Scheuer, 1964).

Mode of translocation in relation to systemic activity

The physiological aspects of translocation are discussed in Chapter 5.

Eradication of an infection at the site of application and limited local movement, such as the protection of the lower surface of the leaf by sprays applied to the upper, may be achieved by diffusion in the free space of the treated tissues. Whether or not more extensive movement, for example from roots to shoots, occurs depends upon the balance between movement in the conducting tissues such as phloem and xylem and any tendency to local adsorption, accumulation behind permeability barriers or preferential partitioning into an immobile phase.

Movement in the xylem will be governed by the flow of the transpiration stream producing an acropetal movement and accumulation in those leaves having the greatest rate of transpiration. This is the type of movement usually observed in most systemic fungicides. It allows the systemic protection of shoot tissues following root application, but foliar application gives only limited local protection. To obtain greater movement from foliar sprays transport in the phloem is required, thus allowing export from mature leaves to the stem and subsequent transportation to root and shoot apices. This type of movement has been difficult to achieve.

A high degree of mobility within the host plant is not always necessary for disease control; this property is more important, for example, when applying a root treatment against a leaf-attacking pathogen than when trying to control a foot-rot or vascular wilt. Chapman (1951) showed that the effectiveness of various compounds applied through the roots differed against the twig-invading Dutch elm disease pathogen and the root-invading Fusarium wilt pathogen of tomato; the differences correlated with the mobility of these compounds in cellulose.

Mode of action of systemic compounds

A systemic compound can modify, and be modified by, the plant tissues with which it comes into contact. It does not necessarily, therefore, reduce disease by direct toxic action against the pathogen and need not itself be fungicidal. The mechanisms of activity that have been postulated are very varied and it is possible for a single compound to have more than one mode of action. Disease can be reduced without rendering the plant tissues any more resistant to attack as, for example, when the growth of the treated host is modified in such a way as to shorten the susceptible phase of development (Kiraly *et al.*, 1962).

Direct activity

This is discussed in detail in Chapter 7. This operates when a fungitoxic chemical accumulates to an effective level at the infection site. The compound actually applied need not itself be fungitoxic *in vitro* as it might be converted to an active compound by metabolism within the plant.

Many compounds that are highly fungitoxic *in vitro* fail to show systemic activity. The requirements for such activity are very exacting, for not only must the compound be antifungal within living plant tissues, a very different environment from that of the standard laboratory tests for *in vitro* fungitoxicity, but it must be selective, damaging the pathogen whilst being innocuous to the host plant.

As has been shown for rimocidin and pimaricin (Oort and Dekker, 1960), certain fungitoxicants may be translocated within the plant and can be detected in tissue extracts by bioassay but yet show no activity against disease. Antagonism by cell constituents (Gottlieb, 1957) and irreversible binding to adsorption sites can render a powerful *in vitro* fungitoxicant inactive *in vivo.*

Achieving the necessary degree of selectivity in toxic action between host and pathogen has proved a major difficulty in the development of systemic fungicides, presumably because of the basic similarity in the biochemistry of fungi and higher plants. It seems likely, as suggested by Kaars Sijpesteijn (1970), that the mechanism of toxicity of a fungicide at the biochemical level determines whether or not it has potential as a systemic.

To achieve a useful degree of systemic activity the active molecule must be reasonably stable within the plant tissues but inevitably most chemotherapeutants will be metabolised. Where the metabolites are inactive this will result in a steady loss of effectiveness with time. This subject is dealt with in Chapter 7 and has also been reviewed by Kaars Sijpestcijn and van der Kerk (1965), Wain and Carter (1967) and Kaars Sijpesteijn (1969).

Indirect activity

It has frequently been postulated that a systemic compound reduces disease indirectly by some mechanism other than direct toxic action against the pathogen. Widely differing mechanisms by which increased resistance to disease is brought about have been suggested (see Chapter 6 and Dimond, 1963a, b; 1965; Grossmann, 1968b) including inactivation of fungal toxins and hydrolytic enzymes, and alterations to the pectins, carbohydrate levels, phenolics and wood morphology of the host plant.

Clearly there is great scope for further investigations into the mode of action of systemics with an indirect effect upon disease development but, as yet, most of the commercially successful systemics seem to work directly against the pathogen. Since those working indirectly do not kill the invading pathogen they tend to suppress symptoms or delay the development of disease rather than effect a cure.

Systemic fungicides versus systemic chemotherapeutants

Strictly a **systemic fungicide** is a compound that is both fungicidal *in vitro* and also shows systemic activity against disease. Frequently, however, the term is used in a wider sense to include all compounds with systemic activity against fungal diseases whether or not they are fungitoxic and work directly against the pathogen *in vivo.* This usage has been criticised (see, e.g., Horsfall, 1956) on the grounds that a non-fungitoxic, indirect-action systemic is not a fungicide. The alternative term of *systemic chemotherapeutant* is preferred by many when referring to all systemic compounds regardless of their mode of action. However, this terminology is open to criticism also since certain of the compounds may well be acting as systemic protectants rather than as therapeutants. Perhaps the best 'umbrella term' for all such compounds is *systemic antifungals* or merely *systemics.*

Although a classification of fungicides in terms of where, when and how they act in controlling disease is both feasible and useful, uptake, translocation and mode of action can be influenced by many factors, as will be apparent in later chapters. Therefore such a system of classification must inevitably be less rigid and precise than one based upon the chemical structure of the antifungal molecule. A successful systemic therapeutant for one disease frequently fails when a different host or pathogen is involved.

Chapter 2

Historical aspects

R. L. Wain and G. A. Carter

Early investigations

Since this topic has been discussed in the reviews of Müller (1926), Roach (1939), Stoddard and Dimond (1949) and Horsfall (1945, 1956), only a brief outline will be given here.

The idea of introducing substances into plants goes back at least to the twelfth century when solid materials such as spices, medicines and colouring matters were inserted under the bark or into a borehole in a tree with the aim of imparting new odours, tastes or colours to the developing fruit. Such experiments were carried out spasmodically for several hundred years, and in the fifteenth century Leonardo da Vinci describes the use of injections of arsenic into fruit trees to render the fruit poisonous. In 1602 an anonymous writer describes methods for improving the flavour and colour of fruits by inserting spices and dyestuffs into the pith; he also mentions the idea of killing 'wormes', i.e. larvae of wood-boring beetles, by inserting a mixture of pepper, laurel, incense and wine into boreholes. Over 150 years later Wilson records the use of liquid mercury inserted into boreholes in branches of trees to combat insect pests. This early work was, inevitably, empirical – there was at that time little knowledge of the uptake and movement of compounds in plants, although pioneer work on movement of sap was performed by Hales in 1726.

During the eighteenth century, when the study of plant physiology developed rapidly, studies were made on the movement of injected dyestuffs and mineral salts. These experiments revealed that certain compounds could be transported readily within the plant, but great differences in the extent of uptake and translocation were observed. A further impetus to these studies was provided by the work of Liebig and other agricultural chemists in their investigations on the mineral nutrition of plants which led to the recognition of nutrient deficiency diseases. In attempts to cure such diseases, and especially chlorosis resulting from iron deficiency, injections of mineral salts were made. This work was pursued by workers in France, Germany, Russia and the USA, culminating in the studies in this country by Roach (1934, 1938, 1939).

Mokrzecki (1903) observed that certain nutrient solutions when injected appeared to alleviate attacks by pests and diseases. Slightly earlier Ray (1901) had injected various liquids, some of biological origin, into leaves via capillary tubes with the specific intention of rendering the leaves immune from disease. In 1906 Bolley reported that injections of copper sulphate, iron sulphate and formaldehyde reduced attacks of *Taphrina* on fruit trees, and ten years later Norton reduced the incidence of *Septoria lycopersici* on tomato plants by injecting copper sulphate solution.

At the turn of the century in Russia and Italy, poisonous substances, particularly potassium cyanide, were injected into plants in the hope of combating insect pests. Work with cyanide was continued in the USA where its movement and persistence in the plant were studied in some detail.

Early this century Brooks and his co-workers began investigations on the silver leaf disease of plum incited by *Stereum purpureum*, and Brooks and Bailey (1919) found that several dyes and disinfectants, when injected into diseased trees, allowed some recovery. Later Brooks and Storey (1923) reported the very high *in vitro* fungitoxicity of 8-quinolinol sulphate towards *S. purpureum*, and this compound also proved effective against silver leaf when injected into the host plant (Roach, 1939).

In the USA Rumbold (1920a, b) made extensive studies on the injection of sweet chestnut trees as a method of curing blight, caused by *Endothia parasitica*. She found that injections of lithium salts could check disease development but the effect was only transient. Later Scherer (1927) and Jacobs (1928) reported the efficacy of injecting a solution of thymol for the control of both fungal and bacterial pathogens.

Throughout most of the early studies on therapy of disease, injection directly into the host plant was the preferred method of application. However, attempts were made to eradicate established infections with surface-applied compounds; Viala (1893) reported that copper sulphate had an eradicant action against anthracnose (*Elsinöe ampelina*) of the vine, and Bolley (1891) controlled scab (*Streptomyces scabies*) on potato tubers with mercuric chloride. Later it was shown that zinc chloride could effectively destroy cankers of fireblight (*Erwinia amylovora*) on pear trees (Day, 1928; McClintock, 1931).

Attempts, too, were made to eradicate pathogens, such as the loose smuts and certain *Helminthosporium* species, carried internally within seeds. Geuther (1895) reported the effectiveness of formaldehyde, Hiltner (1915) of mercuric chloride and Riehm (1914) of organomercurials; during the early 1900s formaldehyde was used extensively in the treatment of cereal seed.

Root applications for the control of foliar pathogens were also employed by some early workers. In 1903 Massee applied copper sulphate to the roots of cucumber plants and reduced the incidence of *Cercospora melonis*, whilst in 1913 Spinks found that the application of salts of lithium to wheat and barley growing in soil or water culture effectively prevented the development of powdery mildew (*Erysiphe graminis*).

A landmark in the study of chemotherapy of plant disease was the publication in 1926 of Müller's monograph *Die innere Therapie der Pflanzen*. In it he summarised much earlier work and described his own injection experiments designed to control insect pests and fungal pathogens. However, despite the publication of this work and the scattered reports of successful chemotherapy of plant disease, little interest in the subject was aroused at that time. Disease control was based on the use of successful protectant fungicides, such as Bordeaux mixture and lime sulphur, and the development of resistant varieties by selection and breeding. The problem of selectively inhibiting one organism within the living tissues of another was thought to be insuperable.

Development from mid-1930s

Interest in chemotherapy was revived in the mid-1930s and it developed rapidly during the next decade. There were several reasons for this. Firstly, there were limitations both in the use of protectant fungicides, which offered little hope of curing infected plants or of controlling vascular diseases, soil-borne pathogens and viruses, and in the use of resistant varieties which frequently lost their resistance rapidly as new races of the pathogen arose. Secondly, many new organic chemicals were now available and these offered more scope than the inorganic salts used by most early workers. Thirdly, the chemotherapy of human diseases, pioneered by Ehrlich, had now been shown to be both possible and strikingly effective.

In 1935 Dogmagk showed that 'Prontosil' could cure bacterial infections in the bloodstream and in the same year Trefouël showed that the active compound, produced from Prontosil within the human body, was *p*-aminobenzene sulphonamide. This chemical became the parent of a whole range of sulphonamide or 'sulpha' drugs. Three years later Hassebrauk protected wheat seedlings from attack by rust fungi by treating them at the roots with *p*-aminobenzene sulphonamide.

A second breakthrough in medical chemotherapy came with the dawn of the 'antibiotic era' when in 1940 Florey and Chain isolated penicillin, which proved outstanding in controlling certain bacterial infections in man. A search for further antibiotics commenced, and chloramphenicol, aureomycin and streptomycin were soon discovered. In 1944 Brown and Boyle achieved a cure of crown gall on several species with applications of crude penicillin, and in 1947 Anderson and Nienow showed streptomycin to be translocated in plants.

The Second World War aided the development of potential chemotherapeutants by stimulating the commercial production of antibiotics. It also stimulated Schrader's work on organic phosphates as poison gases; subsequently these compounds, some of which are freely mobile within plants and non-phytotoxic, were developed into highly successful systemic insecticides.

The development of chemotherapy continued to expand and many hundreds of compounds were examined for therapeutic activity. Progress has been reviewed by many workers, including those listed below:

1946 Zentmyer *et al.*
1949 Dimond *et al.*; Stoddard and Dimond
1951 Horsfall and Dimond (a, b); Horsfall *et al.*
1952 Anderson and Gottlieb; Brian (a, b); Dimond *et al.*; Sharvelle; Wain
1953 Dimond; Wain
1954 Brian; Leben and Keitt
1956 Horsfall; van der Kerk; Rudd Jones
1958 Byrde and Ainsworth; Ford *et al.*; Zaumeyer
1959 Dimond; Dimond and Horsfall; Goodman; Howard and Horsfall; Livingston and Hilborn; Wain
1960 Brian; Oort and van Andel.
1961 Cremlyn; Horsfall; Kaars Sijpesteijn (a, b); van der Kerk; Pridham; Wain *et al.*
1962 Dimond; Goodman (a, b); Rhodes
1963 Dekker (a); Dimond (a, b); Wain
1965 Dimond
1967 Goodman
1968 Grossmann (b)
1969 Byrde
1970 Byrde; Erwin; Fawcett and Spencer
1971 Dekker (a, b); Evans (a, b); van der Kerk (a, b); Oort; Pommer (b); Woodcock
1972 Kirby
1973 Cremlyn; Dekker; Erwin; Evans; van der Kerk

The assessment of therapeutic and systemic activity

As investigations into chemotherapy increased in number, varying methods were employed to detect therapeutic and systemic activity. Rapid laboratory and glasshouse tests were devised for screening large numbers of test compounds, different methods of application and different host/pathogen combinations being used. The choice of test organism was usually made either because of its suitability for routine use or because it was an important pathogen not readily controlled by protectant fungicides.

Seed treatment – Seed-borne pathogens

Studies on the eradication of internally-borne seed pathogens were continued, favoured test organisms being the *Helminthosporium* diseases of cereal seeds, which are carried in the outer tissues of the seed, and *Ascochyta pisi*, which is carried deeper in the tissues of pea seeds. The *Helminthosporium* species could be controlled with organomercurials partly because of their volatility (Leben and Arny, 1952; Arny and

Leben, 1954). The antibiotics antimycin A, helixin B (Leben and Arny, 1952; Leben *et al.*, 1953, 1954; Lockwood *et al.*, 1954) and mycothricin (Rangaswami, 1956) were also found to be effective.

Volatility may also assist in the eradication of *A. pisi* infections with chloropicrin (Kennedy, 1961) and pyridinethiol N-oxide (Kaars Sijpesteijn *et al.*, 1958; Dekker *et al.*, 1958), both of which are effective against this disease. Antibiotics such as GS-1 (Ark and Dekker, 1958), XG (=fungistatin) (Wallen and Skolko, 1950, 1951), cycloheximide (Wallen and Skolko, 1951), pimaricin and rimocidin (Dekker, 1955, 1957; Oort and Dekker, 1960) were also found effective against this pathogen. Another important group of internally-borne seed pathogens are the loose smuts and these were shown to be controlled by a treatment with 'Spergon' (tetrachloro-*p*-benzoquinone) (Porter, 1956) and streptomycin (Paulus and Starr, 1951).

Seed treatment – Seedling diseases

Some investigators have sown treated seed and looked for an enhanced disease resistance in the developing seedling. Bouchereau and Atkins (1950) reported the effectiveness of seed treatment with copper salts against *Ophiobolus miyabeanus* whilst Rodigin and Krasnova (1959) found copper salts effective against leaf rust of wheat when used as a seed dressing, as was boric acid and zinc sulphate. Treatment of rice seed with 2-methyl-1,4-naphthaquinone, 2,4-D, boric acid or copper salts was reported by Akai (1955) as effective in reducing the incidence of *O. miyabeanus*, and Volger (1959) found that treatment of the seeds of *Pinus sylvestris* with TMTD rendered the developing seedlings more resistant towards species of *Pythium* and *Rhizoctonia*. There is no evidence in these studies whether the applied chemical entered the seed and moved directly into the tissues of the seedling or whether it was leached from the seed and taken up by the seedling roots.

Root pathogens

As root-attacking pathogens are difficult to control with protective fungicides, many studies were made on the therapy of diseases caused by soil-borne species of *Fusarium*, *Phytophthora* and *Rhizoctonia*, using chemicals applied as a soil drench. Although some control was often obtained it was difficult to ascertain how much of the effect was due to chemotherapy and how much to direct contact action against the inoculum in the soil.

Stoddard (1952) treated sand-grown plants of *Matthiola* with 8-quinolinol sulphate, lifted them, washed the roots, and replanted them in infected soil. The plants remained healthy, suggesting that the compound was operating within the plant. Previously Stoddard (1951b) had used this technique to demonstrate the therapeutic action of disodium ethylenebisdithiocarbamate against red core (*Phytophthora fragariae*) of strawberry. Zentmyer and Gilpatrick (1960) root-treated avocado seedlings already infected with *Phytophthora cinnamomi* and obtained evidence of a curative effect with 'Dexon' (sodium *p*-dimethyl-amino-benzenediazo sulphonate).

For many crops the preferred method for controlling root diseases would be by the application of a foliar spray containing a chemical which could be absorbed and translocated down to the roots. Successful control, however, has rarely been achieved even under experimental conditions. Zentmyer (1954) sprayed avocado seedlings growing in soil infested with *P. cinnamomi* with 8-quinolinol benzoate and noted a decrease in wilting and an increase in new root growth.

Vascular pathogens

These pathogens remain deep within the plant tissues. They offer one of the greatest challenges to chemotherapy and much work has been done with them, in particular with Dutch elm disease (caused by *Ceratocystis ulmi*) and *Fusarium* wilt of tomato (caused by *Fusarium oxysporum* f. *lycopersici*).

In the early 1900s Rumbold attempted to control chestnut blight by injecting chemicals but with little success. Later Howard (1941) and Howard and Caroselli (1941), working at the Rhode Island Agricultural Experimental Station, studied the bleeding canker of maple (caused by *Phytophthora cactorum*) and found that diamino-azobenzene dihydrochloride injected into diseased trees allowed recovery.

Dutch elm disease appeared in Europe after the First World War and spread rapidly. Several attempts were made to control it by chemotherapy, and both Boudru (1935) and Fron (1936) reported success with 8-quinolinol salts. The disease was introduced into the USA in 1933 and was established there by 1940. In efforts to deal with this situation, research was initiated at the Connecticut Agricultural Experiment Station into various control methods, including chemotherapy. Much of the work done there has been described by Zentmyer *et al.* (1946) and Dimond *et al.* (1949).

Initially over 100 compounds were examined for their toxicity towards the pathogen *in vitro* and the most promising were examined in the field by injecting them into infected elm seedlings. Many compounds were ineffective or phytotoxic but *p*-nitrophenol, hydroquinone and 8-quinolinol sulphate or benzoate showed some promise: 8-quinolinol benzoate, which gave the most lasting effects, was tested in field trials and it was found that soil injection around the feeding roots was a superior method of application to the trunk injections used previously.

To assay potential vascular wilt therapeutants, a technique using *Fusarium* wilt of tomatoes was developed at Connecticut (Dimond *et al.*, 1952). Unlike *C. ulmi*, which invades through the stem, this pathogen invades through the roots, so a method was devised to prevent root-applied compounds from coming into direct contact with the inoculum. Tomato plants were grown in sand and the test compounds applied as a drench; the treated plants were then lifted, their roots washed and deliberately damaged. The injured roots were then dipped into a bud cell suspension of the pathogen and the plants repotted in fresh sand. Subsequent disease development was assessed on the extent of vascular discoloration. Compounds which were effective when applied to sand were later applied to soil-grown plants and it was found that many were inactive. Keyworth and Dimond (1952) showed that chemical as well as mechanical injury to the roots altered their metabolism and reduced their susceptibility to attack by *Fusarium* wilt.

Compounds effective against tomato wilt were also tried against *Fusarium* wilt of carnations and *Verticillium* wilt of eggplants. A high correlation was obtained between the activity shown by compounds towards these three root-invading pathogens, but effective compounds were often inactive against *C. ulmi*. Chapman (1951) attempted to correlate the relative effectiveness of compounds against tomato wilt or Dutch elm disease with their mobility in the plant as measured by their diffusibility on cellulose.

In investigations on the wilt diseases of tomato and carnation (Stoddard and Dimond, 1951; Dimond and Chapman, 1951; Dimond and Davis, 1952, 1953; Davis and Dimond, 1952), several promising chemotherapeutants were discovered, notably 4-chloro-3,5-dimethylphenoxyethanol, 2-norcamphene methanol, n-octadecyltrimethylammonium pentachlorophenate and salts of 2-carboxymethyl mercaptobenzothiazole.

Since these early investigations studies on the chemotherapy of both *Fusarium* wilt of tomato and Dutch elm disease have continued. For example, Davies *et al.* (1954) have found several unsubstituted heterocyclic compounds to be effective against *Fusarium* wilt, whilst Grossmann (1957, 1958b, 1961, 1962a, b, c, 1963, 1968a) has shown that symptoms of this disease can be ameliorated by treatment with dimethyldithiocarbamates, chloronitrobenzenes, sulphonamides and rufianic acid (1,4-dihydroxyanthraquinone-2-sulphonic acid). Thiazolyl mercaptoacetates (Beckman and Howard, 1957), 2,3,6-trichlorophenylacetic acid (Smalley, 1962), halogenated benzoic acids (Geary and Kuntz, 1962) and an aminotrichlorophenylacetic acid (Edgington, 1963) have been shown to exert some control over the development of Dutch elm disease.

Foliar pathogens

Although foliar pathogens are more amenable to control by protectant sprays than are root or vascular pathogens, they have been widely used in studies on chemotherapy. The test method most frequently used is to apply a compound to the roots of a healthy plant growing in soil, sand or water culture and, after a suitable period, to inoculate the foliage and assess the severity of disease development.

Other investigators have measured the eradicant activity of compounds applied to diseased leaves and, to a lesser degree, disease control on untreated foliage following application of the test compounds to selected areas.

Botrytis fabae on broad bean

In early studies on the therapeutic activity of aryloxyalkanecarboxylic acids Crowdy and Wain (1950, 1951) used *Botrytis fabae*, causal agent of 'chocolate spot' of broad bean, as a test organism. Root applications of certain of these acids, notably 2,4,6-trichlorophenoxyacetic acid, reduced symptoms of chocolate spot as shown by a reduction in lesion size rather than lesion numbers. Studies on the activity of structurally-related aryloxy- and arylthio-alkanecarboxylic acids were continued at the Agricultural Research Council Plant Growth Substance and Systemic Fungicide Unit at Wye College (Fawcett *et al.*, 1955, 1957). The systemic activity of 2,4,6-trichlorophenoxyacetic acid against chocolate spot, when applied to the roots of sand-grown broad bean plants, was confirmed and other acids were also found to exert a systemic action, notably 3-phenoxybutyric and 5-phenoxycaproic acids. Although these acids proved to be moderately fungitoxic *in vitro* it was thought that other factors were involved in their activity as therapeutants. *Botrytis fabae* on broad bean has been used for assessing the activity of root-applied (aryloxythio)trichloromethanes (Fawcett *et al.*, 1958), hydroxy-nitroalkanes (Bates *et al.*, 1963), inorganic salts (Carter and Wain, 1964) and *N*-carboxymethyl dithiocarbamic acid derivatives (Heyns *et al.*, 1966).

This host–parasite combination was also used by Crowdy and Davies (1952), who examined various compounds previously reported to be effective therapeutants for wilt diseases. Activity was assessed by measuring the mean lesion diameter; the *in vitro* fungitoxicity of these compounds was also determined. They found that whilst 8-quinolinol derivatives, Helione A, salicylic acid, certain metallic dinicotine salicylates, *p*-aminobenzoic acid and several ureas and thioureas all significantly reduced susceptibility to *B. fabae*, there was no clear correlation between fungitoxicity and systemic activity.

Byrde *et al.* (1953) found that applications in water culture of chlorinated β-naphthyloxyacetic acids and β-naphthols enhanced the resistance of broad beans to chocolate spot. Later Napier *et al.* (1957) reported that root applications of the protectant fungicide 'captan' [*N*-(trichloromethylthio)cyclohex-4-ene-1,2-dicarboximide] also gave protection against chocolate spot and that when this fungicide was applied to the dorsal surface of leaves it protected not only the ventral surface of the treated leaf but untreated leaves above and below. Rombouts and Kaars Sijpesteijn (1958) painted pyridine-thiol-*N*-oxide and its carboxymethyl derivative on to the upper surface of broad bean leaves and obtained protection against *B. fabae* on the untreated surface of the leaflet and also in the twin leaflet of the treated one; lower leaves were also protected to a lesser degree. Root applications of these compounds were not effective.

'Alternaria solani' on tomato

Another foliar pathogen used in early studies on chemotherapy was *Alternaria solani*, causal agent of *Alternaria* or early blight of tomato. As early as 1903 Massee had shown that root applications of copper sulphate were active against this pathogen. Later McNew and Sundholm (1949) reported that when one leaf of a tomato plant was immersed in a suspension of certain substituted pyrazoles, untreated leaves showed an enhanced resistance to *A. solani*; 1-aryl-3,5-dimethyl-4-nitrosopyrazoles were the most effective.

Stubbs (1952) developed a method for the root application of compounds against *A. solani*, using the method of McCallan and Wellman (1943) for inoculation and disease assessment. When applied to the roots of sand-grown tomato plants griseofulvin gave almost complete control; lower levels of activity were shown by 3,5-dimethyl-4-nitrosopyrazoles and 2,4,6-trichlorophenoxyacetic acid. Applying the test compounds to leaves of tomato plants followed by the inoculation of untreated leaves gave variable results. In the same year Dimond and Davis showed that soil applications of the alkali salts of 2-carboxymethyl-mercaptobenzothiazole enhanced the resistance of tomato plants to *A. solani*.

Applying test chemotherapeutants to the roots of tomato plants growing in sand was the method adopted by Fawcett *et al.* (1955, 1957) in their investigation on aryloxyalkanecarboxylic acids, several of which significantly reduced the incidence of *Alternaria* blight. Later (1958) the same workers found certain (aryloxythio)trichloromethanes to be systemically active against *A. solani* as were some inorganic salts (Carter and Wain, 1964).

In 1958 Grossmann applied 1,3,5-trichloro-2,4,6-trinitrobenzene to the roots of tomato plants and reduced the incidence of *A. solani*. Later, in 1961, he found that pre-infection applications of certain sulphonamide drugs including 'Prontalbin' (*p*-aminophenyl sulphonamide) and 'Albucid' (acetyl sulphanilamide), were also partially effective. In the same year Ristich and Cohen showed that soil drenches of 2-pyridinethiol-*N*-oxide were effective against this disease. About the same time Tsao *et al.* (1960) used *A. solani* as a test organism in their examination of the culture filtrates from soil actinomycetes for systemic antifungal activity. Culture filtrates which were active *in vitro* against conidia of *Alternaria solani* were tested further using excised tomato leaves. A culture filtrate from *Streptomyces griseus*, coded B-74, was found to be active in these tests and this material showed systemic activity when applied to intact tomato plants.

Cladosporium cucumerinum on cucumber

A Research Group for Internal Therapy of Plants was established in the Netherlands and here a test for systemic activity was developed using *Cladosporium cucumerinum*, the causal agent of cucumber scab (van Raalte *et al.*, 1955). In this test seven-day-old cucumber seedlings are placed in an aqueous solution of the test compound for two days before being inoculated with conidia of *C. cucumerinum*. They are then kept under high humidity for four to seven days, the severity of disease symptoms being assessed on an arbitrary scale. Alternatively the compound is applied by dipping one cotyledon into an aqueous acetone solution of the compound and then rinsing the treated area prior to inoculating the whole seedling. This test system was used in the intensive investigations on the systemic activity of derivatives of dithiocarbamic acid (van Raalte *et al.*, 1955; Pluijgers, 1959) in which the marked therapeutic activity of *S*-carboxymethyl-*N*,*N*-dimethyldithiocarbamate was discovered. Later investigations showed that DL-serine and DL-threonine (van Andel, 1958), pyridinethiol-*N*-oxide (Rombouts and Kaars Sijpesteijn, 1958), fluorophenylalanine (van Andel, 1962a), thioacetamide and phenylthioureas (Kaars Sijpesteijn and Pluijgers, 1962) and thiosemicarbazides (Pluijgers and Kaars Sijpesteijn, 1966) are also effective in this test system. Earlier (1956) Rich had found that foliar applications of captan to young cucumber plants could give some protection against scab in leaves developed subsequently.

Powdery mildews

Powdery mildew diseases, caused by members of the Erysiphaceae, are found on almost all species of crop plant. Some of the more important are those occurring on vine (*Uncinula necator*), apple (*Podosphaera leucotricha*), cucumber (*Erysiphe cichoracearum* or *Sphaerotheca fuliginea*) and cereals (*Erysiphe graminis*). The cucumber and cereal pathogens have been used extensively in studies on chemotherapy.

Since the mycelium of powdery mildew fungi develops externally upon the leaf surface, foliar sprays do not have to penetrate the plant tissues to be effective eradicants and for this reason most protectant mildew fungicides also show some eradicant activity. Uptake by the sprayed leaves and redistribution within the plant would undoubtedly enhance disease control but only a few studies have been made (e.g. Sharples, 1963) in which the effect of foliar sprays on the disease resistance of unsprayed leaves has been assessed. Most studies on the chemotherapy of powdery mildews have been made using applications to the roots of test plants. Roach (1939, 1942) showed that injections of sodium thiosulphate could protect apple trees against attacks of *P. leucotricha*, and since the early work of Spinks (1913) root applications of inorganic salts have frequently been examined for systemic anti-mildew activity. Salts of copper (Olsen, 1939), cadmium (Sempio, 1938; Meyer, 1950; Tomlinson and Webb, 1960), zinc (Tomlinson and Webb, 1958, 1959, 1960), magnesium (Colquhoun, 1940) and boron (Eaton, 1930) have all been shown to possess systemic activity and there have been repeated reports of the systemic anti-mildew activity of both silicates (Germar, 1934; Wagner,

1940; Minabe, 1951; Grosse-Brauckmann, 1957, 1958) and salts of lithium (Spinks, 1913; Wortley, 1936; Kent, 1941a, b; Vidali and Ciferri, 1951; D'Armini, 1953 and Yarwood, 1959). In 1964 Carter and Wain examined a large number of inorganic salts for their systemic activity against *E. graminis* when applied to the roots of wheat seedlings growing in sand. Most salts were either phytotoxic or ineffective but the marked activity of lithium salts against mildews was again confirmed.

Heyns *et al.* (1966) made root applications of *N*-carboxymethyl-dithiocarbamates to seedlings of wheat, pea, cucumber and apple and found that certain of these compounds showed systemic antimildew activity. Previously Darpoux *et al.* (1958a) found that the manganese salt of pyridinethiol-1-oxide was systematically active against barley mildew and Koopmans (1960) showed that root applications of several substituted triazoles were also effective against this disease; 1-bis(dimethylamido)-phosphoryl-3-phenyl-5-aminotriazole-1,2,4 has been marketed as a systematically active mildew fungicide by N. V. Philips Duphar under the trade name 'Wepsyn'.

Sphaerotheca fuliginea on cucumber has been used by Dekker in his investigations at Wageningen into plant disease chemotherapy. In 1961 he investigated an earlier report that procaine penicillin G was systemically active against cucumber mildew and he showed that the procaine (2-diethylaminoethyl-*p*-aminobenzoate) moiety was responsible for this activity. Procaine hydrochloride, applied to the roots of cucumber plants grown in water culture, protected the leaves against infection and was detected chromatographically in leaf sap. The compound was shown to be effective against several powdery mildews but not other plant pathogens and it was inactive *in vitro*. Microscopic examination of inoculated leaves from treated plants revealed no effect upon the germination of mildew conidia and inhibition of the fungus was only apparent when it was penetrating the host. The hydrolytic products of procaine, *p*-aminobenzoic acid and diethylaminoethanol, proved inactive but several analogues of procaine showed some activity although less than procaine itself. Procaine proved to be disappointing in the field, and in further studies with many procaine analogues and derivatives (Niemann, 1964; Niemann and Dekker, 1966a) none proved better than the parent compound. Following an observation that 3-methyl-*N*,*N*-diethyl-*p*-phenylenediamine is active against cucumber mildew, the same workers examined a series of *p*-phenylenediamines. Marked differences in the systemic activity of structurally-related compounds were found (Niemann, 1964; Niemann and Dekker, 1966b).

The discovery (Dekker, 1963b) that kinetin(6-furfurylaminopurine) was effective against cucumber mildew in leaf disc tests led to the examination of a number of other purines and pyrimidines; five of these proved effective in leaf disc tests but only one, 6-azauracil, was active when applied to roots (Dekker, 1962). This compound was later tried in the orchard against *P. leucotricha* on apple trees. It was effective at low concentrations and gave better control of the disease than some non-systemic fungicides but it damaged the new growth of treated trees (Oort and Dekker, 1964; Dekker and Roosje, 1968).

Root applications of barbiturates have given control of powdery mildew of both marrow (Zaracovitis, 1965) and cucumber (Tempel and Kaars Sijpesteijn, 1967), and in studies on the systemic activity of antibiotics both griseofulvin (Rhodes *et al.*, 1957) and cycloheximide derivatives (Jones and Swartwout, 1961a, b) have shown systemic activity against powdery mildews following root application.

Rusts

Rust fungi have proved one of the most devastating groups of foliar pathogens, attacks of black or stem rust (*Puccinia graminis tritici*) in the wheat lands of Canada and the USA being particularly serious. Many investigations have been made into the chemical control of rust diseases (Dickson, 1959) particularly since the early 1920s when the rust resistance of specially bred wheat cultivars broke down with the emergence of new physiologic races of the pathogen.

The failure to achieve economic control of rusts with protectant fungicides led to many investigations on the chemotherapy of these diseases, particularly using leaf or brown rust, *Puccinia recondita* (*P. triticina*) and stem rust (*P. graminis*) on wheat, *Uromyces appendiculatus* (*U. phaseolicola*) on *Phaseolus vulgaris* and *U. fabae* on broad bean. In most studies either pre-inoculation root applications have been made to plants growing in sand, soil or water culture or the test compounds have been applied as eradicant sprays to infected leaves.

(i) *Root application studies* In 1935 Gigante found that applications of sodium borate to soil in glasshouse and field plots increased resistance of wheat to both *P. striiformis* and *P. recondita*. Subsequent investigators have shown that applications of boron compounds also reduced attacks of *P. graminis* on wheat (Hart and Allison, 1939) and *Melampsora lini* on flax (Heggeness, 1942).

In 1936 Sempio made extensive investigations into the activity of root-applied compounds upon the resistance of *Phaseolus* to *U. appendiculatus* and wheat to *P. striiformis* (*P. glumarum*). Cadmium and nickel salts proved active as did certain alkaloids. In the same year Gassner and Hassebrauk tested 174 compounds for their ability to control brown rust of wheat and barley (*P. hordei*) when applied to the soil surface, rust development being assessed on an arbitrary scale based on the number of sori present. Several compounds including acridine, bromonaphthols, picric acid and sodium *p*-toluenesulphochloramide were effective; the last two compounds were also active when the plants were sown in pre-treated soil, although the chloramide was phytotoxic. Later Hassebrauk (1938) showed that *o*- and *p*-toluenesulphonamide, recently developed for pharmaceutical use, gave complete control of *P. striiformis*, *P. graminis* and *P. recondita* on wheat, *P. hordei* on barley, *P. secalina* on rye and *P. coronata* on oats, although both compounds were more phytotoxic than the moderately effective picric acid. When applied in the field, however, *p*-toluenesulphonamide failed to control yellow or brown rust of wheat and proved phytotoxic (Hassebrauk, 1940).

Hart and Allison (1939) confirmed that root applications of borax, picric acid and both *o*- and *p*-toluenesulphonamides controlled stem rust of wheat but all were phytotoxic; the sulphonamides were more effective in a sandy soil than in a loam.

Hassebrauk (1951) examined other sulphonamide drugs for their systemic activity against cereal rusts, and particularly *P. recondita*. Whilst many of them were inactive, others were highly effective when applied to roots; seed treatment, however, did not enhance the resistance of the developing seedlings to rust, and phytotoxicity was again evident with many of the active compounds. Hassebrauk (1952) later showed that 'Ladogal' (sodium salt of *p*-aminobenzylsulphonoxymethylamide-*N*-*d*-glucoside sulphonic acid), previously shown to be active against *P. recondita*, was also effective against *P. graminis* and *P. striiformis*. The compound remained active in the soil for four weeks after application, but as the humus content increased activity decreased. Infections of *P. recondita* were eradicated by leaf sprays. Another sulphonamide drug, sulphadiazine, was found to be effective against brown rust of wheat by Yamada (1955).

Other compounds which have been shown to be partially effective in controlling cereal rusts when applied as root treatments are salts of lithium (Kent, 1941a, b), 2,4-D (von Witsch and Kasperlik, 1953), ferrous sulphate (Forsyth, 1957) and maleic hydrazide (Samborski *et al.*, 1960).

(ii) *Eradicant studies* Many of the compounds shown to be active against rusts following root application, and in particular the sulphonamides, were subsequently applied as eradicant foliar sprays.

Straib (1941) applied sprays to rust-infected cereal, antirrhinum and flax plants and found picric acid, sulphonamides and acridine to be partially effective but phytotoxic. Mitchell *et al.* (1950) applied several sulphanilamides to wheat as eradicants for stem rust; the sodium salt of sulphadiazine was the most effective and formulation with a wetter or keeping the sprayed plants in humid conditions was shown to enhance activity. Hotson (1952, 1953) also examined sulphonamides as eradicants of wheat stem rust. He applied sprays eight days after planting the seed and the rust susceptibility of the treated seedlings was assessed in terms of their reaction type. Several of the compounds were active; some, e.g. sulphanilamide, gave only a temporary inhibition of rust whilst others, e.g. sulphadiazine, inhibited development throughout the experimental period.

Also in 1953 Livingston examined 179 compounds for their eradicant activity against leaf and stem rust of wheat by spraying plants five days after inoculation. Promising compounds were then applied in the field with knapsack sprayers or by aeroplane. Sulphanilamide was found to retard leaf rust development, but its low solubility prevented its effective application in aerial sprays and sulphanilic acid proved to be more effective. Salts of sulphamic acid were also found to be active in these experiments, the calcium salt being the least phytotoxic. Acosta and Livingston (1955) confirmed that both sodium sulphanilate and calcium sulphamate reduce stem rust on wheat, barley and oats. Sodium sulphanilate was not phytotoxic but the sulphamate markedly reduced the viability of the grain produced from treated crops; its phytotoxicity, however, was found to depend on the stage of growth when it was applied. Further evidence of the effectiveness of sulphanilic acid as an eradicant of *P. graminis* was provided by Campos and Borlaug (1956).

Compounds other than sulphonamides have also been reported as eradicants for rust diseases. Yarwood (1947, 1948) observed that lime sulphur and hydrogen sulphide or hydrogen cyanide inhibit the development of infections of *U. appendiculatus* on leaves of *Phaseolus*, and Ibrahim (1951) found that sprays of 2,4-D retarded the development of *P. graminis* on oats. In a survey by Hotson (1953) picric acid, nitrobenzoic acids, saccharin and methylene blue were found to be effective eradicants of wheat stem rust and, in a similar survey, Livingston (1953) found phenylhydrazine to be effective in the glasshouse although its performance in the field was disappointing. *N*-phenyl-*N*′-3-sulpholanyl hydrazine, however, was found to be a useful therapeutant for rust on *Phaseolus* by Evans and Saggers (1962), and Jaworski and Hoffman (1963) reported various phenylhydrazones to be effective eradicants of wheat leaf rust. Livingston (1953), Silverman and Hart (1954) and Wallen (1955, 1958) tested cycloheximide against both stem and leaf rust of wheat. It was found to show promise both in the glasshouse and in the field although it often proved phytotoxic. Certain salts of cycloheximide, especially the acetate and semicarbazone (Hacker and Vaughn, 1957a, b), gave good and prolonged control of wheat stem rust in the field.

Many other antibiotics have been examined for their eradicant activity against rusts but most proved ineffective. Limited activity has been shown for anisomycin (Zaumeyer and Wester, 1956), F-17 (Preston *et al.*, 1956), oligomycin (Zaumeyer, 1957), P-9 (Davis *et al.*, 1960; Hagborg *et al.*, 1961) and phleomycin (Smale *et al.*, 1961; Purdy, 1964).

Keil *et al.* (1958) applied eradicant sprays in glasshouse tests using leaf rust of rye (*P. secalina*) and several nickel salt–amine complexes were the best compounds tested. Nickel amine complexes applied against wheat leaf rust in the field were also found to be effective by Peturson *et al.* (1958) and Forsyth and Peturson (1959a, b) showed that inorganic nickel salts could also eradicate wheat leaf rust in the field without causing phytotoxicity although the levels necessary for rust eradication proved to be too toxic for use on oats or sunflower. In 1960 these workers showed that nickel salts controlled both stem and leaf rust of wheat in the field and that the timing of their application was less critical than that of the protectant fungicide zineb. A year later,

Keil and Frohlich took out a US patent for the use of nickel compounds as rust eradicants.

Use of rapid laboratory screening techniques

In most earlier studies, made with the object of discovering compounds of immediate practical use in the field, biological activity was measured using entire plants. Other investigators, however, have used more rapid and economic laboratory techniques, such as floating inoculated leaf discs on solutions of the test compounds, as a preliminary test for detecting chemotherapeutic activity. Thus Dekker (1963b) floated cucumber leaf discs inoculated with *Sphaerotheca fuliginea* upon test solutions; kinetin was effective when tested in this manner but was not active when applied to the roots of cucumber plants. Methionine, too, proved active in leaf disc tests but not following root application (Dekker, 1969a). This worker has also used leaf discs from dwarf French bean inoculated with *Uromyces appendiculatus* (e.g. Niemann and Dekker, 1966b).

Excised shoots have been used in translocation studies, in particular using antibiotics (e.g. Alcorn and Ark, 1956) and if treated shoots are subsequently inoculated they can be used to assess the systemic activity of compounds. The large amount of chemical taken up by excised shoots often causes phytotoxicity, and the distribution of toxicity symptoms gives a rapid indication of the mobility of the test compound within plant tissues (Carter, 1968).

Van Raalte (1954) devised a rapid test, using excised petioles, for detecting the translocation of antifungal chemicals. This test has also been used, in a modified form, by Spencer (1957) and Darpoux *et al.* (1958b).

In studies at Wye College on the relationship between chemical structure and chemotherapeutic activity all compounds are examined in a sequential series of tests using leaf discs, excised shoots, foliar sprays and root application in sand, soil or water culture. This test programme yields far more information concerning the properties of the test chemicals than would a single test using entire plants.

The search for new chemotherapeutants

Since the 1930s many thousands of compounds have been examined for potential chemotherapeutic activity but most of these have proved either inactive or too phytotoxic. Some showed promise in glasshouse tests (see Table 2.1) and a few of these were examined in the field, but usually the results were disappointing. The majority of compounds have been examined only by one worker in a single test system but others, e.g. sulphonamides, antibiotics and 8-quinolinol salts, have been tested more extensively. The earliest studies on systemic action were based on screening large numbers of chemically unrelated chemicals but later workers selected compounds for specific chemical or biological properties.

1. Antifungal compounds

Initially, potential chemotherapeutants were selected on the basis of high *in vitro* fungitoxicity coupled with water solubility, but it soon became apparent (e.g. Davis, 1952) that there is rarely a high correlation between *in vitro* fungitoxicity and therapeutic activity.

(a) *Protective fungicides*

In general, commercial protectants have shown little or no therapeutic activity, but some organomercurials have shown limited eradicant activity against both seed-borne (Leben and Arny, 1952; Arny and Leben, 1954; Porter, 1956) and foliar (Hamilton, 1931, 1948; Howard and Sorrell, 1943; Moore, 1966, 1967) pathogens. Dithiocarbamates are one of the most successful groups of protectants and there have been reports of therapeutic activity for ferbam (Taylor, 1953), nabam (Stoddard, 1951a, b), zineb (Taylor, 1953; Rich, 1956) and thiram (Grossmann, 1957; Volger, 1959). There have also been reports of therapeutic activity for captan, not only as an eradicant of seed (Maude, 1966) and foliar (Rich, 1956; Albert and Groves, 1966) pathogens but as showing systemic activity against foliar pathogens after root application (Stoddard, 1954; Napier *et al.*, 1957; Rosen, 1959). Somers and Richmond (1962), however, detected only a very slight movement of root-applied captan into the shoots of broad bean plants. Fawcett *et al.* (1958) examined other compounds containing the thiotrichloromethyl group and found some of them to be systemic; captan itself was not active in their tests.

(b) *Inorganic salts*

Much early work was done using inorganic salts, particularly those of the fungitoxic metals such as copper and mercury. From these studies salts of lithium emerged as showing some promise but in the field they proved relatively ineffective and phytotoxic. Carter and Wain (1964) made extensive studies on the systemic activity of inorganic salts in glasshouse tests but, whilst the activity of lithium salts was confirmed, most compounds proved poorly translocated and often phytotoxic. The most promising application of inorganics in the field appears to be the use of nickel salts as rust eradicants.

(c) *Sulphonamides*

Sulphonamides and related compounds have been examined extensively and their potential use in controlling plant disease was discussed by Rudd Jones (1956). Although they have shown therapeutic activity against facultative parasites (Dimond *et al.*, 1952; Grossmann, 1957, 1961) they are more effective against obligate parasites, particularly rusts. Against these diseases they have proved to be very effective in the glasshouse and moderately effective in the field although they are frequently phytotoxic. Their phytotoxicity can be reduced by applying *p*-aminobenzoic acid, and as this also antagonises their therapeutic activity against rusts, they may act directly against the pathogen as anti-metabolites. The activity of sulphonamides against rusts is fungistatic rather than fungicidal and when treatment ceases then the disease re-establishes itself. Work on their uptake and metabolism by plants has shown that these compounds can be acetylated in plant roots and that their mobility within the plant varies greatly with changes in chemical structure.

Thus Crowdy and his co-workers made studies on the uptake and translocation of certain sulphonamides fed to the roots of seedlings of oat, wheat and broad bean growing in water culture (see Rudd Jones, 1956). Treated plants were inoculated with rust and a relationship was established between the high mobility of sulphanilamide in the plant and its systemic activity as compared with the ineffectiveness of the poorly translocated sulphaguanidine.

(d) *Salts of 8-quinolinol*

8-Quinolinol and its salts are highly antifungal water-soluble compounds which were used extensively in early studies on chemotherapy, particularly the therapy of vascular wilt pathogens. They have shown activity against silver leaf of plum (Roach, 1934; Grosjean, 1951; Bennett, 1962) and Dutch elm disease (Boudru, 1935; Fron, 1936; Horsfall and Zentmyer, 1942; Zentmyer and Horsfall, 1943; Zentmyer *et al.*, 1946; Stoddard, 1946; Dimond, 1947; Dimond *et al.*, 1949) as well as oak wilt (Hoffman, 1952, 1953) and the *Fusarium* wilts of geranium (Stoddard, 1957), tomato (Dimond *et al.*, 1952), carnation (Stoddard, 1951a; Stoddard and Dimond, 1951) and guava (Jain, 1956). Activity has also been reported against *P. cinnamomi* on avocado (Zentmyer, 1954), *Rhizoctonia* on *Matthiola* (Stoddard, 1952), *B. fabae* on broad bean (Crowdy and Davies, 1952) and *Cytospora cincta* on peach and plum (Helton and Rohrbach, 1967). These compounds have been used commercially for the control of Dutch elm disease and carnation wilt. Against the former, 8-quinolinol benzoate gives moderate protection for a year if applied prior to infection but when applied post-infection it does not effect a cure although disease symptoms may be suppressed.

(e) *Fungicide analogues*

The presence of a reactive toxophore in the molecule of a fungitoxicant may render it both poorly translocated and phytotoxic. A derivative of such a molecule, however, may possess improved translocation and therapeutic activity (see e.g. Dimond and Davis, 1953; Hamilton *et al.*, 1956) although *in vitro* fungitoxicity is often lost. Such compounds have been referred to as 'masked fungicides' (Kaars Sijpesteijn, 1961b), the intention being that the original fungitoxic molecule be released within the plant, preferably only at infection sites.

In 1954 van Raalte, using his excised petiole test for demonstrating translocation, showed that the free dimethyldithiocarbamate ion was not readily translocated but that tetramethylthiuram monosulphide (TMTM) was; TMTM was known to liberate the free dimethyldithiocarbamate ion only slowly and it was suggested that the remainder of the TMTM molecule was acting as a protective carrier for the fungitoxic ion. Investigations were therefore made into the systemic activity of compounds with various potential carrier groups attached to the dimethyldithiocarbamate ion (van Raalte *et al.*, 1955) and this led to the discovery of systemic activity in the *S*-carboxymethyl compound whose *in vitro* fungitoxicity, however, was low. Further studies with related compounds (Pluijgers, 1959) did not lead to the discovery of a more effective chemotherapeutant, and later work on the metabolism of these compounds within plants (see Dekhuijzen, 1964) has indicated that the *S*-carboxymethyl compound, which is an active plant growth regulator, works through its effect upon the metabolism of the host plant rather than by liberation of dimethyldithiocarbamate ions.

(f) *Antibiotics of microbial origin*

Since it has been shown that many antibiotics are taken up by and translocated within plants (see Crowdy and Pramer, 1955; Goodman, 1962b; Wain and Carter, 1967) they are of great interest as potential systemics.

The first commercially available antibiotics were antibacterial but, although essentially an antibacterial compound, streptomycin has shown systemic activity against certain fungal pathogens such as hop downy mildew (*Pseudoperenospora humuli*) (Horner and Maier, 1957; Horner, 1963). Other non-fungitoxic antibiotics, notably phleomycin (Småle *et al.*, 1961; Purdy, 1964), have also shown systemic activity against fungal diseases.

Later, antifungal antibiotics were developed, and two of these, cycloheximide (see Ford *et al.*, 1958) and griseofulvin (see Brian, 1960; Rhodes, 1962), have been extensively tested.

Cycloheximide, discovered in 1946, is very active *in vitro* against a wide range of fungi and was released for use in crop protection in 1944–5, being marketed under the name 'Actidione'. It has limited systemic activity but has proved an effective eradicant against *Coccomyces hiemalis* on cherry (Petersen and Cation, 1950; McClure and Cation, 1951; Hamilton and Szkolnik, 1953; Klos and Fronek, 1964), cereal rusts (Livingston, 1953; Silverman and Hart, 1954; Wallen, 1955, 1958), tea rust (Hocking and White, 1965), blister rust on pine (Moss *et al.*, 1960; Moss, 1961), *Cytospora cincta* on peach and plum (Helton and Rohrbach, 1967) and wilt of oak (Phelps *et al.*, 1957, 1966). It is also effective as an eradicant of seed-borne pathogens (Wallen and Skolko, 1951; Krüger, 1960) and as a soil drench against powdery mildew of rose

(Jones and Swartwout, 1961a, b). Its usefulness, however, is limited by its high phytotoxicity. Certain derivatives, such as the acetate, semicarbazone and oxime, have been examined for improved translocation and lower phytotoxicity; the oxime was effective against *C. hiemalis* (Hamilton *et al.*, 1956) and the semicarbazone against wheat stem rust (Hacker and Vaughn, 1957a, b). Foliar injury was reduced if chlorophyllin was added to the spray (Ark and Thompson, 1958; Wilson and Ark, 1958).

Griseofulvin, isolated in the early 1940s, is also active *in vitro* against a wide range of fungi (Napier *et al.*, 1956). Since it is readily translocated in plants (Crowdy *et al.*, 1955, 1956) and has low phytotoxicity it offered much promise as a chemotherapeutant. It proved systemically active against *Botrytis cinerea* (Brian *et al.*, 1951), *Alternaria solani* (Brian *et al.*, 1951; Stubbs, 1952), *Botrytis tulipae* (Aytoun, 1956), *Oidium chrysanthi* (Rhodes *et al.*, 1957), *Pseudoperenospore humuli* (Horner and Maier, 1957) and *Sclerotinia fructigena* (Byrde, 1959). Over 300 derivatives of griseofulvin have been synthesised (Crosse *et al.*, 1964), and some were examined for their systemic activity in plants (Crowdy *et al.*, 1959a, b). None of them however has found commercial use, although griseofulvin itself has been used in Japan for the control of *Sclerotinia mali* and *Mycosphaerella malis*.

In general the many other antifungal antibiotics examined for chemotherapeutic activity have proved less promising. Thus the polyene macrolides appear to be antagonised by plant constituents whilst polypeptide antibiotics are poorly translocated and phytotoxic. Some have shown activity as eradicants of seed infections, e.g. pimaricin and rimocidin (Dekker, 1955, 1957; Oort and Dekker, 1960), helixin (Leben and Arny, 1952; Leben and Keitt, 1952; Leben *et al.*, 1953, 1954; Krüger, 1960) filipin (Krüger, 1960), antimycin (Leben *et al.*, 1954; Lockwood *et al.*, 1954) and mycothricin (Rangaswami, 1956) whilst a streptothricin complex (Sharples, 1963), phytoactin (Phelps *et al.*, 1957, 1966; Moss, 1961; Helton and Rohrbach, 1967), oligomycin (Zaumeyer, 1957) and anisomycin (Zaumeyer, 1957) have shown limited activity against foliar pathogens. Blasticidin-S, a purine antibiotic discovered in Japan in 1958, has a very selective antifungal activity but is therapeutically active against the rice blast pathogen (*Pyricularia oryzae*) and is used commercially for the control of this disease (Misato *et al.*, 1959).

Some investigators have screened culture filtrates from micro-organisms for activity against plant pathogenic fungi. Pridham *et al.* (1956), whilst examining culture filtrates from *Streptomyces* cultures, discovered the F-17 antibiotic complex which is systemically active against rust fungi (Preston *et al.*, 1956; Mitchell *et al.*, 1959). In a similar study Davis *et al.* (1960) discovered the P-9 antibiotic which is also systemically active against rusts and is effective in the field (Hagborg *et al.*, 1961). Rhodes *et al.* (1961), whilst screening soil actinomycetes for activity against the apple scab pathogen (*Venturia inaequalis*) discovered venturicidin, and Tsao *et al.* (1960), also working with soil actinomycetes, tested filtrates for systemic activity and discovered B-74, systemically active against *Alternaria solani*.

(g) *Compounds from higher plants*
Plants are known to contain a wide range of antifungal compounds (see Fawcett and Spencer, 1969; Wain, 1969), many of which, and in particular those of a phenolic nature, have been implicated in the natural resistance to fungal attack shown by most plants. A good example of such a compound is 'wyerone' isolated from broad bean tissue at Wye (Spencer *et al.*, 1957; Fawcett *et al.*, 1965, 1968, 1969; Wain *et al.*, 1961); it shows a wide spectrum of antifungal activity but is much less effective against *Botrytis* species which are pathogens of broad bean (Fawcett *et al.*, 1969).

In addition to preformed antifungal compounds of the wyerone type, since the work of Müller *et al.* (1939) there has been a growing interest in antifungal compounds (phytoalexins) which are produced by plant tissues as a response to infection (see, e.g., Cruickshank, 1963; Fawcett and Spencer, 1969).

Naturally occurring antifungal compounds, whether preformed in healthy tissues or phytoalexins, must be of considerable interest as potential chemotherapeutants. Some, such as phenolic compounds and quinones, are structurally similar to synthetic compounds already known to be active, but many classes of naturally occurring compounds have received little attention. Some have been examined at Wye for their potential as systemic therapeutants (see Fawcett and Spencer, 1970) and some of the Wye compounds are being examined for their capacity to control fungal infections of man.

2. Toxin and enzyme antagonists

There is evidence (see Braun and Pringle, 1959; Ludwig, 1960; Wheeler and Luke, 1963; Deverall, 1964; Pringle and Scheffer, 1964) that toxins secreted by the pathogen *in vivo* are responsible, at least in part, for inducing disease symptoms in the host plant.

Early in the 1900s Rumbold had suggested that toxins produced by *Endothia parasitica* are important in the etiology of chestnut blight and that they might be antidoted with chemicals. Later (Howard, 1941; Howard and Caroselli, 1941) work at Rhode Island Agricultural Experimental Station on the bleeding canker disease of maples (caused by *Phytophthora cactorum*) found evidence that a fungal toxin was important in inducing disease symptoms. This toxin was isolated and it was demonstrated that not only could it be antagonised by diaminobenzene dihydrochloride but also that this compound, when injected into infected trees, led to a reduction in symptoms. Following the work of Howard on bleeding canker, studies were made which established that *Ceratocystis ulmi* also produces toxins capable of causing at least part of the disease syndrome of Dutch elm disease (e.g. Zentmyer, 1942).

Horsfall and Zentmyer (1942) suggested that the activity of 8-quinolinol salts against this disease was brought about by toxin inactivation and pointed out that both this compound and diaminobenzene dihydrochloride contain basic nitrogen atoms in the molecule. They tested malachite green and urea as therapeutants with some success; both compounds also antidoted toxins from *C. ulmi*. In 1950 Feldman *et al.* showed that above pH 6, hydroxyl ions irreversibly inactivate the toxins of *C. ulmi in vitro* and that applications of compounds such as lime around the roots of infected elms mitigated disease symptoms (Feldman and Caroselli, 1951). This work led to the introduction of 'Carolate' as a treatment for Dutch elm disease, this being a mixture of low magnesia lime, urea, salicylic acid and an azo dye.

In vitro antagonisation of fungal toxins has been employed as a screening technique for chemotherapeutants (e.g. Hoffman, 1951, 1952, 1953) but with only limited success. The role of toxins produced in laboratory cultures in the disease syndrome in infected plants has not always been clearly established (see Dimond and Waggoner, 1953). Furthermore toxin antagonists frequently reduce only some of the disease symptoms (e.g. Feldman and Caroselli, 1951) and evidence is available (e.g. Brooks and Brenchley, 1931; Dimond *et al.*, 1949) that a single pathogen can produce more than one toxin.

Extracellular pectolytic enzymes are thought to play a major role in inducing the *Fusarium* wilt syndrome (Wood, 1960), and Grossmann (1962a, b, c, d; 1968a) examined both synthetic chemicals and plant extracts for their ability to antagonise fungal pectolytic enzymes *in vitro*. He then tested the most promising compounds on the plant and found that rufianic acid (1,4-dihydroxyanthraquinone-2-sulphonic acid) reduced wilting although it was less successful in preventing vascular browning or invasion of the tissues by the fungus.

3. Translocated compounds

Since, by definition, systemic compounds must be translocated within the plant a logical approach to the discovery of new systemic chemotherapeutants is the examination of compounds known to be translocated, or structurally related analogues, for their ability to control disease.

Crowdy and Wain (1950, 1951) chose to work with aryloxyalkanecarboxylic acids because these compounds are structurally related to synthetic growth regulators known to be readily translocated in plants (Wain, 1951). By selecting compounds which lacked the structural requirements for growth-regulating activity, they attempted to find chemicals that were readily translocated, non-phytotoxic yet effective against the fungal pathogen.

Crowdy and his co-workers made studies on the uptake and translocation of certain sulphonamides fed to the roots of seedlings of oat, wheat and broad bean growing in water culture (see Rudd-Jones, 1956). Treated plants were inoculated with rust and a relationship was established between the high mobility of sulphanilamide in the plant and its systemic activity as compared with the ineffectiveness of the poorly translocated sulphaguanidine.

4. Compounds working by 'indirect' action

It soon became apparent during studies upon chemotherapy that a number of effective compounds were not fungitoxic *in vitro* and produced no detectable antifungal metabolite *in vivo*. The idea arose, therefore, that such compounds were working 'indirectly' against the pathogen through their effects upon the composition or metabolism of the host.

Thus in 1952 Davis found that several of the most effective chemotherapeutants for *Fusarium* wilt of tomato were non-fungitoxic but had profound effects upon the physiology of the host plant; several, e.g. 4-chloro-3,5-dimethylphenoxyethanol and 2-carboxymethyl-mercaptobenzothiazole salts, inducing formative effects.

Known plant growth regulators, and especially 2,4-D (Ibrahim, 1951; Davis and Dimond, 1953, 1956; von Witsch and Kasperlik, 1953; Natti and Szkolnik, 1954; Akai, 1955; Waggoner, 1956), were extensively studied as possible chemotherapeutants. The effects of treatment with such compounds upon disease resistance varied considerably since, for example, Davis and Dimond (1953) found that certain growth regulators enhanced the resistance of tomatoes towards *Fusarium* wilt whilst Waggoner and Dimond (1952) showed that treatment with maleic hydrazide increased susceptibility to this disease. Furthermore, treatment with growth regulators could enhance the resistance of plants to certain diseases whilst increasing their susceptibility towards others. In an attempt to explain this, Horsfall and Dimond (1957) postulated that protection afforded by growth regulators was related to carbohydrate levels in the plant, some pathogens favouring tissues with a high sugar content and others favouring those with low levels.

Corden and Dimond (1957, 1959) found that the therapeutic activity of naphthalene acetic acids correlated with their ability to inhibit root elongation and in 1960 Corden and Edgington showed that α-naphthylacetic acid (NAA) did not increase the wilt-resistance of calcium-deficient tomato plants. Studies on the nature of pectins in treated plants (Edgington and Dimond, 1959; Edgington *et al.*, 1961) produced evidence that NAA enhanced wilt resistance by rendering the pectins of the host more resistant to attack by fungal enzymes.

In 1957 Beckman and Howard observed that thiazolyl mercaptoacetates were effective against Dutch elm disease only if applied prior to infection and early in the season and that these compounds checked the production of large-vesselled spring wood in treated elm trees. The pathogen penetrates plant cell walls only slowly, and it was suggested (Beckman, 1958) that the modified wood development induced by these compounds acts as a mechanical barrier. Since then, evidence has been

Table 2.1 Compounds reported as showing chemotherapeutic activity

Compound	*Host/pathogen*	*Reference(s)*
Acridine	Rust diseases	Hassebrauk (1938), Straib (1941)
Alkaloids	*Phaseolus/Uromyces appendiculatus*	Sempio (1936)
Alkylmercapto-tetrahydro-pyrimidines	*Phaseolus/U. appendiculatus*	Nickell *et al.* (1961)
Amino acids	Various	see van Andel (1966b)
p-Aminobenzoic acid	Broad bean/*Botrytis fabae*	Crowdy and Davies (1952)
β-Aminoethyl-aryl ketones	*Phaseolus/U. appendiculatus*	Pellegrini *et al.* 1965
	Vine/*Plasmopara viticola*	
Aminotrichlorophenyl acetic acid	Elm/*Ceratocystis ulmi*	Edgington (1963)
Anthraquinones	Tomato/*Fusarium oxysporum*	Fuchs (1964)
Antibiotics (cycloheximide, griseofulvin, etc.)	Various	see Zaumeyer (1958), Goodman (1959), Dekker (1963a, 1969b)
Aryl-3,5-dimethyl-1,4-nitrosopyrazoles	Tomato/*Alternaria solani*	McNew and Sundholm (1949), Stubbs (1952)
	Broad bean/*B. fabae*	Crowdy and Wain (1951)
Aryloxyalkanecarboxylic acids	Broad bean/*B. fabae*	Crowdy and Wain (1951), Stubbs (1952), Fawcett *et al.* (1955, 1957)
	Tomato/*A. solani*	
(Aryloxythio)trichloromethanes	Tomato/*A. solani*	Fawcett *et al.* (1958)
6-Azauracil	Cucumber/*Sphaerotheca fuliginea*	Dekker (1962), Oort and Dekker (1964), Dekker and Oort (1964)
Azo dyes.	Maple/*Phytophthora cactorum*	Howard and Caroselli (1941), Caroselli and Howard (1942)
	Broad bean/*B. fabae*	Crowdy and Davies (1952)
Barbiturates	Cucumber, Marrow/*S. fuliginea*	Zaracovitis (1965), Tempel and Kaars Sijpesteijn (1967)
Benzoic acid	Elm/*C. ulmi*	Zentmyer (1942), Zentmyer and Horsfall (1943), Zentmyer *et al.* (1946)
	Oak/*Ceratocystis fagacearum*	Phelps *et al.* (1966)
Bromonaphthols	Wheat/*Puccinia recondita*	Hassebrauk (1938)
Captan	Various	see text
4-chloro-3,5-dimethylphenoxy-ethanol	Tomato/*F. oxysporum*	Dimond and Chapman (1951)
Chlorogenic acid	Apple/*Venturia inaequalis*	Kirkham (1957)
Chloronitrobenzenes	Tomato/Various	Grossmann (1958a)
Coumaric acid, esters, salts	Apple/*V. inaequalis*	Kirkham and Flood (1963), Kirkham and Hunter (1964, 1965)

Dibenzothiophene	*Pinus banksiana*/*Rhizoctonia solani*	Vaartaja (1955)
Dithiocarbamates	Various	see text
2,4-D	Various	see text
Gibberellins	Potato/*Phytophthora infestans*	Weindlmayr (1963)
	Wheat/*Tilletia foetida*	Király *et al.* (1962)
Hydroquinone	Elm/*C. ulmi*	Zentmyer and Horsfall (1943), Zentmyer *et al.* (1946)
IAA	Tomato/*Verticillium albo-atrum*	Davis and Dimond (1953, 1956), Sinha and Wood (1967)
	Oak/*C. fagacearum*	Phelps *et al.* (1966)
Inorganic salts	Various	see Carter and Wain (1964)
Maleic hydrazide	Wheat/*P. recondita*	Samborski *et al.* (1960)
2-Mercaptobenzothiazole salts	Tomato/*F. oxysporum, A. solani*	Dimond and Davis (1952)
	Elm/*C. ulmi*	Dimond and Davis (1952), Tarjan and Howard (1953), Beckman and Howard (1957), Beckman (1958)
	Oak/*C. fagacearum*	Phelps *et al.* (1966)
2-Methyl-1,4-naphthaquinone	Rice/*Ophiobolus miyabeanus*	Akai (1955)
Naphthyl- and naphthoxy-acetic acids	Broad bean/*B. fabae*	Byrde *et al.* (1953)
	Tomato/*F. oxysporum*	Corden and Dimond (1957)
	Tomato/*V. albo-atrum*	Davis and Dimond (1953, 1956)
Nitrobenzoic acids	Wheat/*Puccinia graminis*	Hotson (1953)
Nitrophenols	Elm/*C. ulmi*	Zentmyer *et al.* (1946)
	Phaseolus/*Erysiphe polygoni*	El-Zayat *et al.* (1968)
N-octadecyltrimethyl ammonium pentachlorophenate	Tomato/*F. oxysporum*	Dimond *et al.* (1952), Davis and Dimond (1952)
Norcamphene methanol	Carnation/*Fusarium dianthi*	Stoddard (1951a)
Organomercurials	Various	see text
Pentachlorophenol	Rice/*O. miyabeanus*	Akai and Kato (1962)
Phenylhydrazines	Wheat/*P. recondita, P. graminis*	Livingston (1953)
	Phaseolus/*U. appendiculatus*	Evans and Saggers (1962)
Phenylhydrazones	Wheat/*P. recondita*	Jaworski and Hoffman (1963)
3-Phenylsydnones	Wheat/*P. recondita*	Davis *et al.* (1959)
	Phaseolus/*U. appendiculatus*	
Phenylthioureas	Apple/*V. inaequalis*	Kuć *et al.* (1957)
	Cucumber/*Cladosporium cucumerinum*	Kaars Sijpesteijn and Pluijgers (1962)
Picric acid	Rust diseases	Hassebrauk (1938), Hart and Allison (1939), Hotson (1953), Straib (1941)

Table 2.1 *continued*

Compound	*Host/pathogen*	*Reference(s)*
Plant growth retardants	Tomato/*V. albo-atrum*	Sinha and Wood (1964, 1967)
	Phaseolus/Sclerotium rolfsii	Tahori *et al.* (1965)
	Blackcurrant/various	Smith and Corke (1966)
Procaine and derivatives	Powdery mildew diseases	Dekker (1961a, b), Niemann (1964), Niemann and Dekker (1966a)
Sodium propionate	Wheat/*P. graminis*	Futrell and Berry (1964)
Pyridinethiol-*N*-oxide	Pea/*Ascochyta* pisi	Kaars Sijpesteijn *et al.* (1958), Dekker *et al.* (1958)
	Phaseolus/Colletotrichum lindemuthianum	
	Barley/*E. graminis*	Darpoux *et al.* (1958a)
	Cucumber/*Colletotrichum lagenarium*	Ristich and Cohen (1961)
	Broad bean/*B. fabae*	
	Peach/*Cytospora cincta*	Helton and Rohrbach (1967)
Pyrogallol	Elm/*C. ulmi*	Zentmyer *et al.* (1946)
8-Quinolinol salts	Various	see text
Rufianic acid	Tomato/*F. oxysporum*	Grossmann (1962a, c, 1963)
Salicylic acid	Broad bean/*B. fabae*	Crowdy and Davies (1952)
	Elm/*C. ulmi*	Tarjan and Howard (1953)
Sulphanilamides, sulphonamides and related compounds	Various, mainly rust diseases	see text
Tetrachlorobenzoic acids	Elm/*C. ulmi*	Beckman (1959)
	Oak/*C. fagacearum*	Geary and Kuntz (1962)
Thioacetamide	Cucumber/*C. cucumerinum*	Kaars Sijpesteijn and Pluijgers (1962)
Thiosemicarbazides	Wheat/*P. graminis*	Fuchs and Bauermeister (1958)
	Cucumber/*C. cucumerinum*	Pluijgers and Kaars Sijpesteijn (1966)
TIBA	Tomato/*V. albo-atrum*	Davis and Dimond (1953, 1956)
	Oak/*C. fagacearum*	Phelps *et al.* (1966)
Triazoles	Barley/*Erysiphe graminis*; Apple/*P. leucotricha*	Koopmans (1960)
Sodium trichlorophenoxy-acetate	Peach/*Cytospora cincta*	Helton and Rohrbach (1967)
Salts of trichlorophenyl acetic acid	Elm/*C. ulmi*	Smalley (1962)
Unsubstituted heterocycles	Tomato/*F. oxysporum*	Davis *et al.* (1954)
Ureas and thioureas	Broad bean/*B. fabae*	Crowdy and Davies (1952)

obtained that 2,3,6-trichlorophenylacetic acid (Smalley, 1962), halogenated benzoic acids (Geary and Kuntz, 1962) and an aminotrichlorophenylacetic acid (Edgington, 1963) exert a control on the development of Dutch elm disease by a similar mechanism.

Amino acids have received considerable attention as possible chemotherapeutants (see van Andel, 1966b). Naturally occurring L-amino acids, other than methionine, have shown little activity although D-isomers are sometimes more effective. Some analogues have shown useful activity and, whilst certain of these, e.g. canavanine, ethionine and fluorophenylalanine, appear to be fungitoxic anti-metabolites which exert a direct action against the pathogen, others, e.g. L-threo-phenylserine, apparently operate through an effect on the host (van Andel, 1966a).

Finally it is interesting that certain fungicide analogues ('masked fungicides') appear to exert their therapeutic effects indirectly rather than by liberation of the free fungitoxic molecule *in vivo* (see Kaars Sijpesteijn, 1961b; van Andel, 1962b; Dekhuijzen, 1964). Thus work on the metabolism of S-carboxymethyl dimethyldithiocarbamate, an active growth-regulator, within plants (see Dekhuijzen, 1964) indicates that its chemotherapeutic effects are not brought about by liberation of effective levels of dimethyldithiocarbamate ions *in vivo*.

Conclusion

Studies prior to the mid 1960s had clearly established both the potentialities and problems in the practical application of chemotherapy and the development of systemic fungicides. The difficulties of achieving the twin requirements of translocatability and selective toxicity towards the pathogen *in vivo* were only too obvious, but the discovery that a chemotherapeutant need not be fungitoxic *per se* greatly widened the range of compounds worthy of examination. Detailed studies such as those of Crowdy *et al.* (1959a, b) with griseofulvin analogues had indicated the complex interaction of many factors which determine the systemic activity shown by structurally related compounds.

As yet, however, no fully effective compound suitable for commercial use in the field had been developed. Even the best discovered by the mid 1960s were insufficiently active, too phytotoxic or too costly although a few, such as 8-quinolinol salts, sulphonamides and certain antibiotics, must be considered 'near misses'.

Chapter 3

Structure – activity relationships

D. Woodcock

The objects of chemical structure–biological activity studies are to discover active basic structures, to establish which functional groups are essential for, or promote high activity, to locate optimal homologues and to complement mode of action studies. Although such work has often been done in a desultory fashion in the past, it is clear that physicochemical parameters such as pK_a, partition coefficient and integrated absorption intensity of the hydrogen bond are being increasingly investigated and are often correlated with fungicidal activity. This has led to limited predictions from partition coefficients using the Hansch relationship, but such forecasts are usually only in a limited context and so far have often remained unpublicised.

In the past decade, the early stages of research and development of systemic fungicides inexorably followed the systemic insecticide lead and explored organophosphorus compounds for activity. Since then a wide variety of organic molecules have been scrutinised for systemic activity, and many interesting data have been published for a large number of pathogens.

In the following account, systemic activity has been interpreted in the broadest possible sense and compounds showing only limited movement, such as in trans-laminar activity, have also been mentioned.

Organophosphorus compounds

For a survey of organophosphorus compounds as fungicides, the review of Grapov and Mel'nikov (1973) is to be recommended, while the text by Fest and Schmidt (1973) is even more comprehensive.

Substituted 1,2,4-triazoles

It was perhaps inevitable that the quest for phosphorus-containing compounds with insecticidal and acaricidal properties should involve heterocyclic nuclei. The compound synthesised by van den Bos *et al.* (1960) from 3-amino-1,2,4-triazole and bis(dimethylamido) phosphoryl chloride, however, not only possessed good insecticidal and acaricidal activity, but was also found to be effective against powdery mildew of barley (caused by *Erysiphe graminis* var. *hordei*). This interesting spectrum of biological activity led to the synthesis of some forty more substituted 3-amino-1,2,4-triazoles containing a phosphoryl group and structural investigations by van den Bos (1960) later established the structure I for these compounds.

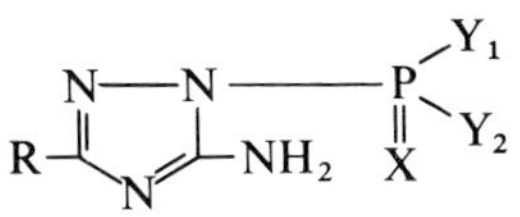

(I)

From the biological results obtained it was clear that the primary requirement for fungitoxicity was the presence of a bis(dimethylamido) phosphoryl group in the 2-position, a thiophosphoryl or other phosphorus-containing substituents being detrimental (Table 3.1). The nature of the substituent in the 5-position was also important, and in the n-alkyl series activity declined after the amyl member. Introduction of a methylene group between a substituent phenyl group and the triazole ring also caused a reduction in activity, as did substitution in the phenyl nucleus. It was noticeable that the insecticidal and acaricidal

Table 3.1 Biological activities of triamphos analogues van den Bos *et al.*, 1960)

Substituents in formula 1							
R	X	Y_1	Y_2	Fungitoxicity†	Insecticidal activity‡	Acaricidal activity‡	Phytotoxicity§
H	O	NMe_2	NMe_2	4	4	4	0
C_5H_{11}	O	NMe_2	NMe_2	5	2	3	2
$C_{11}H_{23}$	O	NMe_2	NMe_2	1	0	0	4
Phenyl	O	NMe_2	NMe_2	4	4	5	0
p-Chlorophenyl	O	NMe_2	NMe_2	3	3	4	1
p-Methoxyphenyl	O	NMe_2	NMe_2	2	1	0	0
Benzyl	O	NMe_2	NMe_2	1	–	4	0
H	S	NMe_2	NMe_2	0	1	1	1
C_5H_{11}	S	NMe_2	NMe_2	2	1	0	3
Phenyl	O	NHMe	NHMe	0	0	0	0
Benzyl	O	NHMe	NHMe	0	0	0	1
H	O	OEt	OEt	0	0	0	1
C_5H_{11}	O	OEt	OEt	0	–	4	0
Phenyl	O	OEt	OEt	0	–	–	4
C_5H_{11}	S	OEt	OEt	0	1	1	2
Phenyl	S	OEt	OEt	0	1	1	1

† O = <1/100 unit of dinocap, 5 = >3 units
‡ MLD O = >1000 ppm, 5 = 10 ppm
§ O = no damage, 5 = leaves dead

activity, as well as the phytotoxicity and mammalian toxicity, paralleled fungitoxicity, and it was significant that several compounds without contact or protectant activity nevertheless showed distinct systemic effects on insects, mites and powdery mildews.

Two compounds (I : R = C_5H_{11} or phenyl) were chosen for further trials and the latter, claimed as the first systemic fungicide, has reached commercial acceptance with the common name triamiphos. Although it has been used for the control of powdery mildew on ornamentals such as roses, on which curative effects have also been reported by Betgem (1960), attempts to control apple powdery mildew by administering it to the trunks and roots have generally failed because of phytotoxicity.

Further analogues have been synthesised (van den Bos *et al.*, 1961, 1966) and the structure–fungicidal activity relationships of the two series (II) and (III) are worth examining.

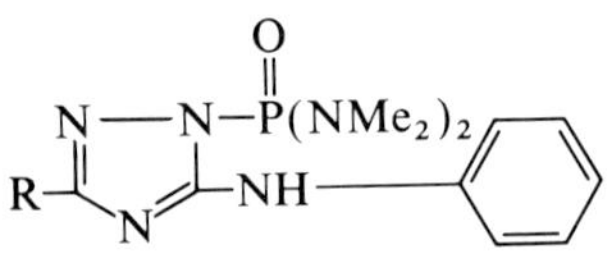

(II)

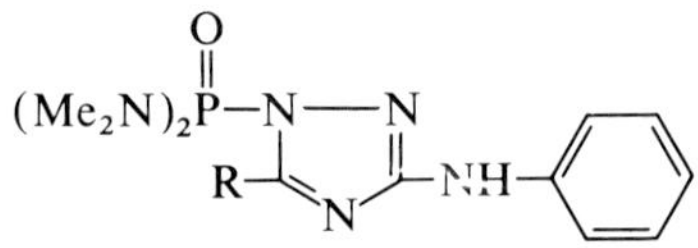

(III)

Fungicidal activity against *Sphaerotheca fuliginea* on cucumber seedlings and *Podosphaera leucotricha* on apple seedlings was examined by Tempel *et al.* (1968) using two tests. In the spray test, leaves were sprayed with aqueous suspensions of the compounds, whilst in the systemic test seedlings were grown in nutrient solution and then transferred to a fresh solution containing the test chemical and kept there for 48 h before inoculation. In another test using apple seedlings, designed to discover whether translocation took place after the leaf application, the test com-

Table 3.2 The activity of 1- and 2-[bis(dimethylamido)phosphoryl]-3-alkyl-5-anilino-1,2,4-triazoles against *Sphaerotheca fuliginea* on cucumber seedlings. Infection in percentages (Tempel *et al.*, 1968)

	N_1 series		**N_2 series**	
R =	Leaf spray test, 1000 ppm	Systemic test, 100 ppm	Leaf spray test, 1000 ppm	Systemic test, 100 ppm
H	30	0	100	100
CH_3	27	0	65	8
C_2H_5	4	7	100	39
i-C_3H_7	0	73	94	40
C_3H_7	29	61	94	100
C_4H_9	0	94	100	100
C_5H_{11}	6	82	—	—
C_6H_{13}	92	100	100	100
Triamiphos	81	86	—	—
Control			100	100

pounds were applied to upper and lower surfaces of one half of the leaf, and the infection on both halves assessed after the incubation period. The most striking feature of this work was the contrast between the activity of compounds with the phosphoryl group in position 1 (N_1 series) and those of the N_2 series: in both tests, with both pathogens, the N_1 members were far superior to their N_2 isomers. Within the active series there was some divergence in the results of the leaf application and systemic tests, and the alkyl chain length appeared to be critical. On cucumber seedlings only the lower members were active in the systemic test, whereas in the leaf spray test the effect was less well defined,

Table 3.3 The activity of 1- and 2-[bis(dimethylamido)phosphoryl]-3-alkyl-5-anilino-1,2,4-triazoles against *Podosphaera leucotricha* on apple seedlings. Infection in percentages (Tempel *et al.*, 1968)

	N_1 series		**N_2 series**	
R =	Leaf spray test, 300 ppm	Systemic test, 30 ppm	Leaf spray test, 300 ppm	Systemic test, 30 ppm
H	31	0	79	56
CH_3	18	13	17	56
C_2H_5	19	30	24	58
i-C_3H_7	15	20	41	71
C_3H_7	8	53	36	70
C_4H_9	3	86	57	94
C_5H_{11}	0	69	—	—
C_6H_{13}	7	58	59	77
Triamiphos	14	11		
Control			100	100

except that activity virtually disappeared with the hexyl member (Table 3.2). With apple seedlings infection was minimal with the n-amyl member (II : $R=C_5H_{11}$) in the leaf spray test, and with the unsubstituted compound (I : R = H) in the systemic test (Table 3.3). Although no translocation from a treated to an untreated leaf took place, lateral movement was demonstrated, and it was clear that the alkyl group again played a critical role, maximum activity being shown by the ethyl, isopropyl and propyl analogues (Table 3.4). It is possible that this is a reflection of optimum partition characteristics for movement in the leaf surface, rather than a distance factor as suggested by Tempel *et al.* (1968).

More recently, a new group of heterocyclic organophosphorus compounds, involving phosphoric esters of pyrazolopyrimidines, have been introduced. *O*-6-Ethoxycarbonyl-5-methylpyrazolo [1,5-*a*] pyrimidin-

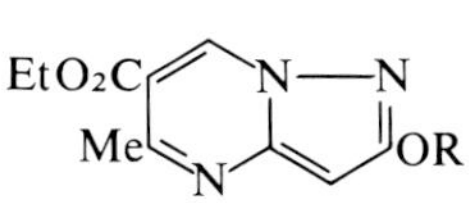

(IV)

Table 3.4 The leaf systemic activity of 1- and 2-[bis(dimethylamido)phosphoryl]-3-alkyl-5-anilino-1,2,4-triazoles against *Podosphaera leucotricha* on apple seedlings. Infection in percentages (Tempel *et al.*, 1968)

	N_1 series		**N_2 series**	
R =	Treated half-leaf 300 ppm	Untreated half-leaf	Treated half-leaf 300 ppm	Untreated half-leaf
H	0	40	44	100
CH_3	0	24	22	76
C_2H_5	0	16	22	82
i-C_3H_7	0	16	—	—
C_3H_7	2	18	46	80
C_4H_9	0	38	38	80
C_5H_{11}	0	44	52	96
C_6H_{13}	0	100		
Control			100	100

2-yl *O,O*-diethyl phosphorothioate [IV : R = $P(S)(OEt)_2$], proposed common name pyrazophos, is absorbed by the foliage and shoots and is translocated in the plant. Root uptake is poor and precludes any possible use by soil application or as a seed dressing. Nevertheless trials have shown that the material is non-phytotoxic to a wide range of crops and excellent control of the powdery mildews *Sphaerotheca humili*, *S. fuliginea* and *Erysiphe graminis* has been obtained. Against powdery mildew of apple, control comparable to that given by binapacryl and dinocap has been obtained, and noticeable improvements in vigour, colour of foliage, crop quality and yield have often resulted (Hay, 1971).

An admirable and comprehensive study of pyrazophos, including metabolism, has recently been made by de Waard (1974). Using the ^{14}C-labelled fungicide, he demonstrated rapid breakdown by *Pyricularia oryzae*, with the formation of the oxygen analogue PO-pyrazophos [IV : R = $P(O)(OEt)_2$] in addition to the des-phosphorylated compound (IV : R = H). The latter might be regarded as the actual fungitoxic principle, since it was active in mycelial growth tests, and in contrast to pyrazophos also tended to inhibit nucleic acid and protein synthesis as well as oxygen uptake. The insensitivity of *Pythium debaryanum* and *Saccharomyces cerevisiae* to pyrazophos, which does not appear to be degraded by them, seems highly significant.

Dialkylphosphorothioates and related compounds

Among the fungal diseases of rice, paddy blast caused by *Pyricularia oryzae* is of major economic importance. Control in the past has been with organomercury compounds such as phenylmercuric acetate and iodide and *N*-phenylmercuric *p*-toluene sulphonanilide, which are being currently criticised on the grounds of hazard to human health. More recently some antibiotics have been used, and pentachlorobenzyl alcohol has also been shown to be particularly specific *in vivo*, although without action *in vitro* (Sumi *et al.*, 1968).

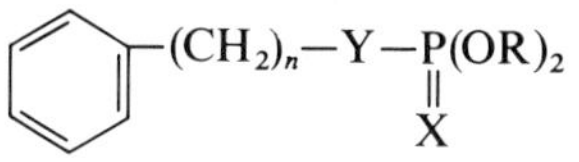

(V)

The benzyl group also features in a group of substituted phosphoric esters (V) recently developed in Japan (Kado *et al.*, 1965) and found to possess excellent systemic properties against rice blast in the field (Yoshinaga *et al.*, 1965). *O,O*-Dialkyl-*S*-benzyl phosphorothioates (V : Y = S, X = O, R = alkyl) were found to be more effective than the corresponding *O*-benzyl phosphates (V : Y = X = O, R = alkyl) and phosphorothionates (V : Y = O, X = S, R = alkyl), and the *S*-benzyl phosphorodithioates (V : Y = X = S, R = alkyl), Fig. 3.1.

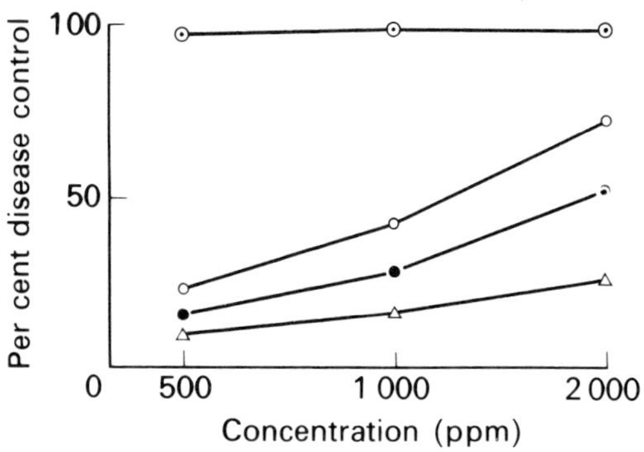

Fig. 3.1 Relation between the per cent control of rice blast on potted plants and the concentration of chemicals (Kado and Yoshinaga, 1969): ⊙ = *S*-benzyl *O,O*-diethylphosphorothioate; ○ = *S*-benzyl *O,O*-diethylphosphorodithioate; ● = *O*-benzyl *O,O*-diethylphosphate; and △ = *O*-benzyl *O,O*-diethylphosphorothionate.

Among a limited number of alkyl groups which were surveyed, greatest activity was shown by the ethyl, propyl and butyl members. Two *S*-benzyl phosphorothioates have reached commercial development – Kitazin (*S*-benzyl *O,O*-diethylphosphorothioate (V : $n = 1$, Y = S, X = O, R = C_2H_5) and Kitazin P – the isopropyl analogue; and two related compounds have also been developed – Hinozan (O-ethyl-*S,S*-diphenyl phosphorodithioate; VI), proposed common name edifenphos, and Inezin (VII: *S*-benzyl *O*-ethyl phenyl phosphorothioate) (Kado and Yoshinaga, 1969).

Adsorption of Kitazin into the rice plant by leaf application and subsequent translocation within the plant, was confirmed by autoradiography using the ^{32}P- or ^{35}S-labelled material. The compound is fairly water-soluble (500 ppm) and it was also readily absorbed through the root system and rapidly translocated in the transpiration stream to the

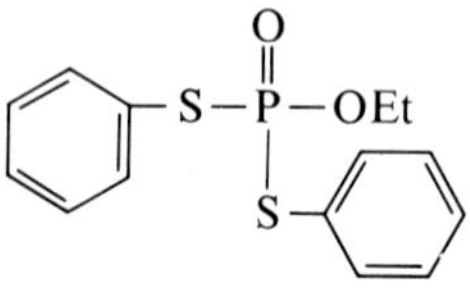

(VI)

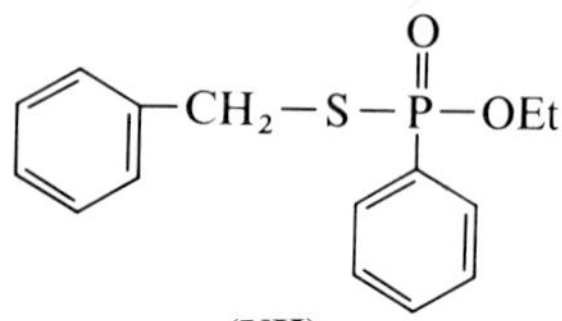

(VII)

site of pathogenic infection. This made possible Kitazin P granules (containing 17 per cent active ingredient), formulated for direct application in the paddy fields, where fungicidal effect is maintained for at least 3 weeks. This is far superior to foliar application, and there is a welcome bonus in the remarkable increase in rice yield, and a shortening of the stalks which reduces the tendency for lodging.

Mode of action studies have been carried out by Misato and his associates. They found that Kitazin at its fungistatic concentration of 55 ppm inhibited the incorporation of ^{14}C-glucosamine into the cell wall fraction in *P. oryzae*, without affecting the mycelial respiration or interfering with protein and nucleic acid synthesis, and they suggested that the site of action of the fungicide might be the cell wall synthesis system (Kakiki *et al.*, 1969). Further confirmatory investigations were carried out using a number of analogues (Maeda *et al.*, 1970). Of the ten compounds tested, the thioates (V: X = O, Y = S) both strongly inhibited mycelial growth *in vitro* and gave excellent disease control on potted rice seedlings (Table 3.5). By contrast, the phosphates (V: X = Y = O), though strong inhibitors of mycelial growth, showed poor blast control and this was thought to be due to breakdown of the compounds – a possibility which seems to be supported by decreasing *in vitro* activity with prolonged incubation. The complete absence of any effect on mycelial growth *in vitro* or any suggestion of disease control *in vivo* reported earlier for the thionates (V: X = S, Y = O) and dithioates (V:

Table 3.5 Fungitoxicity of Kitazin and its analogues (Maeda *et al.*, 1970)

Substitutes in formula V				**Inhibition of mycelial growth†**	**Blast control‡ (pot test)**
R =	X	Y	*n*		
Ethyl	O	S	1	A	A
isoPropyl	O	S	1	A	A
n-Butyl	O	S	3	A	A
Isopropyl	O	O	1	B	D
n-Propyl	O	O	3	A	B
n-Butyl	O	O	1	A	D
isoPropyl	S	O	3	C	D
Ethyl	S	O	1	C	D
n-Butyl	S	S	1	C	D
isoPropyl	S	S	1	C	D
Control				C	D

† A = inhibition ratio > 80%; B = inhibition ratio 60–80%; C = no effect
‡ A = protective value > 80%; D = no effect

X = Y = S) was confirmed. It was also significant that thioates and phosphates, which inhibit mycelial growth *in vitro*, also inhibited ^{14}C-glucosamine incorporation into the cell wall fraction of *P. oryzae*, whereas the non-fungitoxic thionates and dithioates did not do so. This positive correlation strongly indicated that the Kitazin site of action is related to fungal cell wall synthesis, and the accumulation of UDP-acetylglucosamine strongly supports the view that the fungicide affects a stage beyond the formation of the amino sugar derivative. Maeda *et al.* (1970) suggest that all the evidence so far obtained points to penetration of the cytoplasmic membrane as being the site of action of these compounds.

Metabolism studies with edifenphos, Inezin and Kitazin P using *P. oryzae* have been made by Uesugi and Tomizawa (1971a, b) and Tomizawa and Uesugi (1972), who found no differences in the rate and mode of degradation of these compounds by resistant and susceptible strains of the fungus. While *para*-hydroxylation took place in one phenyl nucleus in the case of edifenphos, *meta*-hydroxylation of the benzyl group was a metabolic feature of both Inezin and Kitazin P.

Antibiotics

The translocation of some antibiotics in higher plants and their systemic antifungal activity was first demonstrated some 20 years ago, and the review of the literature up to 1962 by Dekker (1963a) is to be commended.

Griseofulvin

Possibly the most interesting is griseofulvin, a neutral antibiotic first isolated from the mycelium of *Penicillium griseofulvum* by Oxford *et al.* (1939) and more recently recognised as a metabolic product of many *Penicillium* spp. Translocation was observed in lettuce and tomatoes, uptake from aqueous solution through the root system being rapid, with a slower (passive) movement in the shoots in the transpiration stream. Translocation was demonstrated initially by bioassay (Brian *et al.*, 1951) but later the antibiotic was isolated from the tissues of treated plants and identified (Crowdy *et al.*, 1955).

Griseofulvin does not prevent germination of fungal spores, but it has a remarkable effect on fungal development, a concentration as small as 10 μg/ml causing stunting of hyphae, excessive and abnormal branching, hyphal distortion and loss of apical dominance on germinating conidia of *Botrytis allii*. Not all fungi are as sensitive and it was originally thought that griseofulvin exerted this characteristic effect by interfering with the synthesis of cell wall chitin, since fungi having cellulose-based cell walls are unaffected. However, there are a few species of fungi with chitinous cell walls that show no response to griseofulvin

(Napier *et al.*, 1956), and it has been suggested that nucleic acids might be the primary target (McNall, 1960). This is supported by recent work using the highly griseofulvin-sensitive animal pathogen *Microsporium gypseum* (El-Nakeeb and Lampen, 1964a, b; El-Nakeeb *et al.*, 1965a, b). Whatever its *modus operandi* it is relatively non-phytotoxic, but unfortunately slowly degraded in plant tissue and more quickly in soil (Crowdy *et al.*, 1956). This liability makes its antifungal activity of limited practical use in agriculture and horticulture, though it is quite effective against a number of mildews (Brian, 1960). These include the powdery mildew of roses (caused by *Sphaerotheca pannosa*), *Pseudoperonospora cubensis* on cucumber and *Oidium chrysanthemi* on chrysanthemums. It has also controlled some *Botrytis* diseases including grey mould on lettuce, tulip fire and the turf diseases caused by *Fusarium nivale* and *Sclerotinia homeocarpa.* However, although griseofulvin affords some control of these plant pathogens it is no better than, and often inferior to, other established fungicides in general use.

The chemistry and absolute configuration of griseofulvin (VIII: R = Cl), which has asymmetric centres at positions 2 and 6′, have been well established (Grove 1963) and a considerable amount of interesting

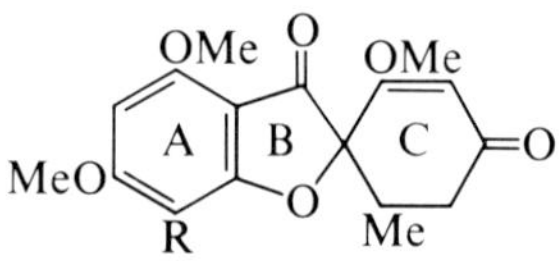

(VIII)

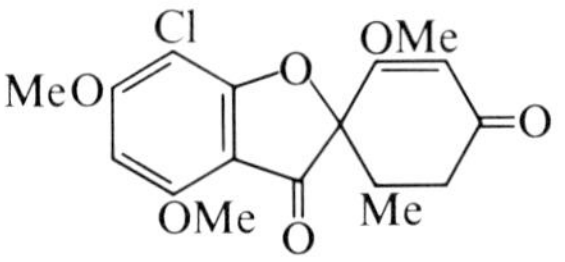

(IX)

structure–activity data is available. The stereochemistry is critical for antifungal activity and the racemic form has only half of the activity of (+)-griseofulvin (VIII: R = Cl) itself. The diastereoisomer (IX), known as epi-(+)-griseofulvin, and all the transformation products with this same stereo-chemistry are inactive. However, some activity is still present after the elimination of the asymmetric 6′-centre, for dehydrogriseofulvin (X) and the synthetic grisenone (XI) are weakly active, though grisan-3,4′-dione (XII), without ring A substituents, is not.

When *Penicillium nigricans* and *P. griseofulvum* are grown in chloride-deficient, and also to some extent in normal media, the dechloro analogue (VIII: R = H) is produced. Addition of potassium bromide to the chloride-deficient medium leads to the production of 7-bromo-7-dechloro-griseofulvin (VIII: R = Br) – the first bromine-containing fungal metabolic product to be characterised. Both analogues are less active

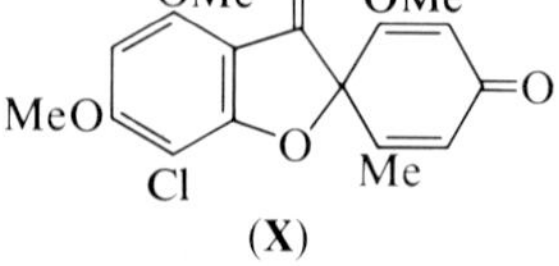

(X)

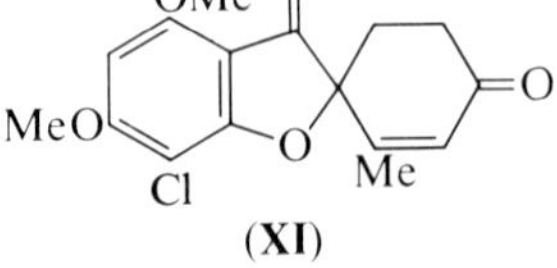

(XI)

than griseofulvin itself, but the (+)-fluoro analogue (VIII: R = F), synthesised by Taub *et al.* (1962), was just as active in the disc plate assay

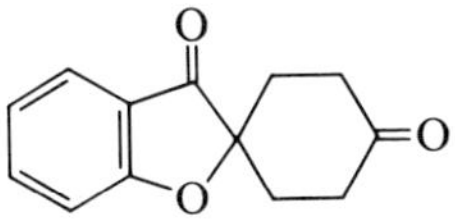

(XII)

using *Botrytis allii.* A free 5-position appears to be essential for high activity, however, for the substitution of a halogen in position 5, instead of position 7, led to reduced activity (Gerecke *et al.*, 1962). Moreover although the 5,7-dichloro compound, geodin (XIII: R = Cl) showed some fungistatic activity, as do many dienones, the typical curling effects on fungal hyphae were absent. A basic requirement would also appear to be the tricyclic system (XIV), synthetic compounds with only a five-membered ring C, such as XV, being inactive. Compounds more active

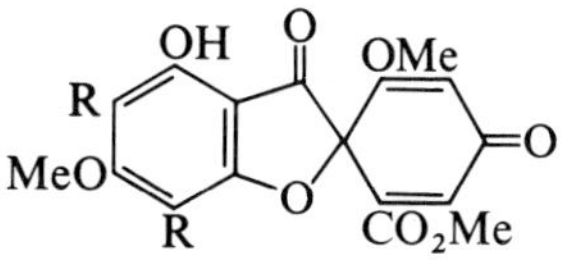

(XIII)

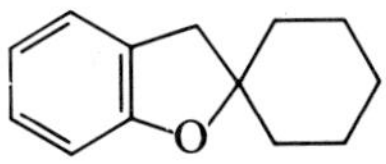

(XIV)

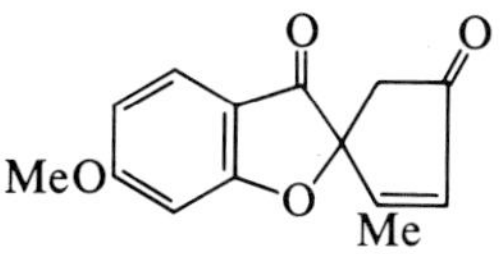

(XV)

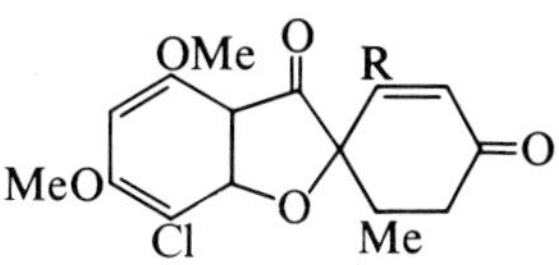

(XVI)

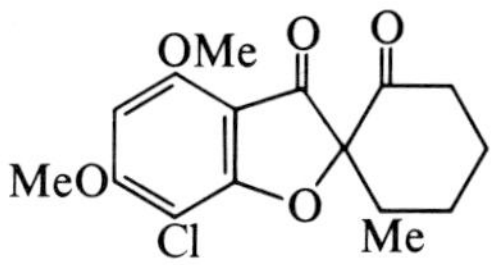

(XVII)

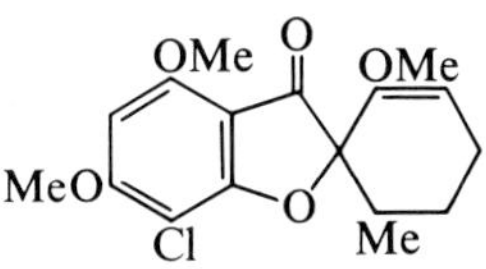

(XVIII)

than griseofulvin can be obtained, however, by increasing the oil–water partition coefficient and homologues such as XVI (R = OPr^n or OBu^n) had 20–50 times the activity of griseofulvin *in vitro* (Crowdy *et al.*, 1959b). On the other hand, a polar group not necessarily in the 4′-position in ring C seems to be desirable for activity since the 3-2′-ketone

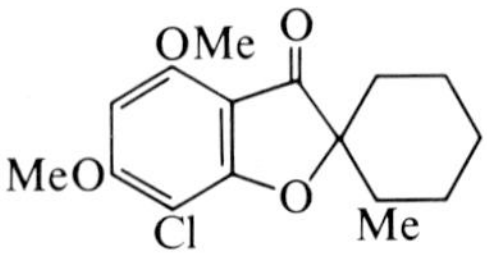

(**XIX**)

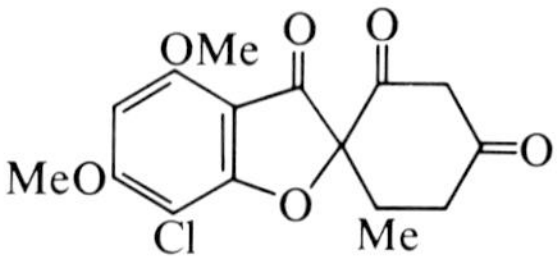

(**XX**)

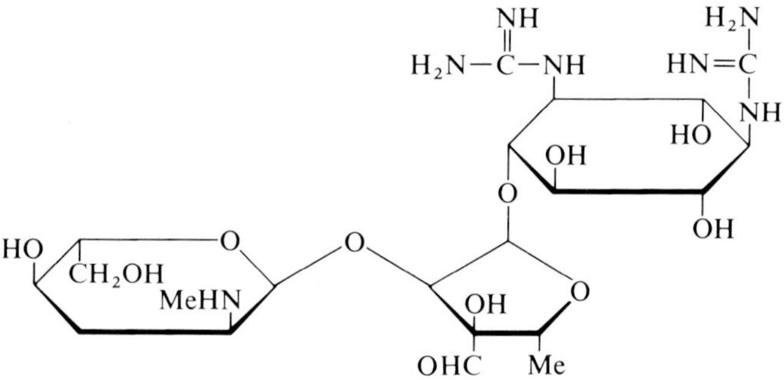

(**XXI**)

(XVII) was active, whereas compounds XVIII and XIX were not. The inactivity of the acidic trione (XX) and the basic griseofulvamine (XVI: R = NH_2) also seems highly significant.

Streptomycin

Other antibiotics have been examined for possible translocation and accumulation in the tissues of higher plants. Thus Mitchell *et al.* (1954) found that of nine antibiotics tested for their ability to protect bean plants against the halo-blight organism, *Pseudomonas medicagensis* var. *phaseolicola* (Burk, Stapp and Kotte), only streptomycin sulphate gave complete protection, although dihydrostreptomycin sulphate and terramycin hydrochloride showed slight activity. This ready translocation of streptomycin (XXI) confirmed previous reports of uptake and translocation of the antibiotic by higher plants (Anderson and Nienow, 1947; Winter and Willeke, 1951) and further confirmation was provided by Pramer (1953) who identified the antibiotic chromatographically in tissue extracts of cucumber leaves after root uptake.

Whilst streptomycin is active against a large number of bacteria, however, it has relatively little effect on anaerobic bacteria, fungi, protozoa and viruses. Thus whilst it has been effectively used as a plant protective agent against such bacterial diseases as fire blight on apples, pears and some ornamental shrubs of the rose family (caused by *Erwinia amylovora*) and against walnut blight (caused by *Xanthomonas juglandis*), its proven systemic properties appear to be of little avail in the control of fungal pathogens. It has however been used to some extent to control early infections of hop downy mildew (caused by *Pseudoperonospora humuli*) and actual translocation in the hop plant was first demonstrated by Maier (1960). It has also some effect on Phycomycetes, such as those causing late blight of tomatoes and potatoes, and even as little as 4 μg/ml of nutrient solution is sufficient to check *P. infestans* completely on tomatoes (Müller *et al.*, 1954; Crosse *et al.*, 1960). The use of streptomycin as a seed disinfectant has also been extensively explored with considerable success and the only disadvantage of such treatments is the possibility of encouraging streptomycin-resistant human pathogens.

Cycloheximide

Also produced by *Streptomyces griseus* is the antifungal antibiotic cycloheximide – possibly more commonly recognised as 'Actidione' – a trade name of the Upjohn Co. It is obtained as a by-product in streptomycin manufacture and is remarkably active against a wide range of fungi and yeasts, but inactive against bacteria. Its use against plant diseases has been rather limited by its phytotoxicity. Cycloheximide, the common name for β-[2-(3,5-dimethyl-2-oxocyclohexyl)-2-hydroxyethyl]glutarimide (XXII), has four asymmetric centres at C-2, C-4, C-6 and C-2′,

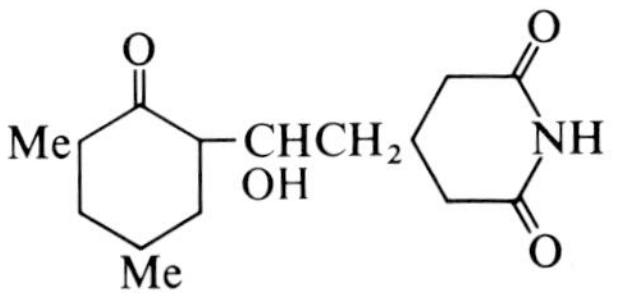

(XXII)

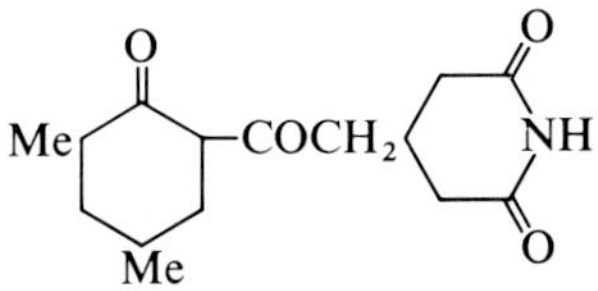

(XXIII)

and a few of the other possible stereoisomers have also been found to be produced by *Streptomyces* spp., including isocycloheximide (Lemin and Ford, 1960), naramycin B (Okuda *et al.*, 1958) and two 'cycloheximide diastereoisomers' (Rao and Cullen, 1960). Isocycloheximide may also be obtained by isomerisation of cycloheximide using deactivated alumina, and it seems likely that neither the carbon atom carrying the secondary alcoholic group nor the C-4 asymmetric centre are involved. Moreover, since there are two isomeric dehydrocycloheximides (XXIII), inversion must be at C-2 or C-6, and because both cycloheximide and isocycloheximide yield the same anhydrocycloheximide (XXIV), it would seem that cycloheximide is 2,6-*trans* whilst isocycloheximide is 2,6-*cis*. Both naramycin B and isocycloheximide appear to be less active than cycloheximide against yeasts, the former only 32 per cent as active

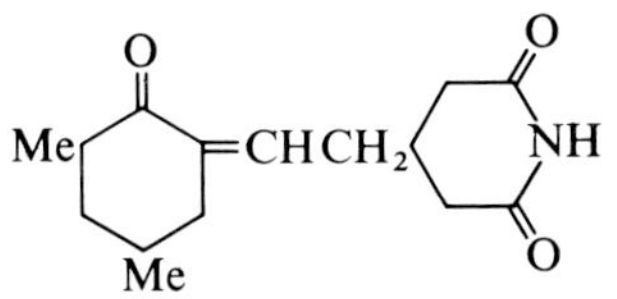

(XXIV)

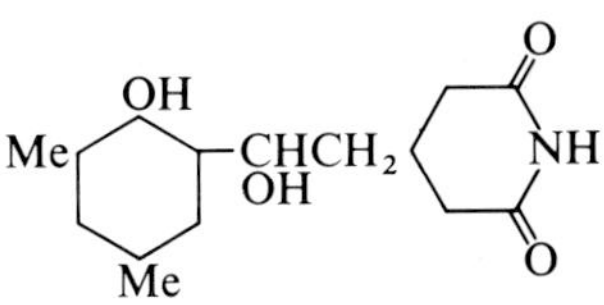

(XXV)

against *Saccharomyces sake* (Okuda *et al*., 1958) and the latter about 30 per cent as active against *S. pastorianus* (Lemin and Ford, 1960). Anhydrocycloheximide, on the other hand, was just as active, but the related compounds dehydrocycloheximide (XXIII) and dihydrocycloheximide (XXV) were much less so (Lee and Wilkie, 1965). It is possible however that the lower activity of these last two compounds may be due to a decrease in the amount entering the cell since they are both insoluble in water, in contrast to cycloheximide and anhydrocycloheximide. Another dehydrocycloheximide, inactone (XXVI), formed by *Streptomyces griseus* also shows no antifungal activity (Paul and Tchelitcheff, 1955).

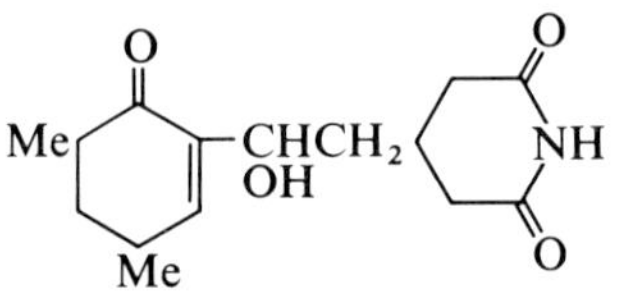

(XXVI)

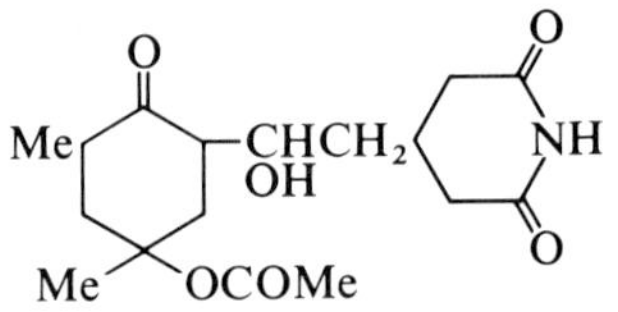

(XXVII)

A number of cycloheximide derivatives have been prepared and assayed against the fungus *Cytospora cincta*. The acetate and acetoacetate were the least fungitoxic, while the oxime, carbazone, thiosemicarbazone and methylhydrazone, although active, were less so than the parent antibiotic (Harvey and Helton, 1962). Whilst these analogues are less fungitoxic *per se*, they are often less phytotoxic and more readily translocated, so much so that use has been made of these attributes in particular applications. Lemin and Magee (1957) using ^{14}C-cycloheximide acetate, have shown that it is taken up by the roots of tomato plants, and thus renders the leaves highly antifungal, though whether because of activity *per se* or because of cycloheximide release is not certain. The oxime shows systemic activity against the cherry leaf spot fungus *Coecomyces hiemalis* Higgins, even protecting new shoots (Hamilton *et al*., 1956), and the semicarbazone confers persistent protection against stem rust in wheat (Anderson and Rowell, 1960). Other compounds closely related to cycloheximide include the antibiotic E-73 (XXVII) produced by *Streptomyces albulus* which has an acetoxy group at C-4 (Rao, 1960), and the streptovitacins formed by a strain of *S. griseus* (Eble *et al*., 1959) which are cycloheximide with the cyclohexyl ring hydroxylated in positions 5, 4 and 3 for streptovitacins A, B and C_2 respectively, together with D – which is ring hydroxylated at an unknown position. These compounds are much less active than cycloheximide and have been studied mainly because of their high antitumour activity.

Kerridge (1958), working with *Saccharomyces carlsbergensis*, first suggested that cycloheximide inhibits both protein and DNA synthesis, and Siegel and Sisler (1963, 1964, 1965) have done much to probe and confirm this. Using a cell-free system of *S. pastorianus* and ^{14}C-labelled leucine, they demonstrated inhibition of this amino acid into protein and suggested that the antibiotic prevents the transfer of aminoacyl-soluble RNA to the ribosomes and formation of the peptide bond. This has been confirmed by both Ennis and Lubin (1964) and by Wettstein *et al.* (1964).

Antibiotics in Japan

Blasticidin

Nowhere is the agricultural usage of antibiotics increasing faster than in Japan (Table 3.6). The first to be developed was blasticidin S

Table 3.6 Agricultural antibiotics used in Japan in 1973 (Mitsato, 1975)

Registration	Antibiotic	Disease	Weight (tons)	Value (1 000 Yen)
Antifungal				
1959	Cycloheximide	Onion downy mildew	3	6 060
1959	Griseofulvin	Fusarium wilt (melon)	1	2 300
1961	†Blasticidin S	Rice blast	4 317	396 986
1965	†Kasugamycin	Rice blast	39 710	3 835 978
1967	†Polyoxin	Rice sheath blight	2 387	1 015 863
1970	†Ezomycin	Stem rot (kidney bean)	0	0
Antibacterial				
1957	Streptomycin	Various bacterial diseases of fruit and vegetables	332	602 377
1964	†Cellocidin	Bacterial leaf blight (rice)	0	0
1964	Chloramphenicol	Bacterial leaf blight (rice)	12	38 478
1968	Novobiocin	Bacterial canker (tomato)	0	0

† Discovered in Japan

(XXVIII), which is produced by *Streptomyces griseochromogenes* (Fukunaga *et al.*, 1955), the structure of which was established by Otake *et al.* (1965). It is active against both bacteria and fungi, and is particularly selective in its antifungal activity. Thus it inhibits growth of *Pyricularia oryzae* at concentrations as low as 5–10 μg/ml, and this has made

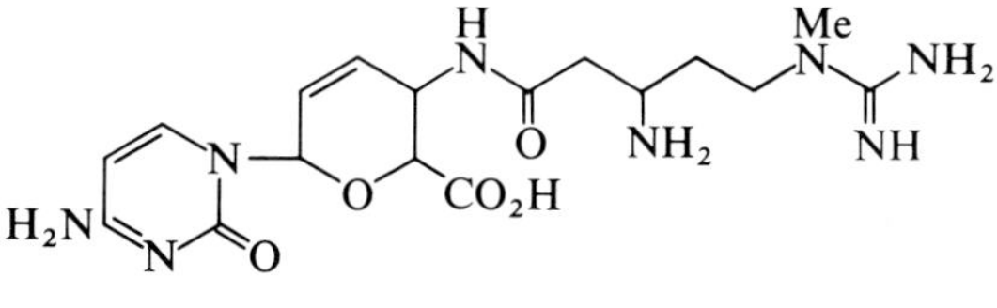

(XXVIII)

the control of paddy blast on rice possible. The antibiotic is up to 100 times as effective as organomercury fungicides (Table 3.7) though con-

Table 3.7 Effect of blasticidin S against leaf blast (Misato, 1969)

		No. of disease spots/pot†	
Fungicide	**Concentration a.i. (mcg/ml)**	Protective effect (spray four days before inoculation	Therapeutic effect (spray one day after inoculation)
Blasticidin S	2	96	19·3
(wettable powder)	5	64	3·7
	10	100	1·5
	20	76	0·2
	50	42	0
	100	11	0
	200	10	0
Phenylmercuric acetate	10‡	48	10·0
emulsifiable concentrate)	20‡	28	7·8
Untreated check	—	86	76·3

† Three pots per treatment, ten seedlings per pot, four leaves per seedling
‡ As Hg

centrations of 20–40 μg/ml may cause chlorotic spots on rice plants. Among a number of derivatives tested, the monobenzylaminobenzyl sulphonate appeared to be least phytotoxic to the rice plant (Asakawa *et al.*, 1963) and this salt is the active ingredient in commercial blasticidin S preparations. Its mode of action against *P. oryzae* is the inhibition of protein synthesis, but no structure-activity information appears to be available.

Its degradation in the rice plant, by sunlight and in the soil, has been examined by Misato *et al.* (1972). Various soil and paddy field microorganisms such as *Pseudomonas marginalis*, *Ps. ovalis* and *Fusarium oxysporum* can cause inactivation, the major metabolite being cytomycin (XXIX) in which cyclisation has resulted in the formation of an additional pyrimidine ring.

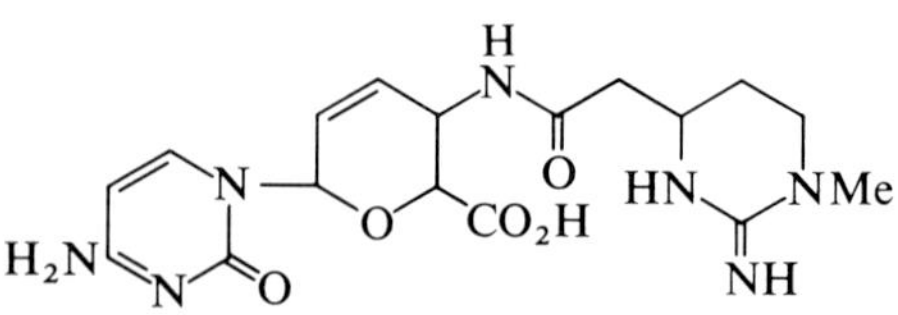

(XXIX)

Though blasticidin S has been used in paddy fields for more than 10 years now, no resistant strain of *P. oryze* has yet been detected (Misato, 1975).

Polyoxins

Another group of antibiotics which also contain the pyrimidine nucleus are the polyoxins (XXX), components of the complex produced by *Strep-*

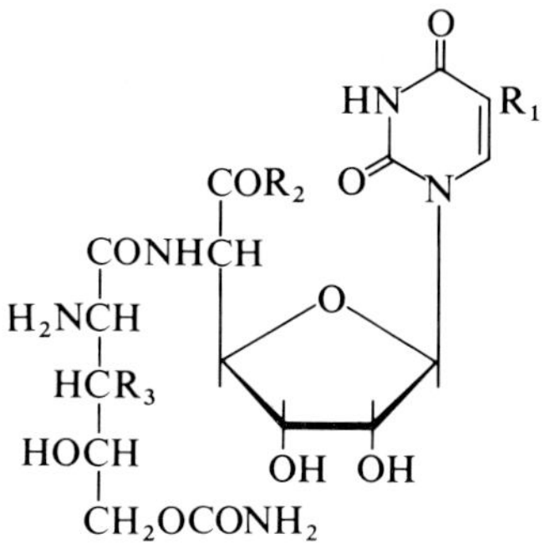

(XXX)

	R_1	R_2	R_3
A	CH_2OH	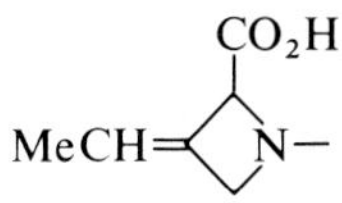	OH
B	CH_2OH	OH	OH
D	COOH	OH	OH
E	COOH	OH	H
F	COOH	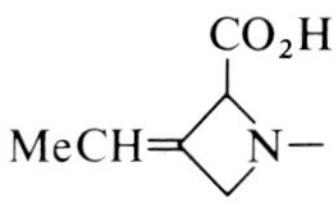	OH
G	CH_2OH	OH	H
H	Me	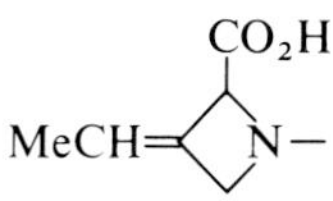	OH
K	H	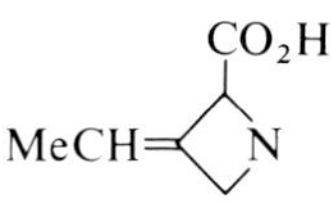	OH
L	H	OH	OH

tomyces cacaoi var. *ascensis* (Suzuki *et al.*, 1965; Isono *et al.*, 1967). These have similar physical and chemical properties but show a marked specificity in their antifungal activity (Table 3.8). Thus of the twelve polyoxins (A–L) characterised so far, only C and I show little activity against phytopathogens. Of the remainder, polyoxin A is effective against brown spot of rice (caused by *Helminthosporium oryzae*), B and G against pear black spot fungus and apple cork spot, whilst D is most effective for rice sheath blight (caused by *Pellicularia sasakii*). Most attention has been devoted to polyoxin D, particularly in mode of action studies

Table 3.8 Antifungal activities of polyoxins (Misato, 1969)

Test organism	A	B	C	D	E
Pyricularia oryzae	3·12†	6·25	>100	3·12	12·5
Cochliobolus miyabeanus	3·12	3·12	>100	6·25	12·5
Pellicularia sasakii	12·5	1·56	>100	<1·56	1·56
Alternaria kikuchiana	50	12·5	>100	50	50
Glomerella cingulata	>100	>100	>100	>100	>100
Physalospora laricina	25	3·12	>100	100	50
Cladosporium fulvum	3·12	1·56	>100	100	25
Fulsarium oxysporum	>100	>100	>100	>100	>100

Test organism	F	G	H	I
Pyricularia oryzae	25	6·25	3·12	>100
Cochliobolus miyabeanus	6·25	3·12	25	>100
Pellicularia sasakii	50	1·56	50	>100
Alternaria kikuchiana	>100	6·25	12·5	>100
Glomerella cingulata	>100	>100	>100	>100
Physalospora laricina	>100	6·25	12·5	>100
Cladosporium fulvum	25	3·12	6·25	>100
Fusarium oxysporum	>100	>100	>100	>100

† M.I.C. (μg/ml)

(Endo and Misato, 1969). It seems clear that this antibiotic selectively inhibits the synthesis of cell wall chitin in *Neurospora crassa* at levels which are comparable to those required for antifungal activity. Kinetic studies in a cell-free system showed that the incorporation of *N*-acetylglucosamine from UDP-*N*-acetylglucosamine into chitin was strongly inhibited. The greater sensitivity to the antibiotic of the chitin synthetase system, compared to the mycelium, seems to point to some kind of barrier to the passage of the toxicant.

Kasugamycin

Two other antibiotics deserve mention, not least because of their obvious potential. Kasugamycin is a water-soluble base obtained from the culture broth of *Streptomyces kasugaensis* (Umezawa *et al.*, 1965), and its structure (XXXI) was established by Suhara *et al.* (1966). It selec-

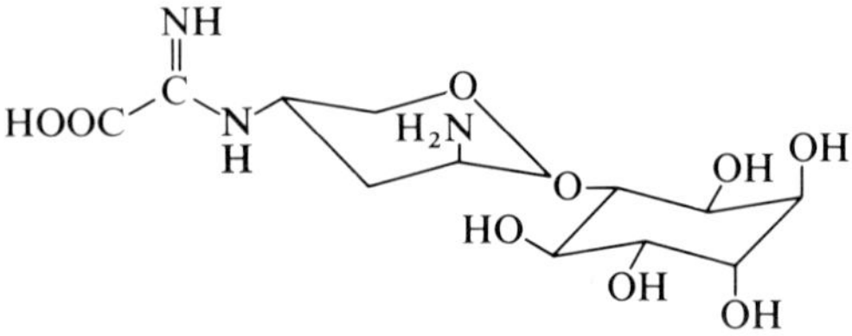

(XXXI)

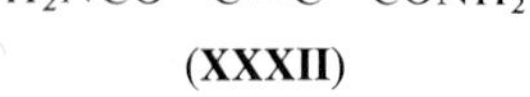

(XXXII)

tively controls *P. oryzae* and *Pseudomonas* spp. and whilst its lowest effective concentration against paddy blast is very similar to that of blasticidin S, it is non-phytotoxic, so much so that heavy infections can be treated safely with high concentrations of kasugamycin for rapid and complete control. Successful paddy blast control for 30 days has also been achieved with the use of seed coated with a 2 per cent wettable powder formulation of the antibiotic.

Unfortunately since 1971 the development of a kasugamycin-resistant strain of *P. oryzae* has become a serious problem, though withdrawal of the antibiotic from use for 1 year in 1973 resulted in a remarkable decrease in the incidence of this.

Cellocidin

Cellocidin (XXXII), an antibiotic produced by *Streptomyces chibaensis* (Suzuki *et al.*, 1958), is unique in its simplicity of structure. This has made synthetic production easy and it is now made from fumaric acid or 1,4-dihydroxybutyne. It has a preventive effect against rice bacterial leaf blight at 100–200 ppm and a study of its effects on the metabolism of *Xanthomonas oryzae* by Okimoto and Misato (1963a, b) suggests that it acts as an inhibitor of the α-ketoglutarate–succinate system.

Polyenes and polyynes

Many more antifungal antibiotics too numerous to mention here are known and the number increases yearly. For example, there are the polyene macrolides which consist of a macrocyclic lactonic ring often containing a polyene chromophore. They are all produced by *Streptomyces* spp. and are highly antifungal *in vitro*, with the exception of those without the polyene chromophore, *e.g.* erythromycin, carbomycin and oleandomycin. A polyene macrolide such as rimocidin (XXXIII) (Bhate

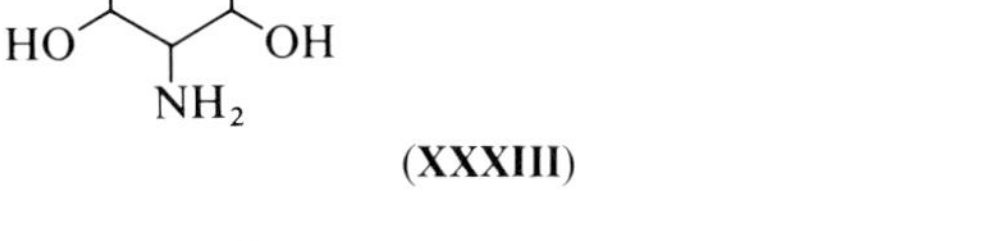

(XXXIII)

et al., 1964), despite its poor aqueous solubility, is nevertheless taken up by broad bean plant roots and can be subsequently extracted from the leaves. Despite this successful translocation no protection was afforded against *Botrytis fabae*, although a comparable concentration of the antibiotic would have inhibited spore germination *in vitro* (Oort and Dekker, 1960). Use of rimocidin as a seed disinfectant is perhaps more promising, for deep-seated infections of *Ascochyta pisi* in pea seeds have been largely controlled by soaking for 24 h (Dekker 1957), whilst a significant reduction of sorghum kernel smut (caused by *Spacelotheca sorghi*) has been obtained by similar treatment of the seed (Krüger, 1959).

Whereas polyene macrolides are formed by Streptomycetes and are almost exclusively antifungal, the non-macrolide polyynes and polyenes are often formed by Basidiomycetes and are both antifungal and antibacterial. Many of them are extremely unstable (e.g. half-life of mycomycin (XXXIV) is only 3 h at 27°C) and for this reason alone, interest in their antifungal activity is largely academic.

$$HC{\equiv}C{-}C{\equiv}C{-}CH{=}C{=}CH{-}CH{=}CH{-}CH{=}CH{-}CH_2CO_2H$$

(XXXIV)

Table 3.9 The systemic fungicidal effect of substituted oxathiins (XXXVI) as measured by their ability to control bean rust (von Schmeling *et al.*, 1966).

R	ppm	% control
Phenyl	12·5	60
	25	95
	50	99
	100	100
2-Methylphenyl	12·5	30
	25	65
	50	90
	100	100
3-Methylphenyl	12·5	30
	25	80
	50	100
	100	100
4-Methylphenyl	125	40
	500	90
	2 000	100
2-Chlorophenyl	500	0
	2 000	90
4-Chlorophenyl	500	0
	2 000	75
4-Ethoxyphenyl	500	10
	2 000	75
3-Ethoxyphenyl	125	100
2,4-Dimethylphenyl	125	100
2-Biphenyl	125	90
	500	96
	2 000	99
n-Butyl	125	25
	500	100
Cyclohexyl	50	60
	100	97
	200	100
Allyl	500	10
	2 000	35
α-Naphthyl	500	0
	2 000	35

Carboxylic acid anilides

Probably the first compound of this class to be shown to be fungicidal was salicylanilide (XXXV) (Fargher *et al.*, 1930) which has been described as a selective protectant fungicide. Despite a close structural similarity to the compounds which follow, systemic properties have never been attributed to it. By contrast the oxathins (XXXVI) which

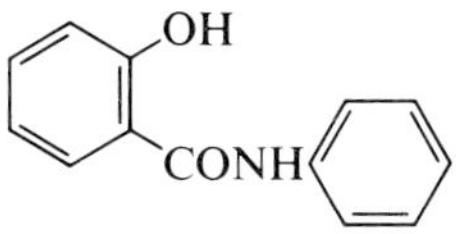

(XXXV)

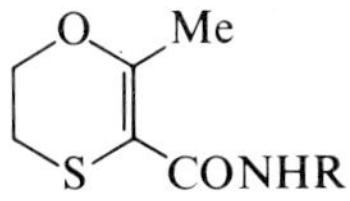

(XXXVI)

were first reported by von Schmeling and Kulka (1966) possess highly selective systemic properties acting mainly on Basidiomycetes – a class of fungi which includes such important pathogens as the rusts, smuts and bunts of cereals and the soil fungus *Rhizoctonia solani* Kühn. Evaluation of these compounds in the control of bean rust (caused by *Uromyces phaseoli typica* Arth.) was carried out using a spray application technique and the results are shown in Table 3.9 (von Schmeling *et al.*, 1966). The most active compound was 2,3-dihydro-6-methyl-5-phenylcarbamoyl-1,4-oxathiin (XXXVI: $R = C_6H_5$), since given the common name carboxin. Any substitution in the aromatic ring of chloro, alkyl or alkoxy groups resulted in reduced fungitoxicity, as did its replacement by alkyl, cyclohexyl or other aryl groups. In soil tests against *R. solani* on cotton the effect of substitution in the phenyl nucleus was less dramatic, and the 5-(2-methylphenyl) carbamoyl analogue was, if anything, slightly more effective against this pathogen than carboxin (Table 3.10).

Table 3.10 Control of *R. solani* on cotton using 20 ppm of substituted oxathiin (XXXVI) in the soil (von Schmeling *et al.*, 1966)

R	% Emergence	% Stand
Phenyl	80	80
2-Methylphenyl	88	88
3-Methylphenyl	76	76
2-Chlorophenyl	80	68
4-Chlorophenyl	52	48
Inoculated soil control	56	16
Uninoculated soil control	76	76

The selective toxicity to Basidiomycetes was confirmed by Edgington *et al.* (1966), who showed that whereas a concentration of 8 ppm or less of carboxin completely prevented the mycelial growth of many Basidiomycetes, other fungal groups such as Phycomycetes, Ascomcetes and

Deuteromycetes were only 50 per cent inhibited by concentrations of 32 ppm or higher. Two exceptions to this fungitoxicity pattern were an isolate of *R. solani* which was not as sensitive as other Basidiomycetes, and the Deuteromycete *Verticillium albo-atrum* which was more sensitive. Against bean rust in greenhouse pot tests, carboxin was found to be inferior to the corresponding sulphone, oxycarboxin (XXXVII), possibly because of its oxidation to the less active sulphoxide (XXXVIII).

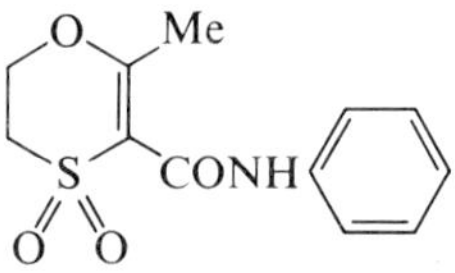

(XXXVII)

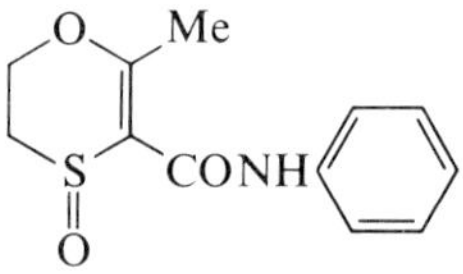

(XXXVIII)

In the field, however, carboxin proved highly effective in the control of loose snut of barley (caused by *Ustilago nuda* (Jens.) Rostr.) whereas oxycarboxin was only moderately successful.

Later work on the uptake of carboxin and oxycarboxin by *R. solani* and *U. maydis*, which are susceptible, and *Fusarium oxysporum* and *Saccharomyces cerevisiae* which are resistant to these chemicals, led Mathre (1968) to suggest that the specificity of the oxathiins may be related to the amount of fungicide taken up and bound to cell organelles.

The stability of carboxin has been examined by Chin *et al.* (1969), who found that oxidation to 2,3-dihydro-6-methyl-5-phenylcarbamoyl-1,4-oxathiin-4-oxide (XXXVIII) took place not only in water and soil, but also in barley, wheat and cotton plants. This change is facilitated by low pH and it is relatively stable above pH 6. There is no evidence so far of any further oxidation to the sulphone and it has been suggested that the energy requirement for this reaction is not available under biological conditions. The sulphoxide has a fungicidal activity against wheat leaf rust (caused by *Puccinia rubigo-vera tritici* (Erikss.) Carleton) *in vitro* at least 5 000 times lower than carboxin, and its production may explain a poor performance of the parent compound in certain conditions. Snel and Edgington (1970) confirmed the uptake of carboxin by bean roots and oxidation to the non-fungitoxic sulphoxide. They also noted fission of the carboxamide linkage in root tissues with the formation of aniline which became bound to plant polymers and also formed highly water-soluble conjugates. The mechanism of action of carboxin in sensitive organisms has been shown to be associated with its ability to inhibit the tricarboxylic acid cycle (Mathre, 1970; Ragsdale and Sisler, 1970). In later work the toxicities of carboxin and seven structurally related compounds (XXXVIII–XLIV) were compared with regard

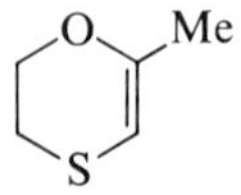

(XXXIX)

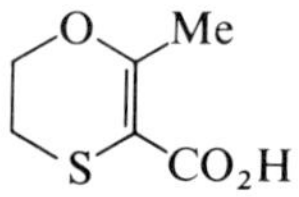

(XL)

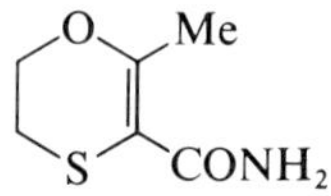

(XLI)

to their ability to affect various metabolic pathways in sensitive fungi (Mathre 1971b). Only the fungitoxic compounds (XXXVII, XLII, XLIII and XLIV) strongly inhibited acetate metabolism and RNA synthesis;

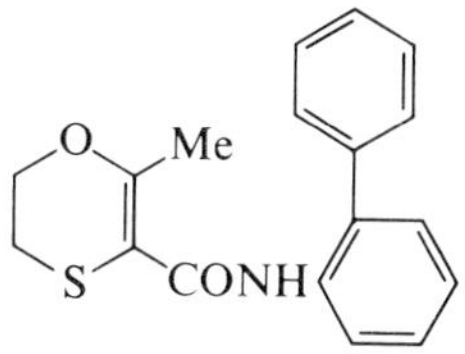

(XLII)

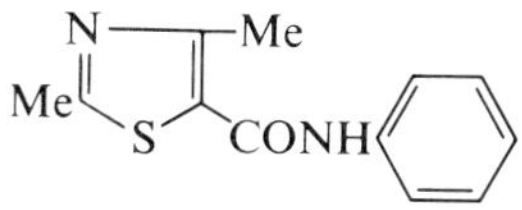

(XLIII)

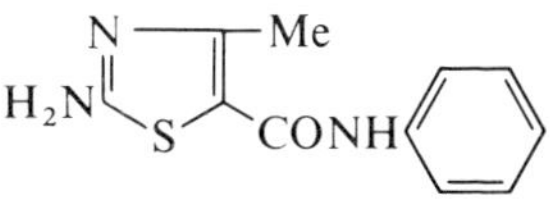

(XLIV)

which seems to indicate a similar mode of action for these compounds, even though two of them are thiazoles (XLIII, XLIV). The poor performance of carboxin sulphoxide (XXXVIII) and carboxin sulphone (XXXVII) against *R. solani* and *U. nuda* is interesting especially in the light of previous results, and with the exclusion of these two compounds, the fungitoxicities of these analogues to *R. solani* correlates well with their partition coefficients determined in a water–octanol system. Significantly, too, elimination of the aromatic ring in compounds XXXIX, XL and XLI destroyed all activity.

Other substituted benzanilides also have an interesting specificity against Basidiomycetes, and while neither of the hetero atoms present in carboxin appears to be critical and a methyl group is often present in the 2-position in active compounds. Thus 2*H*-3,4-dihydro-6-methylpyran-5-carboxanilide (XLV), proposed common name pyracarbolid, containing an *ortho*-methyl group, is also active against rusts, smuts and *Rhizoctonia* spp. (Stingl *et al*., 1970). The early results with the smut diseases of barley and oats (caused by *U. nuda* and *U. avenae* respectively) were particularly promising, activity comparable to that of carboxin being shown (Jank and Grossmann, 1971) (Tables 3.11 and 3.12). 2,5-Dimethyl-3-furanilide (XLVI : R = phenyl, R_1 = H), proposed common name furcarbanil, with only a five-membered heterocyclic ring and again with a methyl group *ortho* to the carbamoyl function has recently been announced by Pommer (1971b) as a new systemic fungicide. It has a much wider spectrum of activity, being particularly effective, not only

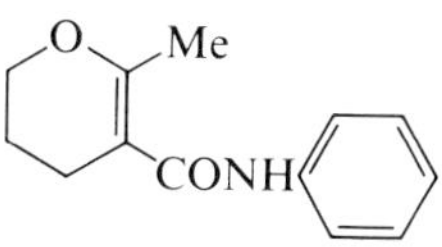

(XLV)

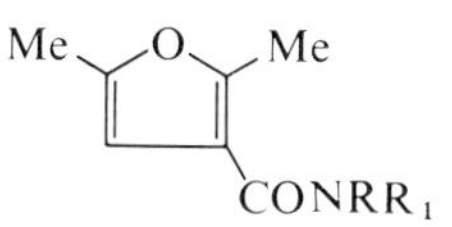

(XLVI)

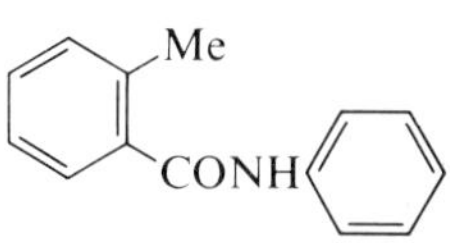

(XLVII)

Table 3.11 Effect of 2*H*-3,4-dihydro-6-methylpyran-5-carboxanilide against loose smut of barley (Jank and Grossmann, 1971)

Fungicide	Application rate, g of active ingredient/100 kg of seed	Smutted ears, %			
		Larker		Pirol	
		Field trial	Pot trial	Field trial	Pot trial
Control	—	20·1[a]	18·7[ab]	28·1[a]	23·7[a]
2*H*-2,3-Dihydro-6-	6·25	12·9[b]	23·6[a]	10·7[b]	25·6[a]
methylpyran-5-	12·5	4·5[cd]	8·3[cd]	6·7[b]	14·8[b]
carboxanilide	25·0	2·3[de]	3·7[de]	0·5[c]	3·2[c]
	50·0	0·4[ef]	1·2[e]	0·3[c]	1·6[cd]
Carboxin	12·5	7.8[bc]	13·5[bc]	7·9[b]	15·4[b]
	50·0	0·3[f]	2·3[e]	0·4[c]	0·3[d]

a–f Treatment means in the same column followed by different letters are statistically different at the 5% level of significance

Table 3.12 Effect of 2*H*-3,4-dihydro-6-methylpyran-5-carboxanilide against loose smut of oats (Jank and Grossmann, 1971)

Fungicide	Application rate g of active ingredient/100 kg of seed	Smutted panicles, %	
		Field trial	Pot trial
Control	—	18·2[a]	29·7[a]
2*H*-3,4-Dihydro-6-	6·25	9·1[b]	5·8[b]
methylpyran-5-	12·5	2·6[c]	1·2[bc]
carboxanilide	25·0	0·2[c]	0·0[c]
	50·0	0·1[c]	0·0[c]
Carboxin	6·25	7·9[b]	1·3[bc]
	50·0	0·1[c]	0·0[c]

a–c Treatment means in the same column followed by different letters are statistically different at the 5% level

against loose and stinking smuts, but also for some *Helminthosporium* and *Fusarium* spp. Its systemic effect was demonstrated in seed and soil treatments by the control of bean rust and it is also being currently evaluated as a seed dressing against *U. nuda* in spring barley and against *U. tritici* in winter wheat. Attempts to improve the fungicidal effectiveness by increasing the lipoid solubility of the molecule were unsuccessful. Thus substitution of a methyl group in the phenyl nucleus reduced fungitoxicity by 30–50 per cent, whilst introduction of a chlorine or bromine atom resulted in up to 100 per cent reduction in activity, Interference with methyl groups of the furane ring was even more disastrous, for the compounds 3- or 5-methyl-3-furanilide were found to be completely inactive when tested against Basidiomycetes.

The implication from Mathre's results is that the carboxylic acid anilide part of the molecule is a prime requirement for activity and this is supported by the fact that similar activity has been reported for a series of substituted benzanilides which do not involve a heterocyclic ring system. 2-Toluanilide (XLVII) (common name mebenil), without any hetero atoms, represents the best of this series for yellow rust control in cereals (Pommer and Kradel, 1969). The importance of the orientation of the methyl group is clearly shown in Table 3.13, as is the inadequacy of the hydroxyl group in the 2-position. More recently 2-iodobenzanilide (proposed common name benodanil) has been shown to be

Table 3.13 The fungicidal activities of substituted benzanilides (Pommer and Kradel, 1969)

	Basidomycetes		**Ascomycetes**		**Phycomycetes**	
Substituent	*R. solani*	Crown rust (oat)	*Aspergillus*	Barley mildew	P. *viticola*	Pythium spp
none	–	–	–	–	–	–
2-Hydroxy	–	–	+	–	+	–
2-Methoxy	–	–	–	–	–	–
2-Methyl	+	+	–	–	–	–
3-Methyl	±	–	–	–	–	–
4-Methyl	–	–	–	–	±	–
2-Trifluoromethyl	+	+	–	–	–	–
2-Chloro	+	+	–	–	–	–
2-Bromo	+	+	–	–	–	–
2-Nitro	+	±	–	–	–	–
2-Amino	±	+	–	–	±	–

+ Effective; ± Less effective; – Insufficiently or ineffective

more active than mebenil and appears to have a greater margin of crop safety. It is also more persistent than the corresponding chloro and bromo compounds. The species specificity of the active compounds of this group is most marked. The use of furcarbanil, as a seed dressing for barley, has been limited to some extent because of the sensitivity shown by some varieties under certain climatic conditions. The cyclohexyl analogue (XLVI: R = cyclohexyl, $R_1 = H$), proposed common name cyclafuramid, however not only gives good control of smuts on barley and wheat, but is safer on the former crop than furcarbanil (Pommer *et al.*, 1971). Further studies by Pommer (1975) involving substitution of the amide proton have revealed interesting structure–activity relationships (Table 3.14). As a result the methoxy analogue of cyclafuramid, 2,5-dimethyl-3-(cyclohexylmethoxycarbamoyl)furan (XLVI: R = cyclohexyl, $R_1 = OMe$) is to be developed as a mercury-free seed treatment.

A study of the systemic activity of another group of carboxanilides by ten Haken and Dunn (1971) has produced even more interesting and valuable structure–activity data. These authors studied the various structural features of oxycarboxin (XXXVII) in relation to its systemic control of broad bean rust (caused by *Uromyces fabae*), using logically modified compounds to assess the contribution of various parts of the molecule. They confirmed the activity of compounds which did not possess hetero atoms and showed that this ring system could in at least one instance be replaced by an open-chain analogue and still retain fungicidal activity. Thus *cis*-crotonanilide (XLVIII) was active, albeit also somewhat phytotoxic, although the corresponding *trans*-isomer (XLIX) was not.

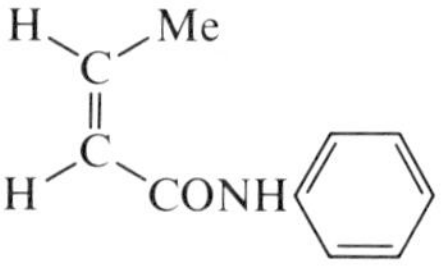

(XLVIII)

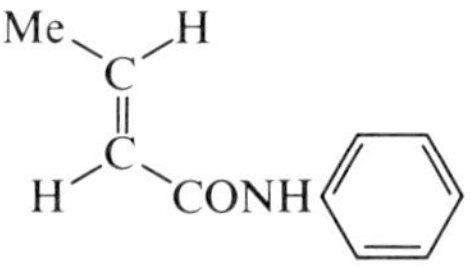

(XLIX)

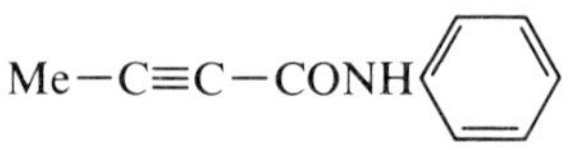

(L)

Reduction of the double bond to give n-butyranilide, however, destroyed activity as did its replacement by a triple bond (L). Ten Haken and Dunn also noted that the double bond may form part of a planar ring system such as benzene, furan, thiazole and oxazole, or of non-planar ring systems like dihydrooxathiin, dihydrofuran and dihydropyran. Their results confirmed the sacrosanctity of a suitable *ortho* group: replacement by hydrogen, ethyl or phenyl groups in 2*H*-2,3-dihydro-6-methylpyran-5-carboxanilide (XLV), or hydroxylation in mebenil (XLVII) caused complete loss of activity.

Table 3.14 The fungicidal activities of *N*-substituted 2,5-dimethyl-3-cyclohexylcarbamoyl furans (Pommer, 1975)

Substituents for XLVI R = cyclohexyl, R_1 =	Ustilago scitaminea	Rhizoctonia solani	Coniophora cerebella
H	+	+	+
CH_3	±	±	+
OH	−	−	−
OCH_3	+	+	+
OC_2H_5	−	−	−

+ very good action ± moderate action − ineffective

The importance of the carboxanilido moiety was also investigated, and in general any interference led to loss of activity (Table 3.15). It is perhaps significant that the function of the —NH— group appears to be more than that of a 'spacer', because replacement by a methylene bridge led to loss of activity, even though the interatomic distances involved were 0.25 nm in both compounds. Ten Haken and Dunn also tested some of the compounds against wheat brown rust (caused by *Puccinia recondita*) and found a good correlation with the bean rust results. In this glasshouse evaluation, they found that certain nuclear substituents in the phenylcarbamoyl part of the molecule had considerable influence on both phytotoxicity and fungitoxicity (Table 3.16).

Table 3.15 Importance of the carboxanilido function in the activity of carboxylic acid anilides against *Uromyces fabae* (ten Haken and Dunn, 1971)

	Foliar application		
Compound	Direct	Systemic	Root drench
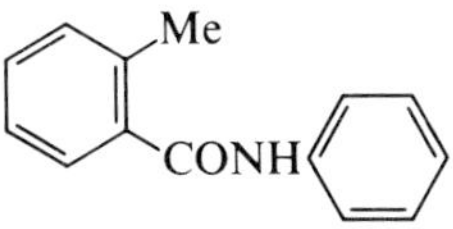	2	2	1
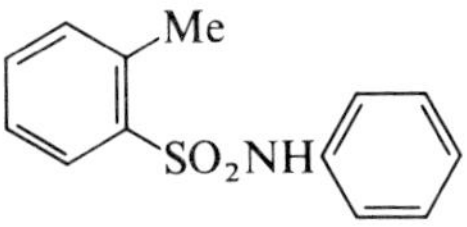	0	0	0
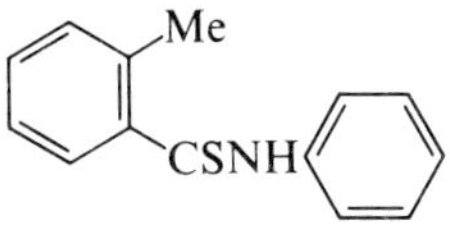	2	0	0
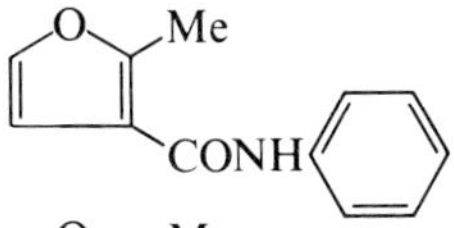	2†	2	1
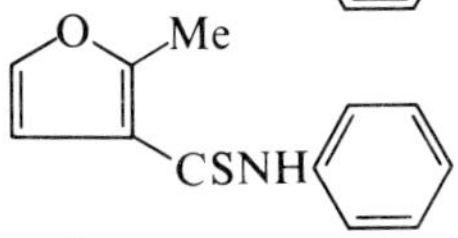	2†	0	0
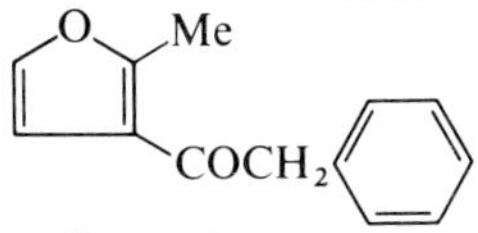	0	0	0
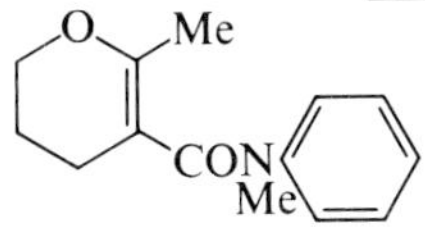	1†	0	0
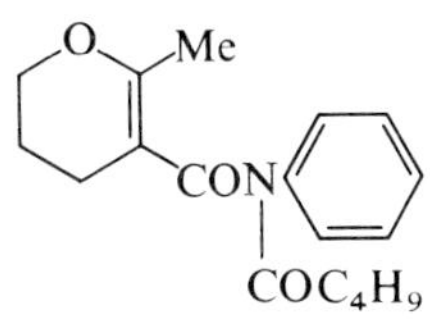	1	0	0

2 80–100% Reduction in rust symptoms; 1 50–80% Reduction in rust symptoms; 0 Insignificant reduction in rust symptoms; † Slight to moderate necrosis

By 1972 it had become obvious that specificity of carboxin resided in the mitochondrial succinic dehydrogenase complex of sensitive fungi such as *Ustilago maydis* (White, 1971; Ulrich and Mathre, 1972), and the structure–activity relationships of some ninety-three carboxin ana-

Table 3.16 Glasshouse evaluation of carboxylic acid anilides against wheat brown rust (ten Haken and Dunn, 1971)

Compound	% Rust control	Phytotoxicity
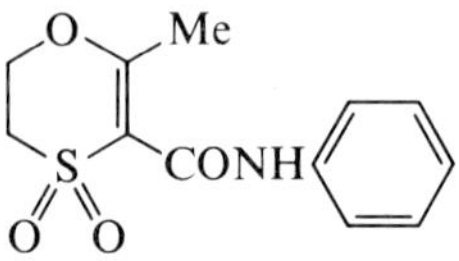	100	1
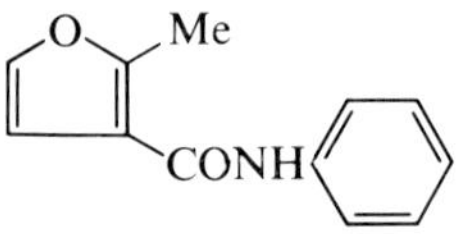	50	2
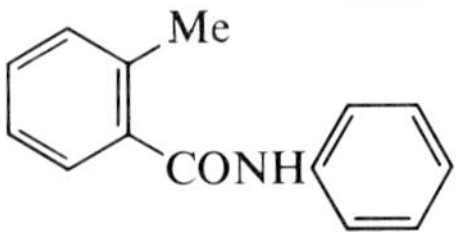	0	1
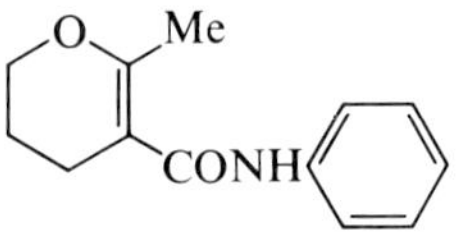	100	3
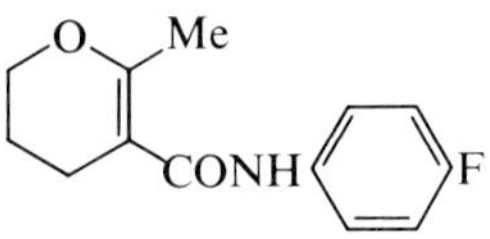	99	1
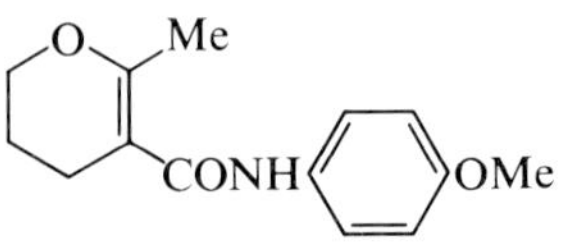	36	0
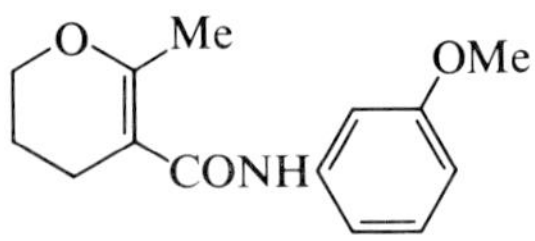	100	2
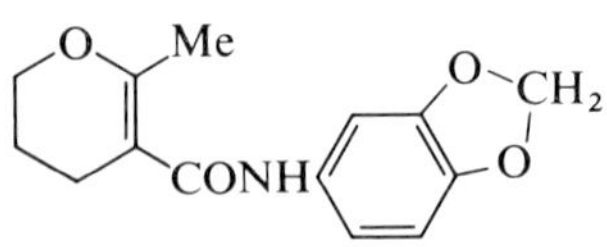	97	0

0 No visible phytotoxity; 1 Slight stunting and chlorosis;
2 Moderate stunting and chlorosis; 3 Severe stunting and chlorosis

logues and related compounds recently reported in detail (White and Thorn, 1975) have confirmed this. With few exceptions, a good correlation was shown to exist between succinic dehydrogenase inhibition and antifungal activity both *in vitro* and *in vivo* (Table 3.17). Compounds

Table 3.17 The effect of carboxamides on succinate-2,6-dichlorophenolindophenol reductase activity in mitochondrial preparations from *C. laurentii* and *U. maydis*, and on the growth of *R. solani* (White & Thorn, 1975)

	I_{50}(μM)		ED_{50}(μM)
Compound	*C. laurentii*	*U. maydis*	*R. solani*
Carboxin (XXXVI: R = phenyl)	0·55	0·5	1·0
Benzanilide	> 500	> 500	> 100
2-Iodobenzanilide (benodanil)	1·0	7·5	—
Mebenil (XLVII)	18	20	10
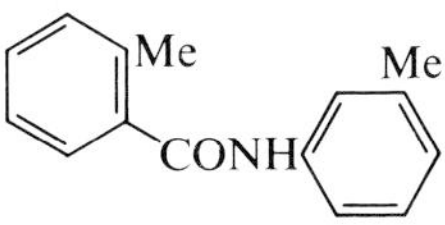	3·9	13	4
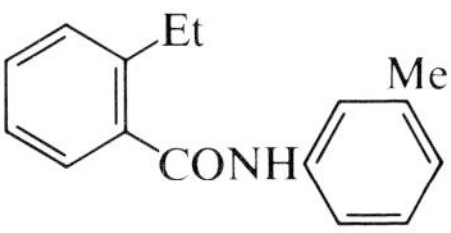	0·5	4·5	2
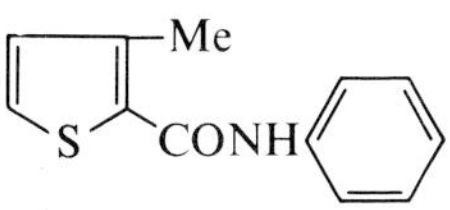	3·3	4·5	3
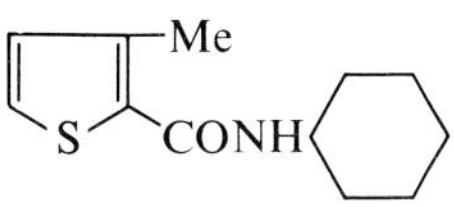	60	20	100
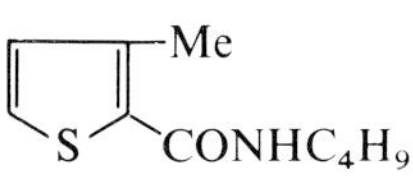	> 500	375	250
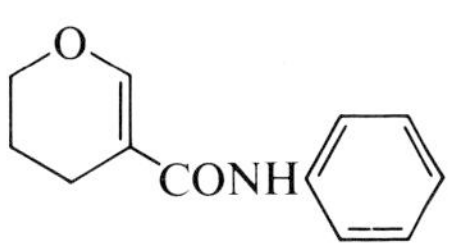	> 1000	> 1000	—
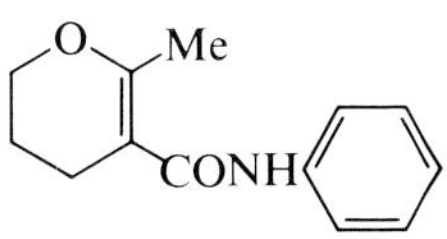 (pyracarbolid)	6·0	5·5	—
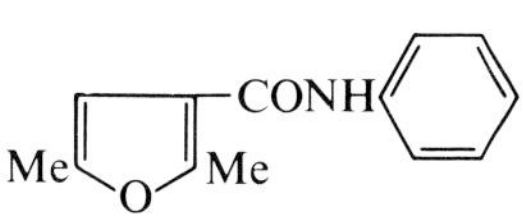 (furcarbanil)	1·9	1·0	—

Table 3.17 *Continued*

Compound	$I_{50}(\mu M)$		$ED_{50}(\mu M)$
	C. laurentii	*U. maydis*	*R. solani*
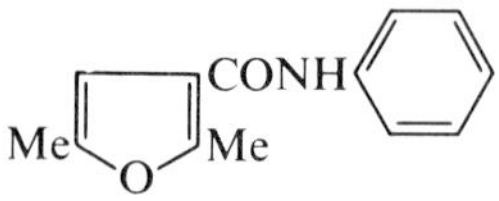 (cyclafuramid)	6·0	1·0	—
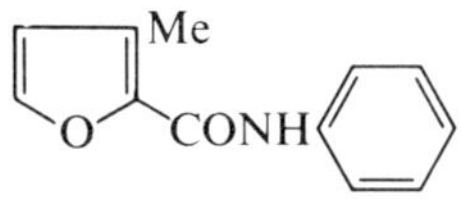	250	150	—

found to be most active in the inhibition of the mycelial growth of *Rhizoctonia solani* were also tested against non-Basidiomycetes, such as *Cladosporium herbarum*, *Monilinia fructicola* and *Verticillium dahliae*, and it was clear that activity was influenced not only by substitution in the aniline ring, but also by carboxy ring substitution. Thus 3-methyl-2-phenylcarbamoylthiophene, 3′-methyl-2-methylbenzanilide and 3′-methyl-2-ethylbenzanilide all had an antifungal spectrum extending beyond Basidiomycetes (Table 3.17). In contrast to an earlier report (Mathre, 1971b), however, no correlation was found between the inhibitory activity of substituted oxathiins and benzanilides, and their calculated partition coefficients.

Heterocyclic compounds

Benzimidazoles

Although 2-aminobenzimidazole (LI: $R_1=R_2=H$) is devoid of fungitoxicity, systemic fungicidal activity was claimed by Klöpping (1960) for various substituted 2-aminobenzimidazoles. This basic work was the prelude to the production of methyl (1-butylcarbamoyl)benzimidazol-2-ylcarbamate (benomyl: LI: $R_1=CONHC_4H_9, R_2=COOMe$) which in addition to eradicant and protectant fungicidal activities and a mite ovicidal action, possesses exceptional systemic fungicidal properties (Delp and Klöpping, 1968). It is active against an impressive list of fungal diseases of fruit and vegetable crops, against *Verticillium albo-atrum* on tomato and cotton and against powdery mildew and black spot of roses. However, it has little effect on diseases caused by *Helminthosporium* spp. (Greenaway, 1972) and by Phycomycetes, such as potato late blight, vine downy mildew and various damping-off diseases. In aqueous solution benomyl has been shown (Clemons and Sisler, 1969) to be rapidly hydrolysed to methyl benzimidazol-2-ylcarbamate (LI: $R_1=H$, $R_2=COOMe$), common name carbendazim (formerly

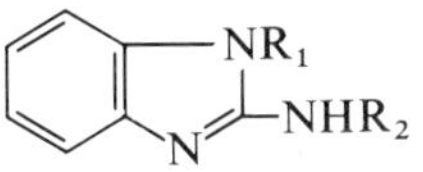

(**LI**)

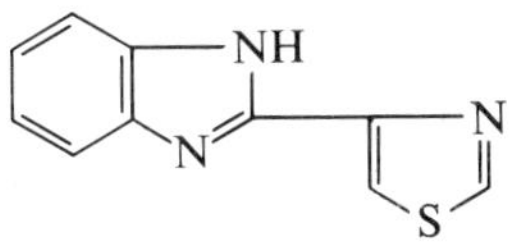

(**LII**)

referred to as MBC), and it is possible that this stable breakdown product is partly or even wholly responsible for fungitoxicity. Both photochemical and microbial breakdown have recently been reviewed (Woodcock, in press).

Although carbendazim was found to be thirty times less toxic to *Saccharomyces pastorianus* Han. (Clemons and Sisler, 1969) it was found to be just as toxic to *Neurospora crassa* Shear and Dodge and *Rhizoctonia solani* Kühn, as benomyl, and the toxicity to these two organisms must be attributed to the breakdown product. The poorer performance of carbendazim against the yeast could be merely a reflection of poorer penetration. Other workers have confirmed the suggestion that the active entity in the plant at sites remote from the point of application of benomyl is carbendazim. Thus, Sims *et al.* (1969) were able to detect it, but not benomyl, in extracts of cotton plants 4 weeks after treatment, while Peterson and Edgington (1969) found that benomyl was completely transformed to carbendazim in 5 days in bean plants. Evidence regarding its mode of action has been provided by Clemons and Sisler (1971) using *N. crassa* and *Ustilago maydis* (DC.) Cda. They showed that though protein and RNA synthesis in *N. crassa* were not inhibited, the action of carbendazim appeared to involve DNA synthesis or some closely related process such as nuclear or cell division.

It seems highly probable that the many analogous compounds which have been synthesised (Loux 1961) will by now have been screened for any sign of improved systemic fungitoxicity.

2-(Thiazol-4-yl)benzimidazole (thiabendazole) (LII), originally introduced as an anthelminthic, has also achieved prominence as a broad-spectrum systemic fungicide effective against many major fungal pathogens (Staron and Allard, 1964; Robinson *et al.*, 1964; Erwin *et al.*, 1968a, b). Translocation in many plants has been reported by bioassays, chromatography and autoradiography, and it seems clear that movement is not only from roots to leaves but also in the less usual reverse direction, and so far there are no indications of any metabolic breakdown *in vivo*. A certain amount of work has been done on its mode of action and like benomyl and carbendazim it appears to be fungistatic. Allen (1969) showed it to be a potent inhibitor of respiration in isolated mitochondria of *Penicillium atroveretum* G. Smith, though the toxicant level required to inhibit whole cell respiration was higher than that needed to inhibit growth. Similarly Kaars Sijpesteijn (1970) reported that 100 ppm – nearly 100 times the growth inhibitory concentration – could not inhibit respiration in *Aspergillus niger*.

Another newly introduced benzimidazole is 2-(2-furyl) benzimidazole (fuberidazole) (LIII). This compound shows considerable promise as a seed treatment against *Fusarium* diseases, such as *F. nivale* of rye and *F. culmorum* on peas (Schuhmann, 1967). Both it and thiabendazole showed good systemic activity, and seed treatments of wheat and barley considerably reduced infections of *Puccinia triticina* Eriksson and *Erysiphe graminis* DC., for several weeks, and also gave extremely effective control of snow mould (*Calonectria nivalis*), being superior to organomercurials at the same concentration.

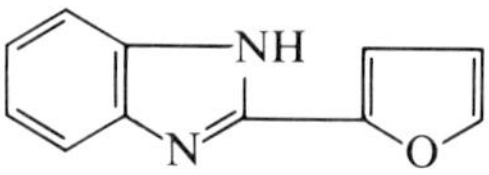

(**LIII**)

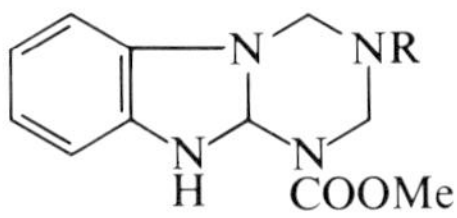

(**LIV**)

Yet another closely related compound methyl 1-(5-cyanopentylcarbamoyl)benzimidazol-2-ylcarbamate (cypendazole; LI: $R_1 = CONH(CH_2)_5CN$, $R_2 = COOMe$), introduced by Bayer A. G. (Daum *et al.*, 1971) is of interest in that not only is it a protectant and eradicant fungicide but it also shows considerable systemic activity. It has a wide antifungal spectrum, being active *inter alia* against apple scab, *Botrytis* spp., powdery mildews, *Cercospora* spp., storage rots, vascular and seedling diseases (Cohick, 1973).

Systemic fungicidal activity is also shown by some substituted triazinobenzimidazoles (LIV) (Härtel and Röchling, 1973) which were active against a number of economically significant pathogens from among the Asco-, Deutero- and Basidiomycetes. Additionally compounds possessing long alkyl substituents were outstandingly effective against a range of Phycomycetes, including *Peronspora* and *Phytophthora* spp. Three of these compounds (LIV: R = isobutyl, n-dodecyl or 1,2-diphenylethyl) have been studied recently by Grossmann and Buchenauer (1975), who showed that in aqueous suspension they were almost completely converted to carbendazim after 12 hours in the dark at room temperature. Carbendazim was also the only fungitoxic metabolite detected in tomato plants after root uptake of these compounds. The dodecyl analogue was exceptional in that it also controlled late blight of tomatoes (caused by *Phytophthora infestans*), and it was thought that this might be due to release of dodecylamine, which is about as active *in vitro* and *in vivo* as compound (LIV: R = n-dodecyl) itself. Significantly, both compounds increased the permeability of plasma membranes, which may be a contributory factor in their fungitoxicity.

The fungitoxicities of benomyl, thiabendazole and fuberidazole *in vitro* have been examined against a wide fungal spectrum and an identical selective antifungal pattern was revealed (Edgington *et al.*, 1971). Thus whilst a similar mode of action for these benzimidazole analogues seems both logical and likely, a report from Sisler (1971) suggests that possession of the same basic ring structure may not always imply the same mode of action. Using *N. crassa* and *U. maydis*, he found that whereas benomyl and thiazbendazole prevented cell multiplication, permitted appreciable increases in dry weight, but had little or no effect on respiration, the trifluoromethyl analogues LV and LVI both prevented increase of cell dry weight and strongly stimulated respiration at

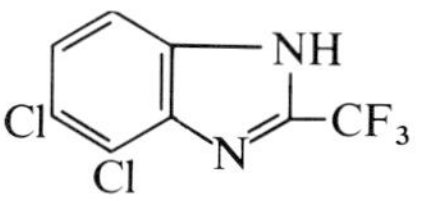

(LV)

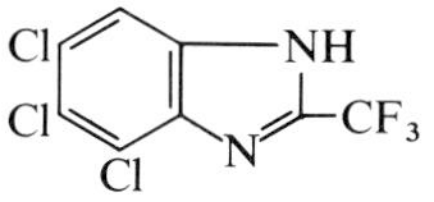

(LVI)

concentrations below or only slightly above those completely inhibiting growth, though higher concentrations did strongly inhibit respiration. Significantly, too, a benomyl-tolerant mutant of *N. crassa* was also tolerant of thiabendazole, but not of compounds LV and LVI, all of which evidence points to a different mechanism of action for the chloro compounds.

Hydroxypyrimidines

Like imidazole, the pyrimidine nucleus is often closely associated with biological activity of one sort or another. An early example was the tetrahydropyrimidines (LVII: R = n-alkyl) some of which were highly active as protectant fungicides (Rader *et al.*, 1952), optimum fungistatic activity occurring against the late blights of tomato and celery when $R = C_{17}H_{35}$. These compounds showed no control of powdery mildews and only moderate control of bean rust. By contrast, the 2-n-alkylmercapto-1,4,5,6-tetrahydropyrimidine hydrohalides (LVIII: R = n-alkyl) show an interesting antifungal spectrum *in vivo*, although they appear to be largely inactive *in vitro* (Nickell *et al.*, (1961). Thus they show systemic activity against rusts and powdery mildews and to some extent against *Fusarium* wilt of tomatoes, 8–14 carbon atoms being required in the alkyl group for maximum effect.

More recently, a growing number of antibiotics which include the pyrimidine nucleus have been found to possess systemic fungicidal activity. These include bacimethrin (LIX) (Tanaka *et al.*, 1962), tubercidin and toyocamycin (Ohkuma, 1961) and pathocidin which is formed by *Streptomyces albus* var. *pathocidus*. The latter, which is highly antifungal, is reported to contain 8-azaguanine (Nagatsu *et al.*, 1962).

A series of phosphorylated pyrimidines related to *O,O*-diethyl-*O*-(2-dimethylamino-4-methylpyrimidin-6-yl) phosphorothionate (LX) de-

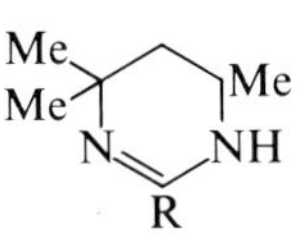

(LVII)

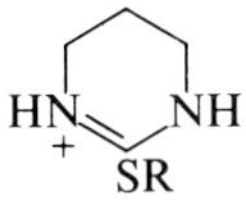

(LVIII)

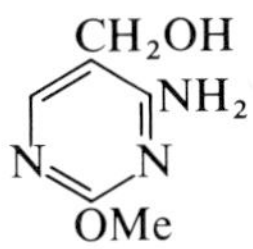

(LIX)

veloped earlier by I.C.I.† for sheep blowfly control, were found by workers at Jealott's Hill Research Station to have protectant fungicidal activity against certain powdery mildews, although the very closely related diazinon (LXI) showed little such activity. This led to further structure–activity work with phosphorylated hydroxypyrimidines,† in which it was found that a wide class of 2-amino-4-hydroxypyrimidines

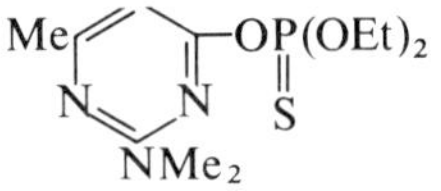

(LX)

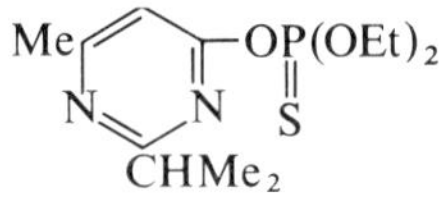

(LXI)

were just as fungicidally active as the derived phosphorothionates. Further synthetic work was therefore concentrated on the synthesis of suitably substituted hydroxypyrimidines, rather than on the relatively more complex organophosphorus derivatives.

Whilst they are generally depicted for convenience as hydroxy compounds, the hydroxypyrimidines exist in solution and in the solid state

Table 3.18 Activity of substituted pyrimidines against cucumber powdery mildew (Snell, unpublished)

Substituent R (Formula LXIII)	Activity at 500 ppm (scale 0–3)		
	Eradicant	Protectant	Systemic
OH	3	3	3
Cl	0	0	0
NH_2	1	0	0
$NHNH_2$	3	3	3
$NHN{=}CHC_6H_5$	3	3	3
$NHNHCOC_6H_5$	0	3	3
Morpholino	2	0	0
SH	3	3	3
$SCO_2C_2H_5$,	3	1	3
$SCH_2CH_2OC_2H_5$	1	1	0

as the tautomeric 1,4-dihydro-4-oxopyrimidines (LXII). This oxygen function at position 4 makes highly active compounds possible. Its replacement by other groups may result in loss of activity (Table 3.18) and where activity was maintained the compounds were not sufficiently superior to warrant further development. The size of the various alkyl

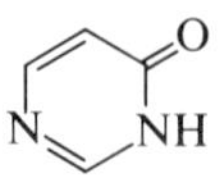

(LXII)

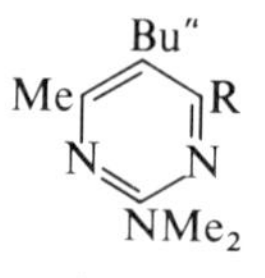

(LXIII)

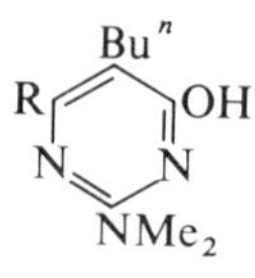

(LXIV)

† British Patents Nos., 1 019 227, 1 204 552, 1 205 000, 1 203 026, 1 129 797, 1 129 563 and others

groups can also affect the level or area of activity and the preferred groups at position 6 are methyl and ethyl. The effect of the 5-substituent provides yet another example of gradual change of biological activity with increasing lipophilicity. Fungicidal activity is at a high level when

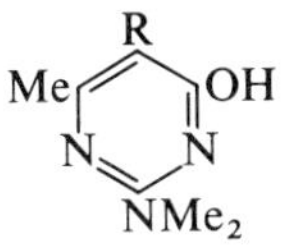

(LXV)

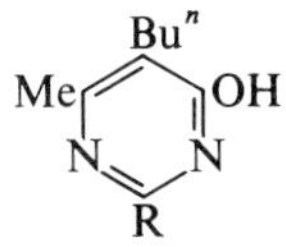

(LXVI)

the 5-alkyl group has from two to five carbon atoms, and n-alkyl groups were somewhat more effective than the corresponding branched-chain compounds (Table 3.19). The alkylamino group in position 2 is also

Table 3.19 Activity of substituted pyrimidines against cucumber powdery mildews (Snell, unpublished)

Substituent R (Formula LXV)	**Activity at 500 ppm (scale 0–3)**		
	Eradicant	Protectant	Systemic
C_2H_5	3	3	3
$i\text{-}C_3H_7$	2	2	0
C_3H_7	3	3	3
$CH_2CH{=}CH_2$	3	3	3
C_4H_9	3	3	3
$i\text{-}C_4H_9$	3	3	3
C_5H_{11}	3	3	3
$i\text{-}C_5H_{11}$	3	3	3
C_6H_{13}	2	2	0
$CH_2C_6H_5$	1	3	3

useful for the highest fungitoxicity. The influence of various 2-substituted amino groups on the systemic activity against cucumber and barley mildews is shown in Tables 3.20 and 3.21. It will be seen that five of the six compounds presented give 90–100 per cent control of the former pathogen when applied as a root drench at 5 ppm and of these

Table 3.20 Systemic activity of substituted pyrimidines† against cucumber powdery mildew at 5 ppm (Snell, unpublished)

Substituent R (Formula LXVI)	**% Leaf surface disease free**
$NHCH_3$	94
NHC_2H_5	100
$N(CH_3)_2$	92
$N(CH_3)C_2H_5$	99
NHC_3H_7	77
$NHCH_2CH{=}CH_2$	91

† British Patent No. 1 182 584

Table 3.21 Systemic activity of substituted pyrimidines† (LXVI) against barley powdery mildew at 50 ppm (Snell, unpublished)

Substituent R (Formula LXVI)	Activity (scale 0–36)
$NHCH_3$	35
NHC_2H_5	36
$NHCH_2CH{=}CH_2$	25
$NHC_3H_7^i$	27
$N(CH_3)C_2H_5$	29

† British Patent No. 1 182 584

the 2-dimethylamino analogue (LXVI: R = NMe_2) was selected for further development with the common name dimethirimol. It is, for example, very effective against cucurbit powdery mildew, and only slightly less effective against cereal powdery mildews. By contrast the 2-ethylamino analogue (LXVI: R = NHEt), common name ethirimol, was somewhat more effective against cereal mildews. Results obtained in glasshouse testing have in general been confirmed in the field (Brooks, 1970), and both dimethirimol and ethirimol have now achieved widespread commercial acceptance.†

Acylation of 4-hydroxypyrimidines is usually unsuccessful, or may lead to the formation of unstable *N*-acyl derivatives. However, acylation of compounds in which a bulky substituted amino group hinders ring access invariably occurs on the oxygen atom (Snell, 1968), and several of the *O*-acyl derivatives (LXVII: R = acyl) have been shown to possess

Table 3.22 Control of apple powdery mildew in glasshouse tests (G. M. Farrell, unpublished)

Compound (LXVII)			Protectant activity (scale 0–3) at concentration (ppm)			
*R*₁	*R*₂	*R*	125	25	5	1
H	Et	SO_2NMe_2	3	3	3	3
H	Pr^n	SO_2NMe_2	3	3	3	3
Me	Me	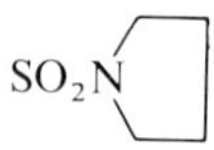	3	3	3	2
H	Et	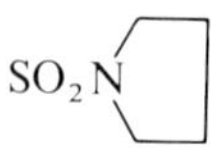	3	3	3	3
H	Pr^n	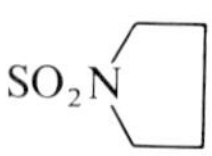	3	3	3	3
Me	Et	COC_6H_5	3	2	0	0
Me	Me	$P(S)(Me)NMe_2$	3	3	0	0
Me	Me	$P(S)(Me)OMe$	3	3	0	0
Me	Me	$P(S)(Me)OPr^n$	3	3	1	0
Me	Me	$P(S)(OPr^n)_2$	3	3	0	0

† British Patent No. 1 182 584

considerable fungicidal activity (Table 3.22).† The superiority of sulphamates over other esters tested was clear. The most active compounds under field conditions were dialkylsulphamates, and 5-butyl-2-ethylamino-6-methylpyrimidin-4-yl dimethylsulphamate (LXVII: $R_1 = H$, $R_2 = Et$, $R = SO_2NMe_2$), common name bupirimate, was chosen for further development against powdery mildews. In addition to strong protectant and eradicant activity against apple powdery mildew, it has marked vapour activity and moves freely within sprayed apple leaves. It is also highly effective against powdery mildew on rose, cucurbit, peach, blackcurrant, vine and strawberry.

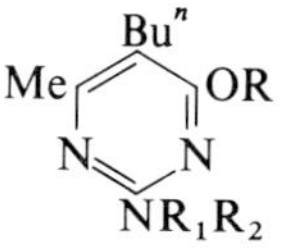

(LXVII)

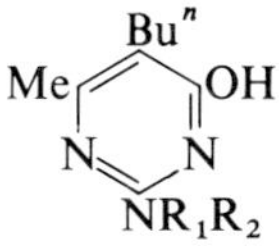

(LXVIII)

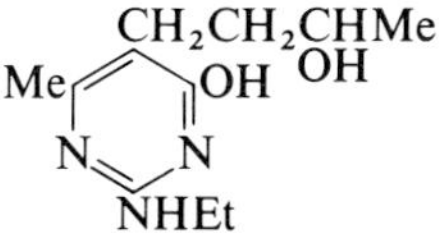

(LXIX)

Autoradiographic studies using ^{14}C-labelled compounds showed that following uptake from the soil they are translocated throughout the plant in the xylem. In contrast to their stability in the soil, however, Hemingway and Cavell have shown that they undergo rapid metabolism in the plant (Slade *et al.*, 1971). Thin-layer chromatographic examination of methanol extracts of cucumber plants which have been treated with ^{14}C-dimethirimol has shown that dimethirimol is dealkylated not only to the desmethyl derivative (LXVIII: $R_1 = H$, $R_2 = Me_3$), which retains fungicidal activity, but also to the amine (LXVIII: $R_1 = R_2 = H$) which is not as active. Both the original fungicide and its primary metabolites react further to form water-soluble conjugates which seem likely to involve glucose, since β-glucosidase brings about at least partial re-conversion. Cavell *et al.* (1971) have demonstrated a very similar metabolic pattern for ethirimol in cereals, dealkylation to the amino compound (LXVIII: $R_1 = R_2 = H$) again taking place. Water-soluble metabolites are also formed and these can be hydrolysed under mild conditions to ethirimol, the amine (LXVIII: $R_1 = R_2 = H$) and the γ-hydroxybutyl analogue (LXIX). Some chromatographic evidence for the formation of ethirimol phosphate has also been obtained. Since some of these metabolites and their conjugates are fungitoxic in their own right, the systemic antifungal activity of these two pyrimidine fungicides must be regarded as a composite effect.

Studies on the mode of action of the hydroxypyrimidines have produced some very interesting initial findings. Bent (1970) has described experiments using segments of wheat leaves, in which the fungitoxicity of dimethirimol has been reversed by the addition of certain chemicals, and it is clear that folic acid, riboflavin and to a lesser extent adenine, thiamine and uracil all act as antagonists. Sampson (1969) has extended the fungitoxicity reversal work using cucumber plants grown in water culture and treated first with ethirimol and subsequently with various

† British Patent No. 1 185 039

potential reversal agents, followed by the spore inoculum. The chemicals which caused reversal in varying degrees, of which the most effective were riboflavin, adenine and guanine, were all active in C-1 metabolism, either as co-factors, and products or C-1 donors. This strongly suggests that the fungicide inhibits a key metabolic reaction. However, despite indications of the inhibition of THFA-directed C-1 metabolism by the hydroxypyrimidines in the reversal experiments, no direct evidence could be obtained using key enzymes, isolated from powdery mildew preparations, though limited evidence has been obtained of interference with a number of pyridoxal-dependent enzymes.

Substituted pyridines and pyrimidines

Structure–activity relationships of a new group of pyridine alkanes (LXXI) and carbinols (LXX), several of which show excellent systemic fungicidal activity, have also been reported (Brown *et al.*, 1967). The pathogens which have been controlled in greenhouse tests include *Erysiphe polygoni* and those causing dollar spot, brown patch and *Helminthosporium* leaf spot in turf grasses. Maximum activity was shown when the pyridine ring was substituted in the 3-position and similar substitution at positions 2 or 4 led to reduced activity. Against *E. polygoni* the substituted methanols and ethanols (LXX) were more active than the alkanes (LXXI) and this activity was not altered by the substitution of amino, chloro, cyano or methoxy groups on the β-carbon atom, although such changes did cause a reduction in the control of the turf pathogens. Excellent control of *E. polygoni* was also obtained with analogues having two aryl substituents on the α-carbon atom and this activity was further increased by substitution, the order being: diphenyl < 4-chlorodiphenyl < bis-(2-, 3- or 4-chlorophenyl). The activity of this type of compound against turf pathogens could be increased by replacement of aryl by cycloalkyl groups, and maximum activity was observed when one or two cyclohexyl groups were present.

(LXX)

(LXXI)

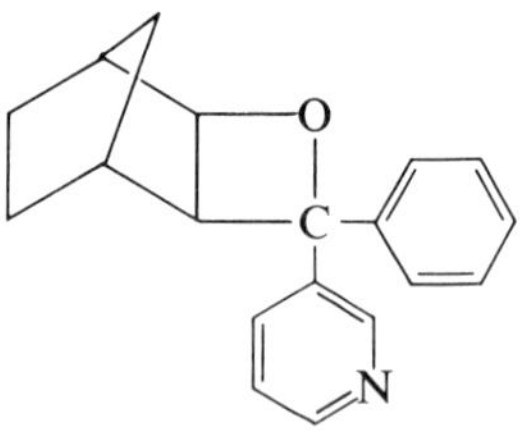

(LXXII)

Bis-(ρ-chlorophenyl)-3-pyridyl methanol (LXX: $R_1 = R_2 = \rho$-chlorophenyl) or parinol was in fact introduced as a new fungicide for the control of powdery mildews, foliar applications at 10 ppm giving >95 per cent control of *Erysiphe cichoracearum* on squash, canteloupe and cucumber (Thayer *et al.*, 1967), and >90 per cent of *Podosphaera leucotricha* on apple, *S. pannosa* on roses and *Uncinula necator* on vines was obtained using 25 ppm. Parinol is of considerable interest because it exhibits translaminar activity – a valuable attribute for the control of pathogens such as powdery mildews which tend to proliferate unhindered on the under surfaces of leaves despite treatment with conventional protectant sprays.

With the related experimental fungicide 4-phenyl-4-(3-pyridyl)-3-oxatricyclo $[4.2.1.0^{2\ 5}]$ nonane (LXXII), no systemic movement of the chemical was observed following soil, stem or leaf applications, but fumigant activity from 100 μg topically applied completely eradicated powdery mildew from the leaf surface of bean plants (Spurr and Chancey, 1970). This is not the first case of fungicide redistribution in the vapour phase, as these authors seem to claim, but a re-discovery of the phenomenon reported first by Yarwood (1950) and more recently by Bent (1967), Hislop (1967) and by Clifford *et al.* (1970).

Two other pyridine analogues were examined by Spurr and Chancey (1970). Diphenyl-3-pyridylmethane (LXXI : $R_1 = R_2 =$ phenyl) also exhibited fumigant activity of the same order as the oxatricyclo compound, but diphenyl-3-pyridylmethanol (LXX: $R_1 = R_2 =$ phenyl) was inactive as a fumigant.

Yet another analogue involves the pyrimidine nucleus. 2.4-Dichloro-α-pyrimidin-5-yl-benzhydrol (LXXIII), common name triarimol, is a

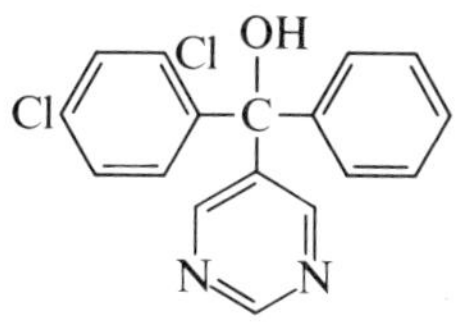

(LXXIII)

broad-spectrum fungicide which is particularly effective for the control of scab on apple (caused by *Venturia inaequalis*) and pear (caused by *V. pirina*) at application rates of 30–40 ppm, whilst the powdery mildews of grape (*Oidium* spp.) cucurbits (*E. cichoracearum*) and rose (*S. pannosa* var. *rosae*) have been controlled at 15–20 ppm (Gramlich *et al.*, 1969). Although it does not prevent spore germination, it stops penetration and arrests any further fungus development within the leaf (Brown, 1970). This curative effect, made possible by its local systemic activity, is a valuable attribute for disease control in the field. In contrast to apple leaves, where movement is limited to the translaminar type, movement in cucumber leaves has been reported as acropetal. Uptake of triarimol from the soil is a slow process and there has been no mention of fumigant activity despite its structural similarity to compounds which have been reported as fumigant in action.

Mode of action studies by Ragsdale and Sisler (1972, 1973) indicated that the primary site of action of triarimol against *Ustilago maydis* was in the sterol biosynthetic pathway preceding ergosterol, and Ragsdale (1975) has recently shown that three reactions in ergosterol biosynthesis are implicated involving the D ring and side chain.

The fungicide S-1358 (LXXIV) from Sumimoto Chemical Co Ltd, which is a derivative of 3-aminopyridine, has also been shown to interfere with sterol biosynthesis, in this case in *Monilinia fructigena* (Kato *et al.*, 1974a, b). It is particularly active against powdery mildews.

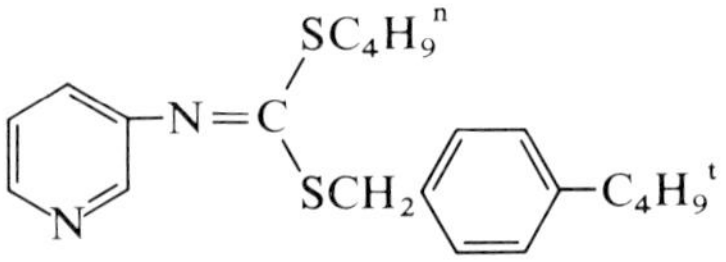

(LXXIV)

Trichloromethyl compounds

Pyridine also provides the nucleus in pyroxychlor (2-chloro-5-methoxy-4-trichloromethylpyridine) (LXXV), first reported by Johnson and Tomita (1966). This compound is of considerable interest because of its activity against *Phytophthora* spp., being the first one reported capable of controlling root fungal infections by foliar application. Thus

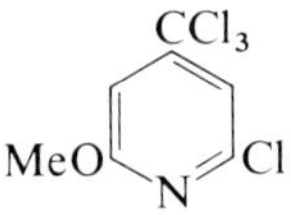

(LXXV)

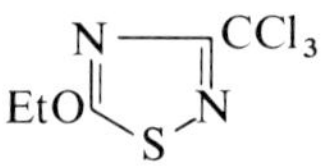

(LXXVI)

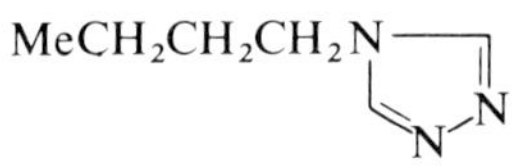

(LXXVII)

the following pathogens are controlled by pyroxychlor applied either as a drench or as a foliar spray: *P. palmivora* on *Diefenbachia picta* (Knauss, 1974), *P. cinnamomi* Rands on rhododendron and *P. megasperma* var. *sojae* Drechs on soybean (Hoitink and Schmitthenner, 1975) and *P. parasitica* on tobacco (Noveroske, 1975a). Against the last pathogen, despite its rapid biodegradation in the soil, it appears to be capable of providing almost season-long protection.

The thiadiazine derivative ethazole (5-ethoxy-3-trichloromethyl-1,2,4-thiadiazole) (LXXVI) was found to be less effective against *P. cinnamomi*, though equally effective but phytotoxic, when used on soybean against *P. megasperma* (Hoitink and Schmitthenner, 1975).

1,2,4-Triazoles

Possibly the most selective fungicidal effect observed to date is that due to 4-n-butyl-1,2,4-triazole (LXXVI), which is the most active of a number of 4-substituted-1,2,4-triazoles against wheat leaf rust (caused by *Puccinia recondita*) (von Meyer *et al.*, 1970). It is effective both as a foliar spray and by root uptake, at concentrations as low as 1 ppm. It fails, however, to control stem rust (*P. graminis*) on the same host, as well as *P. coronata*, *P. hordei* and *Uromyces phaseoli*, at concentrations as high as 1 200 ppm.

By contrast, substitution on the nitrogen atom in position 1 has produced compounds with considerable and varied biological activities. Whilst fluotrimazole (LXXVIII) is useful only as a protectant fungicide, the substitution pattern in triadimefon (LXXIX: $R_1 = Cl$, $R_2 = H$) has resulted in a compound which possesses not only protectant properties,

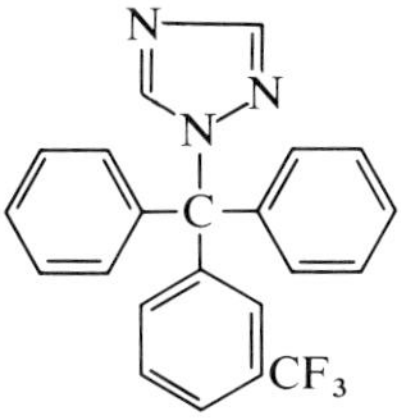

(LXXVIII)

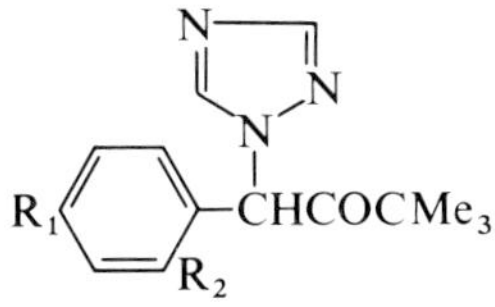

(LXXIX)

but also excellent eradicant and systemic activity against a wide variety of plant pathogens, particularly powdery mildews (Büchel *et al.*, 1975). Despite only weak inhibition of spore germination and mycelial growth *in vitro* of the usual test fungi, its future impact on cereal diseases appears very promising (Frohberger, 1975). Moreover, both acropetal and basipetal translocation have been demonstrated for this compound on cucumbers in greenhouse experiments with *E. cichoracearum* (Kaspers *et al.*, 1975).

The dichloro analogue (LXXIX: $R_1 = R_2 = Cl$) is also a highly effective fungicide but with rather less systemic activity and with an antifungal spectrum orientated rather towards non-mildew fungi. Replacement of the chlorine atom in triadimefon by a phenyl or other aryl group or by cyclohexyl further reduces systemic activity, though the diphenyl analogue (LXXIX: R_1 = phenyl, $R_2 = H$) has a protectant activity equal to or better than triadimefon, giving excellent control of rust fungi on various crops.

N-substituted tetrahydro-1,4,-oxazines

The increase in winter barley acreage in N. Germany, with the consequent escalation of overwintering potential inoculum of *E. graminis*, provided the impetus for the development of this new class of fungicidal compounds based on morpholine, which are both systemic and eradicant in action (Kradel and Pommer, 1967; Pommer *et al.*, 1969). The effects of variations in the nature and size of the *N*-substituent, and of substituents in the heterocyclic ring, on fungicidal activity were examined in several hundred compounds; other structural variations included *N*-oxide and salt formation, and complexing with heavy metals

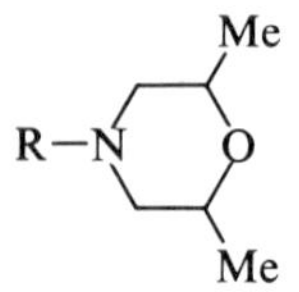

(LXXX)

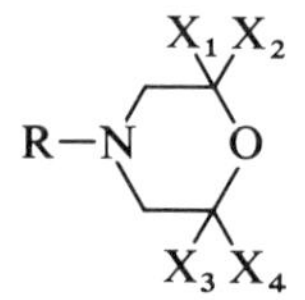

(LXXXI)

(König *et al.*, 1965; Pommer and Kradel, 1967). Table 3.23 shows the influence of the 4-(n-alkyl)-substituents with 9–18 carbon atoms on the fungicidal activity of 2,6-dimethylmorpholine against barley mildew. The effectiveness improves with increasing chain length, reaching a maximum for C_{13} (LXXX: R $=C_{13}H_{27}$) (tridemorph) but unfortunately

Table 3.23 Influence of the *N*-alkyl substituent on the fungicidal activity of 2,6-dimethylmorpholines (LXXX) against barley mildew (Pommer and Kradel, 1967)

	% Active ingredient applied				
Substituent R	0·2	0·1	0·025	0·006	0·0015
			Leaf infection†		
i-C_9H_{19}	§	3	5	9	9
i-$C_{10}H_{21}$	1§	2§	9	9	9
n-$C_{12}H_{25}$	§§	§§	2§	3	7
i-$C_{13}H_{27}$	§§	§§	0	0	1
n-$C_{18}H_{37}$	0‡	0	0	2	3
Cyclohexyl	2	3	4	5	9
Cycloheptyl	2	3	3	3	6
Cyclooctyl	2	2	3	3	4
Cyclododecyl	§	0	0	0	1

†0 to 9. No leaf infection to total infection
‡Slight leaf damage
§ greater damage §§ Total damage

there is a tendency for phytotoxicity to increase at the same time. Higher members still show some phytotoxicity but fungicidal effectiveness is also reduced. With cycloalkyl 4-substituted 2,6-dimethylmorpholines, there is a marked fungicidal effect when the ring substituent contains 7–8 carbon atoms and the distinguishing feature of the 4-cycloalkyl analogues is their low phytotoxicity. 4-Cyclododecyl-2,6-dimethylmorpholine (LXXX: R = cyclododecyl) (dodemorph) was highly effective with only a hint of phytotoxicity at the highest test concentration and its acetate was found to be even more fungicidal than the parent compound.

The effect ot substitution in the heterocyclic ring is shown with 4-cyclooctylmorpholine (LXXXI : R = cyclooctyl; $X_1 = X_2 = X_3 = X_4 =$ H) (Table 3.24). Fungicidal properties change only slightly with increasing alkyl substitution, while aryl substituents result in a marked reduction in activity. The optimum combination of methyl groups (2,6-dimethyl) has been incorporated in both tridemorph and dodemorph.

Table 3.24 Dependence of fungicidal activity on the substituents on the C atoms of the morpholine ring (Pommer and Kradel, 1967)

Substituents (LXXXI: R = cyclooctyl)				**% Active ingredient applied**				
X_1	X_2	X_3	X_4	0·2	0·1	0·025	0·006	0·0015
H	H	H	H	0	1	3	3	5
CH_3	H	H	H	0†	0	2	5	6
CH_3	H	CH_3	H	2	2	3	3	4
CH_3	CH_3	H	H	0	1	2	4	7
CH_3	CH_3	CH_3	H	0	3	3	7	9
CH_3	CH_3	CH_3	CH_3	1	2	3	5	9
CH_3	H	C_2H_5	H	0‡	0†	0	3	9
C_6H_5	H	H	H	0†	2	4	5	7
C_6H_5	H	C_6H_5	H	3	5	9	9	9

†0 to 9. No leaf infection to total infection
‡Slight leaf damage to total damage

Since these compounds show fungicidal activity in plate tests *in vitro*, it seems likely that they are absorbed by the barley plant, either by root or foliage, and are translocated *in toto* throughout the plant, giving protection for a period of 3–4 weeks. Penetration of the leaf tissue by tridemorph is very much slower than root uptake, but after penetration a relatively rapid distribution takes place. Although tridemorph has been used safely on both spring and winter barley there are indications of some phytotoxicity on certain winter wheat varieties. The lower activity of tridemorph against powdery mildew of wheat is probably due to reduced movement within the leaf, and it is also interesting that whereas it is taken up by both surfaces of the barley leaf equally, uptake in wheat is greater on the adaxial surface.

6-Azauracil

6-Azauracil (3,5-dioxo-2,3,4,5-tetrahydro-1,2,4-triazine) (LXXXII) which possesses both cytostatic and antibacterial activity has also been shown to be systemically active, either by foliar or root application, against powdery mildew diseases (Dekker, 1962; Dekker and Oort, 1964). Of eighty-two purines and pyrimidines tested against *Erysiphe cichoracearum* in a floating leaf disc test, only 6-azauracil (at 0·6 ppm) and 2,6-diaminopurine (LXXXIII) as sulphate (at 20 ppm) completely prevented development of mildew; 8-azaadenine (LXXXIV), 8-azaguanine

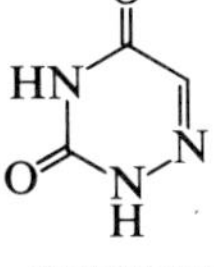

(LXXXII)

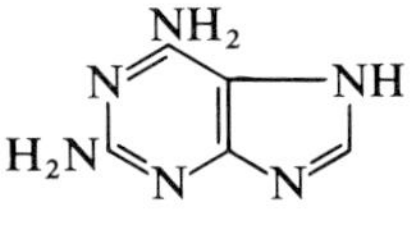

(LXXXIII)

(LXXXV) and dithiouracil (LXXXVI) at 20 ppm also showed a substantial disease reduction. All other purine and pyrimidine derivatives tested were inactive at this concentration and a comparison of these results

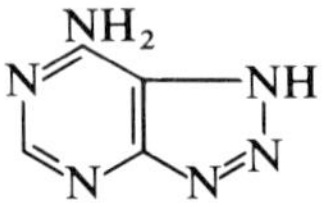

(LXXXIV)

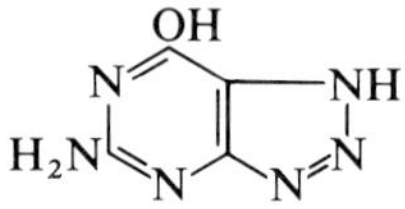

(LXXXV)

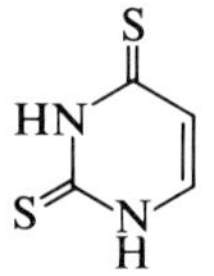

(LXXXVI)

with the occurrence of purines and pyrimidines in ribonucleic acid indicated that only the aza-derivatives of those bases which occur in RNA are active (Table 3.25).

Of the five active compounds, only 6-azauracil was active in the root absorption test, complete prevention of powdery mildew development

Table 3.25 Activity of aza-substituted purines and pyrimidines on cucumber mildew (Dekker, 1962)

Compound	Activity	Bases occurring in RNA
8-Azaadenine	+	adenine
8-Azaguanine	+	guanine
8-Azaxanthine	–	—
8-Azahypoxanthine	–	—
8-Aza-1,3-dimethylxanthine	–	—
8-Aza-2,6-diaminopurine	–	—
6-Azauracil	+	uracil
6-Azathymine	–	—

on the leaves being obtained with 6 ppm. This compound also proved to be active against all other powdery mildews tested by the leaf disc method, including *E. graminis* on wheat, barley and rye and *E. polygoni* DC on beet, peas and lupins. Wheat powdery mildew appeared to be most sensitive and was completely controlled by only 0·3 ppm.

In studies with *E. cichoracearum* on cucumber and *E. graminis* on wheat 6-azauracil did not inhibit germination or penetration of the fungus into the host, but interfered with the growth of the fungus during formation of the first haustorium, which was sufficient to inhibit the pathogen completely. 6-Azauridine and 6-azauridine-5′-phosphate are also active in this way, but their effects, unlike that of 6-azauracil, are not reversed by uracil. In leaf disc tests the uracil precursors, orotic acid, L-dihydro-orotic acid and L-ureidosuccinic acid, also reversed inhibition of cucumber powdery mildew by 6-azauracil. Similar observations were made in leaf disc tests with bean rust (*Uromyces appendiculatus*), cucumber anthracnose (*Colletotrichum lagenarium*) and tobacco mosaic virus.

By contrast, inhibition of bacterial and fungal growth *in vitro* was reversed by uracil, but not by orotic acid, which thus seems to require the presence of the host. These data are consistent with the hypothesis that 6-azauracil is converted to a ribose-containing metabolite which then interferes with pyrimidine biosynthesis by inhibiting orotidylic decarboxylase, and indeed 6-azauracil has been shown to form 6-azauridine in both the cucumber plant and in the pathogen *C. lagenarium.*

Azepines

The azepines (LXXXVII : R_1 and R_2 = alkyl) are rather broad-spectrum fungicides which control some powdery mildews and rusts as well as leaf spot pathogens. Two compounds, 7-dodecylamino-3,4,5,6-tetrahydro-2H-azepine and 4-t-butyl-3,4,5,6-tetrahydro-2*H* azepine hydrochlorides, have been formulated for greenhouse and field trials (Schwinn, 1969). They also display some eradicant activity against

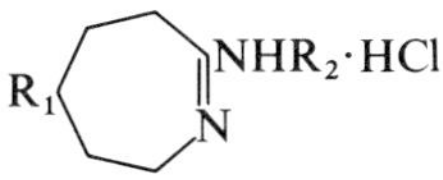

(LXXXVII)

Alternaria solani (treatments up to 24 h after infection), *Uromyces phaseoli* (up to 48 h after infection) and *Erysiphe cichoracearum* (up to more than 72 h). The systemic effect shown by 4-t-butyl-7-declyamino-3,4,5,6-tetrahydro-2H-azepine hydrochloride (LXXXVII : R_1 = t-butyl, R_2 = n-decyl) against cucumber powdery mildew is limited to applications in a nutrient solution and is not observed with soil treatments. The azepines are in general well tolerated by fruit trees but fruit russetting may occur on certain varieties under adverse climatic conditions after application during the period of fruit formation.

Triforine

Yet another heterocycle has been involved recently in systemic fungicide development – piperazine. 1,4-Di-(2,2,2-trichloro-1-formamidoethyl) piperazine (LXXXVIII), common name triforine, has recently been introduced as a foliage fungicide, and it is presumably the best of some sixty-three compounds listed in the original patent (Ost *et al.*, 1969), and prepared from piperazine by condensation with substituted acid amides. It is particularly active against powdery mildews on cereals, apples, cucurbits and ornamentals and against apple scab (Schicke and Veen, 1969). Despite its low aqueous solubility (4 ppm) it was found to control *E. graminis* on rye seedlings by root uptake, and it also acts

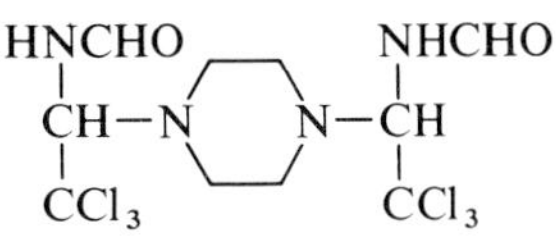

(LXXXVIII)

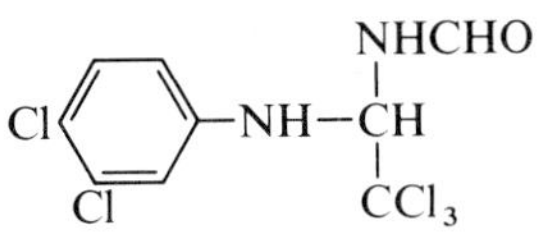

(LXXXIX)

systemically when used as a seed dressing for cereals. Investigations of its absorption and translocation in barley plants using ^3H ring-labelled material have revealed not only the normal upward movement via the xylem, following root absorption, but also a downward movement in the stem to the roots. Less than 1 ppm distributed throughout the plant was found to be sufficient to prevent *E. graminis* (Ost *et al.*, 1971). 1-(3,4-Dichloroanilino)-1-formylamino-2,2,2-trichloroethane (LXXXIX) – proposed common name chloroaniformethan – also contains a formyl-amino group (Malz *et al.*, 1968). It is a specific for certain powdery mildews, being particularly effective in the control of mildew on spring barley (Davis *et al.*, 1971) and has also been used against mildews on wheat and oats. Glasshouse studies showed that the chemical possessed both curative and protective properties in addition to being taken up by roots and showing distal movement within leaves.

In recent structure–activity studies, Carter *et al.* (1972) showed that the heterocyclic portion of triforine could be replaced by certain other groups without substantial loss of activity against powdery mildew on wheat. Replacement of the formyl group by other keto groups however generally destroyed activity, and it was suggested that the Cl_3CCHNH-CHO grouping may be the important active moiety. However, since the compound Cl_3CCH_2NHCHO was found to be inactive in both leaf spray and root drench tests against *Erysiphe graminis* on barley (Brown & Woodcock, 1975a, b), a substituent linked to the methine carbon via a heteroatom may be essential for activity. These latter workers showed that the role of the trichloromethyl group was fairly non-specific, since activity was retained when it was replaced by other groups such as tri-bromomethyl or dichloromethyl. They suggested that the grouping XCHNHCO (where X = substituted nitrogen, oxygen or sulphur) may contain the biologically active centre involved in the toxic step, while other parts of the molecule may influence partitioning to the site of action and exert steric effects. An investigation of quantitative structure–activity relationships in several series of bis-(1-(2,2,2-trichloroethyl) formamides) by the Hansch method has also been reported (Brown & Woodcock, 1975c). It was found that the activity of compounds applied as root-drenches against *E. graminis* on barley correlated with their partitioning properties, as determined by reversed phase t.l.c., and other possible factors governing the activity of these compounds were discussed.

Aliphatic and alicyclic compounds

N-(2,2,2-trichloro-l-methoxyethyl) formamide (XC), which is structurally related to triforine, showed little fungistatic or protectant activity, but was found to control *E. graminis* by root application, though somewhat phytotoxic (Carter *et al.*, 1972). In a subsequent paper in which some hundred related compounds were tested, alkoxy- and alkylamino-

$Cl_3CCH(OMe)NHCHO$

(XC)

$Me_2NCH_2CH_2CH_2NHCOSEt.HCl$

(XCI)

derivatives were found to be more systemically active than the corresponding alkylthio-compounds, and in these series the propyl, butyl and allyl members were more effective than those containing methyl and ethyl groups (Carter *et al.*, 1973).

Ethyl *N*-(3-dimethylaminopropyl)thiocarbamate hydrochloride (XCI), proposed common name prothiocarb, was recently introduced as a systemic soil fungicide capable of being taken up and translocated in plants, and of exerting a fungistatic influence on various *Phycomycetes*, especially *Pythium*, *Phytophthora* and *Peronospora* spp. (Bastiaansen *et al.*, 1974; Kaars Sijpesteijn *et al.*, 1974). Its action is very pH dependent and indications are that it interferes with biosynthesis, though antagonism by cholesterol is rather mysterious. Kaars Sijpesteijn *et al.* (1974) found the corresponding 4-dimethylaminobutylamine to be inactive, but the sulphur-free oxygen analogue appeared to be as active against *Pythium* spp. as prothiocarb itself. Its fate and mobility in biological systems have been examined by Iwan and Goller (1974) using ^{3}H- and ^{14}C-labelled materials.

$CH_3(CH_2)_{10}CONHCHCO_2HCH(CH_3)_2$

(XCII)

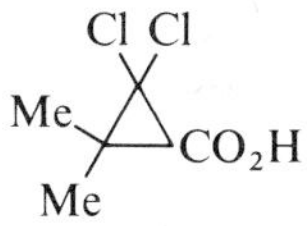

(XCIII)

In a search for naturally occurring substances suitable for use as agricultural chemicals, Misato *et al.* (1972) found that *N*-lauroyl-L-valine (XCII) was very effective against rice blast. Since hydrolysis of the amide linkage would only release the already naturally occurring compounds lauric acid and L-valine, 'residue problems' are obviated.

2,2-Dichloro-2,2-dimethylcyclopropane carboxylic acid (XCIII) is the most effective member so far examined of a group of dichlorocyclopropanes which are also systemically active against rice blast caused by *Pyricularia oryzae* Cav. (Langcake & Wickins, 1975a). They are highly specific in their action, however, being inactive against other host–pathogen combinations, including sheath blight of rice (caused by *Pellicularia sasaki*) and *P. oryzae* on other hosts such as ryegrass or barley. They appear to be more effective on Japonica rather than Indica varieties of rice, and they are most effective by root application, i.e. into the paddy water. Their intrinsic fungitoxicity is low and they appear to act prophylactically rather than curatively. Without any evidence for the formation of a more fungitoxic derivative in the rice plant, it would seem that these compounds act by stimulation of disease-resistance.

Organotin compounds

Although triphenyltin acetate applied to leaves of sugar beet 3 days before inoculation with spores of *Cercospora beticola* failed to show any systemic fungicidal activity (Zolel, 1970), it was later shown that this compound not only possesses translaminar activity, but it can also reduce foliar infection in 1-month-old sugar beet seedlings by way of root uptake (Zolel, 1971). In the latter case, the root system was immersed in a fungicide suspension (10 ppm) for 3 days before inoculation of the leaves with a *C. beticola* spore suspension and incubation in a mist chamber for a similar period. The efficiency of triphenyltin acetate in the field control of *C. beticola* may thus be due in some measure to this moderate systemic effect as well as to its known protectant activity.

The first report of the systemic fungicidal activity of triphenyltin chloride was against *Alternaria tenuis* in sugar beet seedlings (Mukhopadhyay and Thakur, 1972), and it has now been shown to be translocated from the roots to the leaves and leaf sheath of rice seedlings in 24 h (Singh and Sharma, 1973).

Aromatic compounds

Although groups of purely aromatic compounds such as quinones and phenols have furnished effective protectant fungicides without being associated with an organophosphorus residue, a heterocyclic ring system or sulphur atom or atoms, the same hardly applies in the case of systemic fungicides. In the following section only two cases of systemic activity, without associated phosphorus or sulphur atoms, or heterocyclic nuclei are cited. In the first of these (chloroneb) the systemic effect is extremely specific, whilst in the other only trans-laminar or limited systemic activity is shown by the phenols concerned.

Thiourea-based fungicides

A new class of fungicides based on thiourea and containing an aromatic nucleus may however depend on conversion to the benzimidazole ring for their activity. 1,2-Di-(3-alkoxycarbonyl-2-thioureido)benzenes are weakly acidic compounds synthesised by the reaction of *o*-phenylenediamine and isothiocyanoformic esters. The conversion of 1,2-di-(3-ethoxycarbonyl-2-thioureido)benzene (XCIV: R = Et) (thiophanate) and the corresponding methoxy compound, thiophanate-methyl (XCIV: R = Me) to ethyl and methyl benzimidazol-2ylcarbamates (LI: $R_1 =$ H, $R_2 =$ COOEt or COOMe) respectively in aqueous solution, suggested that the fungicidal activity of the thiophanates may well be attributed to these heterocyclic transformation products (Selling *et al.*, 1970) and the inactivity of 1,3- and 1,4-di-(3-ethoxycarbonyl-2-thioureido)-

benzenes, which cannot cyclise in this way, supports this suggestion. Further evidence is provided by the behaviour of the thiophanates in plants and animals. Using both ^{14}C- and ^{35}S-labelled materials, it has been found that cleavage of the C—S bonds takes place readily and

(XCIV)

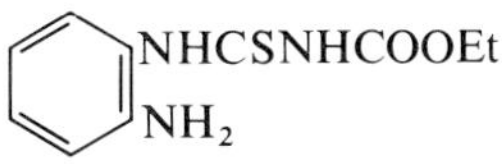

(XCV)

the major resultant metabolites in both plants and animals have been identified as 2-alkoxycarbonylaminobenzimidazoles by co-tlc-autoradiography (Noguchi and Ohkuma, 1971; Noguchi *et al.*, 1971). Additionally, hydroxylation of the aromatic ring of benzimidazole followed by conjugation and subsequent elimination, took place in mice.

Evaluation of the thiophanates in the UK has shown a high degree of systemic activity by root uptake against barley and cucumber mildews, together with a persistence of action, not only for thiophanate, but particularly for the analogous thiophanate methyl and compound NF48 (XCV). There is little systemic activity against apple mildew, however, and the translaminar activity and localised movement towards the leaf apex, shown in barley and cucumber, was also much less marked in apple. These compounds were originally evaluated against tobacco mosaic virus with a view to screening for possible interference with *de novo* synthesis of polynucleotides, and it may be significant that *mono*-alkoxycarbonyl-thioureido benzenes were found to be more active against this organism.

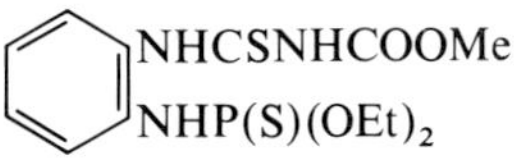

(XCVI)

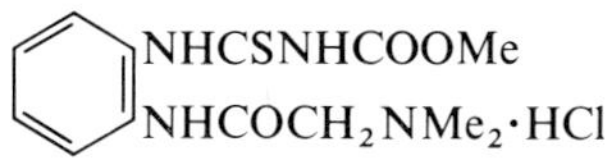

(XCVII)

Two experimental fungicides, also in this class, are of interest. 2-(3-Methoxycarbonylthioureido) *O,O*-diethylthiophosphoranilidothionate (XCVI), from Stauffer Chemical Co, controls rust (caused by *Uromyces phaseoli typica*) and powdery mildew (caused by *Erysiphe polygoni*) at 100 and 10 ppm respectively, when applied as a foliar spray to bean plants in the glasshouse. This compound is also reported to give excellent systemic control of these diseases by root uptake, the corresponding LC_{50}s being 10 and 0·13 ppm respectively. 1-(3-Methoxycarbonylthioureido) - 2 - dimethylaminomethylcarbonylaminobenzene hydrochloride (XCVII), from May and Baker Ltd, is also effective as a foliar spray and/or as a root drench against several diseases, and has been shown to be well distributed and has some prophylactic effect against artificial infections of *Ceratocystis ulmi* when injected into established elm trees.

It is interesting to recall here the systemic activity of phenylthiourea

(XCVIII) which can prevent disease symptoms caused by *Cladosporium cucumerinum* by root treatment of cucumber seedlings with a solution containing 50 ppm (Kaars Sijpesteijn and Pluijgers, 1962). Despite a certain formal similarity with the thiophanates, any further analogy of fungicidal action seems unlikely since phenylthiourea performs very poorly against the pathogen *in vitro*, where a concentration of some 500 ppm is required to inhibit *C. cucumerinum* in the roll culture test. Since the press sap contained less than 20 ppm of phenylthiourea and did not appear to contain any other antifungal derivative thereof, it seems likely that the chemical interferes directly with plant–fungus biochemical relationship or influences plant metabolism in some way (Kaars Sijpesteijn and Sisler, 1968). Chemotherapeutic activity showed a remarkable correlation with polyphenol oxidase – inhibitory capacity and peroxides are also higher in treated than in untreated plants. Increased lignification in treated plants, especially around the penetration points of *C. cucumerinum*, may well be interrelated. There are also indications of increased indoleacetic acid oxidase activity in treated plants and it has been suggested that this leads to increased resistance to cucumber scab (van Andel, 1966a).

(XCVIII)

Substituted thiosemicarbazides

In contrast to phenylthiourea, the closely related 1-phenyl-3-thiosemicarbazide (XCIX) was originally thought to be active *per se*, both *in*

Table 3.26 Antigungal activity of thiosemicarbazides; effect of 4-, 2- and 1-substitution (Pluijgers and Kaars Sijpesteijn, 1966)

Substituent	Minimum concentration causing complete inhibition of growth (ppm)				Systemic activity against cucumber scab after root application
	Botrytis allii	*Penicillium italicum*	*Aspergillus niger*	*Cladosporium cucumerinum*	
—	200	100	200	100	+
4-Methyl	5	5	500	200	+
4-Allyl	5	10	500	200	+
4-n-Butyl	100	50	200	50	+
4-n-Octyl	100	50	100	20	–
4,4-Dimethyl	200	100	200	100	–
2-Methyl	500	500	> 500	50	–
1-Phenyl	5	1	2	1	+ +
1-Benzyl	500	500	500	200	+ +
1-Acetyl	500	500	500	500	–
1-Benzoyl	500	500	500	200	–

vitro and *in vivo* against cucumber mildew, apple scab, potato late blight and *Botrytis* and *Cladosporium* on tomatoes (Pluijgers and Kaars Sijpesteijn, 1966). It has recently been shown, in laboratory trials using *Aspergillus niger*, to be converted to 1-phenylazothioformamide (C) *in vivo* and it is now clear that this is the active agent (Kaars Sijpesteijn *et al.*, 1968). In the light of this discovery the high structural specificity of substitution in thiosemicarbazide (Tables 3.26, 3.27 and 3.28) is very interesting in retrospect. Thus only in those cases where a stable azo-

Table 3.27 Antifungal activity of thiosemicarbazides: effect of aryl substitution (Pluijgers and Kaars Sijpesteijn, 1966)

	Minimum concentration causing complete inhibition of growth (ppm)				**Systemic activity against cucumber scab after root application**
Substituent	*Botrytis allii*	*Penicillium italicum*	*Aspergillus niger*	*Cladosporium cucumerinum*	
1-Phenyl	5	1	2	1	+ +
2-Phenyl	500	500	200	50	+ +
4-Phenyl	200	200	500	100	+
1,1-Diphenyl	500	500	500	500	–
1,4-Diphenyl	500	100	500	20	–
2,4-Diphenyl	500	200	500	50	–
1-(α-Napthyl)	5	1	20	5	+ +
1-Phenylsemicarbazide (for comparison)	500	500	500	500	–

Table 3.28 Antifungal activity of thiosemicarbazides: effect of substitution in 1-phenylthiosemicarbazide (Pluijgers and Kaars Sijpesteijn, 1966)

	Minimum concentration causing complete inhibition of growth (ppm)				**Systemic activity against cucumber scab after root application**
Substituent	*Botrytis allii*	*Penicillium italicum*	*Aspergillus niger*	*Cladosporium cucumerinum*	
1-Phenyl-4-methyl	20	50	100	10	+ +
1-Phenyl-4,4-dimethyl	0·5	50	100	5	+ +
1-Phenyl-4-acetyl	500	500	500	500	–
1-Phenyl-4-benzoyl	500	500	500	500	–
1-(4′-Chlorophenyl)	0·5	0·2	0·5	0·5	+ +
1-(2′,6′-Dichlorophenyl)	50	2	5	1	–
1-(2′,4′,5′-Trichlorophenyl)	20	2	10	1	–
1-(4′-Nitrophenyl)	5	5	50	20	+
1-(4′-Methylphenyl)	2	2	5	1	+ +
1-(4′-Methoxyphenyl)	5	5	20	5	+ +

thioformamide can be expected does the medium become orange-coloured and strong antifungal effects observed (Table 3.29) and it is clear that this requires the presence of a single suitable stabilising aromatic substituent in position 1. The presence of the thione sulphur atom

Table 3.29 Structure-activity relationship of thiosemicarbazides (Kaars Sijpesteijn *et al.*, 1968)

Compound	Minimum concentration (ppm) causing complete inhibition of growth; glucose agar, pH 6·5				
	Botrytis allii	*Penicillium italicum*	*Aspergillus niger*	*Cladosporium cucumerinum*	**Coloration of medium**
1. $C_6H_5.NH.NH.CS.NH_2$	5	1	2	1	+
2. $H_2N.N(C_6H_5).CS.NH_2$	500	500	200	50	−
3. $H_2N.NH.CS.NH.C_6H_5$	200	200	500	100	−
4. $(C_6H_5)_2N.NH.CS.NH_2$	500	500	500	500	−
5. $C_6H_5.CH_2NH.NH.CS.NH_2$	500	500	500	200	−
6. $4\text{-}Cl.C_6H_4.NH.NH.CS.NH_2$	0·5	0·2	0·5	0·5	+
7. $4\text{-}CH_3.C_6H_4.NH.NH.CS.NH_2$	2	2	5	1	+
8. $\alpha\text{-}C_{10}H_7.NH.NH.CS.NH_2$	5	1	20	5	+

would also appear to be essential since the corresponding 1-phenylsemicarbazide, possessing only an oxygen atom, is inactive (Table 3.27).

(XCIX)

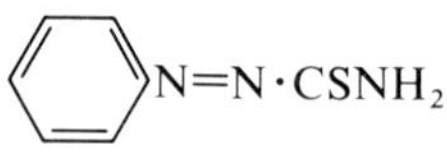

(C)

Phenols and phenol derivatives

1,4-Dichloro-2,5-dimethoxybenzene (chloroneb: CI) was introduced as an experimental fungicide in 1967 by DuPont. Being taken up by root absorption and concentrated in both roots and lower stem, it is useful as a supplemental seed treatment for sugar beet, and as a soil treatment at planting time to control some seedling diseases of cotton and beans. It is highly fungistatic to *Rhizoctonia* spp., moderately so to *Pythium* spp., but gives only poor control of *Fusarium* spp., and is inactive against *Trichoderma* spp.

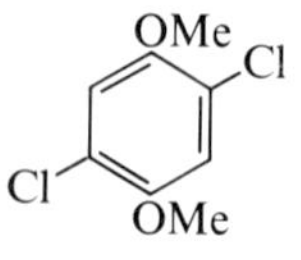

(CI)

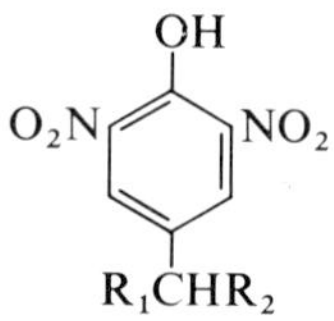

(CII)

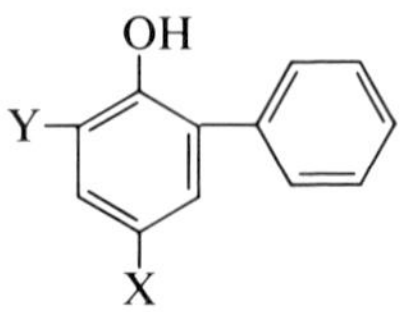

(CIII)

The 4-(1-substituted alkyl)-2,6-dinitrophenols (CII) show an interesting pattern of activity against cucumber and apple powdery mildews in that, in general, their partition and hydrogen bonding characteristics tend to facilitate vapour phase redistribution and also zonal activity, that is protection of large areas of the leaf surface by lateral movement within the leaf surface waxes (Clifford *et al.*, 1971). However, the values of these parameters are such that entry into the leaf, and hence translaminar activity and translocation are precluded. On the other hand, the 4-(1-cyclopropylalkyl)-2,6-dinitrophenols (CII: R_1 = cyclopropyl, R_2 = alkyl) differ in that their partition and hydrogen bonding characteristics are such that some homologues when applied to leaves are extensively translocated resulting in high protectant activities (Table 3.30). Translaminar activity observed with 2-bromo-6-cyclohexyl-4-nitrophenol (CIII: X = NO_2, Y = Br) was also attributed to favourable combinations of these parameters, the isomeric 4-bromo-2-cyclohexyl-6-nitrophenol (CIII: X = Br, Y = NO_2) showing negligible activity (Table 3.31). Since it is generally accepted that nitrophenols act as inhibitors of oxidative phosphorylation, those analogues which are able to penetrate will interfere with energy production in the plant with disastrous consequences, which thus preclude their exploitation commercially. These findings are in agreement with the suggestion of Kaars Sijpesteijn (1969) that all successful systemic fungicides are those which act by inhibition of biosynthetic pathways rather than by inhibition of energy production.

Table 3.30 Activities of 4-(1-cyclopropylalkyl)-2,6-dinitrophenols (CII) against cucumber powdery mildew (Clifford *et al.*, 1971)

Alkyl group	π	A^x	A^y	Protectant activity $10^5 ED_{50}$(M)	Zonal activity
Methyl	0·74	4·6	0·26	17	minimal
Ethyl	0·70	4·4	0·35	22	not tested
Propyl	0·76	4·6	0·47	3	widespread delocalised control and some chlorosis
Butyl	0·79	4·5	0·41	1·5	
Pentyl	0·90	4·1	0·37	1·0	
Hexyl	0·99	3·9	0·33	1·25	
Heptyl	1·07	4·3	0·42	6	

Table 3.31 Activities of 2-cyclohexyl-halogeno-nitrophenols (CIII) against cucumber powdery mildew (Clifford *et al.*, 1971)

Substituents					Protectant activity	Zonal
4	6	π	A^x	A^y	$10^5 ED_{50}$(M)	activity
Cl	NO_2	0·15	2·8	0·07	>40	nil
Br	NO_2	0·09	4·4	0·11	>40	nil
NO_2	Cl	0·66	2·8	0·04	7·8	nil
NO_2	Br	0·78	2·7	negligible	6·2	translocated

It is clear from a review such as this, that a wide range of (often unlikely) chemicals are effective as systemic fungicides. It is also abundantly obvious that while some compounds are freely translocated throughout the plant, others described as systemic, are really only protectant fungicides which are also capable of limited foliar movement. However, such activity can be a valuable adjunct in a protectant fungicide and can play a critical role in protection against certain pathogens. Because of the inevitable intimacy of a systemic chemical with the cells of the host plant, one of the prime requisites of a true systemic fungicide, or indeed of any compound acting systemically, must be a high specificity of action. This is obviously possible with several basic ring systems the structural and stereochemical variations possible with substituents being critical not only for initial uptake and translocation, but ultimately for interference with biochemical processes at enzyme level. While protectant fungicides seem to act by interfering with ATP production by the fungus, the ability to inhibit biosynthesis seems to be common to many systemic fungicides.

Although conventional structure–activity studies based on screening programmes will no doubt continue to supply new fungicides, real hope for the future will surely come from the increased tempo of biochemical research into modes of action.

Chapter 4

Toxicological considerations

D. Woodcock

The prime requirement for the successful utilisation of any biologically active compound in agriculture is that it should be inherently safe both in production and in its application. This safety should extend not only to man, but also to domestic, farm and wild animals, birds, fish, useful plants, beneficial insects and micro-organisms. The metabolism and mode of action of the compound are no less important, and toxicological data for metabolites formed in plants, in animals and in the soil, are becoming increasingly essential.

The toxicity of chemicals can be evaluated in a number of ways using a variety of test animals, birds, fish and insects. Any classification based on acute toxicity to the various experimental subjects is very arbitrary and much depends on other factors. The mean dose in mg/kg of live weight that causes 50 per cent mortality (LD_{50}) can vary enormously between species and the trend with increasing body weight is not always extrapolatable. It is important too that note should be taken of possible penetration routes into the test organism, and compounds of low LD_{50}'s, which are capable of being dermally absorbed or taken in via the respiratory passages are very dangerous indeed.

In the German Democratic and Federal Republics compounds are classified with respect to acute mammalian toxicity into three groups according to their LD_{50}'s: less than 100 mg/kg, 100–300 mg/kg and >300 mg/kg, and compounds of the first group are issued to users only by special permit. In the Soviet Union compounds with LD_{50}'s <50 mg/kg are regarded as 'powerful', compounds with LD_{50}'s from 50–200 mg/kg as 'highly toxic', while LD_{50}'s of 200–1 000 and $>1\,000$ mg/kg are rated as of medium and low toxicities respectively.

In the UK at present there is no classification system, each new active ingredient being dealt with on its merits. However, clearance under the Pesticide Safety Precautions Scheme is only granted (or not) after a detailed study of an enormous amount of toxicological data, and it should be emphasised that acute LD_{50}'s are only a small part of the overall picture.

On the question of classification of pesticides according to hazard much has happened in the past two years. WHO have produced such a scheme based primarily on acute oral and dermal risks of pesticide formulations, though provision is made for other criteria to be taken

Table 4.1 WHO classification of pesticide formulations LD_{50} (mg/kg) rats

Class		Oral		Dermal	
		solids	liquids	solids	liquids
Ia	Extremely hazardous	5 or less	20 or less	10 or less	40 or less
Ib	Highly ,,	5–50	20–200	10–100	40–400
II	Moderately ,,	50–500	200–2 000	100–1 000	400–4 000
III	Slightly ,,	>500	>2 000	>1 000	>4 000

into account (Bates, 1975). In this scheme significantly no compounds are regarded as non-hazardous (Table 4.1).

Fungicides, in general, do not pose any great mammalian hazards and few, apart from organophosphorus-based compounds, will be rated as even moderately hazardous. There are often wide variations within a single class of compound, however, and occasionally some real surprises.

Organophosphorus compounds

The action of organophosphorus compounds in phosphorylating vitally important enzyme systems, and thus inhibiting normal functions, makes them highly active as pesticides. It means, however, that toxicity to vertebrates may also be high and careful structural manipulations are required to produce compounds with suitably moderated mammalian toxicities. Although organophosphorus compounds provided the first systemic insecticides, the logical exploration of this class for antifungal activity has provided relatively few acceptable systemic fungicides. One of these is triamiphos, which was claimed as the first systemic fungicide and which, although possessing an acute oral LD_{50} to male rats of 20 mg/kg, is less toxic as a skin application to rabbits (LD_{50} 1 500–3 000 mg/kg) and mice (LD_{50} > 1 000 mg/kg). However, this compound did not receive clearance under the Pesticide Safety Precautions Scheme on the grounds of its neurotoxicity.

The phosphoric esters of pyrazolopyrimidines recently introduced by Hoechst, which are also effective against powdery mildews, appear to be rather less toxic to mammals than triamiphos. Thus the acute oral toxicities (LD_{50}'s) of O-6-Ethoxycarbonyl-5-methylpyrazolo[1,5-*a*] pyrimidin-2-yl *O,O*-diethyl phosphorothioate[IV: $R = P(S)(OEt)_2$] (pyrazophos), as pure compound and commercial product, are 140 and 370 mg/kg respectively.

The rice blast fungicides, Kitazin and Kitazin P, likewise show only moderate mammalian toxicities with LD_{50}'s of 238 mg/kg (rats) and 660 mg/kg (mice) respectively, being quoted by Mel'nikov (1971).

Antibiotics

Probably the safest systemic fungicide of all is streptomycin, and with its widespread clinical use it is no surprise to find an LD_{50} (acute oral for mice) as high as 9 000 mg/kg. However, the Joint Committee on Antimicrobial Substances is pressing for its withdrawal from agricultural use (Bates, 1975), presumably because of the danger of encouraging proliferation of human pathogens. Griseofulvin also has fairly low toxicity, for Wistar rats of either sex can tolerate a daily dose of 2 000 mg/kg given intraperitoneally, though there is some damage to the seminal and intestinal epithelia. Intravenous injection causes a transient depression of mitosis, most noticeable in the bone marrow, but the rats appear little affected in general condition and recovery proceeds rapidly after 24 h; the LD_{50} is 400 mg/kg. Cycloheximide is considerably more toxic than streptomycin and griseofulvin, acute oral toxicity tests giving the following LD_{50}'s (mg/kg): 133 (mice, 65 (guinea pigs), 2·5 (rats) and 60 (monkeys).

The toxicity of blasticidin is also rather high, its oral LD_{50} for rats being 39 mg/kg. It has also been reported to cause conjunctivitis in humans. By contrast, kasugamycin, the other antibiotic useful in the control of rice blast, has a very low mammalian toxicity. The LD_{50} (oral) for mice is $>$2 000 mg/kg and TLM for carp is $>$1 mg/ml.

The polyoxins are also non-toxic to mammals and fish, the Japanese killifish being unaffected by a 72 h exposure to a concentration of 100 ppm. Oral administration to mice shows an $LD_{50} > 1\,500$ mg/kg, and there appears to be no effect on the human eyes or mucous membranes. The simple molecule of cellocidin is also usually fairly harmless but exhibits a wide range of toxicity to mice, depending on how the antibiotic is applied. Thus the LD_{50}'s for intravenous, oral and cutaneous administration are 11, 89–125 and 667 mg/kg respectively. It has also been reported as being less toxic to fish than DDT.

Carboxylic acid anilides

Members of this class of compound appear to be very safe both dermally and orally. Thus the acute LD_{50} (rats) for carboxin is as high as 3 200 mg/kg and albino rats fed 200 ppm daily for 90 days suffered no detectable symptoms; the acute dermal LD_{50} (rabbits) is 8 000 mg/kg. The corresponding values for oxycarboxin are 2 000 and $>$16 000 mg/kg respectively. The non-heterocyclic mebenil likewise possesses very low mammalian toxicity, the acute oral LD_{50} (rats) being 6 000 mg/kg, with no evidence of skin irritation in laboratory tests. Pyran and furan analogues also provide some very safe compounds. Thus, 3,4-dihydro-6-methylpyran-5-carboxanilide (pyracarbolid) has an acute LD_{50} (rats) of $>$1 500 mg/kg, and in sub-chronic tests the no-effect level was 800 ppm. For 2,5-dimethyl-3-furanilide (furcarbanil), the acute oral LD_{50}

(rats) is >6 400 mg/kg, and this does not appear to alter following the substitution of a cyclohexyl nucleus in place of the phenyl group (cyclafuramid). Of the two substituted thiazoles used in seed treatment, whereas 2,4-dimethylthiazole-5-carboxanilide has a very low mammalian toxicity (LD_{50} 5 600 mg/kg), replacement of one methyl group by an amino group to give 2-amino-4-methylthiazole-5-carboxanilide, results in a dramatic increase in toxicity, the LD_{50} (rats) falling to 141 mg/kg.

Heterocyclic compounds

The substituted benzimidazoles benomyl, fuberidazole and thiabendazole are all apparently very safe compounds. The acute oral toxicity of benomyl against male rats is >10 000 mg/kg and there appears to be no evidence of chronic toxicity in 90-day feeding tests at a dietary level of 2 500 ppm; in 2-year feeding studies with rats and dogs only a low order of toxicity was apparent. There is also no sign of skin irritation and no hint of dermatitis. Substitution of heterocyclic nuclei in position 2 results in slightly increased toxicity but both thiabendazole and fuberidazole containing thiazolyl and furyl substituents respectively, are still virtually innocuous, which is not surprising in the case of the former since it was initially developed as an anthelminthic. LD_{50}'s (acute oral) for thiabendazole are 3 320 (rats), 3 810 (mice) and 3 850 mg/kg (rabbits) and daily doses of 100 mg/kg given orally over a 2-year period induced no toxic symptoms. Fuberidazole had the following LD_{50}'s (mg/kg) for male rats: acute oral, 1 100; acute dermal (7-day exposure), 1 000; acute peritoneal ~100. Five doses per week for 8 weeks at 120 mg/kg caused an increase in liver weight but did not affect other organs.

The breakdown product of benomyl, methyl benzimidazol-2-yl carbamate (carbendazim) which has recently been introduced by BASF as a systemic fungicide in its own right, also seems to be a reasonably safe compound. Approximate median lethal doses (ALD_{50}'s) are >6 400, >8 000 and >8 000 mg/kg for rats, rabbits and dogs respectively, and the 50 per cent formulation examined dermally has ALD_{50}'s of >2 500 and >4 000 mg/kg for rats and rabbits.

The hydroxypyrimidines also appear to be safe compounds. The acute oral LD_{50} (rats) for dimethirimol is more than 4 000 mg/kg and 3-day dermal applications of 400 mg/kg produced neither skin damage nor toxic symptoms. Ninety-day feeding tests at dietary levels of 1 000 ppm also produced no adverse effects in rats or dogs. Ethirimol has the same acute oral LD_{50} for rats and was also harmless in 90-day feeding trials at 200 mg/kg. Other LD_{50}'s (acute oral) are 1 000–2 000 mg/kg (male rabbits); >1 000 mg/kg (female cats); 500–1 000 mg/kg (female guinea pigs); 4 000 mg/kg (hens).

The newly introduced ethirimol derivative, bupirimate (Finney *et al.*,

1975), is also of low mammalian toxicity (Table 4.2). In 90-day feeding tests, no ill effects were detected in rats fed on diets containing up to 1 000 ppm and no appreciable dermal toxicity or sensitisation was observed. It also displays a low toxicity to fish and invertebrates, and no adverse effects have resulted from topical and oral administration to bees at several times the recommended field use dosage rate.

Triarimol, with an acute oral LD_{50} for rats of 600 mg/kg, which places it in the 'slightly hazardous' category, illustrates the necessity for long and comprehensive toxicological testing of candidate pesticides. Although 3-month feeding tests showed that the safe levels for rats and

Table 4.2 Acute toxicity of bupirimate (Finney *et al.*, 1975)

Route	Test species	LD_{50}mg/kg
Acute oral	Rat (female)	*c.* 4 000
	Mouse (female)	>4 000
	Guinea pig (male)	*c.* 4 000
	Rabbit (female)	2 000–4 000
	Pigeon	>2 700
	Quail	>5 200
Acute peritoneal	Rat	2 000–4 000

dogs were 400 and 800 ppm respectively, some longer-term effects have resulted in rejection of the compound by the Pesticides Safety Precautions Scheme.

Triadimefon, recently introduced by Bayer AG, falls in the 'moderately hazardous' class with oral LD_{50}'s for rats of 568 mg/kg (male) and 363 mg/kg (female) respectively. Although no toxic effects were observed in sub-chronic feeding studies over 3 months at levels of 600 and 2 000 ppm in dogs and rats respectively (Frohberger, 1975), longer-term trials will no doubt be scrutinised with care in view of the triarimol experience. No appreciable dermal or inhalation toxicity was evident and no damaging effects were observed on mucous membranes.

The piperazine derivative, triforine, with acute oral LD_{50}'s for mice and rats of >6 000 mg/kg appears to offer little hazard, and it is also reported to be harmless to fish (*Lebistes reticulatus*) at 50 ppm in water, and to bees at 1 000 ppm in feeding experiments. Another compound, chloraniformethan, which also contains the formylamino group, also appears reasonably safe after extensive testing at the Institut für Toxicologie, Elberfeld. Oral toxicity tests showed LD_{50}'s of >2 500, >1 000, >500 and 250–500 mg/kg for male rats, female mice, cats and guinea pigs respectively. In feeding tests, up to 100 ppm in the daily diet for a period of 14 weeks caused no observable changes in rats, and dermal toxicity appeared to be slight (LD_{50} >500 mg/kg). A similar dose applied daily to the clipped dorsal skin of male and female rabbits over a period of 14 days also caused no change in general condition, body weight, blood picture, liver and kidney functions, and application to the ear and conjunctival sac also appeared to have no ill-effects.

Dodemorph must also be regarded as reasonably safe, LD_{50}'s having been reported as 4 800 mg/kg (acute oral) for rats and as 600 mg/kg (intraperitoneal) for mice. Values of 4 500 mm^3/kg for the LD_{50} (acute oral) of the 40 per cent formulation against rats, and of 40 ppm for the LC_{50} against guppies (*Lebistes reticulatus*) have also been given. However, it has been reported to cause severe skin and eye irritation to rabbits, but it does not appear to induce skin sensitivity. Tridemorph is marginally more toxic, the acute oral LD_{50}'s for the 75 per cent (w/v) formulation being 1 270 and 750 mm^3/kg for rats and rabbits respectively. Relative dermal activity is somewhat difficult to assess: the stated LD_{50} (acute dermal) for rabbits is $>$1 000 mm^3/kg. Both these 1,4-oxazines are harmless to bees.

The azepines provide yet another systemic fungicide which has a very low mammalian toxicity. 4-t-Butyl-7-decylamino-3,4,5,6-tetrahydro-2*H*-azepine hydrochloride has an acute oral LD_{50} of 1 100 mg/kg and 1-month feeding studies have shown that a daily intake of 150 mg/kg has no effect on rats nor has 15 mg/kg on dogs. This compound is also non-toxic to honey bees in concentrations up to 0·5 mg.

Aromatic compounds

With an acute oral LD_{50} for rats as high as 11 000 mg/kg, the soil fungicide chloroneb must surely be one of the safest systemic compounds in use. Its dermal activity is also very low (LD_{50} for rabbits $>$5 000 mg/kg) and a 50 per cent aqueous suspension of the 65 per cent wettable

Table 4.3 Acute toxicity of thiophanate-methyl (Noguchi, 1972a) LD_{50} (g/kg)

Species	Sex	Oral	Dermal	Intraperitoneal	Subcutaneous
Mouse (ddy)	M	3·51	>10·0	0·79	>10·0
	F	3·40	>10·0	1·11	>10·0
Rat (wistar)	M	7·50	>10·0	1·64	>10·0
	F	6·64	>10·0	1·14	>10·0
Guinea pig	M	3·64	10·0	—	—
	F	6·70	10·0	—	—
Rabbit	M	2·27	>10·0	—	—
	F	2·50	>10·0	—	—
Dog (beagle)		4·00 (MLD)	>10·0	—	—
Carp	TLM 75·0 ppm				
Rainbow trout	TML 8·8 ppm				

powder caused no irritation to guinea pigs and repeated applications failed to induce skin sensitisation.

A comprehensive toxicological evaluation of thiophanate-methyl has been reported by Noguchi (1972a). This includes not only acute, subacute and chronic toxicity data for several test species but also the results of carcinogenetic, teratogenetic and mutagenic studies. Some LD_{50}'s are shown in Table 4.3.

Residues

The practice regarding statutory pesticide tolerance levels varies from country to country. Thus while elaborate systems of statutory tolerances are operative in the USA, Canada and the Benelux countries, the EEC members Denmark, Ireland and the UK have none. Some idea of the order of permitted levels of systemic fungicides on various food crops is given in Table 4.4.

Table 4.4 Systemic fungicide tolerances in USA (from EPA compendium of registered pesticides 1973, Vol II)

Compound	Crop	Maximum tolerance (ppm)
Carboxin	Barley, wheat	None†
	Cotton	0·2 (seed)
	Winter wheat	0·2 (grain or straw)
Benomyl	Apples	7·0 (fruit) 70·0 (pomace)
	Strawberries	5·0
	Cucurbits	1·0
	Grapes	10·0
	Sugar beet	0·2 (roots) 15·0 (tops)
Thiabendazole	Apples	10·0 (fruit) 33·0 (pomace)
	Bananas	3·0 (fruit) 0·4 (pulp)
	Citrus	2·0 (fruit) 8·0 (dry pulp)
	Pears	10·0

† Non-food use only

From the foregoing brief survey it is clear that systemic fungicides rank among the safest pesticides in current use. Although their metabolites, both enzymic and non-enzymic, and possible conjugates have not been mentioned here, there is no doubt that their toxicology is also of prime importance and will undoubtedly continue to be given exhaustive scrutiny by pesticide manufacturers.

Chapter 5

Translocation

S. H. Crowdy

There are a number of general descriptions of translocation in plants (Richardson, 1975; Peel, 1974; Crafts, 1961) and also more recent reviews of more specialised aspects (Milthorpe and Moorby, 1969; Eschrich, 1970). Data relating specifically to systemic fungicides have been reviewed by Erwin (1973). The movement of chemicals in plants can conveniently be considered in three stages:

(*a*) entry into the free space within the tissues.
(*b*) Movement in the apoplast. This occurs in the non-living parts of the cell, outside the protoplasm, and within the lumen of the dead vessels and tracheids of the xylem, where long distance transport takes place. Apoplastic movement is 'passive' and does not require the expenditure of metabolic energy.
(*c*) Movement in the symplast, which is within the living parts of the cell. This is 'active' and requires the expenditure of metabolic energy. Long distance transport in the symplast system takes place in the phloem, probably in specialised cells, the sieve tubes.

The 'plasmalemma', the membrane bounding the protoplast, separates the symplast from the apoplast. The tissues involved in translocation are described in general textbooks of Botany; in addition there are more detailed monographs on plant anatomy such as Esau (1965).

Entry into the free space within the tissues

Entry

The initial barrier to the penetration of materials into the leaves and aerial parts of the plant is the cuticle and its associated structures. These have been described in a number of reviews (Martin and Juniper, 1970; Hallam and Juniper, 1971; Holloway, 1971; Baker, 1971). The diagram in Fig. 5.1 gives a general idea of the structure of the cuticle; the whole structure is dead and is built on the outer walls of the epidermal cells, which are themselves in contact with the living protoplasm. The layers grade into each other; the cellulose of the cell wall gradually grades into the pectin layer, which may be continuous with the pectin of the middle lamellae of the anticlinal walls of the epidermal cells. Above the pectin layer is a matrix of cutin, a polymer consisting of fatty and hydroxy-fatty acids. At the base of the cutin matrix there seems to be a network of cellulose fibres which fade out in the upper layers. Nearer the surface the cutin contains platelets of wax, which also disappear in the topmost layers of the cuticular membrane. Above the cuticular membrane is commonly a layer of cuticular wax; this may be strikingly

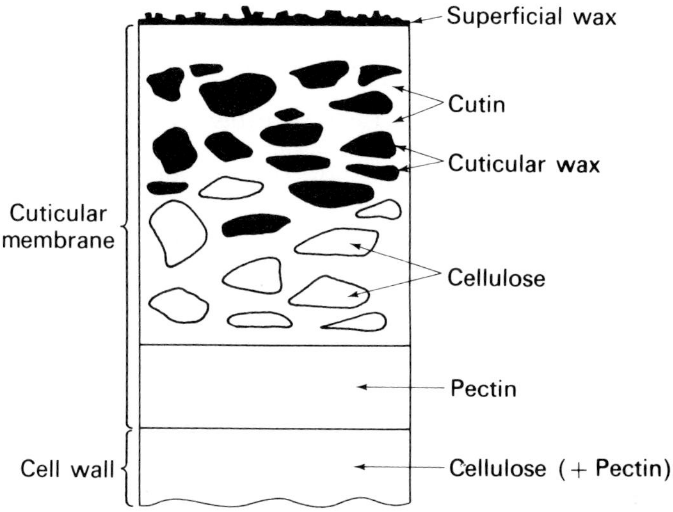

Fig. 5.1 Diagrammatic representation of the plant cuticle. Holloway (1971).

sculptured, exhibiting patterns which are frequently characteristic of different plant species or of different parts of the same species. The main function of the cuticle is to control the loss of water from the aerial parts of the plant; the cutin matrix seems to be permeable to water and the water-proofing is provided by the wax component, mainly the wax platelets embedded in the upper layers of the cutin matrix. The area available for water loss within the cuticle and, naturally, the area available for the entry of aqueous solutions is determined by the separation of these wax platelets, which in its turn depends on the degree of imbibition of the cutin matrix. When the cutin matrix is fully imbibed the separation will be maximal; if the loss of water from the surface exceeds the supply from the roots, the cutin matrix will lose water and contract, bringing the wax platelets closer together and reducing the area available for the passage of water (van Overbeek, 1956). The extent to which this type of regulating system controls water flow has been estimated by Moreshet (1970), who showed that the resistance of sunflower leaves to water loss was inversely proportional to the relative humidity of the surrounding air, rising from an average value of 9·6 mm at a relative humidity of 70 per cent to about 20 mm at relative humidity of 30 per cent. Moreshet also varied the water tension within the plant by varying the water tension in the soil: with a high relative humidity (70 per cent) and a low water tension in the soil (0·2 atm), the leaf resistance was only 1·6 mm, while when the relative humidity was low (30 per cent) and the soil water tension was high (9·9 atm), the resistance rose to 18·7 mm. These data illustrate the very considerable control which the plant exercises over the loss of water from the leaves. Moreshet also calculated the resistance offered by the cuticle in the varied conditions and suggests that cuticular resistance alone rises from about 60 mm at a relative humidity of 70 per cent to about 300 mm at a relative humidity of about 30 per cent.

The freedom with which water and substances dissolved in it will traverse the cuticle is shown by the number of substances which will enter or can be leached from leaves (Franke, 1967; Tukey, 1970); these include a variety of foreign organic chemicals used as insecticides, herbicides and fungicides and natural products such as sugars, sugar alcohols and numerous acids including amino acids. Water solubility seems to be the only absolute requirement for this trans-cuticular movement. The diffusion of systemic fungicides through detached cuticles has been studied in apple (Solel and Edgington, 1973) and coffee (Edgington *et al.*, 1973). The proportion transferred varies with different chemicals, 87 per cent of applied thiophanate-methyl crossed the cuticle as compared with only a trace of dodine. Chemicals applied to the abaxial surface were transferred more readily than those applied to the adaxial surface and transfer was improved by solubilisation and formulating the chemical in oil. It is doubtful whether the channels by which water and solutes enter and leave the leaves can be distinguished by present techniques; however, structures in the cuticle, referred to as ectodesmata, have been described and the recent work related to them has been reviewed (Franke, 1967). It is suggested that the ectodesmata provide an almost direct connection between the leaf surface and the protoplast of the epidermal cells. In section they appear to be outgrowths of the epidermal cell wall reaching into, but never through, the cuticle. In surface view they can be seen as dots which are clustered over the anticlinal walls of the epidermal cells particularly around the stomata and the stomatal accessory cells. These structures are more obvious when solutes are diffusing into the leaf and their distribution appears to be associated particularly with the areas of entry. It can also be shown with tritiated water that the ectodesmata are commonly associated with water loss. Ectodesmata are only revealed when the material is treated with Gilson fixative, which contains mercurous chloride and it has been suggested that the structures have no real existence and are artifacts resulting from the method of treatment, which simply demonstrate the penetration patterns of mercurous chloride (Schönherr and Bukovac, 1970a, b). The balance of evidence seems to support this view. Crowdy and Tanton (1970) and Tanton and Crowdy (1972b) have reported very similar patterns of lead accumulation in leaves treated with the chelate of lead and ethylenediaminetetra-acetic acid, but these have never been associated with cell wall structures which could be mistaken for ectodesmata. However, it should be noted that the technique used would not have revealed structures associated with direct entry from the leaf surface into the protoplasts of the epidermal cells.

Although solutes will enter and leave through the leaves in appropriate conditions, the common pathway by which nutrients and other solutes enter the plants is through the roots. The absorbing surfaces of roots are also covered with a protective layer analogous to the cuticle, but which may consist of precursors of suberin rather than cutin (Martin and Juniper, 1970). The surfaces of older roots are suberised and impervious to water. The initial exclusion of water and passively transferred solutes occurs at an early stage in the development of the tertiary wall of the endodermis. At this stage the plasmodesmata remain functional and the active uptake of ions such as phosphate continues (Clarkson, 1974; Ferguson and Clarkson, 1975). The function of young roots is to absorb water and solutes and the covering to the absorbing zones offers little resistance to water flow to the xylem (Anderson and Reilley, 1968). The main absorbing regions are relatively restricted and may extend only a few centimetres from the root tip. However, the branching of the root system ensures a multiplicity of absorbing points. In the absorbing zone the ratio of surface to volume may be greatly extended by the presence of root hairs, which grow out of the epidermal cells.

The older parts of plants, particularly perennials, are normally covered with bark. This is a heavily suberised multicellular layer, which provides an impervious covering except where the surface is damaged or is penetrated by specialised structures, the lenticels. Aqueous solutions can enter uninjured bark to a small extent and the penetration can be enhanced considerably by scraping or by maintaining a supply of solute on the surface in a gauze pad or by combining both these treatments (Ticknor and Tukey, 1957). There is some evidence that insecticides are systemically distributed after application to the bark (Mitchell *et al.*, 1960). However, it seems unlikely that the bark will in most cases provide satisfactory entry for adequate quantities of systemic fungicides, unlike the roots and the leaves.

Free space

Water and solutes entering either through the roots or the aerial parts of the plant first pass into free space. This can be defined as that part of the plant tissue which communicates freely, by diffusion, with the external environment. The term free space was introduced by Hope and Stevens (1952); the space has also been referred to as the outer space (Epstein, 1955) and the diffusion free space (Johnson and Bonner, 1956). The literature relating to this volume has been reviewed by Kramer (1957) and Briggs and Robertson (1957). Entry into this volume is by diffusion which, in the case of the roots of a transpiring plant, may be assisted by mass flow. It is difficult, if not impossible, to define precisely the free space as an anatomical entity, which precludes direct measurements of the volume. Instead, a number of indirect methods of estimation have been used. One commonly employed is based on ion uptake experiments and is illustrated in Fig. 5.2, reproduced from Briggs and

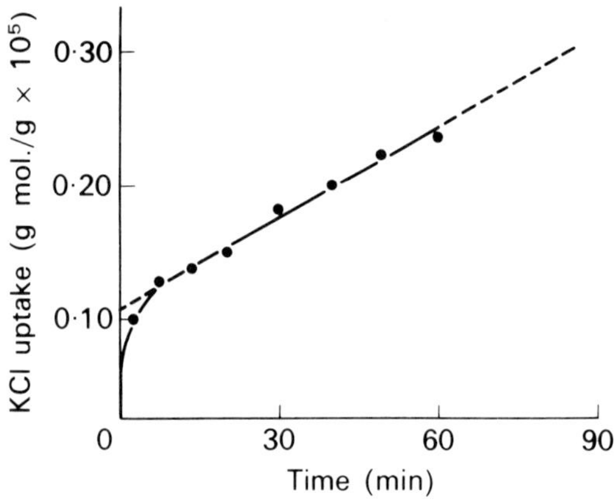

Fig. 5.2 Relation between KCl uptake and time with extrapolation to give free space uptake. Briggs and Robertson (1957).

Robertson (1957). These estimates assume a two-phase uptake; a rapid diffusion into the free space, which is complete in a few minutes, followed by a slow, prolonged accumulation, the amount accumulated being directly proportional to the time of exposure. This accumulation probably reflects the active transfer of the ions from the free space to the cell vacuole. The estimate of the free space volume was obtained by plotting uptake against time and extrapolating the line representing the prolonged steady uptake back to zero time; this provided an estimate U of the amount of salt not already present and diffusing rapidly into the free space. If the concentration of the external solution is C, the estimate of the free space is U/C. The free space estimated in this way includes more than the actual volume of water available for diffusion and has been termed the Apparent Free Space (Briggs and Robertson, 1957). The apparent free space thus includes, in addition to the free space, any systems associated with the free space which interact with the solute such as a Donnan system, which has been considered by Briggs and Robertson (1957) and Pitman *et al.* (1974) or partition into a lipoid phase (Crowdy *et al.*, 1958). The volume of the apparent free space is markedly influenced by the material under study and the conditions in which the experiment was carried out. Some estimates of apparent free space are included in Table 5.1.

The estimates of free space in roots incorporate the uncertainties associated with the concept of apparent free space and, in addition, experimental errors associated with the handling of material. These include failure to make due allowance for the surface film of solute (Levitt, 1957) and adsorption into dead material in the older parts of the roots. A number of these sources of error have been examined systematically by Tanton and Crowdy (1972a) whose conclusions are summarised in

Table 5.1 Estimates of the volume of the apparent free space in various plant tissues (% of tissue volume)

Authority	Tissue	Apparent free space %
Hope and Stevens (1952)	Broad bean roots	13
Butler (1953)	Wheat roots	24·5–33
Epstein (1955)	Barley roots	23
Johnson and Bonner (1956)	Oat coleoptiles	19–22
Kylin and Hylmö (1957)	Wheat roots	27·5
Crowdy *et al.*, (1958)	Wheat leaves	33–65†
Crowdy and Tanton (1970)	Wheat leaves	3–5
Tanton and Crowdy (1972a)	Carrot root discs	4
	Wheat roots	5

† Referred to as Effective Volume

Table 5.2. In this table the uncorrected estimate of apparent free space corresponds to many of the estimates commonly published and agrees reasonably well with the figures published in Table 5.1. A further correction can be applied to the final value in Table 5.2 since the tissue concerned was mainly the cortex and occupied only 75 per cent of the root volume. If allowance is made for this, the apparent free space in the cortex occupies about 5 per cent of the tissue.

Table 5.2 Sources of error in estimates of the apparent free space in roots (% of tissue volume)

Component	Estimate as % of issue volume
Apparent free space uncorrected	18·7
Error due to surface film of solute	7·4
Error due to dead cortical cells	7·5
Apparent free space, corrected	3·8

The concept of apparent free space is not readily interpreted in terms of plant anatomy and it is probably more useful to consider uptake in three stages; the first being diffusion into free space water, the second removal from the aqueous phase of the free space by physical processes and the third metabolically active removal from the first two phases. Phases 1 and 2 comprise the apparent free space. The anatomical entity is the aqueous phase of stage 1 which comprises the free space and seems to be the water of hydration of the cell wall, especially the pectins in the middle lamella. The movement of foreign materials in this volume has been demonstrated with sols of noble metals (Strugger, 1949; Strugger and Peveling, 1961), with the chelate of lead with ethylenediaminetetra-acetic acid (Crowdy and Tanton, 1970; Tanton and Crowdy, 1972a,b), uranyl acetate (Robards and Robb, 1972) and lanthanum (Thomson *et al.*, 1973), all of which give deposits which can be examined in the electron microscope. Since anatomical studies show that the lead chelate is confined to the middle lamella, the estimate of apparent free space of 3–5 per cent of the tissue probably approaches the volume of the free space closely. The sols of noble metals are of special interest since their particles seem to be near the maximum size which will move in the free space; they will move from the shoot to the leaves of *Helxine soleirolii*, but not in *Beta vulgaris* or *Eucalyptus globulus*. Even in *H. soleirolii* the movement of particles with diameters much larger than 5 nm is restricted (Gaff, Chambers and Markus, 1964). The second stage of uptake involves a physical removal of material from the aqueous phase in the free space, possibly by immobile ions or lipoid phases in the cell walls or the protoplasm. Material held in this way will be in equilibrium with the aqueous phase of the free space and, in suitable conditions, can return to it. The prolonged uptake into the cell vacuole is under metabolic control and is in most cases irreversible so that material taken up in the third stage may not be available for translocation.

The distribution of chemical in the free space of the leaves appears to be much the same whether entry has been via the cuticle, or from the roots in the transpiration stream (Crowdy and Tanton, 1970). The leaching of material from plant leaves has already been noted; this loss is adequately described by a diffusion model (Crowdy and Tanton, 1970) and introduces interesting possibilities for re-cycling and re-distributing chemicals applied to the roots.

Since the initial phases of entry are by diffusion, Fick's law will define some of the conditions which will affect entry. This applies particularly to leaves, which will be considered first. The rate of entry is directly proportional to a permeability constant, the area available for diffusion and the concentration gradient; within the limits set by these, the amount of a substance entering a leaf will increase with time until the external and internal concentrations are the same. Since we are considering diffusion from water on the leaf surface into water in the free space, diffusion will be limited to the period while there is liquid spray on the leaves. The permeability of the cuticle does, in fact, vary with external conditions, but this is a factor which is not readily controlled except, possibly, by selecting the time for spraying; the area available for diffusion will be influenced by the extent to which the leaf is wetted and can be maximised by the inclusion of appropriate surface-active agents. The concentration gradient can also be regulated to a large extent by formulation and choice of chemical, since it depends partly on the aqueous solubility of the chemical used and also on the extent to which the solute is removed from the free space in phase 2 of uptake. If removal is rapid and the potential reservoir large, relatively insoluble compounds may enter to quite a large extent. However, retention in phase 2 may impede translocation. The time of exposure can also be increased by formulation, possibly by adding humectants to delay drying. Humectants do not seem to have been used commercially, but glycerol and polyethyleneglycols are reported to have increased the uptake of streptomycin (Gray, 1956; Goodman and Dowler, 1958), various growth regulators (Rice, 1948; Holly, 1956) and possibly zineb (Rich, 1956). Humectants will only be effective when drying is a limiting factor and even then only if the humectant does not adversely affect the phase distribution of the applied chemical.

The same general considerations apply to root uptake, but normally their operation is obscured by the mass flow of the transpiration stream. Most solutions entering the roots are dilute and are not concentrated by evaporation; the transpiration stream prevents accumulation of solute in the free space though material may be removed from the transpiration stream either passively or actively and accumulated in the roots: this has been demonstrated for a number of antibiotics and sulphonamides and examples have been reviewed (Crowdy, 1959). This accumulation can occur in dead root tissue as well as living (Tanton and Crowdy, 1972a) and the material so accumulated may be released when water is substituted for the treating solution, suggesting that in appropriate circumstances the roots could provide storage for a useful chemical (Crowdy and Tanton, 1970).

Apoplastic movement

The plant cell walls provide a system of water pathways outside the living protoplasm which communicates with the environment by free diffusion. Within this system the bulk of the movement is a mass flow from the roots to the leaves in the transpiration stream, via the xylem vessels. This is a passive movement and depends for its motive force on the difference in water potential between the root environment and the leaf environment. Since no living processes are involved, this system will transport poisons (Kurtzmann, 1966), though naturally these can be fatal to the living tissue adjoining the free space of the roots and the leaves. The actual course of the apoplastic pathway starts in the cell walls of the root hairs and is confined to the cell walls and flooded intercellular spaces in the root cortex. The cell wall pathway is blocked at the endodermal cells by an impervious suberised band, the Casparian strip, in the radial walls, which effectively prevents further movement. At this point the solutions moving in the cell walls are diverted into the endodermal cell itself and must pass through the protoplast, which is firmly attached to the Casparian strip. After traversing the protoplast, the solution returns to the cell walls in the vascular cylinder and passes into the xylem vessels through unlignified parts of their walls. The vessels also communicate with each other through the bordered pits (Tanton and Crowdy, 1972a). The protoplast of the endodermis seems to be more permeable to water than to the lead chelate in the solution which was used in these investigations; it was estimated that only about 10 per cent of the lead chelate in the solution presented to the roots crossed the endodermis. There seems little doubt that the protoplast of the endodermis constitutes a major barrier to the movement of chemicals in the transpiration stream and would reward a more detailed

Table 5.3 Concentration of chemicals in the transpiration stream in shoots as a percentage of the concentration supplied

Plant	Chemical	Transpiration conc. as % of treating sol	Ref.
Vicia faba	4,4′-Diaminodiphenyl sulphone	38·3	1
	Sulphanilamide	22·7	1
	Sulphadiazine	14·8	1
	Sulphacetamide	10·0	1
	Sulphathiazole	1·6	1
	Griseofulvin	9·1	2
Triticum aestivum	Lead-EDTA chelate	1·2	3
		1·5	3

Refs. 1 Crowdy and Rudd Jones (1956); 2 Crowdy *et al.* (1956); 3 Crowdy and Tanton (1970)

study. Recent information on the structure and function of the endodermis is provided by Clarkson (1974).

Some information on the extent to which the movement of chemicals is impeded in the roots can be derived from experiments in which accumulation in the shoots is related to water uptake. There is usually a linear relationship between these two and the slope of the linear regression is the average concentration of the transpiration stream. The movement to the shoots can be quantified by expressing the concentration of the transpiration stream as a percentage of the concentration of the solution supplied to the plants. Some data are presented in this form in Table 5.3.

In addition to the filtering effect of the endodermis, there is evidence that retention in the root system is also correlated with an ability to partition into non-polar solvents (Crowdy, Grove and McClosky, 1959). Shone and Wood (1974) reached a similar conclusion after studying the translocation of a number of herbicides and the fungicide ethirimol. Shephard (1973) also recorded the accumulation of systemic fungicides in roots and noted that there was some penetration of membranes. There were marked differences in behaviour between different species, apple and cucumber. Shephard also noted marked differences between the fungicides in their movement through the xylem and release from the leaf veins; these may be related to the activity of the chemicals as systemic fungicides.

Once within the xylem vessels there seems to be little check to the flow of neutral and acidic substances. In contrast basic substances are adsorbed to negative charges on the xylem walls. This effect has been noted in basic dyes (Charles, 1953), basic antibiotics (Crowdy and Pramer, 1955), quaternary ammonium compounds (Edgington and Dimond, 1964; Salerno and Edgington, 1963) and a variety of amino acids (Hill-Cottingham and Lloyd-Jones, 1968). Some of this adsorption is passive and can be reversed by supplying competing cations such as calcium, but there are indications of an active adsorption also (Hill-Cottingham and Lloyd-Jones, 1968).

The xylem vessels carry the transpiration stream to the leaves; within cereal leaves the water in the xylem is retained in the main vascular bundles by the mestome sheath (O'Brien and Carr, 1970), and released through the walls of the secondary vascular bundles into the free space of the leaves, which again is located in the cell walls (Crowdy and Tanton, 1970). Within the cells walls the transpiration stream moves to the cuticle and solutes are deposited at the air–liquid interface, where evaporation occurs. The pattern of this deposition is described by Crowdy and Tanton (1970) and Tanton and Crowdy (1972b). If root treatment is continued for a long time, or if treatment is followed by treatment with water, solutes are washed to the margins of net-veined leaves or to the tips of leaves with parallel veins, where even safe chemicals may attain high enough concentrations to cause a scorch. This pattern of accumulation has been illustrated for benomyl and it can be simulated in filter paper cut to leaf-shaped patterns (Peterson and Edgington, 1970; 1971). The apoplast can also be reached by direct pressure injection into the trunks of trees. Injection of carbendazim and benomyl was as effective as spraying for the control of Dutch elm disease in large urban trees (Alfen and Walton, 1974). Injection of thiabendazole and carbendazim also controlled *Venturia pirina* on pears; the timing of the injection affected the distribution of the fungicide within the tree. The best distribution followed injection in the Autumn (Shabi *et al.*, 1974). The extent to which a fungicide distributed in the apoplast will accumulate in a plant organ will be related to the loss of transpiration water from that organ, which, in turn, is usually related to the frequency of stomata. Peterson and Edgington (1971) have demonstrated this with benomyl, which does not accumulate, or only accumulates to a very small extent, in organs such as tomato fruit, geranium petals or poinsettia bracts, in which stomata are absent or non-functional. This pattern of movement implies that continuous root treatment with a systemic fungicide is needed if a leaf is to remain well protected unless the fungicide is bound in the leaf tissue. It should also be noted that fungicides which enter following leaf application will also travel to the tips and edges in this way.

Symplastic movement

Chemicals which have reached the leaves in the apoplast cannot be redistributed in the plant in the same system except in conditions which occur rarely outside experiments. Movement from the leaves into the main body of the plant occurs in the symplast and requires metabolic energy both to transfer material from the free space into the living parts of the cell and to move material along the symplastic pathways. Movement from the apoplast to the symplast seems in many cases to be associated with specialised transfer cells (Gunning and Pate, 1969). These are located where trans-membrane fluxes might be restricted by adverse surface-to-volume ratios. In the transfer cells the ratios are improved by ingrowths of the cell wall which cause a considerable increase of the surface of the protoplasm which lines it and the interface between the symplast and the apoplast. The living protoplasm forms effectively a continuous network throughout the tissues of the plant, hence the use of the term symplast; protoplasmic connections between adjacent cells are provided through specialised pores, the plasmodesmata. The fine structure of these pores in willow has been described by Robards (1968). Many aspects of the structure and function of the symplast have been reviewed by Clarkson (1974). The symplastic pathway is continuous with the phloem, the vascular tissue in which long-distance transport occurs. The actual cells involved are the sieve tubes, which are long trains of cells communicating with each other through perforated plates which look like and are called sieve plates. Phloem structure and function have been discussed recently by Crafts and Crisp (1971) and by Canny (1973), who have provided an excellent bibliography of literature relating to this tissue. The sieve tubes look initially like normal parenchyma cells but divide early to form a characteristic pair of cells, the sieve tube and the companion cell, which develop together. As the sieve tube develops many of the structures, including the nucleus, degenerate and may disappear, but despite these deficiencies, the sieve tube remains a living cell as long as it is functional. The pores in the sieve plates are derived by enlargement of the pores originally occupied by plasmodesmata (Wark and Chambers, 1965). In contrast to the sieve tubes, the companion cells have normal intracellular structure and dense protoplasm: they are the only cells connected to the sieve tubes with plasmodesmata and these connections are well developed. This close association suggests that the companion cells may have taken over many of the functions of the degenerate protoplast of the sieve tubes.

The sieve tubes are delicate structures which are easily damaged by manipulation and their fine structure is still a matter for debate, since it is always difficult to decide whether the structures seen in the optical or electron microscopes reflect faithfully the conditions *in vivo* or are artifacts. Competing theories of phloem movement require different organisation of the sieve tube trains with the result that proponents of different models find it difficult to agree as to the true status of the artifacts. It seems likely that the full interpretation of the structure of the sieve elements must wait on a better understanding of the mechanisms of phloem movement.

The main function of the phloem is to carry food materials, mainly carbohydrates, from organs in which they are plentiful, 'sources', to organs which require nutrients, 'sinks'. This has been demonstrated by ringing experiments, in which sections of the phloem were cut out or killed by steam and the effect on the distribution of nutrients noted. In its simplest form this implies conveying carbohydrates from the leaves in which they are photosynthesised, to the growing zones in which they are used for tissue building and to storage organs. This is an oversimplification since growing leaves may be net importers of nutrient and may even import and export simultaneously. Equally, carbohydrates may be stored temporarily in various tissues along the route, such as the parenchyma of the phloem. These may act as sinks or as sources depending on the conditions in the tissue and act as a buffer protecting the tissue from extreme fluctuations. In addition to long-distance transport, the phloem must also provide nutrient to the living tissues of the mature plant body, which also implies a capacity for free exchange of nutrients with the surrounding tissues. The concept of sources and sink is useful for describing movement in the phloem provided that the actual tissues involved are defined when necessary.

Material in the symplast is actively accumulated in the sieve tubes against a concentration gradient (Mason and Phillis, 1933) and there are some indications that light is necessary to this loading process. Movement within the sieve tubes is also dependent on metabolic energy; this has been demonstrated by treating zones of the phloem with respiratory inhibitors or by denying them oxygen. General observations on movement have been supported by micro-autoradiography, which makes it possible to locate suitable isotopes in specific cells. The concentration of solutes in the sieve tubes is usually high, much higher than the concentration in the surrounding tissue, so that the contents of the sieve tubes are under pressure and will exude if they are punctured. Phloem exudation can be collected from the cut ends of stems, which provides a general sample of phloem contents, or from aphid stylets, which are commonly located in a single sieve tube and provide a sample from a restricted zone of phloem, possibly from a single train of cells. These methods have provided various data on the contents of sieve tubes and their behaviour. In general the content and concentration of the phloem sap varies with the type of plant used and the environmental conditions, high concentrations being particularly dependent on active photosynthesis. Crafts and Crisp (1971) have summarised data on the composition of phloem exudate. The exudate contains 10–25 per cent dry matter, of which about 90 per cent is sugar. In some species this is entirely sucrose, but other sugars and sugar alcohols may occur. About 2 per cent of the dry matter may consist of mineral nutrients such as potassium, which is present in relatively large amounts, magnesium, calcium and phosphorus. Amino acids may contribute about 0·5 per cent of the dry matter (Tammes and Die, 1966). Hormones and a variety of foreign chemicals including herbicides and insecticides, may also be transported in the phloem. Crafts and Crisp (1971) list 122 chemicals which have been recorded in the phloem of various plants. Data derived from a study of exudates must be treated with some reservations, since the phloem is concerned with distribution as well as transport and this requires a capacity to exchange materials with neighbouring cells. The leakiness of the system, which this implies, confuses the interpretation of translocation patterns.

The actual distribution of the assimilated carbohydrate has commonly been studied by feeding ^{14}C-labelled carbon dioxide or sucrose to leaves and following its distribution by counting or autoradiography of the whole plant, or of tissues, if the aim is location at the cell level. Other phloem-mobile tracers such as ^{35}P, tritium or fluorescein have been used less frequently, but are valuable in special experiments such as those demonstrating bidirectional movement. This work has been summarised by Wardlaw (1968), Crafts and Crisp (1971) and Canny (1973). In general young, growing leaves are net importers of carbohydrate and newly matured leaves are net exporters; the changeover takes place when the leaf is about half its final area. Leaves tend to export to the sinks which are nearest to them, subject always to limitations imposed by the anatomy of the vascular system and the rather restricted movement of the photosynthate around the stem. Thus the upper leaves tend to supply the growing point and the associated young leaves; the lower leaves the roots, while those which are intermediate may export in either direction. This pattern is not absolute and can be varied experimentally by pruning or removing leaves. In some cases photosynthate may be translocated downwards to the roots and then redistributed in the xylem and apoplast to the top of the plant; this can complicate the general picture of distribution. Developing fruits may provide very strong sinks and grapes may accumulate photosynthate from leaves up to 4 m from the cluster (Meynhardt and Malan, 1963). In wheat and rye the source of supply is much more restricted, 80 per cent of the material for grain growth being derived from the flag leaf, stem and ear; in rice the source is rather less restricted and lower leaves may be involved.

Work on patterns of translocation, speeds of movement and possible mechanisms has been reviewed recently by Canny (1973) and Crafts and Crisp (1971). The movement of materials is rapid, probably in the range 0·39–1·09 m h^{-1}, which is too fast for simple diffusion, and can occur simultaneously in both directions. Two general types of mechanism have been suggested to account for this movement, those based on mass flow, which operate best with empty tubes and unimpeded pores, and those which require specialised sieve tube structure or properties. The mass flow hypotheses are historically the oldest and were first clearly formulated by Münch (1927). Basically these envisage a high concentration of soluble carbohydrate in the sources, such as the leaves, and low concentrations in the sinks, where active metabolism occurs, or in storage organs, where soluble carbohydrate is removed from the system either by metabolism or by conversion into insoluble forms. This concentration gradient tends to produce a mass flow of solution from sources to sinks, which would incidentally carry with it other solutes which were present in the phloem. There would only be flow in one direction in a single train of phloem cells, but flow in adjacent trains could be in different directions. There is some evidence for greater turgor pressures at the top of the phloem than at the bottom and it is claimed that the pressure differences are sufficient to maintain the required flow in an unimpeded system. In its simplest form, this model required high pressure differences to account for the speeds and the distances involved and to overcome the resistance of the sieve plates. This model did not call for living processes to maintain the movement between the sources and the sinks although evidence from metabolic inhibitor experiments indicated that active processes were involved in movement throughout the length of the phloem system. Some of these objections have been met by proposing electro-kinetic or electro-osmotic systems, maintained by metabolic energy, to boost the mass flow in the sieve tubes. These essentially involve 'sap-carrying' ions circulating either over an appreciable length of phloem (Fensom, 1957, 1959) or circulating around the sieve plates via either the companion cells (Spanner, 1958) or via adjacent sieve tubes (Spanner, 1970). Potassium, which is in fairly high concentrations in the sieve tubes, has been suggested as the ion involved.

In spite of these modifications, there are a number of observations which throw doubt on the validity of the various mass flow hypotheses. There is evidence from aphid stylet experiments, which indicate bidirectional flow in a single train of sieve tubes in *Vicia faba*. Eschrich (1967) collected honeydew from an aphid feeding on the stem above a leaf treated with fluorescein and below one treated with ^{14}C. Both tracers appeared in the honeydew, presumably derived from a single train of sieve tubes. This experiment did not exclude the possibility of some mixing of materials travelling in adjacent paths and supports rather than establishes bidirectional movement in a single sieve tube. There is also some evidence that the water in sieve tubes is relatively immobile (Gage and Aronoff, 1960; Peel, 1970). In addition Tyree and Fensom (1970) and Canny (1973) find the mass flow models supported by electro-osmotic systems theoretically unsatisfactory. Canny considers the kinetics of a number of alternative model systems and suggests that the most satisfactory is one indicated by Mason and Maskell (1928a, b) in their study of carbohydrate movement in cotton. This suggested that the direction of the transfer of dry material was dictated by the gradient of sucrose concentration in the phloem and that the rate of transfer was directly proportional to this gradient over a wide range of values. This leads to a model which is formally identical to the steady state form of Fick's law of diffusion but with an apparent diffusion coefficient 3×10^4 times greater than the coefficient for the diffusion of sucrose in water. They referred to this process as accelerated diffusion and models of this type are referred to as diffusion analogue models. Diffusion analogue models require specialised structures within the sieve tube to convey the solute and exchange between these structures and a stationary water phase. Suitable structures would be the trans-cellular strands reported by Thaine (1962) in the phloem of *Cucurbita* and *Primula obconica*, which he suggested later were tubules which might carry sucrose solutions by peristaltic contractions (Thaine, 1969). The subsequent lack of agreement as to whether these strands are genuine structures or artifacts as suggested by Esau and her associates (1963) illustrates the difficulties inherent in studying the fine structure of phloem. Jarvis and Thaine (1971) have recently reported strands in quick-frozen material of *C. pepo* confirming the original observations. Milthorpe and Moorby (1969) have reviewed some of the evidence relating to these strands and the studies by Fensom and his associates (1968) and Lee and his associates (1971) on the living phloem of *Heracleum mantegazzianum* indicating a network of micro fibrils, which seem to be in active motion and which carry particles which stain very rapidly when the vital stain, Janus Green, is applied to a cut strand, should also be considered in this connection. Aikman and Anderson (1971) published a theoretical analysis of Thaine's model and concluded that it could function. Further evidence for the presence of microfibrils in living phloem has been derived from *in vivo* studies of *Nymphoides peltata* (Johnson, 1973). Similar structures have been found in the alga *Nitella* where it is suggested that they maintain the protoplasmic streaming (Allen, 1974). It seems unlikely that these structures are tubular and capable of maintaining peristaltic motion. However, it is possible to construct an adequate model of phloem movement based on waves of motion in the microfibrils (Fensom and Williams, 1974).

The source–sink relationships feature in all the models of phloem transport and there is evidence that the sinks are to some extent under hormonal control. The apical bud, which is a typical sink, normally maintains adequate concentrations of indole-3-acetic acid; if the apical bud is removed, the cut end of the stem will only continue to act as a sink when exogenous indole-3-acetic acid is supplied (Booth *et al.*, 1962). In some plants indole-3-acetic acid can be replaced by foreign growth regulators such as naphthylacetic acid 2,4-D or 2,4,5-T (Bowen and Wareing, 1971). Gibberellic acid and cytokinins have no influence on the direction of transport in the absence of indole-3-acetic acid, but when this hormone is supplied, both enhance its effect (Seth and Wareing, 1967; Morris and Thomas, 1968). It is not possible to say how far this effect is directly attributable to the hormone or is indirect and follows hormonal stimulation of growth.

Symplastic and apoplastic transport are physiologically and anatomically distinct within the plant. In most cases movement in these two pathways can be distinguished readily in experiments, since symplastic movement is away from sources towards sinks and can be prevented by treatments which inhibit metabolic activity at the source, or immobilise the phloem, while apoplastic movement takes place passively on the water stream and is directed to zones where water is lost by evaporation, or where it is required to maintain growth or extension (Potter and Milburn, 1970). However, there are circumstances in which solutes will move from a leaf into the body of the plant in the apoplast; then their distribution will have a superficial similarity to symplastic movement; these must be recognised when designing and interpreting experiments. Centripetal movement in the apoplast occurs when the apoplastic pathway is exposed to atmospheric pressure by cutting, abrasion or chemical damage; the tension is released and the solution in the apoplast system sucks back, drawing with it air, or solution, applied to the break. This effect has been used to assist the deep penetration of herbicides into *Convolvulus arvensis* (Crafts, 1933) and in certain conditions may lead to the movement of desiccants into the tubers of potatoes, causing a stem end necrosis (Headford and Douglas, 1967). It is also possible to reverse apoplastic movement by immersing a leaf in water or a solution. This effectively stops evaporation at the immersed surface, which can then supply solution to the main transpiration stream.

Although symplastic movement is mainly concerned with distributing physiologically important chemicals, it will also transport those which are foreign to the plant. The range of chemicals transported is more restricted than in apoplastic movement, since there is a need for active transfer from the apoplast to the symplast; this may involve the ability to use specialised carrier mechanisms. In spite of this restriction, symplastic movement has been demonstrated for a variety of chemicals; these have been listed by Crafts and Crisp (1971). Research has tended to concentrate on chemicals important in agriculture such as herbicides, insecticides and nutrient salts. Movement from the leaves is active, at least at some stage, and does not take place in leaves which have been starved during a period in the dark. This has been demonstrated in the case of 2,4-D, by Hay and Thimann (1956), who also found that exogenous sucrose could be substituted for light while mannitol, arabinose and urea could not. Further, added sucrose had no effect on transport in the light. Ring barking inhibited 2,4-D movement, indicating a symplastic pathway. The sugar is probably needed to provide energy for the transport system; it may also be needed to establish nutrient gradients. Hay and Thimann also showed that excessive 2,4-D inhibited translocation, suggesting that too high a concentration may poison the metabolic processes involved. When foreign chemicals are moving in the symplast, their pattern of distribution indicates that they are moving with the nutrient supply though there are considerable differences in the distribution and in the extent of movement. In general 2,4-D tends to be absorbed during its passage through the plant, while other chemicals such as dalapon (2,2-dichloropropionic acid), amitrole (3-amino-s-triazole) and maleic hydrazide move more freely. Dalapon and maleic hydrazide transfer from the phloem to the xylem, which allows circulation in the symplast–apoplast systems, and are very generally distributed within the plant (Crafts and Crisp, 1971). Foy (1961) even claimed that the effect of dalapon treatment could be carried into a second generation through the seed of wheat. Chemicals, such as dalapon, which are carried in the symplast to the roots may also be released from the roots into the medium surrounding them. This also occurs to a limited extent with 2,4-D, but the effect is more striking with α-methoxyphenylacetic acid (Preston *et al.*, 1954) and 2,3,6-trichloro- and 2,3,4,6-tetrachlorobenzoic acids (Linder *et al.*, 1958), who showed that enough growth regulator was released into the root medium to produce malformations in adjacent plants. Six other assorted di-, tri-, tetra- and pentachlorobenzoic acids and some phenoxy and naphthyloxy acids did not migrate in this way.

The pattern of movement of these foreign chemicals is also markedly influenced by the growth stage and vigour of the plant. Yamaguchi and Crafts (1957) noted that the characteristic retention of 2,4-D in the translocation pathway in *Zebrina pendula* as compared with amitrole and maleic hydrazide was marked in slow-growing plants, but was virtually obscured if growth was rapid. Van der Zweep (1961) was also able to demonstrate an effect of age on the direction in which 2,4-D was translocated in barley, *Hordeum vulgare*, plants. At the two- and three-leaf stages there was good movement to the roots while at the four- and five-leaf stages movement to the roots was much less marked.

Translocation of systemic fungicides

The evidence which has accumulated relating to the translocation of systemic fungicides is much more sparse than that related to the movement of herbicides and insecticides and, in most cases, the reader must infer the pathways involved from data derived from disease control trials. Frequently also disease control or a non-specific bioassay are the only evidence that the chemical is translocated; this inevitably raises doubts as to the identity of the actual chemical moving. The translocated chemical has only seldom been extracted from the treated plant and characterised, though this has been done for the antibiotics griseofulvin and chloramphenicol (Crowdy *et al.*, 1955). The pure antibiotics, recovered from *Vicia faba* after root treatment, were identified by mixed melting point determinations and comparison of their infra-red spectra. More commonly the identity of the translocated compound is based on the similarity of its behaviour on chromatograms to standard pure chemicals. In some cases, such as benomyl and thiophanate, disease control is claimed for the compound applied when it seems likely that the compound actually translocated is a fungicidal metabolite. In most cases the data presented below are supported at least by chromatographic evidence as to the actual chemical translocated or, in the case of some of the antibiotics by bioassays with bacteria specifically requiring the antibiotic for growth. This area has recently been surveyed by Wain and Carter (1967) and Erwin (1973).

Most of the older chemicals, which might be described as near, or not so near, misses showed typical apoplastic movement. These included a variety of antibiotics, of which the most important for the present discussion were probably griseofulvin and streptomycin, whose behaviour has been surveyed briefly by Crowdy and Pramer (1955). Apoplastic movement has also been shown for cycloheximide (Wallen and Millar, 1957) and for certain of its derivatives (Wallen, 1958). A very varied selection of synthetic chemicals have also shown an apoplastic pattern of movement, though in most cases the actual chemical translocated has not been identified. For various reasons only a few of these have even emerged from greenhouse tests before being abandoned. Fungicidal chemicals have only rarely given evidence of symplastic movement, though it is difficult to judge at this stage how far it must be attributed to dilution of not very active chemicals and lack of attention to the patterns of symplastic movement. Crowdy and Wain (1951) provided some evidence of the control of *Botrytis fabae* on *Vicia faba* associated with symplastic movement of 2,4,6-trichlorophenoxyacetic acid. However, the effect was too small to be of commercial interest and the translocated chemical was not identified. Gray (1958) found that the antibiotics streptothricin and pleocidin were translocated from the sprayed leaves of *Phaseolus vulgaris* and would protect younger leaves from infection with *Xanthomonas phaseoli*; he found no such movement with streptomycin. In contrast, Napier *et al.* (1956) reported that a prophylactic spray of streptomycin on the primary leaves of *Phaseolus vulgaris* would prevent infection due to *Pseudomonas medicaginis* as high as the 4th trifoliate leaf for periods of 7–11 days.

The first requirement for symplastic movement is that the chemical should cross the plasmalemma and enter the protoplasm; little effort has been devoted to examining this aspect of translocation with potential fungicides. Pramer (1955, 1956) studied the entry of penicillin, chloramphenicol and streptomycin into the cells of the alga *Nitella clavata*. Penicillin did not enter the cells in detectable quantities, chloramphenicol entered slowly, apparently by diffusion, and attained a concentration of less than half the external solution in 24 h. Streptomycin entered rapidly and accumulated to a concentration of seven times that of the external solution in 18·5 h, which indicated that the uptake was active, a conclusion which was supported by further experiments which examined the effects of competitive ions, respiratory inhibitors and glucose. There is some evidence that lipophilic compounds can diffuse into the vacuoles of the cortical cells of roots where they are available for transport to the shoots (Shone *et al.*, 1974).

The first generation of commercial systemic fungicides also appears to be transported in the apoplast; again, in most cases, this must be inferred from the results of trials designed to study the control of diseases. Peterson and Edgington (1970, 1971) have published a classic series of studies with benomyl and its fungicidally active metabolic product carbendazim, the methyl ester of 2-benzimidazolecarbamic acid (Clemons and Sisler, 1969), which clearly demonstrate that these are mainly transported in the apoplast of *Phaseolus vulgaris*. Benomyl is also translocated in sugar beet, *Beta vulgaris* (Solel, 1970), tomato, *Lycopersicon esculentum*, cucumber, *Cucumis sativus* and watermelon, *Citrullus laratus* (Thanassoulopoulos *et al.*, 1970; Ebben and Last, 1969), seedling of apple, *Malus pumila* and cherry, *Prunus cerasus* (Gilpatrick, 1969) and cotton *Gossypium* sp. (Erwin, *et al.*, 1968a). Thiophanate methyl also appears to translocate in the apoplast in sugar beet (Solel,1970a); since it also has carbdendazim as a fungicidal metabolite (Selling *et al.*, 1970), its behaviour is likely to be similar to benomyl in many respects. Staron and his associates (1966) demonstrated systemic activity following apoplastic movement of thiabenzadole. This compound is also mobile in cotton (Erwin, Sims and Partride, 1968b), sugar beet (Solel, 1970a) and soybean, *Glycine max* (Gray and Sinclair, 1971); in soybean and cotton it appears to translocate unchanged and in the former it tends to remain at the base of the plant (Gray and Sinclair, 1971). The two pyrimidine fungicides, ethirimol and dimethirimol, appear to move unchanged in the symplast to the leaves, where they are subject to degradation. Ethirimol, following root treatment, is generally distributed in cereal leaves, but is retained in the veins of apple leaves: this probably accounts for the failure of the chemical to control mildew, *Podosphaera leucotricha* (Cavell *et al.*, 1971). Triarimol appears to have been used only as a spray, but the eradicant activity of this compound indicates some systemic activity (Gilpatrick and Szkolnik, 1970). The oxathiin derivatives, carboxin and oxycarboxin, have also shown typical apoplastic translocation in cereals (Schmeling and Kulka, 1966) and in other grasses (Hardison, 1967), in cotton (Borum and Sinclair, 1967), tea *Camellia sinensis* (Venkata Ram, 1969) and *Phaseolus vulgaris* (Snel and Edgington, 1970). Comparable evidence has also been presented for the apoplastic movement of tridemorph in barley (Pommer, Otto and Kradel, 1969), triforine in rye, *Secale cereale* (Schicke and Veen, 1969) and three organophosphorus fungicides kitazin and kitazin P in rice, *Oryza sativa* (Yoshinaga, 1969) and pyrazophos against apple mildew (Hay, 1971). Apoplastic transport seldom seems to be specific to a particular plant and, in general, success in treating one herbaceous plant indicates that success can be expected in treating others; this observation may well cover quantitative differences as reported for linuron and atrazine (Walker and Featherstone, 1973), and for fungicides (Shephard, 1973), which the present investigations have failed to reveal. It would be interesting to know if the behaviour of ethirimol in apple is an isolated anomaly, or represents a genuine difference between woody and herbaceous plants.

Most of the experiments noted above provide no evidence that the active chemical moves back from a treated leaf into the rest of the plant. However, it appears that this may occur in some cases. This type of movement has been reported for the benzimidazole derivatives thiabendazole and benomyl, or, more probably, its metabolite carbendazim. Both these fungicides markedly reduced infection on untreated leaves of sugar beet inoculated with *Cercospora* 3 days after the plant had been sprayed (Solel, 1970a). Even more striking results were obtained when sprays of these chemicals, acidified with hydrochloric acid, were applied to cotton leaves. Both sprays were effective against a virulent strain of cotton wilt, *Verticillium albo-atrum*, inoculated into the stem either before or after spraying and fungitoxic materials were detected by bioassay both in leaves above the point of application and in xylem tissue. Material was extracted from the xylem in the stems of plants sprayed three times with these acid formulations and was examined by ultraviolet light spectrophotometry. Thiabendazole was identified in the extract of the xylem of the plant sprayed with this compound, while the fungicidal metabolic product, carbendazim, was identified in the xylem of plants sprayed with benomyl. The solutions sprayed were very acid, pH 1·5–1·7 for benomyl and 2·7–3·0 for thiabendazole, so one must bear in mind the possibility of damage to the apoplast system and suck-back in the xylem although the plants were not visibly damaged (Buchenauer and Erwin, 1971). Solel, Schooley and Edgington (1973) have provided more conclusive evidence of the symplastic movement of benomyl and carbendazim. Carbendazim enters isolated mesophyl cells of both apple and cucumber, possibly by diffusion. Both benomyl and carbendazim will translocate from treated cotyledons or leaves to untreated parts of the plant, including the roots, stem and shoot apex. This translocation is largely prevented if the peripheral tissues of the petiole of the treated leaf are killed by steam. The possibility of xylem movement must also be borne in mind when considering the translocation of certain nitrophenols (El-Zayat *et al.*, 1968). When these are applied to the leaves, they will protect leaves which are formed after treatment. No visible damage appeared to be associated with the treatment, but the chemicals are too phytotoxic to be applied to the roots and the possibility of local damage and xylem suck-back must again be considered. The ability to translocate from treated leaves to younger growth has also been shown by 4-*n*-butyl-1,2,4-triazole, which is effective against *Puccinia recondita* on wheat (Meyer *et al.*, 1970); this chemical can also be used for root treatment. Some basipetal movement has also been demonstrated with 2-chloro-6-methoxy-4(trichloromethyl)pyridine (pyroxychlor); foliar applications are claimed to control black shank of tobacco (Noveroske, 1975) and to delay the development of *Phytophthora* root rot of rhododendron (Hoitink and Schmitthenner, 1975).

On general principles, one would expect to find systemic fungicides which are transported in the symplast and it is possible that the present lack of information in part reflects the difficulty of testing for this ability. Before symplastic movement is assumed, other possibilities must be rigorously excluded and it is also necessary to determine whether the chemical is actively accumulated in the symplast or enters the cells by diffusion, as was suggested for the entry of chloramphenicol into *Nitella clavata* by Pramer (1955), since this could well influence the extent and pattern of translocation.

Conclusion

Most protectant fungicides are general cell poisons and are selective because they remain on the surface of the plant; the fungicidal chemicals are selected and formulated with this in view. When these chemicals penetrate the cuticle, they cause damage since there may be little margin of safety between phytotoxicity and fungitoxicity. A systemic fungicide must co-exist with the cells in the living tissue and this requires an entirely different type of selectivity, which discriminates between the living tissue of the host and of the pathogen. Chemicals showing systemic action must also be chosen on their ability to enter the plant and must be formulated to exploit this ability to the full. These requirements are a specification for an eradicant fungicide and the systemic fungicides now being launched show a remarkable ability to eradicate established infections; with most the recommended uses exploit this feature alone and, with one or two exceptions, the emphasis is on spraying for the control of leaf diseases. Since chemicals which are mobile in the apoplast migrate to the leaf edges and tip, the main leaf surface may only be protected for quite a short time unless a continuous supply of chemical is maintained. When the material is applied to the leaf surface, it may well be necessary to balance entry with retention on the surface and aim at a protectant spray with exceptional eradicant properties. This may prove an unsatisfactory compromise, since the properties required for retention may well militate against entry. Disease eradication implies movement between the host and the pathogen, which will only occur in the absence of an impervious wound barrier isolating the lesion; a better knowledge of this aspect of host–parasite relations would provide a base for planning disease control experiments.

Residue problems with systemic chemicals used in this way may be different to those arising from the use of conventional protectant sprays; they need not necessarily be more intractable. Material on the surface of the leaf would weather off in the usual way and material within the tissue would migrate to the leaf margin and, in due course, leach out of the tissue. A material which has penetrated the cuticle of a fruit might present a more difficult problem than one which is retained on and might weather off the surface. The entry of individual systemic chemicals used as sprays and their leaching from fruit surfaces requires a more detailed study both from the point of view of residues and from possible effects in reducing storage rots.

Treatment through the roots presents a rather different pattern of problems. Assuming that the chemical is stable in the soil, the problem reduces itself to providing sufficient chemical and distributing it so that it is accessible to the roots at the relevant stages of growth. The plant appears to have little capacity to regulate intake through its roots; there appears to be a barrier to movement at the endodermis, but there is no evidence that this responds to conditions in the top of the plant where the translocated material is accumulating. In these circumstances, the proper dose can only be supplied by ensuring a slow and steady release of the chemical into the soil water supplying the transpiration stream. This could be achieved, either by ensuring the correct degree of insolubility or by using a chemical which is adsorbed to the soil and released slowly, such as ethirimol or dimethirimol (Graham-Bryce and Coutts, 1971). The second course has certain possible advantages since desorption from the base exchange system may, to some extent, be related to root activity. A limited mobility in the soil would also be an advantage. An understanding of this system must be based on a detailed knowledge of the interactions between the plant, the soil and the fungicide. The necrotic tissues, which surround living roots, are also capable of accumulating and, later, releasing some chemicals into the transpiration stream (Tanton and Crowdy, 1972a): these might provide a useful store and a buffer against violent fluctuations in the concentration in the soil solution passing through the roots.

Chemicals translocated in the apoplast are not redistributed within the plant; however, they may be leached from the leaves, taken up by the roots and recirculated in the apoplast. This effect has been reviewed by Tukey (1970); the recirculation of nutrients may be of some importance in crops growing in marginal conditions and may reduce toxic accumulations in the leaves. The transfer, in nature, of physiologically active chemicals is illustrated by the interaction between *Camelina alyssum* which produces a toxin which is washed from the leaves and will harm a plant such as flax following root uptake (Grümmer and Beyer, 1960). This process would provide a continuing supply of material to transpiring leaves. Continuing root treatment would be associated with residues in the transpiring parts of the plant; water loss from fruit is probably not great and residues in fruit might not be high. This has been shown for tomato fruit by Peterson and Edgington (1971).

The introduction of chemicals with low inherent phytotoxicity introduces the possibility, which is now being realised, of treating pathogens in the soil and of treating established internal infections such as vascular wilts. If the fungicides are moved in the apoplast, this involves soil treatment, but chemicals distributed in the symplast could be used to control vascular and root diseases, if applied to the leaves. A chemical with the mobility of α-methoxyphenylacetic acid could be sprayed on the leaves, translocated to the roots and released into the soil surrounding them, which could well be the ideal way of treating root diseases. Chemicals distributed in the symplast would open up an entirely different pattern of disease control since they would be transferred from the treated leaves to young growing tissue, possibly reducing the need for repeated spraying. One would expect these chemicals to be distributed in a pattern similar to carbohydrates and this pattern can be profoundly influenced by infection with plant disease; the subject has been reviewed recently by Smith *et al.* (1966). There seems to be a clearly established pattern of accumulation at the lesions of a number of obligate parasites, which occurs both in the host tissue and in the fungal mycelium and may involve retention of carbohydrate in the infected leaf and short-range transport within the leaf to the pathogen. In most cases also there is a general disturbance of the translocation patterns in the plant and active infections may attract photosynthate from adjacent leaves and possibly from a greater distance. In addition to the work reviewed by Smith *et al.* (1969), this phenomenon has been studied further in wheat infected with *Puccinia striiformis* (Siddiqui and Manners, 1971) and with *Ustilago nuda* (Gaunt and Manners, 1971) and in apple scab (Hignett and Kirkham, 1967), who implicated a melanoprotein in polarising the nutrient flow. The mechanism of this disturbance of the translocation pattern has not been elucidated. It is sometimes possible to produce very similar effects by applying growth-regulating chemicals and there is ample evidence that growth-regulating chemicals are produced by fungi and may occur in high concentrations in fungal lesions in higher plants (Wood, 1967): these may well be implicated. Fungicides transported in the symplast with carbohydrates could well accumulate high residues in fruit and storage organs and further work will be aimed at finding out how far the transport of these two groups of chemicals is necessarily related.

Chapter 6

Effects on physiology of the host and on host/pathogen interactions

(the late) A. E. Dimond and S. Rich

Introduction

In broadest terms, systemic compounds act to control disease in one of three ways: they may kill or inactivate the pathogen in the host, they may increase host resistance to infection, or they may interfere with pathogenic processes and, in this manner, block the development of symptoms in the invaded plant. These several mechanisms have been discussed from differing points of view on earlier occasions (Dimond, 1965; Dimond and Horsfall, 1959; Howard and Horsfall, 1959; Grossmann, 1968b; Oort and van Andel, 1960; Woodcock, 1971).

As systemic compounds enter the plant, they encounter a variety of active biochemical systems, and they are often altered in this encounter (see Chapter 8). Some compounds are converted into fungitoxic molecules after relatively minor modification. Others enter as raw materials for biosynthesis of fungitoxic structures. Yet others alter the physiology of the plant or features of its anatomy. Some treatments mimic the mechanisms of natural resistance to disease.

One cannot discuss treatments that affect physiology of the host without also dealing with those that alter pathogenesis. The two kinds of treatment involve different facets of the same relationship; frequently the two are complementary. For example, one treatment may modify host tissue so that is is less readily attacked by a fungal enzyme. Another treatment may inhibit this enzyme, an enzyme that is important in pathogenesis, but not essential for growth or survival of the pathogen. Again, the spread of vascular wilt pathogens through a large host often depends upon migration of propagules with the transpiration stream. The spread of spores through the vascular system can be prevented by the use of treatments that induce tylose and gel formation in conducting vessels of the host. Alternatively, the use of antisporulants or of compounds that inhibit the budding of fungal cells may equally well stop reproduction of the pathogen and spread of its spores through xylem vessels. The one treatment acts upon the host; the other affects pathogenic processes. Either method has the same objective. For such reasons, the discussion that follows may not always be confined to cases where physiology of the host is altered. Rather, discussion will be presented around principles underlying control of disease with systemic compounds.

Although a variety of modes of action for chemotherapy are recognised, the greatest successes to date have been with compounds that are true systemic fungicides. Some of these compounds, in addition to being fungitoxic, also affect the physiology of the host.

Effects of systemic fungicides on the host

Benomyl (Clemons and Sisler, 1969), thiophanate-methyl (Selling *et al.*, 1970), or any other systemic fungicide that can form benzimidazol-2-yl-carbamate (MBC) (carbendazim) may effect changes in and be changed by the host plant.

Carbendazim may be formed in the treated soil or in the plant (Sims *et al.*, 1969) and persists in either for a considerable period of time without further modification (Biehn and Dimond, 1969; Hine *et al.*, 1969; Peterson and Edgington, 1969).

Plants treated with benomyl are altered in turn. Taylor (1970) has noted that benomyl-treated tobacco plants are protected against ozone injury. Other systemic fungicides based on benzimidazole also alter ozone susceptibility of plants. In fact, Pellissier *et al.* (1971) have shown that protection of plants against ozone damage follows treatment with these systemic fungicides or with benzimidazole itself. Plants watered with certain oxathiin compounds are also made resistant to ozone (Rich *et al.*, 1974). Neither benomyl, benzimidazole nor oxathiins react with ozone directly (Rich and Tomlinson, 1970). Therefore, the protective action results from an effect upon the physiology of the plant.

The present evidence is that the fungitoxicity of MBC is caused by mitotic failure (Hammerschlag and Sisler, 1973) resulting from interference with spindle formation (Davidse, 1973). The oxathiins, however, are thought to poison fungi by interfering with electron-transport systems (Georgopoulos and Sisler, 1970) (see also Chapter 7).

It is unlikely that the fungitoxic mechanisms of either MBC or the oxathiins account for their ability to change the physiology of higher plants. Whatever effect they exert is probably on the apoplastic system of the plant (Peterson and Edgington, 1971).

It is now established that benzimidazole-derived fungicides have cytokinin-like properties (Skene, 1972; Thomas, 1974). Cytokinins are also anti-senescent, a property that can reduce ozone injury and its accompanying biochemical changes (Tomlinson and Rich, 1973). This is true not only for benzimidazole-derived fungicides but also for non-fungicides such as phenylurea (Tomlinson and Rich, 1974).

Such compounds may act by modifying either the production or effect of ethylene. Rich (1964), noting the similarity between many symptoms of ozone injury and those produced by ethylene, proposed that the initial injury by ozone may cause the plant to produce ethylene. This was demonstrated by Craker (1971).

We do not know how these systemic fungicides can act as anti-senescent compounds without entering the symplast. Perhaps they do so by modifying the structure of the cellular membranes (Spotts *et al.*, 1974).

Production of fungitoxicants in plants

Natural disease resistance in plants has sometimes been correlated with the presence of fungitoxic barriers. Such barriers may either prevent the formation of appressoria or inhibit invading haustoria.

Lukens (1974) has designed a method for measuring fungitoxicity by the ability of compounds to prevent the formation of appressoria, even though spores germinate in the presence of the toxicant.

More often disease resistance in plants is believed to be associated with the formation of fungitoxic products in the invaded tissue of the host. Some systemic compounds are indeed converted into fungitoxic molecules by the host tissue. An example is the fate of 6-azauracil in cucumber leaves. Dekker (1962) showed that when this compound is introduced into leaves powdery mildew infections fail, even though 6-azauracil is not fungitoxic. Spores germinate and enter leaves normally, but growth of the fungus ceases upon development of the first haustorium.

Uracil reverses the chemotherapeutic action of 6-azauracil. In the plant, 6-azauracil is converted to 6-azauridine and then to the monophosphate (Dekker and Oort, 1964). Applied to plants, the latter two compounds also inhibit haustorial development, but their action is not reversed by uracil. The 6-azauridine monophosphate blocks nucleic acid biosynthesis. These reactions occur in some fungi and in higher plant tissue as well. However, failure of infection from the time when the first haustorium invades a host cell suggests that the host cells have produced the 6-azauridine monophosphate, which then kills the developing haustorium.

6-Azauracil is also effective in preventing development of lesions of cucumber scab, caused by *Cladosporium cucumerinum*, and strains of this fungus are resistant to the chemotherapeutant (Dekhuijzen and Dekker, 1971). In the resistant strain only one-third as much 6-azauracil is converted via 6-azauridine to the monophosphate as is effected by the susceptible strain of the fungus. This is because the resistant strain incorporates orotic acid into RNA regularly but the process is partially blocked in the susceptible strain.

At least superficially, the action of procaine and kinetin on a variety of powdery mildews is similar (Dekker, 1961a, b, 1963b; Dekker and van der Hoek-Scheuer, 1964). Haustorial establishment fails when host cells are invaded. The parallel behaviour suggests that action in these cases also is based upon production of a fungitoxic molecule by host cells.

Phytoalexins are formed in plants under the influence of the invading pathogen. They are also formed as a result of certain types of injury or chemical treatment. Cruickshank (1966) has suggested that in the future chemical control of plant diseases may be based upon systemic treatments that stimulate plants to produce phytoalexins.

The induction of resistance in apple leaves to scab, caused by *Venturia inaequalis*, following injection of DL-phenyl alanine may be an example. Kuć *et al.* reported in 1957 that injection of phenyl alanine into petioles of apple leaves converted their reaction to scab from susceptible to resistant. Subsequently, Holowczak *et al.* (1962) found that phenyl alanine is converted, at least in part, to phloridzin, phloretin and phloretic acid. α-Amino isobutyric acid, which induces a resistant reaction to *Venturia inaequalis* in susceptible apple leaves, is also converted to phloridzin and phloretin (McLennan *et al.*, 1963). Resistance of apple leaf tissue to scab has been associated with the phenolic aglycone, phloretin. This undergoes oxidation and polymerisation as the pathogen enters the leaf and activates the polyphenol oxidase system (Noveroske *et al.*, 1964a, b; Kuć, 1968; Raa, 1968). The intermediate and reactive products of oxidation and polymerisation react non-specifically with enzymes and other proteins and denature them. These reactions are associated with the pin-point lesions associated with a resistant reaction (Kuć, 1968; Raa and Kaars Sijpesteijn, 1968).

The reactions of plant tissue to invasion by a pathogen are similar for a variety of diseases. Phenolic compounds appear and phenol oxidases become activated. The reactions of resistant varieties to invasion are fast and acute, whereas those in a susceptible variety are slow and prolonged, but in a number of respects the reactions occurring in resistant and susceptible varieties are similar (Kosuge, 1969).

The series of responses of tomatoes to invasion by *Fusarium oxysporum* f. sp. *lycopersici* includes the appearance of indole acetic acid, a variety of phenolic compounds, and a very considerable activation of the phenol oxidase system. In resistant varieties, these changes are intense and occur rapidly (Matta *et al.*, 1967). Naphthalene acetic acid, a chemotherapeutant for *Fusarium* wilt, caused a marked activation of the polyphenol–polyphenol oxidase system in susceptible varieties but caused little change in resistant varieties (Matta, 1963). These events suggest how chemical treatment may elicit a non-specific resistance reaction that is closely akin to the hypersensitive reaction.

It is possible that highly specific compounds from plant pathogens may be used to elicit protective responses from host plants. McIntyre *et al.* (1974) were able to confer fire-blight resistance on pear seedlings and shoots by injecting them with a deoxyribonucleic acid preparation derived from both virulent and avirulent isolates of *Erwinia amylovora*. Their experiments indicated that a host response was necessary for protection.

Stomatal closure as a physical barrier

By their action upon stomata, a variety of compounds may create a physical barrier to entry by pathogens that enter through stomates, either exclusively or preferentially. Among compounds that close stomata on leaves are 8-hydroxyquinoline sulphate (Stoddard and Miller,

1962), phenylmercury acetate (Zelitch, 1961), and the alkenyl succinates (Zelitch, 1963). Some of these compounds are themselves directly fungitoxic, whereas others are not. The incidence of infections on bean leaves by *Uromyces phaseoli* was materially reduced following application to foliage of sprays containing non-fungitoxic α-hydroxy succinates that close stomata (Dimond, 1965). The extent to which this principle is useful in disease control in the field needs to be explored systematically.

There are a few cases where open stomata are known to aid the penetration of host plants by fungi. One such well-documented report is the penetration of hop leaves by *Pseudoperonospora humuli* (Royle and Thomas, 1971).

Rich (1963) suggested that closed stomata may impede the extrusion of sporophores through stomata, e.g. *Phytophthora infestans* on potato. Farrell *et al.* (1969) found that stomata of potato leaves infected with *P. infestans* remain open even in the dark and so the sporophores are able to emerge.

Perhaps systemic compounds that can close stomata would be able to protect hops from *P. humuli*, or reduce the sporulation of *P. infestans* on potato.

Systemic compounds that alter growth of the host

The ability of growth regulants to reduce severity of vascular wilt diseases has been known for some time (Davis and Dimond, 1953). Many of these compounds, synthetic as well as natural, have chemotherapeutic activity against a range of pathogens in their hosts. More than one mode of action is now recognised for members of this group.

Alterations in vascular anatomy

No gymnosperm suffers from a vascular wilt disease. In fact, among trees, it is the diffuse porous species that are primarily affected by this group of diseases (Dimond, 1970).

Among species of elms and within varieties, susceptibility to *Ceratocystis ulmi* is correlated with what McNabb *et al.* (1970) have called the vessel group size. Vessel group size is the product of the average vessel diameter and the average number of contiguous vessels in an annual ring.

According to studies of Pomerleau (1970), *C. ulmi* crosses from vessel to vessel only through pit pairs between vessels, when vessels are contiguous. At least during pathogenesis, mycelium of *C. ulmi* is not found in xylem parenchyma or fibres, according to Pomerleau (1970). His findings give the rationale for the concept of vessel group size. This function clearly is related to the ease of invasion by mycelium of the vessels in an annual ring. If this thesis is correct, then any factor that changes the vessel group size in a tree should modify its resistance to vascular wilt diseases such as Dutch elm disease and oak wilt.

Perennial hosts occasionally recover naturally from vascular wilt diseases. When elms recover from Dutch elm disease, for example, the pathogen continues to grow in the originally invaded vessels in an annual ring but does not cross into vascular tissues of subsequently developed rings (Banfield, 1968). Natural recovery of American elms from infections by *C. ulmi* occurs frequently in young trees and rarely in large ones. The reason for this relationship is believed to be associated with the thickness of the annual ring, the number of vessels that must be crossed in a year to keep pace with the growth of the tree. It is also related to the size and distribution of vessels, both across the annual ring and longitudinally.

The action of certain growth regulants as chemotherapeutants in Dutch elm disease and oak wilt is related, probably, to their ability to modify vessel group size. Amino trichlorophenylacetic acid retards development of Dutch elm disease, according to Edgington (1963). When applied early, before the onset of the main flush of cambial growth, this treatment induced rapid cell division, and caused formation of a layer of dense, small, starch-filled cells between earlier and later developing vessels, an effect that significantly reduced vessel group size and, at the same time, reduced severity of disease in inoculated trees.

Trichlorophenylacetic acid applied to oaks affected the characteristics of sapwood and altered susceptibility of trees to oak wilt, caused by *Ceratocystis fagacearum*, according to Venn *et al.* (1968). Young vessels that were formed prior to the time of treatment rapidly filled with tyloses. Wood that formed subsequently consisted of fibres, mainly, and fibre tracheids of small diameter, together with xylem parenchyma. Symptoms rarely progressed following treatment. This treatment also altered vessel group size.

Alteration of the lumina of vessels

The fungi causing vascular wilt diseases spread within their hosts by producing spores or bud cells, which are transported through xylem vessels. On lodging, these spores germinate and establish new centres of infection. In this way, a large host plant can become extensively invaded more rapidly than it could through mycelial growth alone. This process occurs in a variety of vascular wilt diseases in both annual and perennial hosts (see Beckman, 1964; Dimond, 1970). For example, Pomerleau and Mehran (1966) have isolated spores of *C. ulmi* from shoots and leaves of American elm, harbouring Dutch elm disease, while Talboys (1962) had earlier called attention to the same process in the case of *Verticillium* wilt of hop.

When wilt-inducing pathogens invade through roots, the process of spore production and migration may pass through several cycles before the plant as a whole becomes invaded. Beckman *et al.* (1962) have observed how spores of *Fusarium oxysporum* f. sp. *cubense* in the plate region of banana are carried to a cross wall, become lodged, germinate and grow through pores, then produce more spores, which are dislodged and carried to the next obstruction. Beckman *et al.* (1961) have correlated resistance to Panama disease with the rapid development in xylem vessels of tyloses and gels in response to the presence of the pathogen. When these obstructions are formed quickly and are both stable and persistent, spores cannot migrate and the host cannot be invaded by the pathogen (Beckman, 1964). Tomato cultivars resistant to *F. oxysporum* f. sp. *lycopersici* show this same response (Beckman *et al.*, 1972).

In Dutch elm disease, a similar situation prevails. In addition to the correlation of susceptibility in elms with vessel group size, Elgersma (1970) has correlated susceptibility with conductivity of vessels in 2-year-old shoots and related conductivity to length and diameter of vessels. Conversely, resistance is related to lower conductivity, and Elgersma (1967) has noted how spore transport is prevented in resistant elms by the rapid deposition of gum in vessels.

The vascular gels contain both pectic substances and hemicelluloses (Beckman, 1969a, b; Beckman and Zaroogian, 1967). Tyloses are formed simultaneously and provide a stable network for holding gels.

Hyperauxiny is characteristic of wilt-diseased plants as is true of some other diseases (Beckman, 1964); and the tyloses that abound in vessels of oaks suffering from oak wilt have been attributed to the effects of auxins (Beckman *et al.*, 1953). In Panama-diseased banana, Mace and Solit (1966) have isolated indole acetic acid and have experimentally demonstrated its ability to induce the formation of tyloses in vessels.

The foregoing provides a background for explaining a second type of action of growth-regulating compounds as chemotherapeutants in the vascular wilt diseases. They may inhibit growth of sapwood, as Beckman has shown (1958) when auxin supplies are normal, and by doing so, change the vessel group number. But in addition, they induce tylose formation and reduce the conductivity of vessels before infection takes place. Thus, in Dutch elm disease, Smalley (1962) reported the occlusion of large vessels by tyloses following treatment of trees with 2,3,6-trichlorophenyl-acetic acid and suggested that the protection of trees given by this treatment is at least partially attributable to its preventing movement of spores through vessels. After inoculation, pretreated elms developed tyloses in from 2 to 4 days, whereas tyloses in untreated but inoculated trees took from 7 to 9 days, according to Brener and Beckman (1968). This treatment delayed formation of spring wood also. Similarly, Venn *et al.* (1968) reported that oak treated with this same compound developed heavy tylosing in vessels existing prior to treatment and a reduction in diameter and numbers of vessels formed after treatment. Severity of disease was reduced.

Halogenated benzoic acids have also been employed to reduce Dutch elm disease and oak wilt on an experimental basis. Geary and Kuntz (1962) noted a delay in formation of spring vessels, along with reduction in severity of oak wilt. However, the chemotherapeutic effects and phytotoxicity have been erratic following this treatment.

In herbaceous plants, growth-regulating compounds that induce tylose formation have likewise reduced the severity of vascular wilt diseases. 2-Chloroethyltrimethylammonium chloride (cycocel) had little effect upon growth of *Verticillium albo-atrum in vitro* but induced tylose formation in tomato plants and reduced severity of disease in treated plants (Sinha and Wood, 1964). 3-Indoleacetic acid, and α-naphthalene acetamide had similar effects and reduced invasion of stems by the pathogen (Sinha and Wood, 1967).

Modification of cell walls

Among plant diseases, those causing rotting of tissues involve the degradation of cell walls by hydrolytic enzymes that are produced by pathogens. In addition to the rots, a wide variety of other diseases causes decomposition of cell walls. The involvement of pectic and cellulolytic enzymes in pathogenesis is now well established (Bateman and Millar, 1966). A variety of hydrolytic enzymes is produced during pathogenesis that reflects the wide variety of chemical bonds and the variety of carbohydrate monomers that are represented in cell walls (Albersheim *et al.*, 1969).

As the composition of host cell walls varies with stage of host maturity and with species, so also must the hydrolytic enzymes produced by different pathogens vary. The nature of some of these enzymes has been reviewed by Bateman and Millar (1966) and by Wood (1967). Two types of enzymes effect major changes in the molecular weight of pectic substances: the polygalacturonases and the pectate lyases. The former tend to be inhibited when their substrate is largely in the form of calcium pectate or when Ca^{++} is present in quantity. The pectate lyases require calcium (Bateman and Millar, 1966). Thus, a pathogen that degrades pectin primarily by means of polygalacturonase should damage host tissues according to the level of calcium present.

This has proven to be so. *Fusarium oxysporum* f. sp. *lycopersici* produces polygalacturonase both *in vitro* (Waggoner and Dimond, 1955) and in the infected tomato plant (Patil and Dimond, 1968; Mussell and Green, 1968). Tomato plants that are calcium-deficient are severely attacked by *Fusarium*, whereas plants grown on a normal calcium supply suffer less (Edgington and Walker, 1958).

How growth regulators act as chemotherapeutants in *Fusarium* wilt began to be apparent when Corden and Dimond (1959) correlated the chemotherapeutic effectiveness of a series of naphthalene-substituted aliphatic acids with their activity as growth regulants by each of several criteria. Chemotherapeutic activity was highly correlated with the ability of compounds to inhibit root elongation, but with no other index of growth hormone activity. This result suggested that the compounds acted upon pectins in cell walls by causing increased calcium bonding, which in turn decreased root elongation. Corden and Edgington (1960) then showed that the chemotherapeutic effect of naphthalene acetic acid was calcium-dependent. Calcium-deficient plants were just as susceptible to wilt whether or not they had been treated with naphthalene acetic acid!

The effects of calcium nutrition and treatment with naphthalene acetic acid upon the pectins in cell walls were then examined (Edgington *et al.*, 1961). Although total pectins were not altered, calcium-deficient plants contained more water-soluble pectins than plants grown on a normal calcium supply. Plants treated with naphthalene acetic acid contained less water-soluble pectin than normal plants. The pectic compounds of calcium-deficient plants were readily attacked by pectolytic enzymes from the tomato wilt *Fusarium*, but those from naphthalene acetic acid-treated plants on a normal calcium supply were resistant to enzymic hydrolysis. Thus, susceptibility of plants to *Fusarium* wilt and susceptibility of their pectic components to enzymic attack were well correlated. Evidently, when applied in growth-inhibiting concentrations, growth regulants reduce methoxylation of pectic substances and, in the presence of a normal supply of calcium, components in cell walls are highly cross-bonded with calcium bridges which are polygalacturonase-resistant. In calcium-deficient plants, this cannot occur, so the pectic components of cell walls are readily hydrolysed.

This interpretation is consistent with what is known of the role of auxins in maintaining plasticity of cell walls in healthy plants, a property long associated with effects on pectins. Bonner (1961) has offered evidence that its site of action is the same as for calcium-dependent rigidity.

Ethionine has shown chemotherapeutic activity on several occasions, Zentmyer *et al.* (1962) showed its action against root rot of avocado by *Phytophthora cinnamomi*, and Moje *et al.* (1963) demonstrated that methionine antagonised this effect. Action by ethionine and its reversal by methionine (or the converse in appropriate host–pathogen combinations) can be interpreted as an effect upon cell walls, much as auxins can modify the pectins in cell walls. Methionine in the presence of auxins donates methyl groups to pectins in wall formation (Cleland, 1963a) and ethionine prevents this donation of methyl groups (Cleland, 1963b). Thus, for diseases in which cell wall attack is by comparable mechanisms, ethionine or auxin could produce similar effects. Jones and Woltz (1969) have provided the data to confirm this by showing that ethionine reduces the severity of *Fusarium* wilt of tomato. Edgington *et al.* (1961) showed that naphthalene acetic acid reduced *Fusarium* wilt of tomato through its action upon pectins of the cell wall.

Other growth effects

The correlation between growth-regulant activity and chemotherapeutic activity has been used as a basis for seeking new compounds that may be useful in control of plant diseases. A number of compounds have been found in this way, but as yet their mode of action is not known. The action of salts of 2-benzothiazolyl thioglycolate against

Fusarium wilt and Dutch elm disease is a case in point (Dimond and Davis, 1953). These compounds produce formative effects in the host. Whether they stimulate tylose formation or act in some other way is not known. The same is true of the morphactins (methyl, n-butyl, and n-heptyl esters of 9-hydroxyfluorene-9-carboxylic acid and the 2-chloro-substituted analogues). Buchenauer and Grossmann (1969) have reported moderate reductions of *Fusarium* and *Verticillium* wilts and of late blight of tomato as a result of treatment, but the mode of action is not known.

In three instances, at least, the use of growth retardants has reduced disease severity. The amount of brown spot, caused by *Alternaria longipes*, was reduced in tobacco to which maleic hydrazide had been applied for sucker control (Ramm *et al.*, 1962). Van Andel (1966a, b, 1968) has reported a reduction in cucumber scab, caused by *Cladosporium cucumerinum*, as a result of applying compounds that reduced auxin through increasing IAA oxidase activity. Among them were L-threo-β-phenyl serine and 2-chloroethyltrimethylammonium chloride. In addition, several unnatural D-amino acids retarded root extension in assays and reduced cucumber scab (van Andel, 1962b).

This evidence suggests that growth regulants and retardants affect cucumber scab and *Fusarium* wilt of tomato oppositely. Growth-promoting substances tend to reduce wilt and growth retardants, such as ionising radiation or maleic hydrazide greatly increase severity of disease (Waggoner and Dimond, 1957).

Beckman *et al.* (1962) suggested a mechanism by which growth regulants could block the invasion of wilt-inducing fungi. They demonstrated that in bananas with *Fusarium* wilt, vascular parenchyma cells begin to grow again in response to the presence of the advancing fungal hyphae. The renewed growth either crushes the small protoxylem cells or causes the large metaxylem cells to invaginate. This, in turn, helps to seal the infected vessels and to wall off the infection.

The severity of powdery mildews is reduced by treatment, not only with ethionine, but with methionine also. In fact, L-methionine is the only natural amino acid having a chemotherapeutic effect (van Andel, 1966a). Dekker (1969a) reported poor development of powdery mildew (caused by *Sphaerotheca fuliginea*) on cucumber leaves, an effect confirmed by Covey (1971a) with apple powdery mildew, caused by *Podosphaera leucotricha*.

In similar fashion L-methionine prevents root rot of peas, caused by *Aphanomyces euteiches*, and ethionine is also active (Papavizas and Davey, 1963). The basis of this action remains unexplained. L-Methionine also reduces root rot of sugar beet, caused by *Aphanomyces cochlioides* (Winner, 1966).

The apparently diverse behaviour of amino acids and their analogues in reducing severity of disease or in negating effects induced by an antagonist surely arises because of the many ways in which these compounds can act in biological systems. In each disease, two organisms are involved, a host and a pathogen, and observed response can be from either organism. Depending upon the mechanism of pathogenesis, response to a given amino acid will vary. Impose, in addition, response to natural (L) and unnatural (D) amino acids, and the multiplicity of possible responses increases yet more.

The amino acids can be grouped in families that share a common pathway of biosynthesis. A more orderly picture of amino acid interrelations, for bacteria, at least, has been presented by Dawes and Large (1968). Sands and Zucker (1971) have compared phytopathogenic with saprophytic pseudomonads in respect to their amino acid metabolism. Pathogenic species proved frequently to have impaired biosynthetic pathways, whereas saprophytic species rarely do. Toxic effects of an administered amino acid may be related to a faulty step in the biosynthetic pathway. A host plant can compensate in its own amino acid supply for the faulty pathway in a pathogen. If the compensation by the host can be blocked, then therapy is possible.

Treatments that block pathogenic processes

Antisporulants

Blocking the spread of propagules of vascular wilt pathogens through xylem vessels has already been mentioned as a potential method of chemotherapy. Two approaches are possible. One can induce tyloses and gums in vessels to prevent ready movement of spores through vessels. Alternatively, one can inhibit the development of propagules by fungi through use of antisporulants. Horsfall and Lukens (1968) have explored the possibility of disease control through using antisporulants to reduce the aerial spread of plant pathogens in the field and have sought compounds that reduce spore formation *in vitro*. Howard and MacHardy (1969) have discussed this principle in reference to vascular wilt organisms. Biehn (1972a) has examined the extent to which the antisporulant *p*-fluorophenylalanine prevents bud cell formation by *Ceratocystis ulmi* with therapy in mind. Keen *et al.* (1971) have examined the effectiveness of various antisporulants, including 5-fluorodeoxyuridine, on cultures of *Verticillium albo-atrum*.

The idea of controlling disease through controlling the amount of available inoculum is appealing, and with the perfection of systemic antisporulants, a new method of control through chemotherapy will be available.

Inactivation of pectolytic enzymes

Just as maceration of tissues can be reduced by means of treatments that make cell walls more resistant to enzymic hydrolysis, so can maceration be reduced through compounds that inactivate these wall-splitting enzymes. Grossmann (1958b) sought to inactivate the pectolytic enzymes present in culture filtrates of *F. oxysporum* f. sp. *lycopersici*. He found a number of phenolic compounds, among them rufianic acid (1,4-dihydroxyanthraquinone-2-sulphonic acid), that would do this *in vitro* (Grossmann, 1962a). Rufianic acid is also effective *in vivo* as a chemotherapeutant against *Fusarium* wilt of tomato (Grossmann, 1962b). This work demonstrated the feasibility of chemotherapy by enzyme inhibition and offered a means of disease control on a clearly demonstrable scale *in vivo*. It was also a necessary part of the demonstration of the role of pectolytic enzymes in pathogenesis of the wilt diseases.

In principle, the inhibition of macerating enzymes is also a means for controlling soft rots and such leaf-spotting diseases as depend upon hydrolysis of cell wall polysaccharides. Grossmann (1968a) found that rufianic acid reduced maceration of plant tissue by *Rhizoctonia solani* and that non-fungicidal inhibitors of pectolytic enzymes also prevented infection of tomato by *Alternaria solani*.

The phenolic compounds in plant extracts often auto-oxidise and polymerise to form highly reactive products. These react non-specifically with proteins generally, a process that leads to inactivation of pectolytic and other enzymes. Byrde (1956, 1957) had called attention to this as a natural process, accounting for resistance to rotting in some cider apples and Cole (1958) in this way accounted for lack of detectable pectolytic activity when some cultivars of apples are attacked by rotting fungi. Grossmann (1962c), in his assays, noted loss of pectolytic activity when plant extracts containing phenolic compounds had auto-oxidised. Byrde (1963) has traced the course of this idea and the role of natural phenolic compounds in enzyme inactivation. Products of chlorogenic acid or caffeic acid, arising from the action of potato phenolase, were shown to inactivate polygalacturonase of *Verticillium albo-atrum* (Patil and Dimond, 1967a).

Beckman *et al.* (1974) proposed that phenolic infusion at infection sites is the final step that restricts the advance of wilt-inducing fungi. Not only can these compounds act as fungitoxicants, but they may also stabilise and protect cell-wall membranes from further degradation and penetration by the pathogen.

In many diseases, plant tissues exhibit an activation of phenol oxidases, and phenols become oxidised and polymerised. Such reactions are now believed to be involved in disease resistance, the hypersensitive reaction, and, to some degree, in phytoalexin formation. These reactions occur both in susceptible and resistant tissues, but are far more rapid in resistant tissue and generate fungitoxic materials. The response can be evoked by pathogens, non-pathogens and certain chemical and physical treatments. Hopefully, such treatments will one day be used in disease control to convert susceptible into hypersensitive plants.

Regulation of enzyme production by chemotherapy

The role of hydrolytic enzymes in breakdown of host cell walls is now recognised to be an important aspect of pathogenesis. Historically, attention was first focused on the pectolytic enzymes and their role in maceration of tissues.

As attention was directed experimentally toward these enzymes, the conditions favouring induction and repression began to be stressed. Deese and Stahmann (1960) reported the higher polygalacturonase activity when various *formae* of *F. oxysporum* were grown on tissues from susceptible hosts than when grown on tissues from plants that are resistant to these *formae*. Deverall and Wood (1961) compared pathogenesis of *Botrytis cinerea* and *B. fabae* in respect to polygalacturonase production under comparable circumstances.

Repression of polygalacturonase synthesis was explored in *F. oxysporum* f. sp. *lycopersici* in the presence of glucose or galacturonic acid (Patil and Dimond, 1967c). In contrast, polygalacturonase production by *V. albo-atrum* showed end-product repression but the enzyme was produced in considerable amounts when glucose or galactose were present (Patil and Dimond, 1967b; Mussell and Green, 1970).

Patil and Dimond (1968) then examined the effect of applying regulators of enzyme production to infected plants. Glucose, 2-deoxy-D-glucose, and monogalacturonic acid all interfered with production of polygalacturonase by *F. oxysporum* f. sp. *lycopersici*. Each was introduced into the vascular system of tomato plants previously inoculated with *Fusarium*. For treated plants and inoculated controls, the amount of polygalacturonase activity and the rate of advance of disease symptoms were compared. In plants treated with a repressor, less polygalacturonase was detected and symptoms increased at a slower rate than in controls. Plants treated with glucose showed fewer disease symptoms than plants treated with water, despite the fact that the former contained more culturable mycelium.

Such studies are being extended to other types of enzymes that are involved in degradation of cell walls. Thus, Keen *et al.* (1970) have examined the conditions affecting induction and repression of β-galactosidase synthesis in *V. albo-atrum*. As such studies are extended to a wider range of enzymes involved in the attack of cell walls, greater variety of treatments will be available for interfering with this process of pathogenesis.

The diversity of wall-splitting enzymes produced by a given pathogen is greater than was formerly realised, thanks to the work of Albersheim *et al.* (1969). He has shown how enzymes are induced in a pathogen in a prescribed sequence, according to the sequence of carbohydrates in wall components of the host. As tissues in a growing host plant develop, the components of its cell walls change and the sequence of enzyme induction must also change. This relationship is thought to account for the short-lived susceptibility of juvenile tissues to certain pathogens (Nevins *et al.*, 1968). If the wall-digesting enzymes are not induced in the correct sequence, pathogenesis is impaired.

These ideas suggest that attack of cell walls could be inhibited were it possible to interfere with processes of enzyme induction and especially to interfere with the sequence.

Synthesis of a variety of wall-hydrolysing enzymes can be regulated. Both cellulase and polygalacturonase are inhibited by the presence of glucose in *F. oxysporum* f. sp. *lycopersici*. Polygalacturonase synthesis is affected oppositely by glucose in *F. oxysporum* f. sp. *lycopersici* and *V. albo-atrum* (Patil and Dimond, 1967b, c; Mussell and Green, 1968). These differences aid in accounting for susceptibility of plants to infection by organisms, according to the amount of sugar in their tissues. Horsfall and Dimond (1957) have discriminated between 'high sugar' and 'low sugar' diseases, according to how sugar levels in tissue affect susceptibility to infection by different pathogens. In general, the rusts and powdery mildews are best able to infect tissues that are high in sugar, whereas low-sugar diseases are caused by *Helminthosporium* and *Alternaria* spp. These differences are consistent over a variety of circumstances. When pathogenesis involves pectolytic or cellulolytic enzymes to a considerable extent and when these enzymes are consistently repressed in synthesis by the presence of glucose or sucrose or other sugar, the regulation of enzyme synthesis dictates a low-sugar disease. Moreover, the end product of hydrolysis of cellulose will repress synthesis of polygalacturonase or of any other enzyme repressed by glucose. When cell-wall destruction is unimportant and when pathogenesis depends upon the synthesis of other enzymes that are not affected by sugars, high sugar levels do not interfere with pathogenesis.

End-product repression of synthesis of wall-attacking enzymes and its absence in the regulation of synthesis of enzymes in other species of pathogens accounts for some, but not all, high-sugar diseases. It does account for how *Botrytis cinerea* and *V. albo-atrum* (Patil and Dimond, 1968) can attack tissues containing more sugar than those attacked by *B. fabae* (Deverall and Wood, 1961) or *F. oxysporum* f. sp. *lycopersici* (Patil and Dimond, 1967c). But it fails to explain why diseases caused by the obligate parasites, the rusts and powdery mildews, are high-sugar diseases as a group.

Two low-sugar diseases have been controlled by chemotherapy by taking advantage of the hypothesis. Lukens (1970) noted that the melting-out disease of turf, like all diseases caused by species of *Helminthosporium*, behaves as a low-sugar disease. He has used this information to alleviate *Helminthosporium vagans* on Kentucky bluegrass turf. High cutting, elimination of shading or use of glucose sprays on turf increased reducing-sugar levels in leaves and reduced severity of melting-out disease. Kaars Sijpesteijn (1961a) has offered evidence in a second case. Cucumber scab, in her studies, behaved as a low-sugar disease, just as *Alternaria* does. The chemotherapeutant carboxymethyldimethyl dithiocarbamate in treated seedlings reduced sugar levels and increased resistance to infection by *Cladosporium cucumerinum*. When seedlings were treated with both sugar and the same chemotherapeutant, they remained susceptible to infection.

Higher plants can produce enzymes that lyse the cell walls of fungi. Pegg and Vessey (1973) found that both healthy and Verticillium-infected tomato plants contain chitinase. Chitin is not a constituent of higher plants but is an important component of the cell walls of many fungi. Dixon and Pegg (1969) had shown earlier that the mycelium of *V. albo-atrum* actually disappears from invaded vessels of tomato plants that had some resistance to *Verticillium* wilt. Pegg and Vessey discovered that infection by *V. albo-atrum* did cause a significant increase in the chitinase activity of the host tomato, and that this chitinase is able to lyse the mycelium of the pathogen. However, they could not correlate this increase with host resistance.

Wargo (1975) reported that many healthy deciduous trees contain chitinase and β-1,3-glucanase and that both enzymes are able to lyse the cell walls of *Armillaria mellea.*

Abeles *et al.* (1971) detected trace amounts of chitinase and β-1,3-glucanase in uninjured leaves of healthy beans (*Phaseolus vulgaris*). When these leaves were exposed to ethylene for 3 days, the activity of both enzymes increased about 50-fold. In fact, chitinase made up 4 per cent of the soluble protein of the ethylene-treated leaves. Abeles *et al.* proposed that this process served to protect plants from pathogenic fungi.

Another link in this relation was offered by Valent and Albersheim (1974). They demonstrated (*a*) that an endo-β-1,3-glucanase produced by beans in response to ethylene can lyse the cell walls of the anthracnose fungus, *Colletotrichum lindemuthianum*; and (*b*) the pathogen can produce a protein that inhibits the bean glucanase but not an endo-β-1,3-glucanase that is secreted by the pathogen.

Because the production of ethylene by higher plants can be mediated by a number of chemical treatments (Abeles, 1973) these studies offer interesting possibilities for the use and design of chemotherapeutants.

Counteracting toxins in plant disease

To an increasing degree, specific symptoms associated with plant disease are being ascribed to the action of toxins originating in the pathogen. Toxins have in a number of cases been isolated from cultures of the pathogen and from diseased plants, and, in a few cases, they have been identified.

The idea that toxins can be neutralised or antidoted and so bring about chemotherapy is an old one (Howard and Horsfall, 1959). For example, in infections by *Pseudomonas tabaci* on tobacco leaves, chlorotic haloes surround the lesion and the chlorosis has been attributed to a toxin on numerous occasions. Lovrekovich and Farkas (1963) called attention to the similarity between development of chlorosis and of altered protein metabolism in leaf tissue and processes of senescence, and they demonstrated that these changes in infected leaves were reversible by administration of kinetin. Here is a potentially useful illustration of therapy. But there are few cases where therapy, based on toxin inactivation, has been achieved in the field. This possibility remains to be considered, however, especially when a specific toxin is known to play a critical role in the induction of disease.

An example where a toxin plays a critical role in disease involves pyricularin, the toxin of *Pyricularia oryzae*, the cause of rice blast. Tamari *et al.* (1966) found that ferulic acid detoxifies pyricularin and increases resistance of the rice plant to blast. Extracts of treated plants also inactivated pyricularin. Pyricularin appears to act by suppressing the critical hypersensitive reaction of host tissues, and ferulic acid in the rice plant causes respiratory changes that are not related to the hypersensitive reaction. Under such circumstances, the neutralisation of a toxin can make a critical difference in the success of an infection.

The host-specific toxins illustrate a second case where toxins play a critical role in disease induction – to the extent that Scheffer and Pringle (1967) refer to them as primary determinants of disease. As yet, among the various diseases where host-specific toxins are known to participate, no compound has been reported that will neutralise the effect of these toxins or suppress their synthesis by the pathogen. These possibilities could prove useful in combating such important diseases as southern corn blight, caused by *Helminthosporium maydis* or *Helminthosporium* blight of oats, caused by *H. victoriae*.

Future prospects

The foregoing discussion has dealt with many instances where disease can be alleviated through treatments that are not in themselves fungitoxic. Primarily, these treatments interfere with one or more phases of the process of pathogenesis. These treatments generally delay the development of disease, but in only a few instances has treatment effected a cure.

Postponing an epidemic sometimes makes the difference between success and failure in harvesting an annual crop. Among perennial plants, such as shade trees, cure of disease is more pertinent than alleviation of symptoms.

A systemic fungicide can eliminate infection in a plant and can effect a cure. A systemic chemotherapeutant can only slow the progress of the pathogen or make the host resistant to its invader. As yet, systemic fungicides would seem to hold the key to control of disease by chemotherapy.

Chapter 7

Effects on fungal pathogens

A. Kaars Sijpesteijn

Introduction

Knowledge of the mechanism of action of fungicides is not only important from a purely scientific point of view but it seems also a necessary prerequisite for a proper evaluation of the general toxicology of these compounds. In addition, knowledge of the mode of action may in many cases help to elucidate the mechanism by which resistant mutants differ from the normal sensitive strain. For obvious reasons the fundamental study of the mode of action of antifungal agents used in plant protection has for a long time lagged far behind that of antibacterial agents in medicinal chemotherapy. In this latter field especially the antibiotics have been the subject of numerous investigations (e.g. Gottlieb and Shaw, 1967; Franklin and Snow, 1971; Corcoran and Hahn, 1975).

This chapter attempts to deal with a variety of compounds in a comparative way.

The description of the effect of a fungicide on a fungus requires morphological and physiological as well as biochemical observations. Generally we may distinguish at least three different methods by which fungicides can exert their action.

(*a*) *Inhibition of energy production,* or in other words, of ATP production; for instance by inhibition of respiration or by uncoupling of oxidative phosphorylation. The energy-producing processes are located partly in the cytoplasm and partly in the mitochondria. Strong inhibition of such processes will eventually have a fungicidal effect.

(*b*) *Interference with biosynthesis,* i.e. with processes leading to the production of new cell material required for growth and maintenance of the organism. Biosynthesis of low molecular weight compounds such as amino acids, purines, pyrimidines and vitamins takes place in the cytoplasm, whereas protein synthesis proceeds mainly on the ribosomes, and DNA- and part of RNA-synthesis in the nucleus. Inhibition of biosynthesis sometimes has only a fungistatic effect.

(*c*) *Interference with cell structure,* for instance by affecting permeability of the cell membrane resulting in leakage of the cell contents. This effect may be fungicidal.

One has of course to distinguish between primary and secondary effects of a toxicant. Thus, increase of permeability of the cell membranes will lead eventually to inhibition of respiration; alternatively, inhibition of energy production will result in a decrease of biosynthesis for lack of ATP and often also for lack of necessary intermediates normally produced during respiration. Hence it is of paramount importance to detect which after addition of the fungicide is the earliest effect on the fungus.

When a new fungicide becomes available for study a variety of experimental results will be required to build up a picture of how the compound might act. Primary questions are the level of activity, the antifungal spectrum, the influence of pH and composition of the medium on activity. It must be determined whether respiration is inhibited and it is useful to investigate whether certain compounds, as for instance growth factors, amino acids, lipids, –SH compounds, can act as antagonists of the fungitoxic agent. One needs to know whether the agent exerts a fungistatic or a fungicidal effect. This may be different again whether tested under conditions of growth or non-growth, as in a nitrogen-free medium. Investigations are needed to determine at which stage the compound acts, e.g. on spore germination, or only when hyphal growth starts. Moreover, a study of the morphology and cytology of the inhibited cells can give useful information. The structure of the antifungal agent can sometimes give a clue as to its mode of action; there is the question whether the fungicide applied is itself the actual toxic agent or whether the true toxicant is formed non-enzymatically in the medium or by metabolic activity of the fungus.

The above broad survey of activities of an antifungal agent constitutes only the first phase of investigation; a careful analysis of its results usually leads to indications of the possible primary site of action. This hypothesis has to be checked in a second phase of study in which one tries to relate the antifungal effect more definitely to interference with a particular enzyme system or a specific function of certain cell constituents.

A third phase may reveal the more molecular aspects of the interference of the toxicant with the target system. This is a field in which very little progress has yet been made, but Davidse's work (1975a, b) with carbendazim seems to be a most encouraging example.

Since at higher concentrations further processes may gradually become inhibited, it is advisable to use low growth-inhibiting concentrations when trying to elucidate the primary mode of action. In such studies intact cells should be used as much as possible, one of the reasons being that in cell-free preparations conditions, e.g. the concentration of the toxicant at the site of the action, may deviate considerably from those in intact organisms.

Obviously all these studies can often be carried out more efficiently with saprophytes than with parasites. For systemic compounds that act exclusively on obligate parasites such as mildews, rusts and to some extent smuts, the *in vitro* study of their action can only be of limited scope.

In the past few years knowledge of the mode of action of systemic fungicides has been expanded at a vast rate but at the same time new compounds have appeared. Interestingly, so far most new compounds reveal new mechanisms again.

As we shall see, the effect of a fungicide on two different fungi is frequently not the same, slight deviations or additional effects being often observed. These differences, which presumably are mostly due to inherent differences in the normal physiology of the organisms studied, stress the importance of comparative work involving more than one organism. Needless to add, the study of antifungal activity is conversely making a continuous contribution to the elucidation of the fundamentals of fungal physiology, knowledge of which is still limited.

The following sections deal with the *in vitro* effects on fungi of systemic fungicides, including those used in practice and some others. Sufficient knowledge seems available at present to make an attempt to group the different compounds largely according to their mode of action as was tentatively done for fungicides in general (Kaars Sijpesteijn, 1970).

A far more detailed treatise of this material will be presented in the forthcoming volumes on *Antifungal Compounds* by Siegel and Sisler (1977).

Table 7.1 Minimum inhibitory conc.† (ppm)

	Botrytis cinerea	***Penicillium italicum***	***Aspergillus niger***	***Cladosporium cucumernium***
Benomyl	0·02	0·05	0·5	0·2
Carbendazim	0·05	0·05	0·5	0·2
Thiophanate-methyl	0·2	0·5	20	2
Thiabendazole	0·2	0·1	2	0·5
Fuberidazole	0·5	0·5	10	2
Carboxin	20	100	100	10
F427‡	20	5	10	1
Dimethirimol	> 500	> 500	> 500	> 500
Ethirimol	> 500	> 500	> 500	> 500
Triarimol	2	2	20	0·2
Tridemorph	5	200	> 500	10
Triforine	> 500	> 500	> 500	100
Chloroneb	> 500	> 500	> 500	> 500
Pyrazophos	> 500	> 500	> 500	> 500
Kitazin	100	500	500	200
Cycloheximide	20	20	100	10
Imazalil	0·2	0·005	0·2	0·02
Triadimefon	50	200	> 500	20
Pyroxychlor	> 500	> 500	> 500	50
Prothiocarb	> 500	> 500	> 500	> 500

† On glucose mineral salts agar pH 6·8, three days after inoculation with conidial suspension

‡ See p. 134

It seems useful to give, for comparison, the minimum concentrations of a variety of compounds inducing complete inhibition of visible growth on glucose mineral salts agar, pH 6·8 (Table 7.1), in the standard roll-culture test (Pluijgers and Kaars Sijpesteijn, 1966).

The chemical structures of the various compounds dealt with are given in other chapters.

Interference with respiration

Carboxin and oxycarboxin These oxathiins display a remarkable selectivity in their antifungal spectrum. This was shown for carboxin (Edgington *et al.*, 1966), as well as for its sulphone, oxycarboxin, and for various other derivatives (Snel *et al.*, 1970). Fungi belonging to the Basidiomycetes are particularly sensitive, e.g. loose smut of barley, bean rust and *Rhizoctonia solani.* Of fungi belonging to other groups only *Verticillium albo-atrum* and *Monilia cinerea* f. *americana* (= *Monilinia fructicola*) were inhibited by low concentrations (Edgington and Barron, 1967). The same authors reported that the antifungal spectrum of the carboxin analogue, 2,3-dihydro-6-methyl-5-(2′-diphenylyl)carbamoyl-1,4-oxathiin (F 427), is much wider than that of the other oxathiins, since apart from inhibiting Basidiomycetes it also is highly active on representatives of other groups of fungi. This observation would support the view that the selectivity in the oxathiins is not connected with their mode of action but rather depends on differences in their availability at the sites of action, caused by selective absorption or permeation.

Studies by Ragsdale and Sisler (1970) revealed that in a medium lacking a nitrogen source, low growth-inhibiting concentrations of carboxin strongly inhibit glucose oxidation of *R. solani* and *Verticillium albo-atrum.* With *Ustilago maydis* and *Neurospora crassa* glucose oxidation was little or not inhibited if glucose was the substrate, but acetate oxidation was strongly affected by concentrations which cause almost complete growth inhibition. For *U. maydis* and *N. crassa* such concentrations are 2 and 50 ppm respectively with glucose as a carbon source, but ten times less is needed when acetate is the substrate. Work along the same lines by Mathre (1970) disclosed a strong inhibition of CO_2 evolution from acetate and pyruvate in *R. solani* and *U. maydis.* All evidence obtained pointed to a block in the tricarboxylic acid cycle.

In *R. solani* and *V. albo-atrum* lack of energy supply may be the primary cause of growth inhibition. *Ustilago maydis* and *N. crassa*, however, apparently do not require the tricarboxylic acid cycle for oxidation of glucose or are able to bypass the blocked site. Growth inhibition of these organisms on a glucose medium is thus due not to lack of energy but presumably to a shortage of necessary TCA intermediates required for growth.

Further studies specifying the site of inhibition were made by Mathre (1971a) and White (1971): low growth-inhibiting concentrations had a strong effect on respiration of mitochondrial preparations of *Ustilago maydis* at or close to the site of succinate oxidation. Inhibition of succinate-coenzyme Q reductase, but not of NADH-coenzyme Q reductase, was reported by Ulrich and Mathre (1972). The same authors observed that only when intact mitochondria were used did carboxin inhibit the succinic dehydrogenase system in *U. maydis*; when the enzyme was solubilised and no longer bound to structural units the fungicide proved ineffective. They concluded that the main site of action of carboxin, and equally of oxycarboxin, lies between succinate and coenzyme Q in the intact succinate–ubiquinone reductase complex of the electron transport system. This is in line with the observation of Georgopoulos *et al.* (1972) that a mutant of *U. maydis* which is highly resistant to carboxin on either glucose or acetate as a substrate possesses a modified succinic dehydrogenase system. Later work by Georgopoulos *et al.* (1975a) supported this view and the authors speculated that carboxin might interfere with the catalytic function of a specific nonheme Fe-S component of the above mentioned enzyme complex. A similar model is proposed by Lyr *et al.* (1975b).

A second site of inhibition was discovered in *U. maydis*. It is in the cyanide-insensitive, alternate pathway for electron transport which according to investigations by Sherald and Sisler (1970, 1972) operates in certain fungi. This pathway, in contrast to the normal cytochrome pathway is tolerant to antimycin A, an antibiotic which inhibits electron transport between cytochromes b and c. A group of mutants which, on glucose but not on acetate, appeared slightly more resistant to carboxin than the wild type at the same time proved extremely sensitive to antimycin A and cyanide (Georgopoulos and Sisler, 1970). These mutants seem to lack the alternate pathway.

Although the relation between the two sites of inhibition by carboxin in *U. maydis*, as well as several other aspects, still needs further clarification, succinate oxidation is obviously the main site of inhibition.

Working with *Ustilago hordei*, Ben-Yephet *et al.* (1975) largely confirmed the observations made with *U. maydis*. A marked effect of carboxin on phosphate uptake accompanied the inhibition of succinate oxidation in this former organism. It is not clear yet whether the effect on phosphate uptake is an independent effect or a consequence of the inhibition of substrate oxidation. In mutants of *U. hordei* which were somewhat less sensitive to the fungicide, not only was the succinic dehydrogenase system less sensitive, but other factors accounted for resistance as well. According to Lyr and co-workers (1972) the selectivity by carboxin with regard to a variety of yeasts is related to their lipid content rather than to differences in uptake. The relative insensitivity of *Saccharomyces cerevisiae* coincides, however, with a relative insensitivity of its succinic dehydrogenase (Ulrich and Mathre, 1972).

Mebenil and *pyracarbolid* These compounds, related to the oxathiins, are effective against diseases caused by Basidiomycetes, which include *R. solani* and rusts (Pommer and Kradel, 1969), and smuts (Jank and Grossman, 1971). They appear to act like oxathiins, since low concentrations strongly inhibit succinic dehydrogenase of *U. maydis* and *Cryptococcus laurentii* (White and Thorn, 1975).

Interference with protein synthesis

Cycloheximide This antibiotic acts on a wide variety of fungi including the Phycomycete *Pythium debaryanum* (Whiffen, 1950), but it often shows a great difference in activity towards closely related organisms. Respiration is only slightly inhibited in most organisms. A thorough study of its mode of action was made by Sisler and Siegel (1967). Using *Saccharomyces pastorianus* they found that low fungitoxic concentrations (3·16 ppm) specifically inhibit protein synthesis at the stage where in a complex with the ribosomes amino acids are transferred from tRNA into polypeptides. The precise mechanism and site of action of cycloheximide is still obscure. Although DNA synthesis is also strongly inhibited, protein synthesis is regarded as the primary basis for toxicity. This conclusion finds support in the observation that cell-free protein-synthesising systems of resistant strains of *S. cerevisiae* are resistant to the toxicant whereas those from sensitive strains are sensitive.

In contrast to *S. pastorianus* and *S. cerevisiae*, *S. fragilis* is highly insensitive to cycloheximide. This is not due to lack of penetration of the compound or to metabolic conversion, since cell-free protein-synthesising systems from the latter species are far less sensitive to the compound than those of the former. Cross-over experiments with the 60 *S* and 40 *S* sub-units of ribosomes from the sensitive *S. cerevisiae* and the insensitive *S. fragilis* show that sensitivity to cycloheximide is associated with the 60 *S* ribosomal unit (Rao and Grollman, 1967). These apparently differ in the sensitive and insensitive organisms.

A detailed discussion of the work on cycloheximide is given by Sisler (1969).

More recent work by Timberlake and Griffin (1974) raises the question whether in the Oömycete *Achlya bisexualis* cycloheximide may also have a direct effect on RNA synthesis, because this process is inhibited much more rapidly than would be expected from inhibition of protein synthesis alone.

Blasticidin *Pyricularia oryzae* is particularly sensitive to this antibiotic, a pyrimidine derivative which, unlike most compounds dealt with, also inhibits growth of certain bacteria (Misato, 1967). Since mycelial growth is affected by lower concentrations than spore germination, blasticidin can be applied to rice plants after infection has occurred. The compound is somewhat phytotoxic. Respiration of glucose is 50–60 per cent inhibited both by growth-inhibiting concentrations (0·1 ppm) and by far higher concentrations. At 1 ppm the incorporation of ^{14}C-amino acid into the protein fraction of intact mycelium and of cell-free extracts of *P. oryzae* was completely inhibited; nucleic acid synthesis was undisturbed. Since amino acid activation and the transfer of activated amino acids to soluble RNA were less inhibited than overall incorporation, Misato and co-workers concluded that the probable site of action is the final step of protein synthesis taking place on the ribosomes. The rather serious inhibition of respiration will presumably also be of influence on inhibition in as far as it limits energy production.

In strains of *P. oryzae* which had become resistant to blastidin, this resistance could be ascribed to a difference in permeability of the cells, because in cell-free extracts protein synthesis was as sensitive as in those of the normal strain.

In contrast to *P. oryzae*, *Pellicularia sasakii* is highly tolerant to blasticidin; protein synthesis both in whole cells and in cell-free extracts was little affected by the compound. Therefore in this case insensitivity to blasticidin is not due to impermeability of the cell membrane but probably to an inherent insensitivity of the processes leading to protein synthesis.

No experiments on cross resistance between cycloheximide and blasticidin are known so far; also it is unknown whether both compounds have the same mode of action.

Kasugamycin This antibiotic is an aminoglycoside which selectively inhibits *P. oryzae* and some bacterial species. It is active only at low pH and inhibits protein synthesis *in vivo* as well as in cell-free systems, nucleic acid synthesis being undisturbed (Tanaka *et al.*, 1966; Tanaka, 1975). Okuyama *et al.* (1971) describe the compound as a selective inhibitor of initiation in bacterial protein synthesis.

Interference with nucleic acid synthesis and nuclear and cell division

Benzimidazole and thiophanate fungicides

This group of systemic fungicides comprising benomyl, carbendazim, thiabendazole, fuberidazole and thiophanate-methyl arouses much interest. Table 7.1 illustrates their high activity *in vitro*. The compounds lack antibacterial activity but are active towards a great many fungi including the powdery mildews.

Benzimidazoles Investigations by Bollen and Fuchs (1970) of the antifungal spectrum of benomyl reveal a striking selectivity in action within the group of the Ascomycetes: whereas many species studied are highly sensitive, those belonging to the Porosporae or the Annellosporae proved to be quite insensitive. The Oömycetes and other Phycomycetes were also insensitive.

Edgington *et al.* (1971) obtained similar results and reported, moreover, a great similarity in the antifungal spectrum of benomyl, thiabendazole and fuberidazole. No experimental data were given, however, for the latter two compounds.

Clemons and Sisler (1969) showed that in water benomyl decomposes very rapidly to methyl benzimidazol-2-ylcarbamate (carbendazim). Since moreover the antifungal effect of benomyl and carbendazim is similar, benomyl is assumed to act only after conversion into this latter compound, which is also itself used in practice.

Similarity in structure and antifungal spectrum of carbendazim and thiabendazole has led many investigators to suggest a common biochemical mode of action, the benzimidazole moiety being the active part of the molecule. The same may be expected for fuberidazole, but data on its antifungal spectrum are very limited so far.

The finding of Bollen and Scholten (1971) that a benomyl-resistant strain of *Botrytis cinerea* isolated from benomyl-treated cyclamen was also resistant to thiabendazole and fuberidazole, supports the idea of the common mode of action. Similarly, benomyl-resistant isolates of *Penicillium brevicompactum* and *P. corymbiferum* were observed by Bollen (1971) to be resistant to thiabendazole and fuberidazole. On the other hand Bartels-Schooley and MacNeill (1971) reported on a fuberidazole-resistant mutant of *Fusarium oxysporum* obtained by ultra-violet irradiation, which was normally sensitive to benomyl and thiabendazole. No quantitative data are given on the degree of resistance. Van Tuyl *et al.* (1974) made the interesting observation that of 203 thiabendazole-resistant strains of *Aspergillus nidulans* obtained by u.v. irradiation eight proved exceptional in not being also resistant to benomyl; they even were more sensitive to it than the wild strain. This phenomenon could be clarified by the work of Davidse (1975a, b) described below.

Thiophanates In their performance against plant diseases thiophanate-methyl and thiophanate show a striking similarity to benomyl (Aelbers, 1971; Matta and Gentile, 1971); moreover, their antifungal spectrum closely parallels that of benomyl (Bollen, 1972). We suggested that thiophanate-methyl might be transformed into carbendazim, the toxic principle of benomyl (see p. 139) and indeed in tap water, buffer or sterile nutrient medium thiophanate-methyl slowly formed carbendazim (Selling *et al.*, 1970; Vonk and Kaars Sijpesteijn, 1971) (as illustrated by Fig. 7.1). Antifungal activity as well as the amount of carbendazim formed increased with the age of a solution of thiophanate-methyl; moreover, fungitoxicity as well as the rate of carbendazim formation (see Fig. 7.2) strongly decreased with pH. From this correlation we concluded that the antifungal effect of thiophanate-methyl depends entirely upon the formation of carbendazim. Similarly, the antifungal effect of thiophanate could be assigned to formation of ethyl benzimidazol-2-yl carbamate (EBC) which is slightly less fungitoxic than carbendazim.

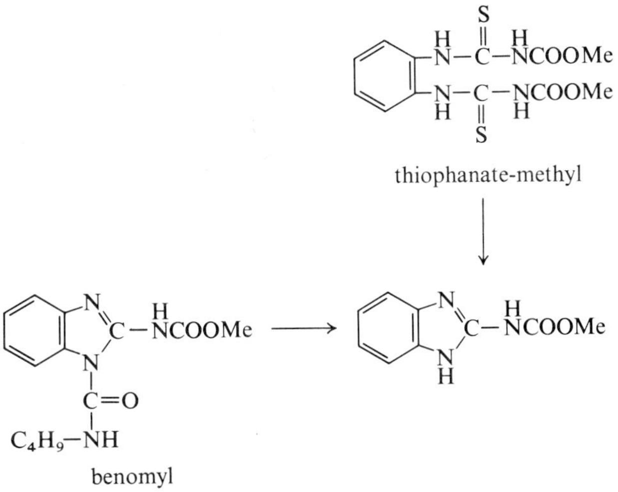

Because fungitoxicity of both benomyl and thiophanate-methyl apparently depends on formation of the same compound, carbendazim, it is obvious that benomyl and thiophanate-methyl have in fact the same mode of action. The observation that the transformation into carbendazim of benomyl proceeds far more rapidly than that of thiophanate-methyl, even at pH 7, explains the stronger fungitoxicity of the former compound. In view of this common fungitoxic principle benomyl-resistant fungal strains should of course also be resistant to thiophanate-methyl. This was in fact observed by Bollen and Scholten (1971) for

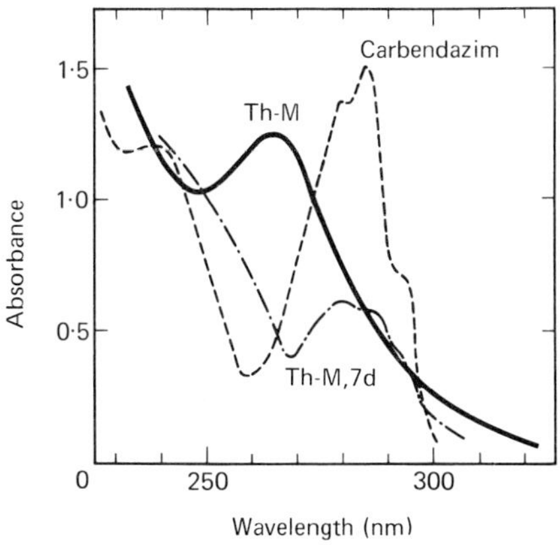

Fig. 7.1 Ultra-violet-spectra of thiophanate-methyl (Th-M), carbendazim and of thiophanate-methyl shaken for 7 days in sterile glucose-mineral salt solution (Th-M, 7 d.).

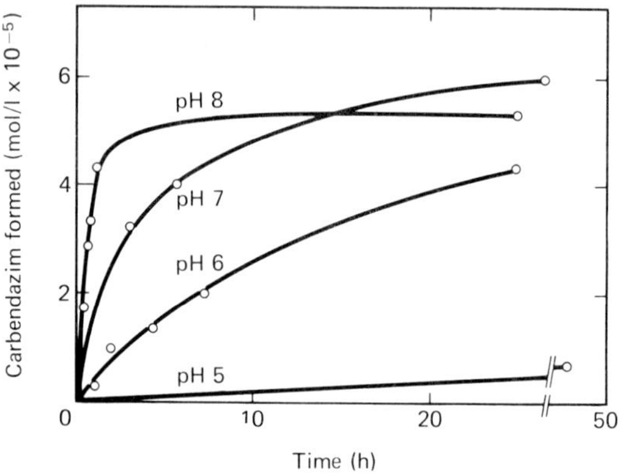

Fig. 7.2 Influence of pH on the formation of carbendazim from thiophanate-methyl in phosphate buffers at 70°C.

Botrytis cinerea and by Bollen (1971) for *Penicillium brevicompactum* and *P. corymbiferum*.

In contrast to the above, Noguchi *et al.* (1971) reported that the thiophanates, though active in combating plant disease, are inactive *in vitro* against fungi. No mention is made, however, of the concentrations examined nor of the pH of the medium used. It seems remarkable that, although the authors observed transformation of the thiophanates into carbendazim and EBC in phosphate buffer pH 7·0 as well as in plants (Noguchi and Ohkuma, 1971), and incubation in rice sheath sap was found to activate the former two compounds (Noguchi *et al.*, 1971), neither carbendazim nor EBC was recognised by the authors as the actual fungitoxic principle.

From the above observations and those of Matta and Gentile (1971) that in contact with plant tissue or sap the thiophanates become more fungitoxic than they are in water, one must conclude that the rate of carbendazim or EBC formation increases under these conditions. Autoclaved sap did not influence fungitoxicity. These results point to a metabolic conversion of thiophanates which is superimposed on the conversion taking place in water. Also fungi were found to increase the rate of carbendazim formation from thiophanate-methyl (Vonk and Kaars Sijpesteijn, 1971).

Further details on metabolic conversion are given in Chapter 8.

Mode of action of benzimidazole and thiophanate fungicides

As shown above, benomyl and thiophanate-methyl have the same biochemical mode of action, both being converted into carbendazim. Although this is true in general, it was later discovered that the action of benomyl on a very limited number of fungi differed from that of carbendazim.

For thiabendazole and fuberidazole the mode of action is generally assumed to be quite similar to that of carbendazim in view of the similarity in structure and antifungal spectrum as well as the observed cross-resistance with carbendazim.

Benomyl The activity of benomyl was first believed to be entirely due to formation of carbendazim, the mode of action of which is described below. Later it became clear that by this conversion a second antifungal breakdown product is also formed, the importance of which must be discussed.

In comparative studies with benomyl and carbendazim Hammerschlag and Sisler (1972, 1973) noted that benomyl inhibits growth of *S. pastorianus*, *S. cerevisiae* and *U. maydis* at considerably lower concentrations than does carbendazim. Moreover, in these organisms benomyl, in contrast to carbendazim, at its MIC value strongly inhibits respiration; also the morphology of the inhibited cells deviated markedly from that of carbendazim-treated cells.

The problem of this different behaviour of benomyl and carbendazim was solved by demonstrating that in solution benomyl not only gives rise to carbendazim but also to a volatile, selectively fungitoxic product which was recognised as butyl isocyanate $C_4H_9{-}N{=}C{=}O$ (Hammerschlag and Sisler, 1973). This compound is a strong inhibitor of oxidation of glucose and acetate for the above organisms. Since it acts at a lower concentration than carbendazim, its effect is predominant in this instance.

Thiols were reported to reverse the effect of benomyl on *S. cerevisiae* (Mailman *et al.*, 1971); this now seems explicable in view of the chemical reactivity of thiols with butyl isocyanate.

Carbendazim The first thorough study of the action of carbendazim was made by Clemons and Sisler (1971). Carbendazim (1 ppm) allowed spore germination of *Neurospora crassa*, but germ tube development was prevented and, while the dry weight increased normally for about 6 h before slowing down, enlarged and distorted cells were formed. Whereas carbendazim evoked these lethal effects in a medium supporting growth, the cells remained viable when incubated in phosphate buffer. Also oxidation of glucose by conidia in buffer solution was not inhibited by carbendazim (1 ppm). Uptake of ^{14}C-labelled carbendazim from a 2 ppm solution was complete within 0·5 h but the amount taken up or bound to the cells was very low.

With *U. maydis* at 8 ppm results appeared quite comparable to those described so far for *N. crassa*. The number of cells did not increase on incubation with this concentration of carbendazim. All these observations pointed to interference with growth phenomena.

Incorporation studies with ^{14}C-phenylalanine and 2-^{14}C-uridine by conidia of *N. crassa* and sporidia of *U. maydis* revealed a very strong effect on DNA synthesis which in *N. crassa* showed a delay of 3–4 h.

Protein and RNA synthesis were not yet inhibited after 8 h incubation of *N. crassa*, but in *U. maydis* their inhibition rapidly followed that of DNA synthesis. More or less similar effects were observed for *S. pastorianus* (Hammerschlag and Sisler, 1972).

These phenomena, which suggested interference of carbendazim with DNA synthesis or some closely related process such as nuclear or cell division, were further clarified by the investigations of Hammerschlag and Sisler (1973) with *U. maydis* and *S. cerevisiae* and of Davidse (1973) with the filamentous fungus *Aspergillus nidulans*.

With the aid of synchronous cultures the former could prove that the primary process inhibited in both organisms by carbendazim is not DNA synthesis but mitosis. Growth of carbendazim-treated cells terminated with the formation of doublets (joined mother and daughter cells) which contained only one single nucleus; neither nuclear division nor cell separation took place. In untreated cells one nucleus is present in each of the doublet components just before cell separation occurs. Interference with mitosis also explains why in *U. maydis*, where DNA synthesis occurs immediately after cell division, almost no DNA is synthesised after addition of carbendazim, whereas in *S. cerevisiae*, in which in contrast the DNA always doubles just before mitosis, the total DNA content still doubled before inhibition of further DNA synthesis took place. *Neurospora crassa* conidia apparently behave like *S. cerevisiae* in this respect. Hammerschlag and Sisler (1973) suggested interference of carbendazim with the mitotic spindle in the same way as known for colchicine.

The investigations of Davidse (1973) fully confirmed the former observations. Whereas carbendazim (4 ppm) allowed the dry weight of young mycelium of *A. nidulans* to increase more or less uninhibited for the first 4 h, the increase in number of nuclei per germ tube was suppressed immediately and completely by addition of the toxicant. The DNA content increased slowly until the double amount was present, much in the same way as described for *N. crassa* and *S. cerevisiae*. Moreover, abnormal mitotic phases were observed. The author also suggests for carbendazim an antimitotic effect as colchicine has, namely by formation of a complex of carbendazim with a subunit of microtubuli, preventing the normal assembly of microtubule subunits into spindle fibres; this would lead to inhibition of spindle formation.

In further investigations Davidse (1975a, b) compared the carbendazim-binding capacity of macromolecules of a normally sensitive *A. nidulans* (wild strain) with that of a highly resistant mutant strain R which had been obtained by u.v. irradiation. As a third strain he used the mutant 186 which, like strain R was resistant to thiabendazole but unexpectedly was even more sensitive to carbendazim than the wild strain. Genetic analysis of the two mutants had shown that mutation was in the same gene in both (Van Tuyl *et al.*, 1974). Cell-free extracts of these three strains were incubated for 2 h with 2-^{14}C-carbendazim and subsequently subjected to gel filtration on Sephadex G-100 columns. The eluate of the two sensitive strains showed radioactivity in a macromolecular fraction with a molecular weight of approximately 110 000 daltons; this label could be ascribed to bound carbendazim. In contrast, mycelial extracts of strain R did not display carbendazim-binding activity. Further characterisation of this binding substance strongly indicated its identity with fungal tubulin. The different effect of carbendazim on growth of the mutants may be based on different changes in affinity of the receptor site of tubulin to carbendazim, caused by mutation of its coding gene. With regard to thiabendazole, the affinity of the receptor site should be about equal in both mutants.

For sensitive and resistant strains of *Penicillium* species, similar differences in carbendazim-binding activity were observed. Moreover, the finding that mycelial extracts of *Alternaria brassicicola* and a *Pythium* species were devoid of carbendazim-binding activity seems to provide a reasonable explanation for the observations that carbendazim is unable to inhibit growth of organisms belonging to these genera.

Cytological details of the antimitotic effect of carbendazim on *Botrytis cinerea* were recently described by Richmond and Phillips (1975).

Strong inhibition of respiration of *Fusarium oxysporum* by benomyl was reported by Decallonne *et al.* (1975). Since this applies only to concentrations far exceeding those preventing growth, the observations do not bear a relation to the primary mode of action which will be that of carbendazim.

The work of Hastie (1970) revealed that benomyl can cause genetic instability, since low concentrations evoke haploidisation of diploids of *Aspergillus nidulans*. The same ability to induce genetic segregation was reported by Kappas *et al.* (1974) for benomyl, carbendazim, thiophanate-methyl and thiabendazole; it may be due to the fungicides causing non-disjunction.

Indications for a mutagenic effect of carbendazim as well as fuberidazole were obtained by Seiler (1972) for *Salmonella typhimurium*. Equally Dassenoy and Meyer (1973) presume a mutagenic effect of benomyl since in a limited experiment mutants with nutritional deficiencies were obtained when spores of *Fusarium oxysporum* were incubated under growth conditions with sublethal concentrations. More critical work of larger scope is obviously required for a proper assessment of these mutagenic effects.

Thiabendazole Although thiabendazole is generally believed to act like carbendazim as an antimitotic agent, it is surprising that no direct evidence such as that presented for carbendazim has been reported. However, its antifungal spectrum, its cross resistance with carbendazim, its causation of genetic instability (Kappas *et al.*, 1974) as well as the following results all support this hypothesis.

Gottlieb and Kumar (1970) reported a greater effect on elongation of germ tubes and on mycelial growth than on spore germination of

Penicillium atrovenetum and other fungi. Germ tubes showed malformation (Koch, 1971).

There was no inhibition of glucose oxidation by washed suspensions of *Aspergillus niger* (Kaars Sijpesteijn, 1970) or *Cladosporium cucumerinum* treated for a 24-h period with 100 ppm of thiabendazole, which is about 100 times the growth-inhibiting concentration. Similar observations were made by Lyda and Burnett (1970) for *Phymatotrichum omnivorum*. Yet Allen and Gottlieb (1970) concluded from their experiments with *Penicillium atrovenetum* that inhibition of respiration probably is the primary effect of thiabendazole; their data so far have not received confirmation.

Fuberidazole As regards the mode of action of fuberidazole, no inhibition of respiration of washed suspensions of *Aspergillus niger* in glucose phosphate buffer containing 100 ppm of this compound over a 16-h incubation period was found (Kaars Sijpesteijn, 1972). No further data are available. In view of its structure and observations with resistant strains it appears for the time being most likely that fuberidazole has the same mode of action as carbendazim and thiabendazole.

Other compounds

Chloroneb

Hock and Sisler (1969a) found that this seed disinfectant inhibits growth of *R. solani* and *Sclerotinia rolfsii* at a level of 5–8 ppm, whereas it is less active on *N. crassa* and *S. pastorianus*. A further study revealed that this selectivity is not based on selectivity in uptake or on metabolic conversion of the compound (Hock and Sisler, 1969b). Both *S. pastorianus* and *Scl. rolfsii* proved unable to metabolise chloroneb, whereas *R. solani* and *N. crassa* were able to do so. Growth inhibition by chloroneb of heavy mycelial suspensions of *R. solani* could even be overcome by quick metabolic conversion of the compound, 2,5-dichloro-4-methoxyphenol being formed.

A study of the mode of action of chloroneb by the same authors (1969a) showed that at a level of 20 ppm respiration of *R. solani* is not inhibited. Incorporation of thymidine into the DNA fraction of the hyphae was 65–90 per cent inhibited by 8 ppm of the compound. Incorporation of phenylalanine into protein and of uridine into RNA were far less inhibited.

A study by Tillman and Sisler (1973) with *Ustilago maydis* revealed further details. Chloroneb at 8 ppm kills the sporidia in a nutrient medium but not in phosphate buffer. Whereas under growth conditions the number of sporidia remains unchanged, dry weight as well as protein, RNA and DNA content of the cultures still increased for a few hours. In contrast to carbendazim-treated cells, in synchronised chloroneb-treated sporidia the DNA content still doubles and nuclear division proceeds normally, but this leads to binuclear cells as cell division does not take place. Thus the primary effect is not on DNA synthesis or on mitosis; its true character is, however, still unknown.

Interference with enzymatic processes in the cell wall has been suggested by the authors. The fact that a mutant with slight resistance to chloroneb is also more resistant to aromatic hydrocarbon fungicides like hexachlorobenzene, 2,6-dichloro-4-nitroaniline (dicloran), pentachloronitrobenzene (quintozene) and others is intriguing. For the latter compounds an effect on the polysaccharides in the cell wall was observed by Macris and Georgopoulos (1973) but it may not be the primary effect. Recently an effect of chloroneb, quintozene and dicloran on the genetic stability of a diploid strain of *A. nidulans* was reported by Georgopoulos *et al.* (1976); this would be suggestive of a primary action of these compounds on nuclear processes.

6-Azauracil In studying the action of 6-azauracil, Dekker and Oort (1964) found that it has systemic activity against powdery mildew of cucumber, as well as against *C. cucumerinum* on cucumber and several other diseases but it never reached the stage of practical application.

Its structure raises the question whether its mode of action is related to that of dimethirimol and ethirimol. A comparison of the properties of these compounds with those of azauracil demonstrates, however, that this cannot be so. Extensive investigations (Dekker and Oort, 1964) have shown that azauracil undergoes metabolic conversion into azauridinemonophosphate. This latter compound inhibits the enzyme orotidylic decarboxylase, which catalyses the decarboxylation of orotidinemonophosphate to form the nucleotide uridinemonophosphate.

Its action is reversed by uracil, which can act as an alternative precursor for uridine monophosphate. 6-Azauracil does not inhibit spore germination, but strongly suppresses the development of haustoria in the host plant. These facts as well as the fact that dimethirimol-resistant cucumber mildew is not resistant to azauracil (personal communication of Dr J. Dekker) clearly demonstrate the different mode of action of azauracil as compared with dimethirimol and ethirimol.

Interference with synthesis and function of cell walls and membranes

Polyoxin D The antibiotic polyoxin D is a pyrimidine derivative which inhibits growth of *Cochliobolus miyabeanus*, *Alternaria kikuchiana*, *Pyricularia oryzae*, *Neurospora crassa* and other fungi. Sasaki *et al.* (1968) found that it inhibits neither respiration nor synthesis of nucleic acids or proteins of the former fungus. Incorporation of ^{14}C-glucosamine into cell walls was, however, strongly inhibited.

The biochemical mode of action was studied more closely by Endo *et al.* (1970), using *N. crassa*. A concentration of polyoxin D which severely affects growth (100 ppm) inhibited the incorporation of ^{14}C-glucosamine into chitin while UDP-*N*-acetylglucosamine accumulated. The rate of synthesis of chitin from ^{14}C-UDP-*N*-acetylglucosamine by cell-free preparations of the fungus was also strongly decreased. In such preparations the compound was even far more effective than *in vivo* possibly owing to the presence of a penetration barrier in intact mycelium (Endo and Misato, 1969). Thus the antifungal action of polyoxin D appears due to inhibition of the enzyme chitin synthetase. The antibiotic is a competitive inhibitor for UDP-*N*-acetylglucosamine in this reaction; obviously the structural resemblance between these two compounds leads to an antimetabolite effect. Similar results were obtained with *P. oryzae* (Ohta *et al.*, 1970; Hori *et al.*, 1971). Later results from these workers afforded still more detailed information on the structural resemblance between the polyoxins and the natural substrate with which it competes (Hori *et al.*, 1974). The antibiotic interferes in a similar way with chitin synthesis in *Mucor rouxii* (Bartnicki-Garcia and Lippman, 1972), and in *S. cerevisiae* (Bowers *et al.*, 1974).

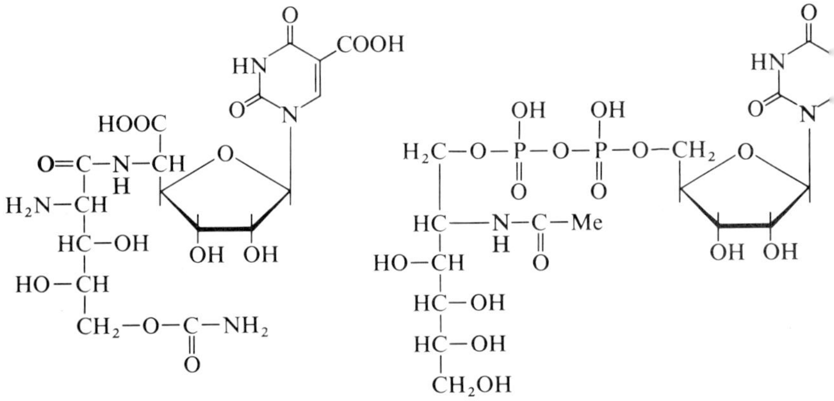

polyoxin D　　　　UDP-N-acetylglucosamine

No information is available whether the Oömycetes which lack chitin in their cell walls are insensitive to polyoxin D. Endo *et al.* (1970) report that even several fungi species with a very high content of chitin are insensitive to polyoxin D, but possibly the compound does not reach the site of action in these organisms.

The antifungal mode of action of polyoxin D parallels in certain respects the antibacterial mode of action of penicillin, for this compound inhibits the final step in the biosynthesis of murein, a constituent of bacterial cell walls. A structural resemblance of penicillin with an intermediate in murein synthesis is supposed to account for this effect.

When *Cochliobolus miyabeanus* was grown in the presence of polyoxin D it formed giant, osmotically sensitive, protoplast-like structures (Endo *et al.*, 1970), a phenomenon somewhat similar to that obtained with bacteria grown in the presence of penicillin.

Triarimol This pyrimidine derivative is active against a variety of fungi including powdery mildews; *Pythium* spp. proved rather sensitive *in vitro* (Desai *et al.*, 1975). *In vivo* the action is curative (Brown and Hall, 1971). Even at 500 ppm it does not inhibit oxygen uptake of *Aspergillus niger* in glucose phosphate buffer (Kaars Sijpesteijn, 1972). Houseworth *et al.* (1971) found no inhibition of CO_2 production from glucose or acetate for this organism and germ tube development is affected rather than spore germination.

Thorough investigations by Ragsdale and Sisler (1972, 1973) using *Ustilago maydis* revealed that 2 ppm terminated sporidial multiplication after the first division which was complete after 3·5 h, yet growth continued and abnormal sporidia were formed. Neither glucose nor acetate oxidation was affected; DNA, RNA and protein synthesis were only slightly decreased during a 9·5 h period. However, the quantity and nature of lipid components were markedly altered during a 5·5-h growth period. Thus incorporation of ^{14}C-acetate into ergosterol and another sterol was inhibited more than 90 per cent already during the first hour whereas others appeared or increased. Furthermore, the quantity of free fatty acids was higher and their ratio was altered after 5 h. The primary site of action of triarimol is presumed to be in the sterol biosynthetic pathway at a point preceding ergosterol. Almost certainly an effect on synthesis and function of membranes should result from this. Since fungitoxicity is not reversed by addition of ergosterol, deficiency of this compound may not be the only cause of growth inhibition.

In a recent study Ragsdale (1975) specified that triarimol inhibits three reactions in ergosterol biosynthesis by *U. maydis*, all involving the D-ring and the side chain of the molecule; inhibition of demethylation at C-14 appears to be the principal site.

Sherald *et al.* (1973) made the remarkable observation that triarimol-resistant mutants of *C. cucumerinum* also exhibited resistance to the structurally unrelated fungicide triforine. These two agents have a similar antifungal spectrum and show the same abnormally swollen germ tubes. Moreover, of a variety of compounds antagonising the antifungal effect of triarimol on the above fungus, most did so for triforine as well: β-carotene, vitamin A, farnesol, progesterone, testosterone. The mechanism of this antagonism is not yet known. It is most remarkable that in the case of *U. maydis* none of these compounds appeared to be antagonistic to triarimol, whereas Tween 20 and Tween 40 reversed the action against both fungi, and oleic acid only that on *U. maydis*.

The conclusion that triarimol interferes with sterol biosynthesis fails to explain the observation of Desai *et al.* (1975) that *Pythium* species which do not possess sterols are rather sensitive. The same applies to our observation (unpublished) that gram-positive bacteria are strongly inhibited. Another mechanism may apply here.

Recently a close derivative, fenarimol, has been introduced as an experimental fungicide.

Triforine Triforine shows activity against powdery mildews on cereals, apples and cucurbits, as well as against rusts on cereals, and apple scab (Schicke and Veen, 1969). According to Fuchs and Drandarevski (1973) there is no correlation between *in vitro* and *in vivo* activity; the compound acts *in vitro* on a variety of Ascomycetes, Basidiomycetes and Fungi Imperfecti but not on Phycomycetes. Haustoria formation by barley mildew was impaired *in vivo* (Schlüter and Weltzien, 1971). No inhibition of respiration of *C. cucumerinum* in glucose phosphate buffer by 500 ppm was observed (Kaars Sijpesteijn, 1972).

As mentioned earlier, Sherald *et al.* (1973) observed that triarimol-resistant mutants of *C. cucumerinum* were also resistant to triforine. Other similarities in the action of both fungicides were also revealed which suggested that both compounds, although structurally unrelated, might have a common mode of action. Indeed triforine inhibited ergosterol biosynthesis in *N. crassa* at a low growth-inhibiting concentration (50 ppm). Ergosterol antagonised the fungitoxicity to *C. cucumerinum* of triforine but not of triarimol. Several other compounds did, however, reverse the activity of both fungicides.

A recent study with *Aspergillus fumigatus* by Sherald and Sisler (1975) showed no influence of triforine on respiration and revealed an inhibition of DNA, RNA and protein synthesis by 10 ppm starting after 2 h simultaneously with an inhibition in the increase of dry weight. Ergosterol synthesis was, however, already nearly completely inhibited in the first hour after treatment. The gross effects of triforine and triarimol on lipid synthesis by *A. fumigatus* were similar; also triarimol-resistant mutants of this fungus proved resistant to triforine. There is strong evidence that both fungicides inhibit sterol C 14-demethylation, thus interfering with membrane development. Triarimol also inhibits a further site in ergosterol biosynthesis.

S-1358 This new fungicide is active against powdery mildews; of a series of fungi studied *in vitro*, the Phycomycetes, Basidiomycetes and a variety of Ascomycetes and Fungi Imperfecti were, however, insensitive (Kato *et al.*, 1975b). While spore germination of *Sphaerotheca fuliginea* was unimpaired, abnormally swollen hyphae developed; there was no distortion as evoked by benomyl.

A fungitoxic vapour was observed which may be due to the compound itself or a conversion product.

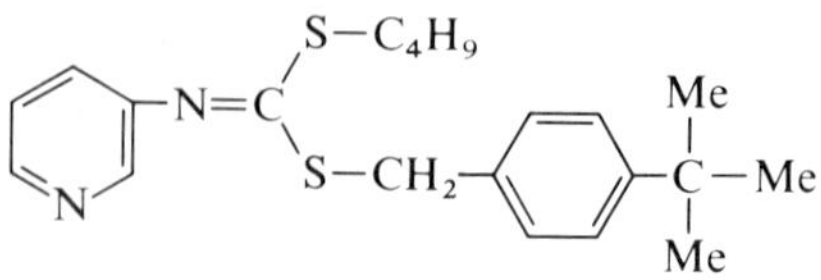

S-1358

Kato *et al.* (1974) described formation of abnormal germ tubes of *M. fructigena* after spore germination has occurred. Low growth-inhibiting concentrations do not inhibit respiration and only slightly affect incorporation of nucleic acid precursors and amino acids. Two hours after treatment of growing mycelium the increase of dry weight slowed down. This was preceded by a sharp decline of the incorporation of acetate into 4-desmethyl sterols like ergosterol. Evidence is given that S-1358 blocks C 14-demethylation reactions (Kato *et al.*, 1975a; Kato, 1977) pointing to a mode of action similar to that of triarimol and triforine. The observation that two mutants of *C. cucumerinum* resistant to triarimol and triforine were also resistant to S-1358 strengthens this hypothesis (Sherald and Sisler, 1975).

Kitazin P, edifenphos and Inezin These synthetic organophosphorus fungicides are used mainly against fungal diseases of rice. Similarity in their mode of action is assumed in view of structural relationship as well as cross-resistance of slightly resistant isolates and similarity in other effects.

Kitazin P inhibits growth of *P. oryzae*, a particularly sensitive fungus, at 50 ppm (Yoshinaga, 1969; Maeda *et al.*, 1970). Mycelial respiration, protein or nucleic acid synthesis are not affected, but the incorporation of ^{14}C-glucosamine into chitin is impaired by this concentration. It was, moreover, observed that UDP-*N*-acetylglucosamine accumulated in the cells under these conditions. Fungitoxicity of analogues ran parallel to their ability to inhibit ^{14}C-glucosamine incorporation.

These results might suggest an inhibition of chitin synthetase by Kitazin. However, Maeda *et al.* (1970) found that not only chitin formation is impaired but also the incorporation of ^{14}C-glucose into cell wall glucans markedly decreased in treated mycelium. Therefore they suggest that the primary action of Kitazin is not on chitin synthetase itself. They rather believe that it limits the outward permeation through the cytoplasmic membrane of the substrates for chitin synthetase and glucan synthetase. Since the formation of glycolipids in the membranes was also markedly inhibited in treated mycelium, the authors hypothesise that this may be the primary cause of the impairment of chitin synthesis and growth inhibition, glycolipids being necessary for membrane transport of chitin precursors. Since Kitazin can inhibit acetyl cholinesterase, interference with esterases connected with glycolipid metabolism may well be the actual mode of action of this organophosphorus fungicide.

Edifenphos according to de Waard (1972) also caused inhibition of chitin synthesis and this was confirmed in a further study with *P. oryzae* for edifenphos and Kitazin, the ethyl analogue of Kitazin P, by the same author (1974). At their growth-inhibiting concentration both compounds also caused leakage of ^{32}P-phosphate from mycelium. It is supposed that the fungicides act primarily on cell membrane permeability, the other effects, including interference with chitin synthesis, being consequences of this. The observation that both compounds are effective against *Pythium debaryanum*, an Oömycete lacking chitinous walls, would agree with this hypothesis.

Uesugi *et al.* (1974) made the remarkable observation that isolates of *P. oryzae* which are somewhat less sensitive to the above fungicides than the wild strain proved in contrast more sensitive to *N*-methyl-*N*-phenyl-phosphoramidates.

Imazalil This experimental fungicide shows very high activity *in vitro* (Table 7.1) as well as in the field, in particular against *Penicillium* diseases of citrus fruit. Against such diseases Laville (1974) found curative activity; spore germination as well as mycelial growth of *Penicillium* species were inhibited and the compound was active against benomyl-resistant strains.

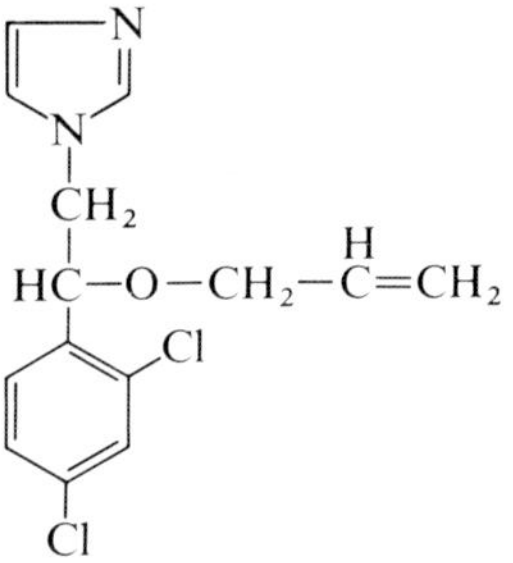

imazalil

So far there is no direct information on the mode of action of imazalil but structural similarity with the antimycotic compound miconazole may suggest a similarity in mode of action. Miconazole is highly active on a wide variety of fungi as well as on Gram-positive bacteria (van Cutsem and Thienpoint, 1972). In *Candida albicans* its primary effect appears to be on the function of the plasmalemma resulting amongst others in a selective decrease in uptake of nutrients (Van den Bossche, 1974). In many respects it resembles the action of another fungicidal imidazolyl derivative, clotrimazole (Iwata *et al.*, 1973; Van den Bossche *et al.*, 1975).

Preliminary studies with imazalil on *Penicillium italicum* by Siegel *et al.* (1977) support this hypothesis.

Mode of action still uncertain

Of the following compounds the mode of action is still unknown or the scarce observations available are insufficient to place them into any of the foregoing groups. They appear to be mainly agents with activity against powdery mildews or against Phycomycetes. The former can only be studied properly if they also act on fungi growing *in vitro*. With Phycomycetes the study is complicated by the fact that methods to investigate these organisms are little developed yet; their physiology, moreover, is still poorly known.

Dimethirimol and ethirimol The pyrimidines dimethirimol (Elias *et al.*, 1968) and ethirimol (Bebbington *et al.*, 1969) have been recommended respectively for the control of cucumber powdery mildew caused by *Sphaerotheca fuliginea* and of cereal powdery mildews caused by *Erysiphe graminis*. They have low mammalian toxicity and act selectively on mildews only.

According to Bent (1970) they inhibit spore germination of mildew of wheat and barley *in vitro*. Complete suppression of spore germination was achieved by 50 ppm of either compound but 5 ppm was quite effective (cf. also Table 7.3). In contrast, a concentration of 50 ppm did not affect germination of conidia of *Venturia inaequalis* or of uredospores of *Puccinia recondita* and the germination of conidia of *Botrytis fabae* and of zoosporangia of *Phytophthora infestans* was reduced by only 7 and 26 per cent, respectively. Mildew of vine (*Uncinula necator*) was less sensitive to the compounds than the other powdery mildews tested. Whereas cucumber mildew proved more sensitive to dimethirimol than to ethirimol, the reverse is true for wheat and barley powdery mildew.

In a study of the effect of various metabolites on fungitoxicity of the two pyrimidine fungicides Bent (1970) observed that inhibition by dimethirimol (10 ppm) of *in vitro* spore germination of the wheat mildew fungus was markedly antagonised by the incorporation of riboflavin or folic acid (100 ppm) into the solution of the fungicide (Table 7.2). Slighter reversal of activity was found by addition of guanine, thiamine,

Table 7.2 Effects of metabolites on the systemic action and on the fungitoxicity of dimethirimol in tests on wheat powdery mildew

Chemical treatment	Disease on leaf segments† (% of disease on untreated segments)		Germination of spores *in vitro* (% of germination of untreated spores)	
Dimethirimol alone	0	0	3	4
+adenine	9	—	12‡	—
+folic acid	64	31	38	34
+guanine	0‡	—	12	—
+riboflavin	88	97	47	8‡
+thiamine	13	—	24	—
+thymine	0‡	—	28	—
+uracil	17	—	24	—
None	100	100	100	100
Dimethirimol concentration (ppm)	13	10	10	10
Metabolic concentration (ppm)	1 300	100	100	100
Disease on untreated segments (% area)	100	63	—	—
Germination of untreated spores (%)	—	—	22	31

† Floated on solutions of the chemicals
‡ No significant reversal ($P = 0{\cdot}05$); all other treatments gave a significant decrease in dimethirimol activity

thymine and uracil. Somewhat similar effects were observed *in vivo*. With ethirimol the same observations were made. The effect of riboflavin could be interpreted as a photochemical inactivation of the fungicide. It was not stated whether folic acid also reverses when used in a lower, more physiological range of concentrations.

Present evidence points to an interference with tetrahydrofolic-acid-directed C-1 transfer reactions essential for a wide range of cellular components, as for instance purines (cf. Calderbank, 1971; Slade *et al.*, 1971). The effect of the two pyrimidine fungicides on enzymes involved in tetrahydrofolate metabolism was examined by Sampson (1969). No inhibition was found, however, of dihydrofolic acid reductase or of N^5N^{10}-methylene tetrahydrofolate reductase in enzyme preparations of cucumber powdery mildew. Limited evidence suggests an inhibition of some pyridoxal-requiring enzymes. Clearly more work will have to be done to reveal the true nature of the action of these fungicides.

Bupirimate, the recently developed sulphamate ester of ethirimol, doubtless acts on the same systems as the above pyrimidines.

Triamiphos Triamiphos or Wepsyn is active against powdery mildew of barley and of apple (Koopmans, 1960). Microscopic observations by Magendans and Dekker (1966) revealed that on treated barley plants the main barrier to the development of powdery mildew lies at the stage of haustorial formation. The compound only slightly reduces germination of the conidia placed on glass slides.

De Waard (1971a) developed a technique which facilitates the *in vitro* study of germination of powdery mildew conidia. It is based on his observation that a very high percentage of the conidia of cucumber and barley mildew germinates when dusted onto a cellophane membrane

Table 7.3 Percentage of germination and estimated percentage of collapse of conidia of *Sphaerotheca fuliginea* on cellophane membranes, laid on an agar medium, mixed with systemic and non-systemic fungicides. Percentage of germination based on control as 100%

	10^{-6} M		10^{-5} M		10^{-4} M	
Concentration	germ.	coll.	germ.	coll.	germ.	coll.
Karathane	48	0	0	>90	0	>90
Acrex	15	0	1	0	1	0
Morestan	1	>90	1	>90	1	>90
WP 155	94	0	88	0	39	20
WP 356	92	0	69	0	21	5
PP 149[1]	2	0	2	0	0	0
PP 675[2]	11	0	0	0	0	0
Benomyl	62	0	60	0	57	30
Hoe 2873[3]	57	32	39	37	17	47

Note. The concentrations of the pure compounds in the formulations used were 22·5%, 98%, 25%, 25%, 25%, 80%, 10%, 50% and 100% respectively

[1] ethirimol [2] dimethirimol [3] pyrazophos

which had been placed on agar in a petri dish. The medium used was Czapek-Dox agar to which 3 per cent sucrose and 0·1 per cent yeast extract had been added. The effect of fungicides could be studied by incorporating them into the agar (de Waard, 1971b). As his results show (Table 7.3), neither Wepsyn (WP 155) nor the related compound WP 356 inhibits germination of conidia at low concentration. The same was observed by Koopmans (1960) for another derivative of triamiphos and by Dekker and Van de Hoek-Scheuer (1964) for procaine, a systemic antimildew compound.

The fact that these compounds are only active on mildew in a stage where they rely entirely on the plant for further development has led to the suggestion that they act by changing host metabolism, thus rendering the plant more resistant to mildew, or by disturbing the intimate relation between host and parasite which is imperative for growth of mildew.

Tridemorph This agent performs well against powdery mildew of barley (Pommer and Kradel, 1967). Pommer. Otto and Kradel (1969) described its antimildew action as eradicant. Schlüter and Weltzien (1971) found inhibition of haustorial formation of barley mildew on treated barley leaves.

Tridemorph has also activity against a wide variety of other fungi (Bergmann *et al.*, 1975), and we observed, moreover, growth inhibition of gram-positive bacteria by 10 ppm of the compound. According to Fisher (1974) mycelial growth is affected at far lower concentrations than spore germination.

The long alkyl chain of tridemorph might suggest interference with the function of fungal cell membranes, but convincing experimental evidence is lacking so far. Low growth-inhibiting concentrations did not inhibit glucose oxidation by *C. cucumerinum* in phosphate buffer over a 16-h period; a concentration of 100 ppm caused slight inhibition (Kaars Sijpesteijn, 1972).

Bergmann *et al.* (1975) observed, however, for the yeast *Torulopsis candida* 30–60 per cent inhibition of respiration at the ED 50 value for growth inhibition (*c.* 25 ppm). Permeability was not influenced at this concentration, and incorporation experiments with ^{14}C-valine and ^{14}C-adenine did not reveal a primary effect on protein or nucleotide synthesis. Antimycin A has a strong synergistic influence on the inhibition of respiration by tridemorph. From these results the authors conclude that there is a primary effect of this latter compound in the respiration chain.

Microelectrophoretic studies of Fisher (1974) did not reveal an influence of tridemorph on membranes of protoplasts of *Botrytis fabae* and *N. crassa* but at a concentration below that causing growth inhibition the incorporation of ^{14}C-histidine by such protoplasts was considerably inhibited within 1 h; higher concentrations were required to inhibit their respiration. These results led the author to suggest that tridemorph may act by inhibition of protein synthesis. Experiments with intact cells are indispensable for a further evaluation of this hypothesis.

Buchenauer (1975) observed increased leakage of electrolytes at the ED 50 value for *Verticillium albo-atrum*. Since this effect was not displayed by the far less sensitive fungus *Fusarium oxysporum* he does not regard it as a primary mode of action. The author also made the surprising observation that the insensitive Oömycete *P. debaryanum* became sensitive to tridemorph (10 ppm) when cholesterol (25 ppm) was added to the medium; thus the fact that *Pythium* normally does not possess sterols may explain insensitivity.

The diverging observations on the effect of tridemorph on various fungi so far do not afford a convincing picture of its mode of action; possibly different mechanisms are operative.

Triadimefon This triazole derivative is an experimental fungicide effective especially against powdery mildews (Grewe and Büchel, 1973).

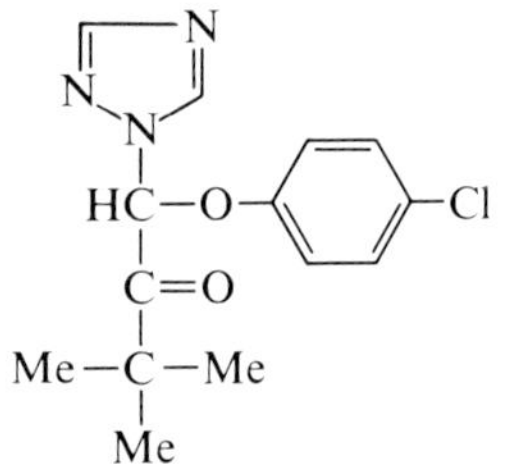

triadimefon

According to Frohberger (1975) it is also very effective against several other diseases, but *in vitro* the action of triadimefon on spore germination and mycelium growth is very weak. No explanation is available so far for this anomaly.

Büchel *et al.* (1975) presume that the agent might act by a selective effect on the cell-membrane. Experiments by Buchenauer (1976) revealed inhibition of ergosterol biosynthesis in *Ustilago avenae*, 1·5 ppm causing 80 per cent inhibition after 8 h while another sterol accumulated. This result is reminiscent of the action of triarimol.

Pyrazophos This organophosphorus compound, like triamiphos, is highly effective against powdery mildew of cereals and cucumber (Mariouw Smit, 1969). Experiments with barley mildew showed an effect on appressorial formation and penetration into the plant (Schlüter and Weltzien, 1971). By the technique described above de Waard (1971b) observed that low concentrations affect *in vitro* the germination of conidia of cucumber and barley mildew, thus suggesting a different mode of action from triamiphos.

Further investigation by de Waard (1974) revealed the high sensitivity of *P. oryzae* as compared with many other fungi. A mode of action different from that of Kitazin and edifenphos is indicated since there was no effect on chitin synthesis or on membrane permeability. Also a resistant strain showed no cross-resistance with Kitazin or edifenphos (M. A. de Waard, personal communication). The author proved that a phosphorus-free metabolite, 6-ethoxycarbonyl-2-hydroxy-5-methyl-pyrazolo (1,5-a) pyrimidine (PP), is responsible for the fungitoxic effect (cf. also de Waard, 1975). This metabolite is quickly formed in mycelium of *P. oryzae* but not by *P. debaryanum* and *S. cerevisiae*, which may explain their insensitivity. For formula see Chapter 8.

PP is highly fungitoxic at pH 4; at higher pH values slow penetration into the fungus may limit activity. Tested at concentrations somewhat above those limiting growth PP strongly inhibits respiration by mycelial suspensions of glucose in phthalate buffer pH 4·0. Oxygen uptake by mitochondrial preparations was severely impaired, and an influence on protein, DNA and RNA synthesis was also observed. At present insufficient information is available to decide which is the primary effect limiting growth.

Another metabolite of pyrazophos is its *O*-analogue, PO-pyrazophos. Its fungitoxicity is somewhat higher than that of pyrazophos itself but this is ascribed to its more rapid metabolic conversion into PP.

Studies by Georgopoulos *et al.* (1975b) confirmed for *U. maydis* that PP is the actual toxic agent, since low growth-inhibiting concentrations of this compound killed the cells within 40 h in a nutrient medium.

Ethazole This compound finds application as a seed disinfectant against *Pythium* and *Phytophthora*.

Lyr *et al.* (1975a) observed *in vitro* activity against a variety of fungi,

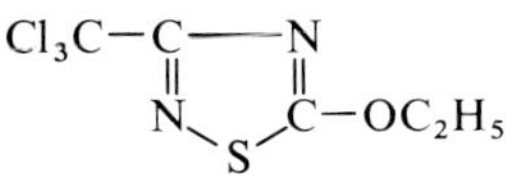

ethazole

the Phycomycetes being most sensitive. A study with *Mucor mucedo* revealed inhibition of spore germination as well as of hyphal growth. Respiration and nucleic acid synthesis were unimpaired, but alterations in ultrastructure became rapidly visible; changes in lipid synthesis were also observed. Evidently further work will be necessary to determine its primary site of action.

Prothiocarb Prothiocarb,

$$(CH_3)_2NCH_2CH_2CH_2NH\underset{\underset{O}{\|}}{C}SC_2H_5.HCl,$$

controls diseases caused by *Pythium*, *Phytophthora*, and other Phycomycetes (Bastiaansen *et al.*, 1974). It also has a curative effect.

In vitro experiments showed selective activity on the Oömycetes, a group of fungi which instead of chitosan and chitin possess cellulose and glucans as main cell wall constituents; Zygomycetes were not affected (Kaars Sijpesteijn *et al.*, 1974). Fungitoxicity is strong at pH 7 but negligible at pH 5·3. This fact as well as selectivity may be a consequence of selective penetration into the cells. Spore germination of *Pythium splendens* is not inhibited and short hyphae are formed before growth stops. Also treated mycelium still continues to grow for some time. Prothiocarb does not inhibit respiration, and under growth conditions does not kill the fungus, suggesting interference with a biosynthetic process. Cholesterol, although not a cell constituent of *Pythium*, at low concentration antagonises the action of prothiocarb to a certain extent.

At relatively high pH prothiocarb slowly releases ethyl mercaptan. This compound is not regarded as the active principle since on *Pythium* the carbamate analogue,

$$(CH_3)_2NCH_2CH_2CH_2NH\underset{\underset{O}{\|}}{C}OC_2H_5.HCl$$

is about as active as prothiocarb itself.

Recently we observed the action on *Achlya radiosa* to differ from that on *Pythium*; it could be ascribed to ethylmercaptan released (Kerkenaar and Kaars Sijpesteijn, 1977).

Pyroxychlor This experimental fungicide is very active mainly against *Phytophthora* diseases (Hoitink and Schmittbrenner, 1975).

In vitro it inhibits *Phytophthora parasitica* and *Pythium ultimum* at about 10 ppm, but the compound appears to be far more effective when applied to plants or to soil than *in vitro*. It is suggested therefore that a metabolite might be the actual active fungicide (Noveroske, 1975a, b). No further information is available so far.

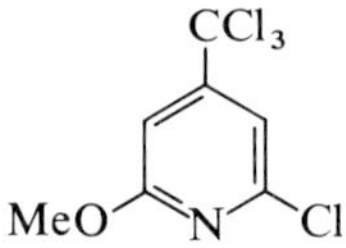

pyroxychlor

2,2-Dichloro-3,3-dimethyl cyclopropane carboxylic acid This compound effectively controls *P. oryzae* on rice plants but *in vitro* fungitoxicity is weak or even negligible depending on the composition of the medium. There is a long delay in fungitoxic effect which could not be ascribed to conversion, since the compound proved stable during the experiment (Langcake and Wickins, 1975a).

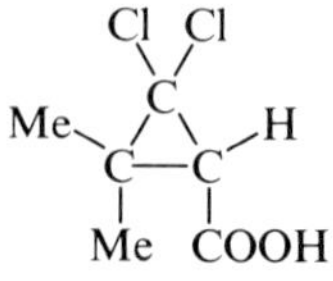

2,2-dichloro-3,3-dimethyl-
cyclopropane carboxylic acid

In vivo activity is believed not to be due to the weak fungitoxic properties of the compound; it could, however, be related to a strong enhancement of peroxidase activity which is thought to increase the resistance of the host plant to the disease (Langcake and Wickins, 1975b).

General remarks on mode of action, selectivity and resistance

In the foregoing pages a bird's-eye view has been given of present-day knowledge of the effect of systemic fungicides on their target organisms, the fungi. Obviously very much work remains to be done in this rapidly developing field.

Mode of action of systemic fungicides

As far as the biochemical modes of action are known, it is apparent now that most systemic fungicides affect biosynthetic processes and that the oxathiins, and some related compounds, may so far perhaps be the only compounds which act by inhibiting respiration. Kitazin and related compounds may interfere with cell, or more particularly membrane structure, but this is far from certain.

In contrast most protectant fungicides act by inhibition of processes involved in energy production. This at least applies to the dialkyldithiocarbamates, the bisdithiocarbamates, captan, dinocap, triphenyltin acetate, chlorothalonil and phenylmercuric acetate.

A possible explanation for this very low incidence of inhibitors of energy production among the systemic fungicides may be that such inhibitors, as for instance inhibitors of respiration, or uncoupling agents, if systemic, would tend to be equally toxic for plant tissues as for fungal tissues. In contrast inhibitors of biosynthesis which act primarily under conditions of growth might often be far more effective on the fast growing parasitic fungi than on the slow biosynthetic processes going on in most plant cells. Moreover, inhibitors of energy production, on the whole, appear less selective in their antimicrobial spectrum than inhibitors of biosynthesis. Strikingly also antibacterial agents used for medical purposes do not act by inhibition of energy production. The same is true for systemic insecticides used in plant production.

Obviously the oxathiins are exceptional in that they are systemic fungicides which do inhibit respiration and yet are not too phytotoxic to be used. But it must be remembered that also on fungi these compounds have rather selective activity, and even on sensitive fungi they do not act always by impairment of energy production but sometimes by lack of intermediate products required for synthetic processes. Mathre (1971a) has, moreover, revealed that the mitochondrial systems of Pinto bean plants are far less sensitive to these compounds than those of fungi. This may explain the absence of strong phytotoxicity.

The study of the biochemical mode of action appears to be almost a prerequisite for the understanding of selectivity of action and of the mechanisms of the resistance which has been observed with certain fungicides, although metabolic conversion of the compounds by the fungal cell or impermeability of the cells may also be a factor of importance. Conversely observations on selectivity of action and on the mechanism of resistance may help to establish the true nature of the biochemical mode of action. The following will show how much interwoven these features are.

Selectivity of action and resistance

Whereas protectant fungicides generally show a broad antifungal spectrum, selectivity in antifungal action is far more common among the systemic fungicides. Little is known as yet of the basis of this selectivity, which frequently bears a relation to the taxonomy of the fungus involved. In the case of the benzimidazoles absence of carbendazim-binding capacity of microtubular protein corresponded to lack of sensitivity of the fungus. One may expect fungi-lacking chitin to be insensitive to polyoxin in view of the interference of this compound with chitin biosynthesis, but no reports on the antifungal spectrum are available so far.

Lack of activity of cycloheximide and of blasticidin on certain fungi has been related to an inherent insensitivity of the system which in sensitive organisms is affected, namely protein synthesis.

The lower activity of chloroneb on certain fungi could not be ascribed to differences in the rate of metabolic conversion of this compound. Pyrazophos, however, acts only on fungi which have the ability to convert this compound into the active agent PP.

Carboxin on the whole is more active on Basidiomycetes than on Ascomycetes and Fungi Imperfecti. In this case differences in uptake by the cells appeared to account for selectivity. This view is supported by the observation that by introduction of one phenyl group a compound with a far wider spectrum was obtained. It seems possible that differences in penetration and accumulation frequently account for selectivity of fungicides.

Most anti-powdery mildew agents appear to act only on these obligate parasites, whereas compounds inhibiting growth of other fungi are frequently inactive towards powdery mildews. This exceptional situation of the mildew fungi may partly be due to their rather strongly lipophilic character; on the other hand several anti-mildew agents, e.g. triamiphos, are supposed to act primarily on the host plant, the effect on the parasite being only a consequence of this.

For many years the available systemic fungicides appeared to lack almost all activity with regard to Phycomycetes, of which especially the Oömycetes comprise most of the phytopathogens. The reason for their insensitivity may be manifold: lack of inherent sensitivity, lack of chitinous walls, lack of penetration.

Recently, however, compounds acting selectively on this group have also made their entry; for example, ethazole, prothiocarb and pyroxychlor. Presumably, because of the urgent need for such compounds, primary screening directed to activity against diseases caused by Phycomycetes is now receiving more emphasis.

The above observations enumerate various mechanisms by which selectivity in antifungal action may be brought about. These are: absence of a vulnerable mechanism, inherent insensitivity of the system affected in sensitive organisms, metabolic conversion of toxicant to inactive product or of an inactive compound into the actual toxicant, lack of sufficient penetration and accumulation of the compound in the cells.

Selectivity in action is also met when resistant mutant strains of a sensitive species emerge. The mechanisms on which selectivity between sensitive and mutant strains can depend may in principle be the same as described above for selectivity between different species: thus lack of binding site in the case of carbendazim, inherent insensitivity of protein synthesis in the case of cycloheximide, decreased permeability of the cells in the case of blasticidin and lack of metabolic conversion into the toxicant in the case of 6-azauracil. This subject will be dealt with in far more detail in Chapter 9.

Since many biochemical mechanisms in plants, animals as well as micro-organisms show basic similarities, it is not surprising that fungicides often show wider biocidal activity. Thus tridemorph and blasticidin have antibacterial activity; benomyl is known as an acaride; benomyl, carbendazim and thiabendazole are anthelminthics; triamiphos, pyrazophos and Kitazin have relatively high mammalian toxicity, possibly because they inhibit cholinesterase activity. Cycloheximide, an inhibitor of protein biosynthesis in higher plants and in animals, is a specific inhibitor of 80 *S* ribosomes and does not act on 70 *S* ribosomes such as occur in bacteria, and consequently has no antibacterial activity. From the point of view of selectivity polyoxin D is an ideal fungicide in that its antifungal action depends on inhibition of chitin synthesis, a process only known to occur in certain fungi and lower animals. Its low mammalian toxicity may be due to this selective activity.

This chapter has shown that studies on mode of action, selectivity, resistant strains, metabolism and structure–activity relation are all closely interconnected. It is hoped that a thorough understanding of the mode of action of fungicides will contribute in future to as wide as possible an evaluation of their toxicological hazards. A better understanding of primary mechanisms of action should ultimately contribute to the development of new active compounds.

Chapter 8

Metabolism

J. W. Vonk and A. Kaars Sijpesteijn

Introduction

Systemic fungicides are subject to a number of natural processes which will alter their chemical structure. Nowadays the term metabolism is generally used to designate such alterations. This chapter attempts to discuss the present knowledge of the metabolism of systemic fungicides.

In view of the toxicological and environmental implications of the use of crop protection agents, it is important to have thorough knowledge of the chemical and biochemical conversion of systemic fungicides. In most countries therefore data are required by the registration authorities, and agrochemical manufacturers are requested to submit these. In this way industrial research in the last decade has broadened our knowledge of metabolism of systemic fungicides, but also a substantial amount of research work has been performed by independent groups, often on more fundamental aspects of the problems.

In recent years improved techniques have become available for such studies. The use of radio-actively labelled compounds has become more or less a matter of routine, and techniques like high-pressure liquid chromatography (HPLC) and mass spectrometry coupled to gas chromatography (GCMS) have made possible both qualitative and quantitative analysis of metabolites.

Generally we may distinguish between two main processes by which the structure of systemic fungicides can be altered: firstly, purely chemical break-down processes and secondly, the metabolic systems present in plants, animals and micro-organisms.

Among chemical, non-enzymic reactions spontaneous hydrolysis and oxidation may be mentioned, e.g. in the formation of carboxin sulphoxide in soil. Irradiation of fungicides with sunlight, either in the presence or absence of biological material, may lead to photochemical reactions.

By contrast, most reactions occurring in the metabolic systems of organisms are enzyme-catalysed. When the substrate specificity of an enzyme is wide enough, it may accept a foreign organic molecule as a substrate. The biochemical conversions which systemic fungicides undergo in this way can be roughly grouped as oxidation, reduction, hydrolysis and conjugate formation. Examples may be found in the sections devoted to the individual compounds. *Systemic* fungicides due to their very nature are brought into close contact with the metabolic system of the plant.

The following sections deal with the metabolism of most systemic fungicides used in practice. No attempt has been made to discuss this matter in full detail, since a recent, more exhaustive, treatment is available (Kaars Sijpesteijn *et al.*, 1977). Throughout the chapter the discussion of the data for a particular systemic fungicide is grouped according to observed pathways and metabolites. The nature of the transforming agent is indicated in the figures. The abbreviations of Fig. 8.1 are used throughout.

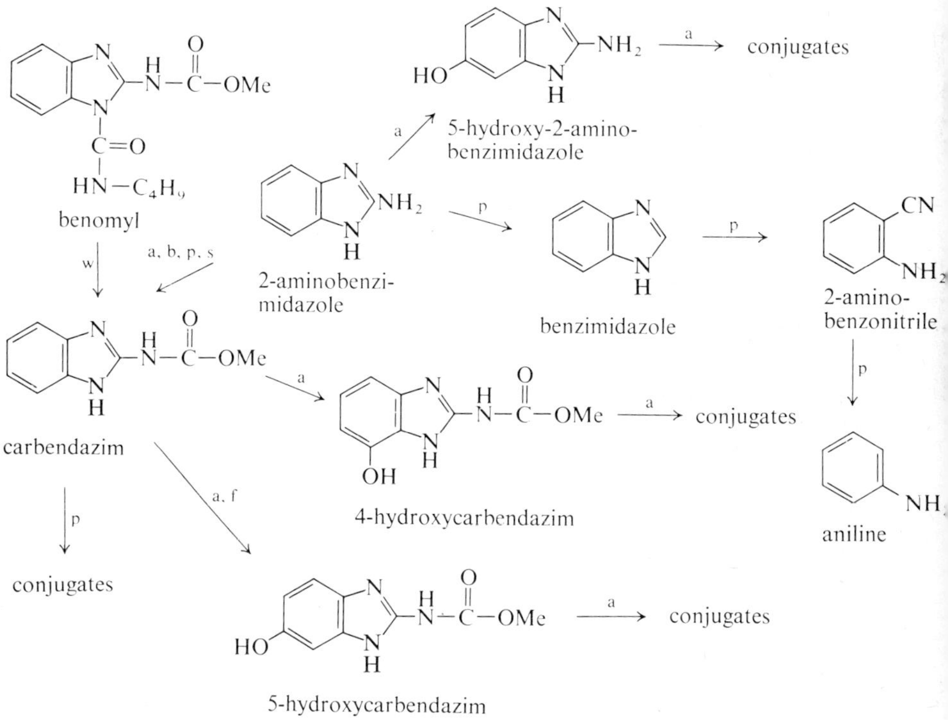

Fig. 8.1 Known metabolic pathways of benomyl and carbendazim a: animals; b: bacteria; e: enzymes; f: fungi; l: light; p: plants; s: soil; w: water.

Benzimidazoles and related compounds

Benomyl and carbendazim

In dilute aqueous solution or in organic solvents benomyl rapidly loses the butylcarbamoyl group, forming methyl benzimidazol-2-ylcarbamate (carbendazim, MBC) (Clemons and Sisler, 1969; Chiba and Doornbos, 1974). This compound is also formed from benomyl in microbial cultures, soil, animals and plants. When, however, applied to plants by spraying, benomyl is obviously more stable because it constitutes a major part of the spray deposit for extended periods (Baude *et al.*, 1973).

Carbendazim is rather resistant to attack by plants and micro-organisms and is persistent in soil.

Siegel (1973) found no label of [ring-2-^{14}C]-carbendazim in products of intermediary metabolism after 88 days in strawberry plants and only 1·5 per cent label of [ring-2-^{14}C]-carbendazim was respired as $^{14}CO_2$ after 14 days in cucumber plants (Solel *et al.*, 1973). However, although the benzimidazole nucleus thus appears stable, small amounts of the non-fungitoxic 2-aminobenzimidazole were observed by Siegel (1973) and certain products resulting from cleavage of the benzimidazole nucleus, viz. 2-aminobenzonitrile and aniline, were detected by Rouchaud *et al.* (1974) after 2 months in melon plants, in addition to benzimidazole. These latter products might be the result of photochemical decomposition.

Hydroxylated derivatives of carbendazim have not been reported in plants, although evidence was obtained for the presence of *O* or *N* conjugates (Siegel, 1973). In animals a rapid hydroxylation at the 4 and 5 position followed by the conventional mechanisms of conjugation, viz. glucuronide and sulphate formation, was observed (Gardiner *et al.*, 1974; Douch, 1973). 2-Aminobenzimidazole, together with 5-hydroxy-2-aminobenzimidazole and its conjugates, were also detected (Douch, 1973).

2-Aminobenzimidazole was also formed by degradation of either carbendazim or benomyl in soil. Carbendazim is rather persistent in soil, the half-life of the total benzimidazoles in benomyl-treated soil being 0·5–1 yr and the formation of 2-aminobenzimidazole is slow (Baude *et al.*, 1974). Since sterilisation abolishes the rate of disappearance of carbendazim, micro-organisms are involved in breakdown (Helweg, 1973; Siegel, 1975). Fleeker *et al.* (1974) and Siegel (1975) have reported the formation of $^{14}CO_2$ from [ring-^{14}C]-carbendazim, thus indicating that breakdown of the benzimidazole nucleus does finally occur in soil. Several organisms were isolated from treated soil which degraded benomyl or carbendazim (Helweg, 1973; Weeks and Hedricks, 1971; Fuchs and Bollen, 1975) and 2-aminobenzimidazole and 5-hydroxy-carbendazim have been identified as metabolites. The latter compound resulted from metabolism by *Aspergillus nidulans*. Davidse (1976) suggested that this hydroxylation is connected with melanogenesis, a process in which polyphenoloxidase plays an important role. In animals hydroxylation of carbendazim is probably performed by a mixed function oxidase (Douch, 1973).

Figure 8.1 summarises all known pathways for benomyl and carbendazim.

Thiabendazole

Since thiabendazole was first developed as an anthelminthic agent, its metabolism in animals was studied at an early stage. Like benomyl and carbendazim, thiabendazole is hydroxylated in animals, but only 5-hydroxythiabendazole and its conjugation products have been reported as metabolites (Tocco *et al.*, 1965) (Fig. 8.2).

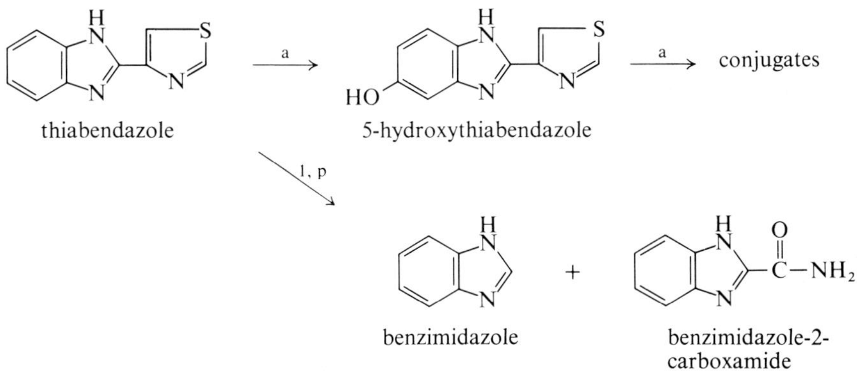

Fig. 8.2 Metabolic pathways of thiabendazole.

The metabolism of thiabendazole in plants, soil and micro-organisms has not been studied as much as the metabolism of benomyl, probably because its practical application is less extensive. Erwin *et al.* (1970) and Wang *et al.* (1971) have studied the fate of [ring-^{14}C]-thiabendazole in cotton plants and speculated that conjugates of metabolites were formed. The unchanged parent compound was also detected in the plants after uptake via the roots (Gray and Sinclair, 1971). In pepper leaves thiabendazole disappeared faster than carbendazim (Ben-Aziz and Aharonson, 1974). After leaf application on sugar beets Jacob *et al.* (1975) found no transformation of thiabendazole under artificial light, whereas on plants exposed to sunlight small amounts of benzimidazole and benzimidazole-2-carboxamide were formed, obviously by photodecomposition. A carboxamide was also reported as a metabolite in animals of cambendazole, an anthelminthic related to thiabendazole (VandenHeuvel *et al.*, 1974).

There is no information on the breakdown of thiabendazole by pure cultures of micro-organisms. In several soils investigated, however, thiabendazole was slowly broken down (Aharonson and Kafkafi, 1975).

Fuberidazole

No information is available on breakdown of fuberidazole by plants and micro-organisms. The metabolism in animals was studied by Frank

(1971), and as with the other benzimidazole fungicides the 5-hydroxylated compound and its conjugates were formed. The furan ring is cleaved to yield the γ-hydroxybutyric acid derivative of benzimidazole (Fig. 8.3).

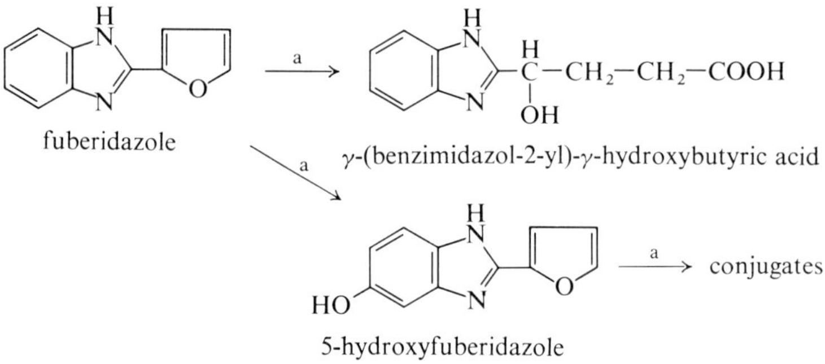

Fig. 8.3 Transformation of fuberidazole in animals.

Thiophanates

Ring cyclisation of thiophanate-methyl and thiophanate yields the alkyl benzimidazole carbamates (carbendazim and its ethyl analogue, respectively), which are responsible for their activity. This phenomenon is discussed in detail in Chapter 7. Only the fate of thiophanate-methyl will be dealt with here, that of thiophanate not being essentially different (Fig. 8.4).

Although a slow formation of carbendazim from thiophanate-methyl

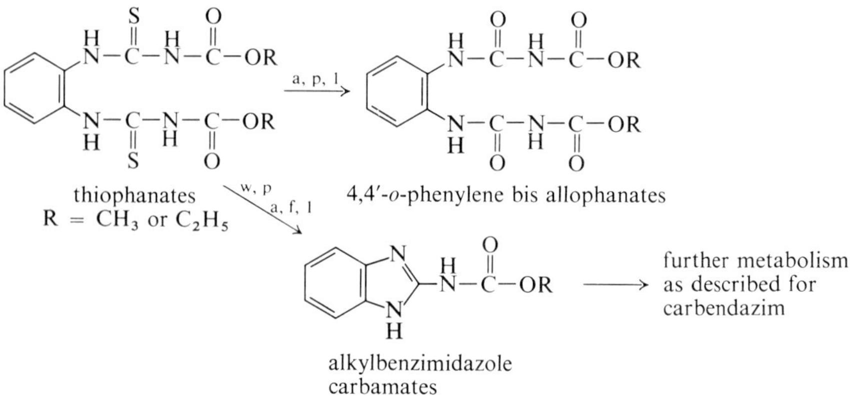

Fig. 8.4 Metabolic pathways of thiophanates.

in aqueous media was observed (Vonk and Kaars Sijpesteijn, 1971), this conversion was found to proceed much faster in several organisms. Thus, the formation of carbendazim from thiophanate-methyl has been reported in plants (Soeda *et al.*, 1972a), in fungi (Vonk and Kaars Sijpesteijn, 1971; Yasuda *et al.*, 1973), in animals (Noguchi, 1972; Douch, 1974) and also in soil where it was the major metabolite (Noguchi, 1972). The biological formation of carbendazim in fungi is probably mainly responsible for the fungitoxic effect of thiophanate-methyl.

Buchenauer *et al.* (1973) observed photochemical formation of carbendazim from thiophanate-methyl on the surface of cotton leaves, and Matta and Gentile (1971) reported that tissue fragments of different plants accelerated this conversion. Studies with potato tuber cell-free extracts revealed that intermediary *o*-quinones formed by polyphenoloxidase activity may catalyse this conversion (Vonk and Mihanović, 1976). Studies with liver microsome preparations indicated that in animals the cyclisation of thiophanate-methyl into carbendazim is catalysed by a mixed-function oxidase (Douch, 1974). Any carbendazim formed from thiophanate-methyl can be expected to be eventually metabolised further.

Besides carbendazim, small amounts of dimethyl-4, 4′-*o*-phenylene-bis-allo-phanate were formed on leaf-treated plants and evidence was obtained for the photochemical nature of this conversion (Soeda *et al.*, 1972b).

Oxathiins and related compounds

Carboxin and oxycarboxin

The oxathiins are more liable to conversion than the benzimidazole-derived fungicides. Carboxin is readily oxidised to the non-fungitoxic S-oxide (carboxin sulphoxide) in various plant species, animals and soil (Chin *et al.*, 1970a, b; Snel and Edgington, 1970). Small amounts of the sulphone (oxycarboxin) were sometimes found as well. The formation of carboxin sulphoxide also occurs slowly in water, but inside organisms this proceeds much faster. Oxidation of an S atom to an S-oxide is a common type of conversion in organisms.

A large amount of carboxin is bound in barley and wheat plants as insoluble complexes, probably with lignin. According to Chin *et al.* (1971) this lignin complex contains the intact carboxin molecule. In contrast, Snel and Edgington (1970) postulate that in bean plants the amide linkage is hydrolysed and the resulting aniline is bound to the plant polymers. The presence of free aniline in oxycarboxin-treated plants has been reported (Newby and Tweedy, 1970). Recently the *p*-hydroxyphenyl derivative of carboxin was detected as a metabolite of carboxin in barley plants by Briggs *et al.* (1974). Since this compound is readily bound to lignin, it obviously escaped the attention of earlier authors. Strangely enough, Briggs *et al.* could detect only traces of the sulphoxide. The same discrepancy was observed for the metabolism of carboxin in animals. Whereas Chin *et al.* (1971) reported the formation of the sulphoxide in rats. Waring (1973) could not detect the latter compound, but instead glucuronides of the *o*- and *p*-hydroxyphenyl derivatives were found in rat and rabbit urine. Waring has questioned the assignment of the sulphoxide structure by the earlier authors, who might in view of identity of R_F values have confused the sulphoxide with the phenol structure.

In soil carboxin was eventually oxidised to the sulphoxide, but no hydrolysis took place. The sulphoxide formation also occurred in sterilised soil and was therefore not due to microbial action (Chin *et al.*, 1970b). Of several micro-organisms tested Lyr *et al.* (1974) found that

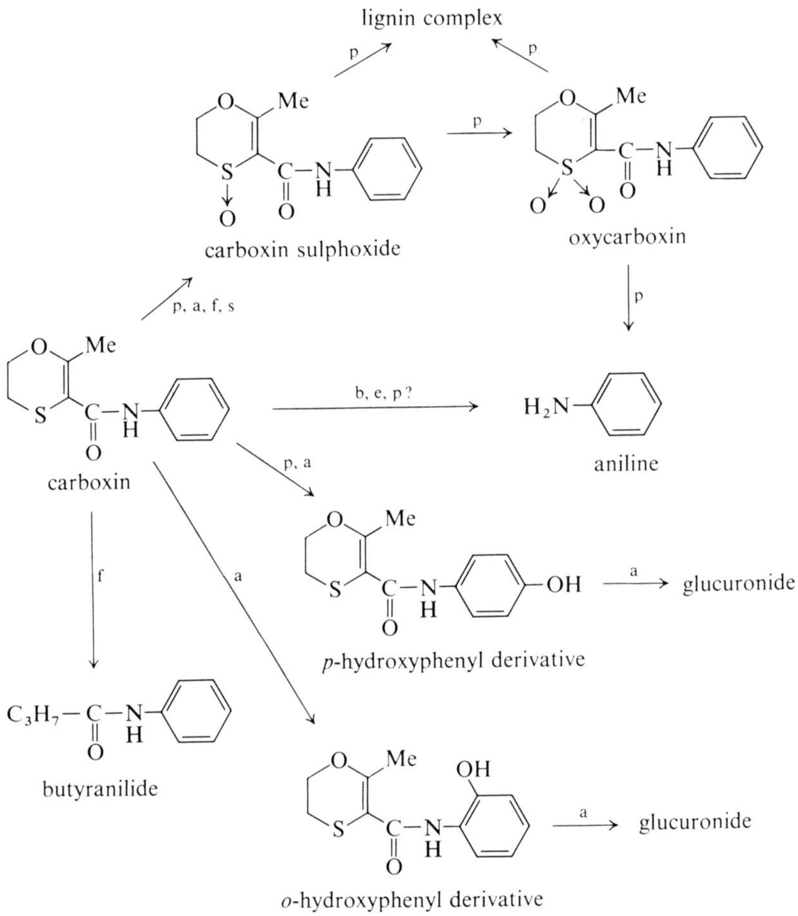

Fig. 8.5 Metabolic pathways of carboxin and oxycarboxin.

only *Ustilago maydis* could form the sulphoxide from carboxin. Moreover, isolated mitochondria from fungi could oxidise carboxin. Photochemical sulphoxide formation occurred in the presence of flavin compounds (Lyr *et al.*, 1974) and by the action of ultra-violet light (Wolkoff *et al.*, 1975).

Ring cleavage of carboxin with the formation of butyranilide by a fungus was reported by Wallnöfer (1969), whereas hydrolytic cleavage of carboxin was found in cultures or enzyme extracts of *Bacillus sphaericus* (Wallnöfer and Engelhard, 1971; Engelhard *et al.*, 1973).

Figure 8.5 gives the reported pathways of carboxin and oxycarboxin.

Mebenil, 2,5-dimethyl-3-furanilide and pyracarbolid

Nothing has been published about the fate of mebenil in plants. It is transformed into its *p*-hydroxyphenyl derivative however by the fungus *Rhizopus japonica* (Wallnöfer *et al.*, 1971) (Fig. 8.6). In contrast, the

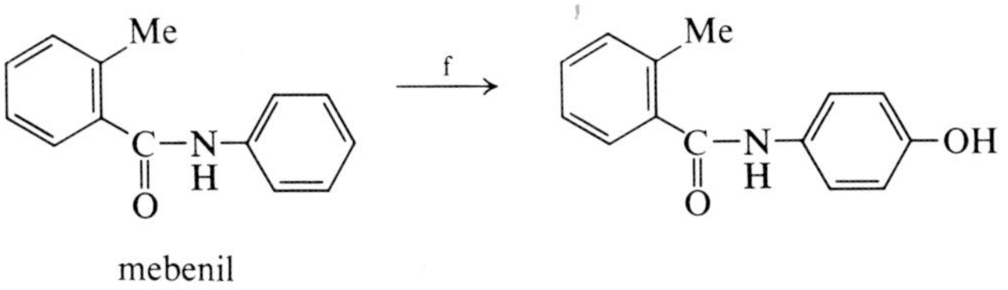

Fig. 8.6 Transformation of mebenil in fungi.

methyl groups of the related compound 2,5-dimethyl-3-furanilide are hydroxylated by several Phycomycetes, including *R. japonica* (Wallnöfer *et al.*, 1972) (Fig. 8.7). It is not known if a *p*-hydroxyphenyl derivative is also formed. Oeser *et al.* (1975) have reported that pyracarbolid is metabolised in wheat into three compounds. One was tentatively identified as 2,6-dihydroxy-3-phenylaminocarbonyl-hex-2-ene, the result of ring splitting of the dihydropyran ring (Fig. 8.8)

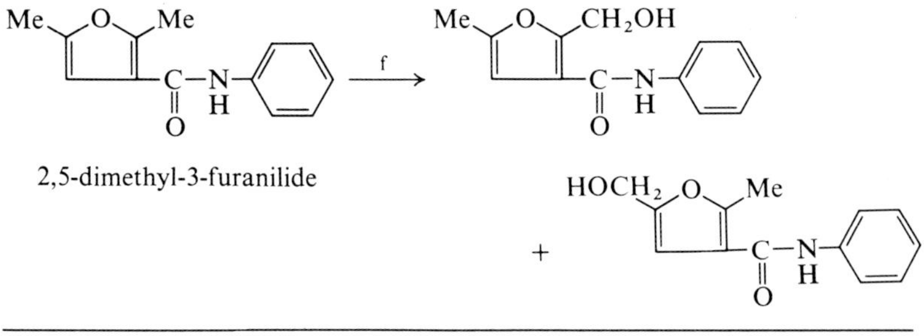

Fig. 8.7 Transformation of 2,5-dimethyl-3-furanilide in fungi.

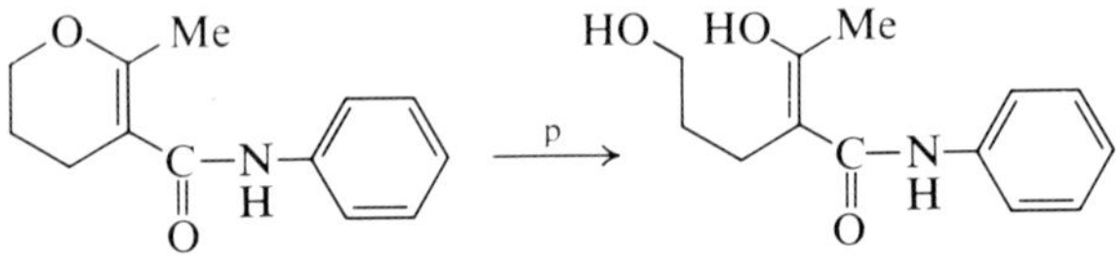

pyracarbolid

Fig. 8.8 Transformation of pyracarbolid in wheat.

Organophosphorus esters

Pyrazophos

The fungus *Pyricularia oryzae*, but not *Pythium debaryanum* and *Sacharomyces cerevisiae*, was able to metabolise pyrazophos into 6-ethoxy-carbonyl-2-hydroxy-5-methylpyrazolo[1,5-a]pyrimidine (PP) and the oxygen analogue of pyrazophos (PO-pyrazophos); water-soluble, non-identified metabolites were observed as well (de Waard, 1974) (Fig. 8.9). The formation of an oxygen analogue is well known for organophosphorothionate insecticides. The β-glucoside of PP was detected as a metabolite on cucumber leaves, whereas glucuronide and sulphate conjugates of PP were formed in rats. Possibly, PO-pyrazophos was present in small amounts in wheat leaves between the 4th and 10th day after

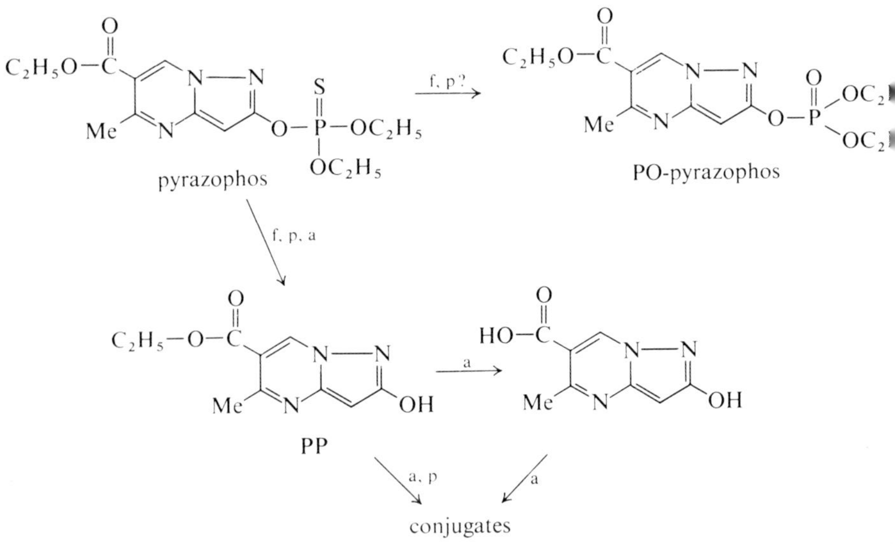

Fig. 8.9 Metabolic fate of pyrazophos.

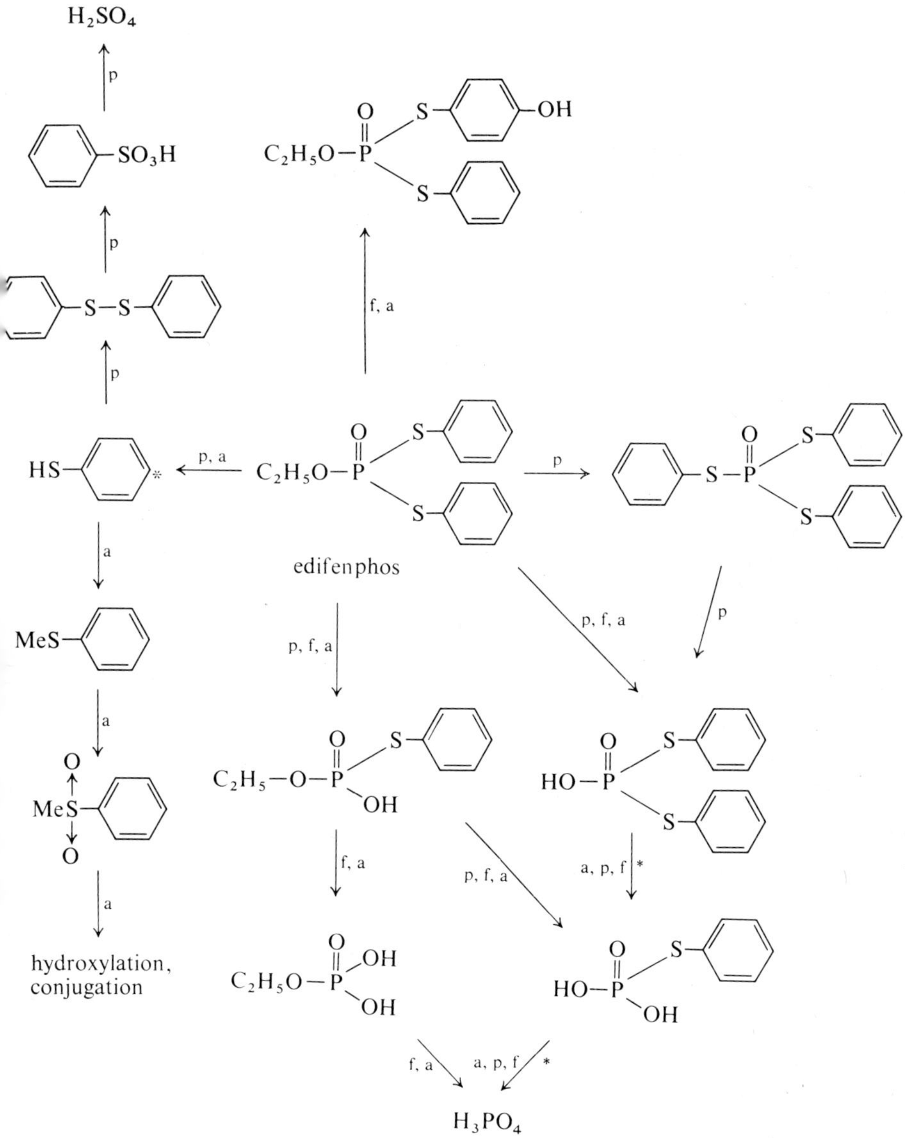

Fig. 8.10 Degradation of edifenphos. * = proposed pathway or compound.

pyrazophos treatment. PP was further metabolised in rats to yield the carboxyl derivative (Gorbach *et al.*, 1975).

Kitazin P, edifenphos (Hinosan) and Inezin

A number of related organic phosphate esters, active against rice blast disease, have been developed in Japan. Extensive and careful studies of their metabolic fate in plants, fungi, soil and animals have been carried out. Thus the fate of edifenphos in rice plants was studied by Ueyama *et al.* (1973), in *P. oryzae* by Uesugi and Tomizawa (1971), in animals by Uesugi *et al.* (1972) and Ueyama and Takase (1975). The fate of Kitazin P in rice plants was studied by Yamamoto *et al.* (1973), in *P. oryzae* by Tomizawa and Uesugi (1972), in soil by Masuda and Kanazawa (1973). The fate of Inezin in rice plants was studied by Endo *et al.* (1970), in *P. oryzae* by Uesugi and Tomizawa (1972). Uesugi *et al.* (1972) have reviewed the literature on degradation of these organic phosphate fungicides.

It would take too much space to discuss the fate of all these compounds in detail here. As an example, a schematic representation is given of the breakdown of edifenphos in Fig. 8.10.

The ester linkages of the compound are hydrolysed successively and the resulting intermediates are subject to further breakdown, yielding finally phosphoric acid and sulphuric acid. Most pathways run along parallel lines in plants, fungi and animals. However, a *p*-hydroxyphenyl phorothionate (P=S) formation by isomerisation was observed for between molecules (transesterification) was only observed in plants.

Similar routes of degradation apply to Kitazin P and Inezin. An interesting difference from edifenphos is, however, that hydroxylation by *P. oryzae* occurs at the *meta* position instead of at the *para* position. Phosphorothionate (P=S) formation by isomerisation was observed for Kitazin P metabolism in plants. Moreover, the metabolite *O,O*-diisopropyl hydrogen phosphorothioate was methylated to *O,O*-diisopropyl *S*-methyl phosphorothioate, a mechanism resembling the biosynthesis of methionine from homocysteine.

$$(i\text{-PrO})_2\overset{\overset{\displaystyle O}{\|}}{P}\text{-SH} \longrightarrow (i\text{-PrO})_2\overset{\overset{\displaystyle O}{\|}}{P}\text{-SCH}_3$$

Chloroneb

Chloroneb has slight systemic activity, as it accumulates in the roots and lower parts of plants. The main metabolic pathway is the demethylation to 2,5-dichloro-4-methoxyphenol. This occurs in plants (Rhodes *et al.*, 1971; Thorn, 1973), animals (Rhodes and Pease, 1971) and by several fungi under conditions of active growth (Wiese and Vargas, 1973) (Fig. 8.11). It is noteworthy that some fungi are also able to perform the reverse reaction, viz. formation of chloroneb from 2,5-dichloro-4-methoxyphenol. In plants and animals 2,5-dichloro-4-methoxyphenol was conjugated, a β-D-glucoside being formed in the former (Thorn, 1973). Further minor conversion products isolated from

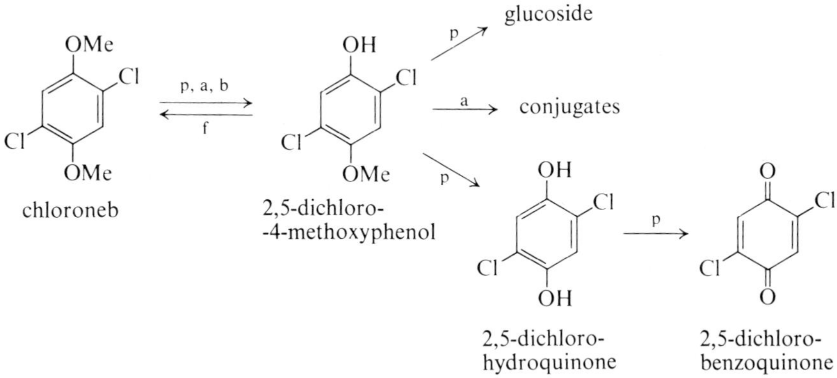

Fig. 8.11 Metabolic fate of chloroneb.

plants were 2,5-dichlorobenzoquinone and 2,5-dichlorohydroquinone (Rhodes *et al.*, 1971).

In soil chloroneb has a half-life of 3–6 months, the products formed not being known (Rhodes *et al.*, 1971). Little is known about its metabolism by bacteria, and of three species investigated only one was able to demethylate chloroneb (Wiese and Vargas, 1973).

Pyrimidines

Ethirimol and dimethirimol

Dimethirimol and ethirimol are translocated from the roots to the above ground parts of barley and cucumber plants (Calderbank, 1971; Slade *et al.*, 1971). The fate of these compounds in plants as far as elucidated has been summarised in Fig. 8.12. Metabolism of both compounds follows a pattern similar to that observed in animals, viz. *N*-dealkylation, hydroxylation of the butyl group and conjugate formation (Bratt *et al.*, 1972). In plants, the hydroxylation appears to be a relatively minor pathway as compared to conjugate formation. The conjugates formed in plants are glucosides and probably phosphates, while in animals glucuronides were the only conjugates identified.

The *N*-demethylation of dimethirimol occurs very quickly to give the still fungicidally active *N*-monomethyl derivative in cucumber plants. The loss of the second *N*-methyl group occurs more slowly and gives rise to the relatively inactive 2-amino derivative, which is also formed from ethirimol in barley plants.

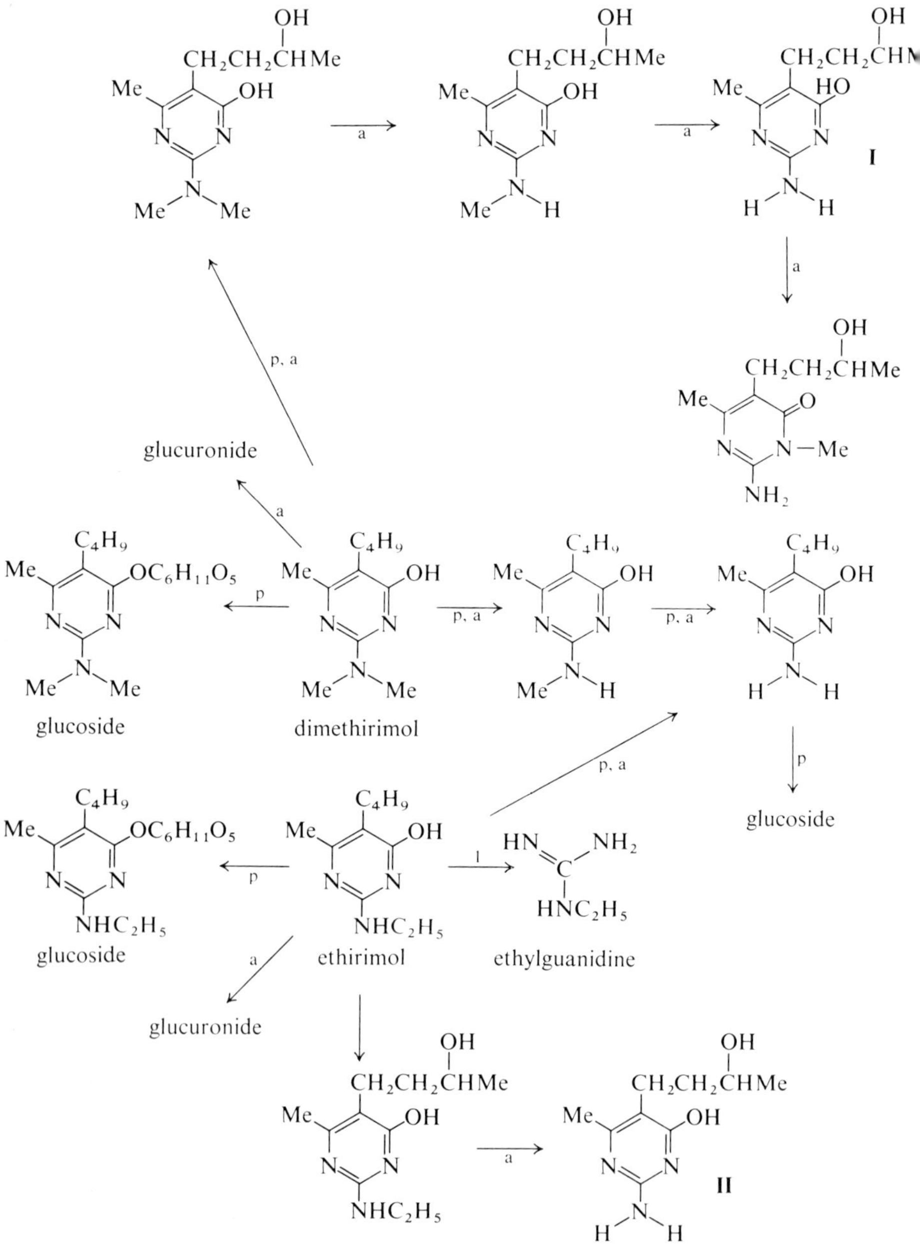

Fig. 8.12 Fate of dimethirimol and ethirimol. N.B. Compounds I and II are identical.

Ethirimol is broken down in soil, the rate of degradation being dependent on its type (Hill and Arnold, 1973). Cleavage of the pyrimidine ring was indicated by $^{14}CO_2$ evolution from appropriately labelled parent compound, but no other metabolites were identified.

In aqueous solution ethirimol is photolytically degraded into ethylguanidine (Teal and Cavell, 1975).

Bupirimate

On the surface of apple leaves and fruits bupirimate is readily converted into ethirimol, which itself again undergoes further degradation (Teal and Cavell, 1975).

Tridemorph

Tridemorph (2,6-dimethyl-4-tridecylmorpholine) consists of two diastereoisomers, in which the methyl groups have a different spatial arrangement. The tridecyl group is probably highly branched (Kost and Gurfinkel, 1975).

In soil tridemorph is slowly transformed into tridemorph-*N*-oxide and 2,6-dimethylmorpholine; $^{14}CO_2$ was evolved from the labelled compound (Otto and Drescher, 1973). There were also indications of the formation of hydroxylated tridemorph in barley plants (Waring and Wolfe, 1975), whilst in rats there was some evidence for a hydroxylated metabolite (Hawkins *et al.*, 1974) (Fig.8.13).

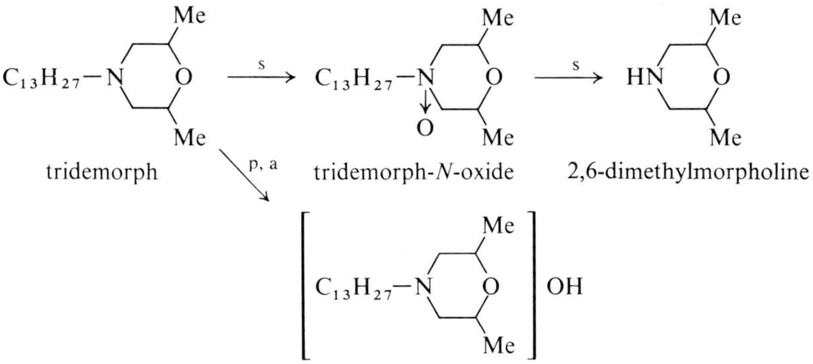

Fig. 8.13 Metabolic pathways of tridemorph.

Triforine

Only limited data on the metabolism of triforine are available.

In dilute aqueous solution the trichloroethyl formamido groups were successively converted into glyoxyl groups which then split off to yield

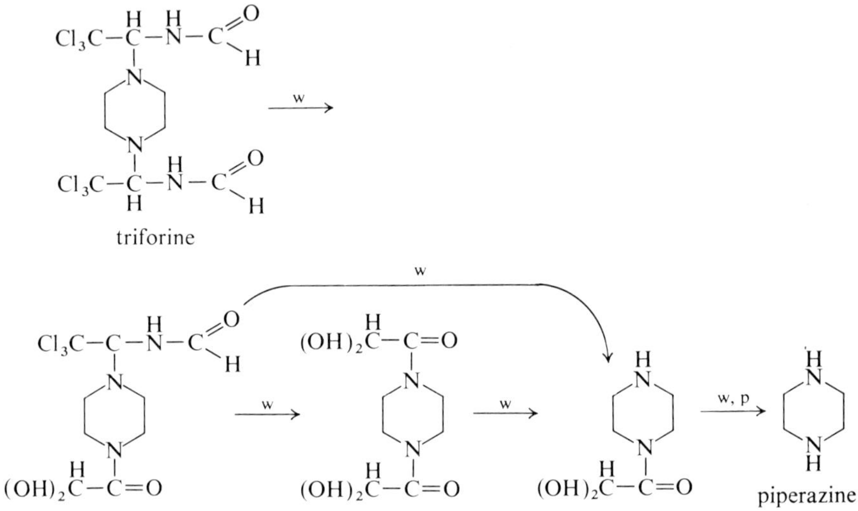

Fig. 8.14 Degradation of triforine.

piperazine (Fuchs and Ost, 1975) (Fig. 8.14). In plants triforine is broken down and the presence of at least four non-fungitoxic metabolites could be demonstrated, one of which was identified as piperazine (Bruchhausen and Stiasni, 1973). The three unknown metabolites might be identical with the degradation products in aqueous solution (Fuchs and Ost, 1975).

Prothiocarb

Prothiocarb is degraded in soil. The formation of $^{14}CO_2$ from labelled product was observed. Two metabolites in soil have been tentatively identified, viz. *N*,*N*-dimethylpropylene diamine and the demethylated product *S*-ethyl *N*-(3-methylaminopropyl)thiocarbamate (Iwan and Goller, unpublished). The second metabolite may be of microbiological origin (Pieroh *et al.*, 1975).

In soy beans no difference was observed in migration between propyl-^{3}H and ethyl-^{14}C-labelled prothiocarb. This may indicate that breakdown is slow in this plant species (Iwan and Goller, 1975).

Final remarks

This survey of metabolic transformations obviously shows great gaps in our knowledge. One may, however, assume that much knowledge gathered by industrial research has remained unpublished.

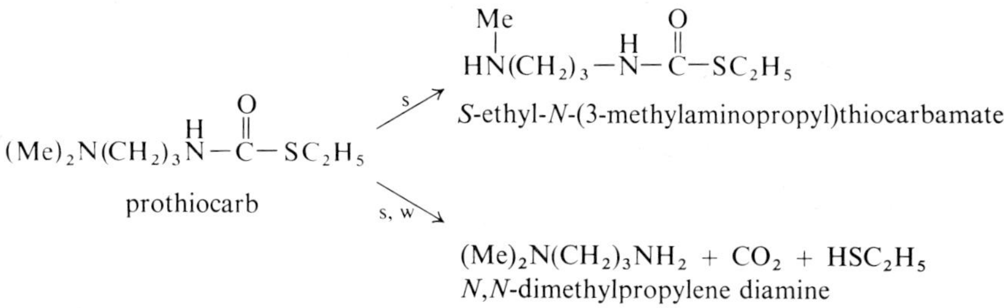

Fig. 8.15 Breakdown of prothiocarb.

After conversion of a fungicide has led to inactive compounds the effect on the fungus has come to an end; the above survey may, however, have shown that breakdown of the compounds in most cases does not by any means go to completion. In contrast, the skeleton of the molecule often remains more or less intact and this fact may stimulate the search for compounds which after having done their job are subject to extensive decomposition.

Chapter 9

Resistance

J. Dekker

Introduction

Living organisms possess the ability to adapt to changing environmental conditions. This is illustrated in the history of life on earth; the evolution of living matter would have been impossible without this property. This adaptation may occur also to changes, brought about by man, such as the introduction of biocides in the environment. Well known is the adaptation of bacteria to antibiotics, which became apparent shortly after the application of these biocides for control of bacterial diseases in humans. Insects and mites also have adapted rapidly to certain new organic insecticides, which has caused problems in the control of plant pests. In contrast to this, adaptation of fungi to agricultural fungicides has until recently been observed only rarely in practice. In view of this, the opinion was expressed that fungi might be less liable to adapt to biocides than bacteria or insects. From experience with several new organic chemicals with selective action on fungi, it is now known that this assumption is no longer tenable. Reduced sensitivity of fungi to various of these fungicides has now been encountered frequently, both in the field and in the laboratory.

Organisms which adapt to a toxicant by a reduction of their sensitivity are called 'resistant' or 'tolerant'. Both terms are often used interchangeably, when describing the relation between a fungus and a fungicide, although they have a different meaning with respect to the relation between a host plant and its fungal parasite. In the latter case the plant may resist attack or be susceptible to it, and after the attack has succeeded may prove tolerant or sensitive. In the former case it seems of little practical relevance whether the fungal cell 'resists attack' by preventing the fungicide from entering, or is insensitive after penetration has taken place. Moreover the mechanism of resistance or tolerance is not known for many fungicides.

In order to avoid confusion with the use of resistance and tolerance in reference to the host, the term 'insensitivity' may also be used.

In this chapter attention will be paid to genetic, biochemical and epidemiological aspects of acquired resistance and to measures aimed at avoidance or reduction of acquired resistance in practice.

Occurrence of acquired resistance

To conventional fungicides

Since Millardet, in 1882, discovered the effect of Bordeaux mixture against downy mildew on vine, copper compounds have been used on a large scale against many fungal plant diseases. Later on, these fungicides were to a large extent replaced by dithiocarbamates, which have now been in use for several decades, and by various other conventional fungicides.† These fungicides have rarely given serious problems with respect to acquired resistance and there are only a few reports on development of tolerance in practice. Failure to control *Pyrenophora avenae* (*Drechslera avenae*) on oats by seed treatment with organic mercury compounds has been reported from Scotland (Noble *et al*., 1966) and various other countries. Development of resistance in *Sclerotinia homoeocarpa* on turf grass to cadmium-containing compounds (Cole *et al*., 1968) and to dyrene (Nicholson *et al*., 1971) was observed in the USA. Application of dodine, after 10 years of use for control of apple scab in New York State, USA, became less effective in 1969 (Szkolnik and Gilpatrick, 1969). Development of tolerance in the causal organism, *Venturia inaequalis*, was demonstrated: ascospores from orchards with inadequate scab control were two to three times less sensitive to dodine than from orchards where dodine performed satisfactorily (Gilpatrick and Blowers, 1974). Development of resistance in practice has also been reported for several members of the so-called aromatic hydrocarbon group, such as diphenyl (Harding, 1964), pentachloronitrobenzene (Shatla and Sinclair, 1963), hexachlorobenzene (Kuiper, 1965) and dicloran (Locke, 1969; Webster *et al*., 1970).

For *in vitro* studies of resistance to conventional fungicides, the so-called 'training' has been the preferred method. It involves exposure of the fungus, usually as mycelium, to sublethal concentrations of the fungicide in an agar medium, with subsequent transfer to media containing slightly higher concentrations of the chemical. A high degree of resistance to these fungicides was seldom obtained in this way, and acquired resistance usually disappeared after transfer of the fungus to a medium free of toxicant. An example is the tolerance to organic mercury in *Pyrenophora avenae*, obtained by training this fungus on media containing the toxicant (Greenaway and Cowan, 1970). Another method consists of seeding fungal spores on an agar medium which contains a normally inhibitory concentration of the fungicide. Survivors which have acquired resistance, usually by mutation, will then be selected by this fungicide. Before plating out of the spores on fungicide containing medium, they may first be exposed to u.v. or other mutagenic treatments in order to increase the mutation rate. Also this method was not very successful with conventional fungicides, except for a number

† The adjective 'conventional' will be used for those fungicides which show little or no penetration into the plant tissue.

of compounds belonging to the aromatic hydrocarbon group. In laboratory experiments mutants of *Hypomyces solani* f. sp. *cucurbitae* were obtained resistant to chloronitrobenzenes (Georgopoulos, 1962, 1963) and to diphenyl (Vomvoyanni and Georgopoulos, 1966), and mutants of *Ustilago maydis*, resistant to chloroneb (Tillman and Sisler, 1971). Most of these fungicides are rather selective in action and are used only to a limited extent.

Considering the long period during which conventional fungicides have been in commercial use, the number of cases in which emergence of resistance has become a practical problem is rather small, and the practical significance not great. Reviews on development of resistance have been published by Ashida (1965) for inorganic fungicides and by Georgopoulos and Zaracovitis (1967) for organic fungicides.

To systemic fungicides

Before the introduction of systemic fungicides in practice, it was already known from laboratory experiments that resistance in fungi to selective fungicides, namely various antibiotics and antimetabolites, could arise readily. Resistance to the antibiotic antimycin A in *Venturia inaequalis* was demonstrated by Leben *et al.* (1955) and different types of resistance to 6-azauracil in *Cladosporium cucumerinum* by Dekker (1967, 1968). Soon after application of systemic fungicides was started in practice, reports appeared about increasing failure of some of these compounds to provide disease control. In most cases this could be attributed to development of resistance. Benomyl resistance in cucumber powdery mildew was reported within 1 year after introduction of this fungicide (Schroeder and Provvidenti, 1969). Thereafter many other reports were published on development of fungicide resistance.

A dramatic example is the rapid development of tolerance in *Sphaerotheca fuliginea* to the pyrimidine derivative dimethirimol (Milcurb). This compound was introduced in Holland in 1968 and used on a large scale for control of powdery mildew in cucumber greenhouses in the following year. A suspension of this fungicide, applied to the soil around the base of the plant, provided continuous protection of the whole plant for several weeks. The absence of residues on the leaves and the low cost of this labour-saving technique also contributed to the success of this systemic fungicide. In the autumn of 1969, and more so in the spring of 1970 the results obtained were less satisfactory in some greenhouses. Experiments with powdery mildew from sites of poor control and from sites where dimethirimol had never been used, revealed that fungicide-tolerant strains had developed (Bent *et al.*, 1971). As a consequence the use of dimethirimol for control of cucumber powdery mildew had to be abandoned.

A second example of fungicide resistance has been encountered in the cyclamen culture in Dutch greenhouses, where *Botrytis cinerea* causes a serious heart rot. As conventional fungicides performed unsatisfactorily against this disease, the infected plant parts had to be removed by hand, a labour-intensive and therefore costly procedure. Application of benomyl provided an excellent solution to this problem as a 0·15 per cent spray with the formulated product (Benlate 50 per cent W.P.) kept the plants free from the disease for several weeks. The enthusiasm of cyclamen growers was tempered, however, when after a short period of use benomyl appeared to be less effective than originally. From diseased plants a strain of *B. cinerea* was isolated which tolerated much higher concentrations of benomyl than the wild type

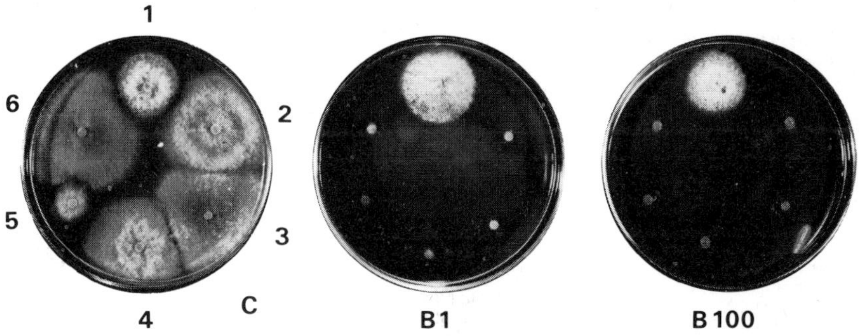

Fig. 9.1 Acquired resistance of *Botrytis cinerea* to benomyl. Growth of six isolates on PDA without fungicide (C) and with 1 (B1) and 100 (B100) μg/ml Benlate, 50 per cent W.P.; 1: resistant isolate from cyclamen. 2, 3, 4, 5, 6: sensitive isolates from cyclamen, dahlia, lettuce, sunflower and soil, respectively (from Bollen and Scholten (1971), *Neth. J. Pl. Path.* 77, 86)

fungus. Even at 1 000 ppm of benomyl there was not a complete inhibition of growth, while the wild type fungus was eliminated at 0·5 ppm of benomyl in the medium (Bollen and Scholten, 1971) (Fig. 9.1). The

Table 9.1 Development of resistance to benomyl (Benlate 50 per cent W.P.) in *Sphaerotheca fuliginea*. Disease assessment after seven days, using a scale from 0 (no mildew) to 5 (covered with mildew). R = resistant, S = sensitive

Leaf discs floating on solution							Roots in nutrient solution					
	0	0·1	1	10	100	500 ppm		0	0·5	5	50	500 ppm
S	5	0	0	0	0	0	S	5	5	0	0	0
R	5	5	5	5	3	0	R	5	5	5	4	0

tolerant strains showed cross resistance to the benomyl-related compounds thiabendazole and fuberidazole, and also to thiophanate-methyl. Cross-resistance to the latter compound, which is not a benzimidazole derivative, can be understood, since we know that both benomyl and thiophanate-methyl are converted to methyl benzimidazol-2-yl-carbamate (MBC), which is the actual fungitoxic compound (Selling *et al.*, 1970).

Often the fungicide resistance is of a high level. From a cucumber greenhouse in Holland, where benomyl had been applied against powdery mildew, a strain of *Sphaerotheca fuliginea* was obtained which, in a leaf disc test, tolerated concentrations of benomyl which were a hundred to a thousand times as high as those tolerated by the sensitive fungus (Table 9.1).

During recent years numerous other cases of acquired resistance have been reported from many parts of the world, predominantly to benzimi-

Table 9.2 Acquired resistance to systemic fungicides in practice

Benomyl
Botrytis cinerea on cyclamen (Bollen and Scholten, 1971), chrysanthemum (Watson and Koons, 1973), lettuce and tomato (Miller and Fletcher, 1974), raspberry and strawberry (Jarvis and Hargreaves, 1973) and cucumber (Iida, 1975)
Cercospora apii on celery (Berger, 1973)
C. arachidicola on peanut (Littrell, 1974)
C. beticola on sugarbeet (Georgopoulos and Dovas, 1973)
Cercosporidium personatum on peanut (Clark *et al.*, 1974)
Colletotrichum musae on banana (Griffee, 1973)
Erysiphe graminis on *Poa pratensis* (Vargas, 1973)
E. cichoracearum on cantaloup (Paulus *et al.*, 1972)
Fusarium oxysporum f. sp. *gladioli* on *gladiolus* (Magie and Wilfret, 1974)
Penicillium corymbiferum on lily (Bollen, 1971)
Sclerotinia homoecarpa on turf (Warren *et al.*, 1976)
Septoria leucanthemi on daisy (Paulus and Netzer, 1974)
Sphaerotheca fuliginea on cucumber (Schroeder and Provvidenti, 1969)
S. humuli on strawberry (Iida, 1975)
Venturia inaequalis on apple (Wicks, 1974)
Verticillium malthousei on mushroom (Wuest *et al.*, 1974)
V. dahliae on strawberry (Anonymous, 1973a)
Thiabendazole
Penicillium digitatum on citrus (Harding, 1972)
P. italicum on citrus (Harding, 1972)
Thiophanate-methyl
Botrytis spp. on egg plant (Tezuka and Kiso, 1975)
Erysiphe cichoracearum on egg plant (Iida, 1975)
Penicillium italicum on citrus (Iida, 1975)
P. fructigenum on citrus (Iida, 1975)
Sphaerotheca humuli on strawberry (Iida, 1975)
Venturia inaequalis on apple (Iida, 1975)
Oxycarboxin
Puccinia horiana on chrysanthemum (Abiko *et al.*, 1975)
Dimethirimol
Sphaerotheca fuliginea on cucumber (Bent *et al.*, 1971)
Ethirimol
Erysiphe graminis on barley (Wolfe, 1971)
Polyoxin
Alternaria kikuchiana on pear (Shimada *et al.*, 1973)
A. mali on apple (Iida, 1975)
Kasugamycin
Pyricularia oryzae on rice (Miura *et al.*, 1973)

dazole compounds and thiophanates, but also to several other new systemic fungicides. Examples are given in Table 9.2.

Origin of resistance

Resistance to fungicides may emerge as a consequence of genetic changes in the fungal cell. Although fungicide tolerance due to non-genetic factors has often been observed, the physiological and biochemical aspects of this phenomenon have been relatively little studied. Partridge and Rich (1962) accepted adaptive enzyme formation as the most probable origin of increased tolerance of *Penicillium notatum* to glyodin and captan. As this type of tolerance seldom reaches a high level, and is usually rapidly lost again after transfer to a fungicide-free medium, it seems of limited practical importance. Therefore no special attention will be paid to it in this chapter.

For studies of genetic aspects of acquired resistance use is made of fungi which allow genetic analysis, such as *Aspergillus nidulans* (perf. *Emericella nidulans*), *Neurospora crassa* and *Saccharomyces cerevisiae*. In addition to these fungi, experiments are carried out with plant pathogens, such as *Venturia inaequalis*, *Ustilago maydis*, *U. hordei* and *Hypomyces solani* (*Nectria haematococca*). The most extensive studies on development of fungicide resistance have been carried out with members of the aromatic hydrocarbon group, benzimidazole derivatives and related compounds, and oxathiins.

From crosses made between resistant and sensitive strains of fungi it has been shown that resistance to these fungicides may develop through mutation of any one of a number of Mendelian factors. Five loci for resistance to aromatic hydrocarbons have been identified in *Nectria haematococca* (Georgopoulos and Panopoulos, 1966) and two in *Aspergillus nidulans* (Threlfall, 1968). Resistance to benomyl, induced by u.v. irradiation, has been studied in nine strains of *Aspergillus nidulans* (Hastie and Georgopoulos, 1971). From crosses with the wild type fungus and subsequent ascospore analysis, it was concluded that benomyl tolerance in each strain was determined by a single gene mutation. The presence of two genes for resistance was demonstrated, located on linkage groups VIII and II. Van Tuyl (1975b) confirmed this and in addition discovered a third gene for resistance, located on linkage group VII. Induced resistance in *Ustilago hordei* to carboxin appeared to be controlled monogenically or polygenically (Ben-Yephet *et al.*, 1975) and in mutants of *U. maydis* resistant to carboxin three genes for resistance were shown to be present (Georgopoulos *et al*, 1975a; Van Tuyl, 1975a). In mutants of *Hypomyces solani* f. sp. *cucurbitae* four unlinked loci for resistance to dodine were found (Kappas and Georgopoulos, 1970), and in mutants of *Aspergillus nidulans* five loci for resistance to imazalil (Van Tuyl, unpublished).

The level of the resistance may or may not be different for different loci and further influences may result from the existence of more than one resistant allelomorph at each locus and also by the presence of modifying genes. The studies on resistance to aromatic hydrocarbons and to dodine in *Nectria haematococca* and to cycloheximide in *Neurospora crassa* (Vomvoyanni *et al.*, 1971; Vomvoyanni, 1974) have provided good examples of the existing possibilities in this respect.

When more genes, each of which confers resistance, are brought together into the same haploid nucleus, resistance may be additive, as was found for resistance to dodine in *Nectria haematococca* (Kappas and Georgopoulos, 1970), to imazalil in *Aspergillus nidulans* (Van Tuyl, unpublished) and to benomyl in *A. nidulans* (Van Tuyl, 1975b), or non-additive, as was demonstrated for aromatic hydrocarbons, where double mutants of *Hypomyces solani* f. *cucurbitae* appeared phenotypically indistinguishable from the single tolerant strain (Georgopoulos, 1963).

Resistance may be dominant, semidominant or recessive. Carboxin resistance was found semidominant in *Ustilago maydis* (Georgopoulos *et al.*, in the press), but dominant in *U. hordei* (Ben-Yephet, 1974); in *Aspergillus nidulans* two genes for carboxin resistance were semidominant and one dominant (Van Tuyl, 1975a). In benomyl-resistant isolates of the latter fungus all three genes for resistance appeared semidominant (Van Tuyl, unpublished). In *Saccharomyces cerevisiae* three genes for resistance to cycloheximide were semidominant, two were recessive and one dominant (Wilkie and Lee, 1965). Resistance to pentachloronitrobenzene in *A. nidulans* appeared recessive (Threlfall, 1968). Mutations for tolerance will be immediately expressed in haploid pathogens, and in diploid pathogens, when tolerance is dominant to sensitivity. When tolerance is recessive in diploid pathogens, the expression of resistance may be delayed until after segregation.

Heterokaryotic phenomena may also play a role in development of fungicide resistance. An example is resistance of *Botrytis cinerea* to dicloran (Webster *et al.*, 1970).

On media containing this fungicide tolerant sectors appeared frequently in the colonies. No training was necessary to obtain high levels of resistance. The authors express the opinion that heterokaryosis may serve as a mechanism in maintaining low levels of tolerance to dicloran in a multinucleate fungus such as *B. cinerea*. Dicloran resistance might emerge when the number of nuclei carrying resistance increases due to selection pressure exerted by the chemical. Evidence for such a phenomenon was also presented by Meyer and Parmeter (1968), who obtained tolerance in *Thanatephorus cucumeris* to pentachloronitrobenzene after heterokaryotic association from homokaryons both of which were intolerant.

Mutation for resistance to a fungicide may occur spontaneously, i.e. not deliberately induced by treatment with a mutagenic agent. The majority of these mutants will disappear from the fungal population, unless environmental factors act in favour of the mutant. This will be the case when a fungicide is present which discriminates between the mutant and the wild strain, thus exerting a selective action. It seems plausible that this is the most common way of emergence of fungicide resistance *in vitro* or in the field.

Indications were obtained that, in addition, some fungicides may also exert mutagenic activity. It has been reported that benzimidazole and thiophanate fungicides induce instability in diploid cells of *Aspergillus nidulans* (Hastie, 1970; Kappas *et al.*, 1974). The presence in these cells of recessive or semidominant genes for benomyl resistance may, after exposure to this fungicide, result in resistant mitotic segregants (Hastie and Georgopoulos, 1971). The genetic effect of benomyl on *Fusarium oxysporum* might possibly be explained in the same way (Dassenoy and Meyer, 1973). So far, no unequivocal evidence has been obtained that benomyl induces point mutations in fungal plant pathogens. The possibility that certain systemic fungicides may exert mutagenic action cannot be excluded, however, and should be examined carefully for newly developed fungicides.

Mechanism of resistance

The interesting question arises as to how a genetically determined resistance becomes expressed in the fungal metabolism. Elucidation of the mechanism of resistance will often require knowledge about the mechanism of action of the fungicide concerned.

Studies of various experimental antifungal compounds with selective action, such as antibiotics and antimetabolites, and research on systemic fungicides which have been recently introduced in practice, have greatly increased our understanding of fungicidal action. Fungicides may interfere with the metabolism in the fungal cell at various sites. Some examples of fungicidal action and acquired resistance are presented in Table 9.3.

Table 9.3 Sites of action of fungicides and some examples of acquired resistance in fungi

Site	Interference with	Fungicide	Acquired resistance
Cell wall	chitin synthesis	polyoxin D	*Alternaria kikuchiana*[1]
Protoplast-membrane	permeability	dodine	*Venturia inaequalis*[2]
		nystatin	*Saccharomyces cerevisiae*[3]
Mito-chondria	energy production	oxycarboxin	*Puccinia horiana*[4]
		antimycin A	*Venturia inaequalis*[5]
Ribosomes	protein synthesis	kasugamycin	*Pyricularia oryzae*[6]
		cycloheximide	*Neurospora crassa*[7]
Nucleus	nucleic acid metabolism	6-azauracil	*Cladosporium cucumerinum*[8]
	functioning of spindle	benomyl	Various fungi (see Table 9.2)

[1] Hori *et al.* (1974) [2] Szkolnik and Gilpatrick (1969) [3] Patel and Johnson (1968) [4] Abiko *et al.* (1975) [5] Leben *et al.* (1955) [6] Helser *et al.* (1971) [7] Vomvoyanni *et al.* (1971) [8] Dekker (1968)

Table 9.4 Types of resistance to fungicides

Fungicidal activity *in vitro*	Decreased permeability	Detoxification (a) no conversion into toxic compound (b)	Decreased affinity at site of action	Circumvention (a) compensation (b)	Type of resistance
+	+	. . . .	. . . .	. . . .	I
−	−	+	. . . .	. . . .	IIa, IIb
		−	+	. . . .	III
			−	+	IVa, IVb
				−	no resistance

It appears that fungi possess the potential for a wide range of metabolic variation resulting in resistance to those fungicides which interfere at specific sites with the fungal metabolism. The fungal cell may be changed in such a way that the fungicide does not reach the site of action. This may be due to a decreased permeability of the protoplast membrane, or to an increased capacity of the fungus to detoxify the fungicide. When the fungicide does reach the site of action, resistance may be due to a decreased affinity at the site of action to the fungicide or to other changes in the fungal metabolism, resulting in a compensation for the inhibiting effect or a circumvention of the blocked site (Table 9.4).

It seems plausible that resistance to general plasmatoxicants, to which many of the conventional fungicides belong, must be explained by a decreased permeability of the protoplast membrane or by detoxification before the site of action has been reached. Resistance to selective fungicides might be due also to the other mechanisms mentioned above. Examples of each of these resistance mechanisms are given below.

I. Decreased permeability

Resistance in pathogens to both the antibiotics blasticidin-S and polyoxin-D has been attributed to a decreased permeability of the fungal protoplast membrane to these antibiotics.

Strains of *Pyricularia oryzae* were obtained in laboratory experiments, which tolerated a concentration of 1 000–4 000 μg/ml of blasticidin-S in the medium. This antibiotic is known as an inhibitor of protein synthesis. From studies with cell-free extracts of *P. oryzae* it appeared that it inhibits protein synthesis in extracts of resistant strains as well as in those of sensitive strains. It was concluded that in the intact cell the antibiotic is unable to reach the sites of protein synthesis, possibly due to decreased permeability of the protoplast membrane (Huang *et al.*, 1964).

The antibiotic polyoxin-D is known to interfere with chitin synthesis by competitive inhibition of the enzyme chitin synthetase (Hori *et al.*, 1971). Polyoxin-resistant strains of *Alternaria kikuchiana*, the cause of pear black spot, were isolated from orchards, where disease control by this antibiotic failed (Nishimura *et al.*, 1973). Experiments with cell-free extracts of the fungus showed that the antibiotic also inhibits the chitin synthetase in extracts from the resistant strains; it was assumed that resistance was due to decreased permeability of the protoplast membrane (Hori *et al.*, 1974).

IIa. Increase of detoxication

Detoxication of a fungicide may occur by a modification of its molecule or by binding to cell constituents, with a concomitant loss of fungicidal activity after entrance into the cell. Evidence has been obtained that naturally-occurring insensitivity of a fungus to a fungicide may be due to this phenomenon. *Fusarium oxysporum* f. *lycopersici*, which is resistant to ascochitin, appears to be able to reduce this antibiotic to the less fungitoxic dihydroderivative (Nakanishi and Oku, 1969). When comparing fungi resistant to quintozene (pentachloronitrobenzene), such as *Fusarium oxysporum* f. *niveum* and f. *lycopersici*, with quintozene-sensitive fungi, such as *Rhizoctonia solani*, the same authors (1970) found a positive correlation between quintozene resistance and ability to convert this fungicide into the much less fungitoxic compounds pentachloroaniline and pentachlorothioanisole. That not only a difference in sensitivity between various fungi, but also between strains of one fungus might be explained by this phenomenon, was indicated by the finding that the same correlation was shown for two strains of *Botrytis cinerea* (Nakanishi and Oku, 1970). Therefore it seems quite possible that acquired tolerance might also be attributed to detoxication.

Studies with dodine-resistant isolates of *Nectria haematococca* f. sp. *cucurbitae* also indicate that acquired resistance might be due to detoxication; an increased capacity of tolerant strains to convert dodine into less fungitoxic compounds was observed (Kappas and Georgopoulos, 1970).

Development of resistance due to inactivation of the fungicide by binding to cell constituents has been suggested for organic mercury resistance of *Drechslera avenae*, but the results obtained by various authors have been contradictory (Greenaway, 1971; Ross and Old, 1973a, b; Ross, 1974).

IIb. Decrease of conversion into fungitoxic compound

As a counterpart of detoxication the fungus may convert an inactive compound into a fungicide, which is called lethal synthesis. This appeared to be the case with a 6-azauracil (AzU)-resistant strain of *Cladosporium cucumerinum*, obtained after exposing conidia to ultra-violet irradiation. For antifungal activity AzU has first to be converted to 6-azauridine-5′-phosphate (AzUMP), which is synthesised by this fungus from AzU via 6-azauridine (AzUR) in the following way:

$$\text{AzU} \xleftrightarrow[\text{uridine phosphorylase}]{} \text{AzUR} \xleftrightarrow[\text{uridine kinase}]{} \text{AzUMP}$$

The final product, AzUMP, inhibits one of the enzymes of the pyrimidine biosynthesis *de novo*. The resistant strain was unable to convert AzU into AzUR, so that the toxicant AzUMP was not formed. This type of acquired resistance was apparently due to a loss of activity of the enzyme uridine phosphorylase. No resistance was observed when AzUR or AzUMP were administered to the fungus (Dekker, 1967). A second AzU-resistant strain showed a decreased conversion of AzUR into AzUMP (Dekhuijzen and Dekker, 1971).

Evidence has been obtained that, for fungitoxic action, pyrazophos has to be converted into 2-hydroxy-5-methyl-6-ethoxycarbonyl-pyrazolo(1,5-a)pyrimidine. In that case a decreased ability of a fungus to perform this conversion might result in resistance. In laboratory experiments, however, no resistance to pyrazophos could be induced by u.v. irradiation (de Waard, 1975).

III. Decreased affinity at the site of action

When a fungicide reaches the site of action without being detoxified, tolerance of the fungus may be due to lack of affinity for the inhibitor at the reactive site. Various examples of this type of resistance are available. It has been shown that methylbenzimidazol-2-yl carbamate (MBC) (carbendazim), which is considered to be the main toxic principle of benomyl and thiophanate-methyl, inhibits the separation of chromosomes at mitosis (Hammerschlag and Sisler, 1973; Davidse, 1973). Many cases have been reported of development of resistance in fungi to these fungicides, either in the field or after u.v. irradiation in the laboratory. The mechanism of action of carbendazim and the nature of the acquired resistance to this fungicide have been studied in detail by Davidse (1973, 1975a, b), using *Aspergillus nidulans* as a test organism. He found that the antimitotic activity of carbendazim in *A. nidulans* resembled that of colchicine in human cell cultures. The latter compound binds to protein subunits of the spindle microtubules, thus preventing assembly of the spindle and mitosis. In experiments with cell-free extracts of *A. nidulans* carbendazim appeared to be bound to some macromolecule, which was very probably identical with microtubular protein. The strongest binding was found with a supersensitive mutant and no binding at all with a resistant mutant. The close correlation between binding of carbendazim to the macromolecule and sensitivity to carbendazim indicates that this binding is responsible for antimitotic activity. Development of resistance may then be ascribed to a slight change in the protein subunits of the spindle due to mutation, resulting in a decreased affinity for carbendazim.

It is known that the antifungal activity of the antibiotic cycloheximide is due to inhibition of protein synthesis on the ribosomes. Cooper *et al.* (1967) found that in tolerant strains of *Saccharomyces cerevisiae*, obtained from sensitive organisms by gene mutation, resistance was located in the ribosomes. It seems possible that, by gene mutation, the composition of the ribosome has been slightly altered in such a way that affinity to cycloheximide has decreased without impeding the capacity of the ribosome to synthesise proteins.

Development of resistance in *P. oryzae* to the antibiotic kasugamycin, an inhibitor of protein synthesis, might be attributed to a change in affinity of the ribosomes to this antibiotic. The nature of resistance to kasugamycin, however, has not yet been studied in fungi, but only in bacteria (Helser *et al.*, 1971).

When a fungicide primarily acts upon one particular enzyme, a slight change in this enzyme as a result of gene mutation may be the cause of loss of affinity to the fungicide without loss of its normal capacities, thus conferring resistance. Such a case has been demonstrated by Lewis (1963) in a study with the fungus *Coprinus lagopus* and ethionine. This compound is an antimetabolite of methionine and interferes with the so-called 'methionine-activating' enzyme. In the resistant mutant this enzyme had been changed in such a way that it lacked affinity to ethionine, but not to methionine. This could be explained as a result of a mutation in the structural gene, which forms the messenger RNA coding for this enzyme.

IVa. Circumvention

If a fungicide blocks a reaction at one site in the fungal metabolism, the fungus may adapt to this situation by shifting its metabolism in such a way that the blocked site is bypassed.

It is known that antimycin A acts upon the electron transport in the respiratory chain between cytochromes *b* and *c*. Tolerance of *Ustilago maydis* to this antibiotic could be attributed to a shift in the electron transport at a site preceding cytochrome *b* to an alternate terminal oxidase (Georgopoulos and Sisler, 1970). This alternate pathway is apparently of considerable value to *U. maydis* for growth in the presence of the antibiotic, which is demonstrated by the fact that a mutant, which lacked this alternate pathway, appeared sensitive to antimycin A. This example concerns a mutation from a normally tolerant fungus to a sensitive one, and it is not known whether the reverse might be possible in a normally sensitive fungus. The resistance mechanism of antimycin-A-tolerant strains of *Venturia inaequalis*, obtained by Leben *et al.* (1955) after irradiation of conidia with ultra-violet light, has not been investigated. It is of interest to note that the antimycin-A sensitive mutant of *U. maydis* was resistant to the oxathiins carboxin and oxycarboxin. This indicated a relation between the action of the antibiotic and these compounds. In the meantime it has been shown that the oxathiins inhibit respiration by preventing the transport of electrons in the mitochondrial respiratory chain, and that their main site of action lies between succinate and coenzyme Q in complex II (Georgopoulos *et al.*, 1975a).

IVb. Compensation

When a particular enzyme is the primary site of attack by a fungicide, increased tolerance may be the result of the ability of the organism to compensate for the effect of the inhibition; for example, by an increased production of the inhibited enzyme. An example of this phenomenon is strain R II-1 of *Cladosporium cucumerinum*, which is resistant to 6-azauracil. As has been explained earlier, 6-azauracil, after conversion into its nucleotide 6-azauridine monophosphate, inhibits primarily one enzyme in pyrimidine biosynthesis, namely, orotidine monophosphate decarboxylase. The activity of this enzyme was determined in the resistant strain and the wild type fungus, using carboxyl-labelled orotidine monophosphate, according to a method described earlier (Dekker, 1968b). It appeared that the resistant strains produced at least three times as much of the enzyme than the sensitive fungus (Dekker, 1971b). As a consequence the resistant fungus needs more of the fungicide to be inhibited, i.e. is more tolerant.

Cross-resistance

The term 'cross-resistance' is used when a change in one genetic factor results in resistance to different fungicides. It usually occurs only between compounds with a similar mechanism of action. In rare cases a mutation may increase tolerance to one fungicide and decrease it to another one. This phenomenon, which is called 'negatively correlated cross-resistance', has been reported for phosphoramidate and phosphorothiolate fungicides in *Pyricularia oryzae* (Uesugi *et al.*, 1973). An

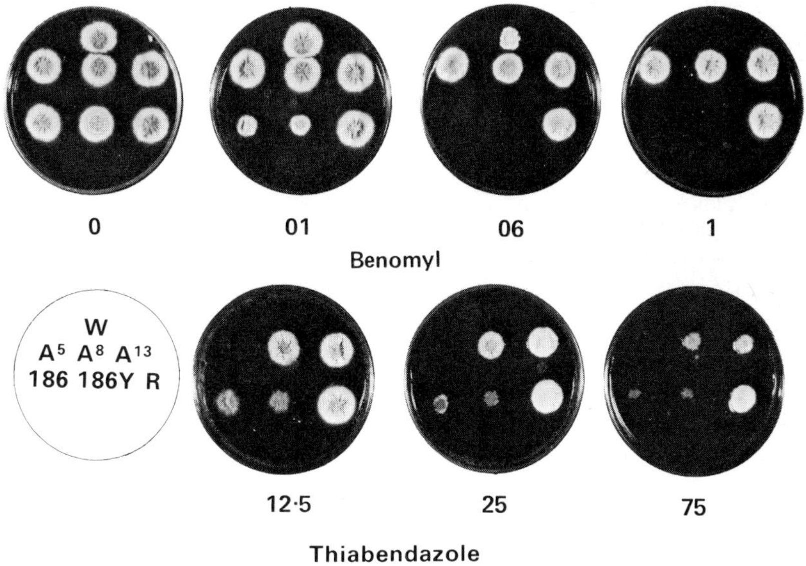

Fig. 9.2 Cross-resistance and lack of cross-resistance in strains of *Aspergillus nidulans*, isolated from medium containing benomyl (A_5, A_8, A_{13}) or thiabendazole (186, 186Y, R); w = wild type fungus. Colonies replicated on medium containing increasing concentrations of benomyl or thiabendazole, in ppm.
A_8, A_{13}, R: resistant to both compounds.
A_5: resistant to benomyl only.
186, 186Y: resistant to thiabendazole only, extrasensitive to benomyl.

example of cross resistance between thiabendazole and benomyl, and of exceptions to this rule, is shown in Fig. 9.2.

Mutation for resistance *in vitro*

The frequency with which fungicide-resistant strains emerge *in vitro* depends on the type of fungicide, the nature of the pathogen and environmental conditions. For an estimate of the potential of a fungus to develop fungicide resistance, use is often made of chemical or physical mutagenic agents. In this way little or no genetic resistance has been found to conventional fungicides, such as dithiocarbamates and metal toxicants. Most of these fungicides are 'multisite inhibitors', which react at many sites in the fungal cell. Mutation leading to changes at all sites involved is unlikely or impossible and a significant increase in tolerance due to changes in the permeability of the protoplast membrane or detoxication of the fungicide does not seem to occur frequently with these fungicides. The situation is, however, different for fungicides which act primarily at a specific site in the fungal metabolism. In several cases it was shown that mutation in one gene may confer resistance. This raises the question whether all site-specific fungicides eventually will encounter resistance. The rate of mutation for resistance to these compounds may depend on the number of mutations required for a certain level of resistance and the mutability of the genes at the loci concerned. It seems, moreover, of importance to determine how far mutation for resistance influences the vitality of the mutant; if the organism possesses little or no flexibility for change at the site of attack, mutants may tend to be lethal.

Among the fungicides with specific action are compounds which interfere with biosynthesis of vital compounds, energy production, permeability, mitosis, etc. The antifungal antibiotics cycloheximide and kasugamycin are selective inhibitors of protein synthesis. Resistance to these compounds can be readily induced in various fungi, and appears associated with the functioning of the ribosome (Cooper *et al.*, 1967; Helser *et al.*, 1971). By mutation for resistance the protein-synthesising apparatus apparently changes in such a way that it shows a decreased affinity for the inhibitor without concomitant loss of the capacity to synthesise proteins. The situation seems different for a number of antibiotics which interfere specifically with the replication or transcription of DNA, such as bleomycin, phleomycin, lomofungin, edeine, thiolutine, phytoactin, actinomycin and related compounds; mutational modifications in fungi resulting in resistance to these compounds seem to be rare (Dekker, 1977). It might be that the organism allows little or no change at these sites, so that corresponding mutations tend to be lethal.

The oxathiins carboxin and oxycarboxin are inhibitors of respiration. In the laboratory, mutants of *Ustilago maydis* and *U. hordei* were readily obtained (Georgopoulos *et al.*, 1975c; Ben-Yephet *et al.*, 1974). In *Aspergillus nidulans* the frequency of mutation for resistance was one in $5{\cdot}10^6$ conidia, surviving doses of u.v. irradiation, which killed 95 per cent of the conidia.

Inhibitors of mitosis in fungi are carbendazim (MBC), the fungitoxic principle of benomyl and thiophanate-methyl, and the antibiotic griseofulvin. From 14×10^8 conidia of *A. nidulans*, of which 6 per cent survived u.v. irradiation, 211 benomyl-resistant colonies were obtained (Van Tuyl *et al.*, 1974). Ultra-violet irradiation of *Fusarium oxysporum*, which killed 95 per cent of the spores, yielded one benomyl-resistant mutant per $4{\cdot}6 \times 10^5$ survivors. It has been shown that benomyl resistance may occur as the result of a single gene mutation. No resistance, however, has been induced in fungi to griseofulvin.

Several antifungal compounds are known to interfere with permeability. Among these are a number of polyene macrolide antibiotics, the antifungal activity of which is related to the affinity to certain sterols in the protoplast membrane (Lampen *et al.*, 1962). Mutations leading to alterations in sterol patterns may confer resistance (Molzahm and Woods, 1972), but the degree of resistance is generally low. Although some of these antibiotics, such as nystatin and pimaricin, have been used for medical purposes for a number of years, no resistance problems have emerged; it seems that increased tolerance is associated with diminished virulence (Hamilton-Miller, 1974).

In some cases acquired resistance to a fungicide is accompanied by reduced vigour. Mutants of *Cladosporium cucumerinum* resistant to triforine, an inhibitor of sterol biosynthesis, have been reported to be less virulent (Fuchs and Viets-Verweij, 1975). Moreover spores of triforine-resistant mutants showed a decreased ability to germinate, which was counteracted by addition of ergosterol (Fuchs, 1976).

From the data available it is clear that the rate of emergence of resistance and its level may vary for different types of specific fungicides; probably also the vigour of resistant mutants may, at least in some cases, be related to the mode of action of the fungicide concerned.

It is not yet clear how far the nature of the pathogen is of importance with respect to development of fungicide resistance. It seems that, when mutation for resistance occurs readily in some fungi, it may be expected also in other fungi; from studies at our laboratory it appeared that resistance to benomyl could be induced in all fungi tested (Van Tuyl, 1975b). The readiness with which such mutations are expressed may, however, depend to some extent on the type of fungus. This will occur immediately in haploid pathogens and in diploid pathogens when tolerance is dominant. Expression of resistance may be delayed when this is not the case and further with pathogens in a dicaryotic or multinucleate phase.

Emergence of resistance in the field

Laboratory experiments with or without the use of rays or other mutagenic agents may provide an estimate of the potential of a fungus for mutation towards resistance to a fungicide. They do not, however, answer the question whether resistant mutants will survive and become economically significant under field or greenhouse conditions. If a mutation for resistance has a pleiotropic effect, resulting in increased tolerance concomitant with a decreased virulence, problems in practice will be delayed or may not emerge. In this connection it is of interest to note that isolates of *Pyricularia oryzae*, resistant to the antibiotic blasticidin-S, were readily obtained in the laboratory, but never from the field (Nakamura and Sakurai, 1968; Uesugi *et al.*, 1969). Kasugamycin-resistant strains of this fungus, obtained in the laboratory, were reported to be less infectious on rice plants (Ohmori, 1967), and for several years this antibiotic has been used for control of rice blast without resistance problems. In recent years, however, a decreasing effect of the antibiotic in certain areas appeared due to development of resistance (Misato, 1975).

In our work with 6-azauracil and *Cladosporium cucumerinum* two out of fifteen resistant strains had lost their pathogenicity to cucumber plants, while a few others appeared less virulent. Most of the resistant strains, however, were as virulent as the wild type fungus. Tolerance to pentachloronitrobenzene in *Sclerotium rolfsii* was not associated with loss in pathogenicity (Weigun Yang and Lung-Chi Wu, 1971). Also in polyoxin-tolerant strains of *Alternaria kikuchiana*, isolated from the field (Sakurai and Shimada, 1974), or in dodine-tolerant strains of *Nectria haematococca* (Kappas and Georgopoulos, 1971), a relation between resistance and pathogenicity was not found. Wolfe (1971) concludes that most of the recorded examples of resistance do not necessarily seem to be associated with changes in pathogenicity, assessed in laboratory studies, and that the potential might therefore be said to exist for the evolution of forms which possess the combinations of tolerance and pathogenicity necessary for successful competition with sensitive forms of the fungus.

Although virulence and other factors which contribute to survival and frequency of resistant mutants in the field, such as infection chance, speed of colonisation and reproduction rate, can be investigated in laboratory experiments, these do not provide all the information necessary to evaluate the competitive ability of a resistant population in the field. This is argued by Wolfe (1971) in the following way. 'The theory of genetic homoeostasis indicates that, although particular individuals with a characteristic such as fungicide tolerance may be selected, the whole population does not easily shift in this direction since there are many characteristics in the population which are held in a complex balance in the existing environmental situation. Thus, if the new selection pressure is relaxed, or incomplete, the population may tend to shift back to its previous state of balanced adaptation.'

Recently some interesting examples of this phenomenon have been reported. Although kasugamycin-resistant and sensitive isolates of *Pyricularia oryzae* showed no differences in mycelium growth on agar, sporulation and in size and number of lesions after inoculation (Miura *et al.*, in the press), the resistant populations in the field dropped sharply after application of the antibiotic was stopped (Misato, 1975). Similar observations were made with benomyl-resistant strains of *Cercospora beticola* on sugarbeet in the USA (Ruppel, 1975) and with a polyoxin-resistant population of *Alternaria kikuchiana* (Kohmoto *et al.*, 1974). A decline of the benomyl-resistant population of *Sphaerotheca fuliginea* (see also Table 9.1) was found in greenhouse experiments with cucumber. Plants were inoculated with various mixtures (90 : 10, 50 : 50 and 10 : 90) of conidia from the resistant and the sensitive fungus; after sporulation the newly produced conidia were used for inoculation of a new batch of healthy plants. After this procedure had been repeated five times, the resistance had almost disappeared from the 90 : 10 mixture, and completely from the other mixtures (Dekker, unpublished).

Although a decline of the resistant population in absence of the fungicide may be the rule, the possibility exists that mutants arise which can compete with sensitive forms. It seems, for example, that the benomyl-resistant population of *Cercospora beticola* on sugarbeet in Greece, 3 years after application of benomyl was stopped, has not yet declined (Georgopoulos, in the press). Even the emergence of resistant strains with a greater virulence than the wild type fungus cannot be excluded. Evidence for this was obtained in studies of benomyl-resistance in *Verticillium malthousei*, a mushroom parasite (Bollen and Van Zaayen, 1975).

Wolfe (1975) has further pointed out that low initial fitness of resistant mutants may be altered, either through selection pressure or through evolution of improved fitness by selection of modifiers, or both.

Besides the properties of the resistant mutants, the type of disease may be of importance with respect to development of resistance problems in practice. It seems plausible that resistance may spread more rapidly in a population of a heavily sporulating pathogen than in that of a slowly growing parasite which does not produce aerial spores, e.g. *Ceratocystis ulmi* in elm (Brasier and Gibbs, 1975). If, moreover, it concerns a type of disease where the pathogen is difficult to eliminate by a fungicide, competition by the wild type fungus may counteract the spread of the resistant mutants. This might explain why control of strawberry wilt with benomyl and thiophanate-methyl has not shown evidence of increased tolerance, although fungicide-resistant strains of *Verticillium dahliae* were obtained in laboratory experiments (Anonymous, 1973a).

Development of fungicide resistance may also influence the microbiological balance in the soil or on the plant surface. In experiments on control of Botrytis rot on cyclamen with benomyl it was found that plants infected with benomyl-resistant forms of *B. cinerea* were more severely diseased than plants in untreated greenhouses. This is not necessarily due to an increased virulence of the resistant strain, but might also be caused by the inhibition of benomyl-sensitive antagonistic fungi. It was noticed that after prolonged benomyl treatment disease incidence again appeared reduced in some plants. From these plants

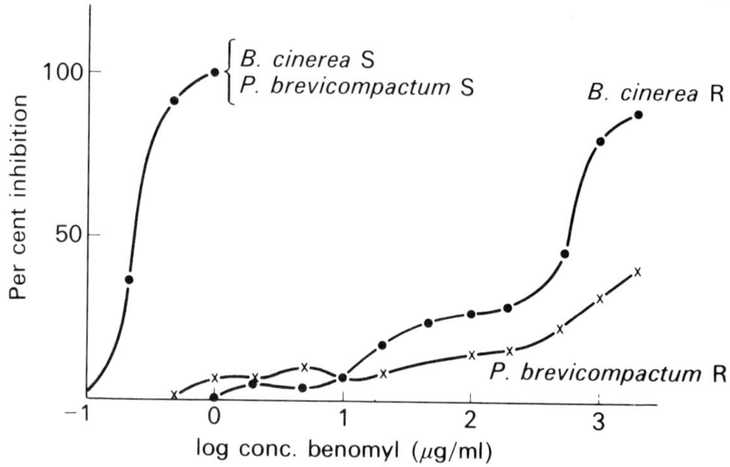

Fig. 9.3 Inhibition of mycelial growth of benomyl-sensitive (S) and benomyl-resistant (R) strains of *Botrytis cinerea* and *Penicillium brevicompactum*, isolated from cyclamen.

Bollen (1971) isolated two strains of *Penicillium brevicompactum*, non-pathogenic to cyclamen, which also had acquired resistance to benomyl. Both strains tolerated a concentration of 1 000 ppm of benomyl in an agar medium, while growth of the wild type fungus was completely inhibited at a concentration of 0·5 ppm (Fig. 9.3). When tested on malt agar, a striking antagonism between benomyl-resistant strains of *B. cinerea* and *P. brevicompactum* was obtained. This antagonism, which was shown by a growth-free zone between the two fungi, was stronger in the presence of high concentrations of benomyl than without this fungicide in the medium. This indicates that the use of selective fungicides may interfere with the relation between both pathogenic and non-pathogenic micro-organisms.

Similar phenomena probably also occur in the soil, as the activity of plant pathogens is influenced by the presence of other micro-organisms, which compete for food and space, or produce antibiotic substances. The existing balances may be disturbed by the introduction of selective fungicides, to which some components of the soil microflora may develop resistance more readily than others.

As little information is yet available on ecological and epidemiological aspects of fungicide resistance, studies in this area should be encouraged.

Manipulation of fungicide resistance

Search for new fungicides

It has been argued above that the possibility of emergence of fungicide-resistant forms of fungi depends among other factors upon the mode of action of the fungicide. Elucidation of the mechanism of action usually requires a time-consuming investigation. Further, it is impossible yet to predict whether a fungus may acquire resistance to a fungicide, solely on the basis of its chemical structure. It is nevertheless of great value to obtain an impression of the potential of fungi to develop resistance to a promising chemical before it is released for practical use. Ultra-violet irradiation or other mutagenic treatments may be helpful in this respect; if such treatments do not result in emergence of resistance, it seems unlikely that it will occur under practical conditions. If resistance does emerge in these experiments, greenhouse experiments may provide further information about the properties of the resistant mutants.

In the search for new chemicals for control of fungal plant diseases, it seems worthwhile also to pay attention to compounds which, though not fungicidal *in vitro*, nevertheless enhance the resistance of the host plant indirectly. This artificially induced resistance may impose a lower degree of selection pressure on the fungus population than the more highly selective systemic fungicides. Possibly this might be compared to the so-called 'horizontal' or polygenic resistance of a crop to a fungal pathogen, which is not easily broken down by a new strain. It has been shown that various experimental systemic compounds may protect plants in this way (see Chapter 6). An example is procaine hydrochloride, which after introduction into cucumber plants provides protection against powdery mildew in spite of the fact that it is not a fungicide *in vitro* (Niemann and Dekker, 1966a). It scarcely inhibits germination of conidia of *Sphaerotheca fuliginea*, nor elongation of germ tubes *in vitro* (Van't Land and Dekker, 1972). At present, however, no information about the potential of fungi to acquire resistance to this type of systemic compounds is yet available.

Application of fungicides

When using systemic fungicides, to which resistance in fungi may develop, the build-up of a resistant population of the pathogen in the field may be influenced by the way in which these fungicides are applied. Application of high doses of a fungicide will of course reduce the rate of emergence of resistance, but, if tolerant strains have appeared, frequent application of high doses of the same compound will maintain a high selection pressure, which will promote the build-up of a resistant population. Also the method of application may play a role with respect to continuation of the selection pressure. Especially after in-furrow or seed-treatment of a selective fungicide, the chemical may be present at lethal or sublethal concentration for an extended period. Also mutants with a low level of resistance may be selected, when the fungicide gradually decreases due to breakdown and dilution.

Thus the persistence of a fungicide or its persistent use may be important factors in relation to selection of resistant mutants. In view of this, discontinuance of use of ethirimol on winter barley has been advised in order to increase the proportion of the non-selective versus the selective periods in the asexual cycle, so that homoestasis or stabilising selection could better operate (Wolfe and Dinoor, 1973).

When development of fungicide resistance may be expected, application of a combination of fungicides should be considered. In this case compounds should be used with different mechanisms of action in order to avoid failure of control by the phenomenon of cross-resistance. A combination of different compounds will at least delay the appearance of resistance. The integrated use of benomyl and dimethirimol for control of cucumber powdery mildew did not result in emergence of forms resistant to both compounds (Ebben and Spencer, 1973). Even the combination of a systemic fungicide with a conventional one seems useful, since the latter may reduce the sporulation and spread of mutants which have become resistant to the systemic compound. The problem of acquired resistance probably would be solved if combinations of compounds could be found with negatively correlated cross-resistance, so that mutants resistant to one of the fungicides would be eliminated by the other compound, and the reverse. Uesugi and co-workers (1969, 1973) described negatively correlated cross-resistance in *Pyricularia oryzae* between phosphoramidate and phosphorothiolate fungicides. Indications that possibilities in this direction may exist were also obtained with other fungicide combinations (Georgopoulos and Sisler, 1970; Van Tuyl *et al.*, 1974).

The selection may be increased undesirably by application of a fungicide, to which resistance may develop, in such a way that the sensitive population of the pathogen is eliminated in a geographically or artificially (greenhouse) isolated area, so that competition with the resistant forms is decreased sharply.

In order to reduce the emergence of resistance in the field, Wolfe (1971) also advocates the use of fungicides preferably on host varieties which have at least a moderate degree of disease resistance.

The question arises whether, by stopping the application of a systemic fungicide to which resistance has developed, resistance will disappear in the fungal population after a period of time, so that the same fungicide might be used again. As this problem has not yet been studied with respect to fungicide resistance, it is of interest to consider the results obtained in experiments with insects, which have become resistant to new organic insecticides. These led Keiding (1967) to the conclusion that a high level of resistance may persist for a long time and will usually decrease only partially when the selection pressure is relaxed. Even if a population shows an apparently complete reversion of resistance, the potential of a new development is usually much higher than before the pesticide was used, which indicates that resistance, at least against insecticides, is essentially a 'one-way street'. If this should hold true also for fungicides, it should, however, not exclude the possibility of 're-use' of a fungicide as a secondary compound, following the use of a new primary compound, to help reduce the pressure for emergence of isolates tolerant to the new compound.

Conclusion

The overall rate of mutation towards fungicide resistance depends on the type of fungicide which is used. Resistance to conventional fungicides, most of which are multi-site inhibitors, does not develop easily. With respect to systemic fungicides, most of which are site-specific, the rate of mutation for resistance varies widely for different fungicides. In some cases resistance may be acquired readily by single-gene mutation; in other cases, however, mutation for resistance seems to be rare or absent. Normally resistance arises 'spontaneously', followed by selection of the tolerant forms by the fungicide.

Whether emergence of resistant mutants will lead to resistance problems in practice depends on the properties of resistant mutants, the type of disease, the persistence and method of use of the fungicide and environmental conditions. Fewer problems are likely to arise when increased tolerance is accompanied by lowered virulence or vigour. When using fungicides, to which resistance in fungi may develop, a continuous selection pressure by one type of fungicide should be avoided, which might favour the build-up of a resistant population. Matters of consideration in this respect are, among others, the doses of the fungicides, the frequency with which they are applied, the methods of application, discontinuation of the use of a particular compound and the possibilities of combining fungicides with different modes of action. Use of a new fungicide on a large scale in practice should be discouraged before the potential of fungi to develop resistance to the compound has been tested in the laboratory, or the properties of resistant mutants have been studied on the plant either combined or alternately. The use of the fungicides should be restricted in space and time, as much as economic disease control allows, and the method of application carefully considered, so that selection pressure is not increased unnecessarily. The question remains, however, whether a combination of different chemicals or a restricted use as mentioned above will appear feasible and economical.

Chapter 10

Efficient use of systemic fungicides

(the late) E. Evans

In agriculture the effective use of any treatment implies both economic and technical satisfaction. It is therefore proposed to discuss both these subjects under the headings of,

economic considerations
application technology
future strategy

Economic considerations

In practical terms the first decision relating to the use of any fungicide is that of whether or not to apply fungicide at all. This entails distinguishing between the 'damaging threshold' and 'economic threshold'. The former is defined as the minimum level of disease that affects yield while the latter is that at which a disease control programme would have been economically worth while. The farmer's criterion for economic satisfaction is based on the formula (Headley, 1973):

$$\frac{B}{C} \geqslant 1$$

where B is the expected benefit from treatment, C is the total cost of application.

In this context, the economic threshold is that level of disease at which the benefit/cost ratio would have been at least 1. This is a vast oversimplification of the practical situation relating to plant diseases which until the introduction of systemic fungicides were almost wholly controlled by prophylactic treatments of various kinds at a time when the disease was not even known to be present in the crop.

The problems of assessing crop losses are extremely complex (James, 1974) and the conversion of these findings to monetary values even more so, as the price of any crop will vary throughout the year. Economists are well aware of these changes and recognise that while the demands for certain crops such as fruit, coffee and potatoes are inelastic and not greatly affected by prices, those of cereals are elastic and tend to increase as prices are lowered (Ordish, 1962).

Treatment costs

In discussing the debit side of the cost/benefit equation it should be realised that total treatment costs involve more than that of the chemical itself, and that in practice the aim is always to reduce this value as much as possible. This is achieved by:

selecting the cheapest effective product;
using the minimum necessary dosage;
making the least number of treatments;
using the most convenient method of application.

The cheapest products are not necessarily the most effective, as most of the new materials, including systemics, are only introduced when they possess some technical advantage over those already available. As the costs of development of all new pesticides are very high and rapidly increasing (van Tiel, 1975) it follows that the newer products are also more expensive.

The minimum necessary dosage is usually that recommended by the manufacturer as being the dosage which was experimentally found to be reliable in terms of disease control and crop safety, with corresponding yield and quality improvements, Since this is an experimental value it cannot be predicted except in general terms, noting that when systemic and non-systemic fungicides are compared as protectants, e.g. in the control of apple scab, the systemic materials are found to be more active weight for weight of ingredient (Evans *et al.*, 1971). However, when used curatively for the control of apple mildew the effective dosages are very similar.

The number of treatments necessary is a function of both the method of application and the dosage recommended. Thus while a seed dressing treatment and a foliar spray for the control of cereal mildew may not be technically comparable this is a perfectly valid practical comparison, with advantages and disadvantages for both techniques.

Since infection conditions often occur very erratically the use of routine spray schedules of any fungicide sometimes entails unnecessary applications being made. In order to avoid this, various spray forecasting systems have been developed for different diseases. Even so, a farmer with a large area may find it difficult to treat all his crop between one infection period and the next, and the only kind of product that would then be acceptable would be one that could destroy the pathogen after infection had occurred, i.e. a curative fungicide. It is perhaps significant that most present-day systemic fungicides can act in this fashion and it is suggested that this aspect of their activity should be exploited wherever possible in order to limit the number of applications made.

The methods of application used in a given situation vary according to the biology of the pathogen concerned, the physical–chemical properties of the active ingredient and the kind of biological activity it possesses. This subject is discussed later in this chapter where it will be noted that the chosen method is based on a combination of technical and economic considerations.

Technical/economic benefit

The precise benefit from a given treatment is always difficult to predict both in technical and economic terms. However, Watts Padwick (1956) recognised three different categories of loss, listed respectively as due to killing diseases, debilitating diseases and diseases that cause direct yield and quality losses at harvest and in store. Of these, the killing diseases, usually root or general systemic pathogens, are undoubtedly the most damaging in economic terms particularly on perennial crops, since a dead plant involves a lost capital as well as a loss of yield during the replanting period. Prior to the introduction of systemic fungicides such problems could not be corrected although they might have been avoided by careful management. By comparison it is now sometimes economically worth while and technically possible to control certain soil borne pathogens such as Fusarium wilt of carnations (Scholten, 1970).

Most of the economically important foliar plant diseases are best described as debilitating organisms that attack the stem or leaves of the plant and thus reduce the amount of functional green tissue that may contribute to yield. The losses incurred are not simply relative to the amount of fungus present, the site and time at which infection occurs being equally important (James, 1974). For these reasons it is extremely difficult to assess the economic importance of any pathogen; the epidemic intensity of which also varies from place to place and from season to season.

This variability makes treatment decisions extremely difficult and introduces a risk element into all calculations (Gilmour and Fawcett, 1973) based on past experience rather than factual information relating to the epidemic under consideration (James, 1974).

The third category listed by Watts Padwick concerned crop losses due to yield and quality reductions at harvest and in store. These can be quite considerable depending on the immediate pre-harvest environmental conditions as well as those in store. By this stage the crop will have already absorbed most of the growing and harvesting costs and thus reached its maximum value to the grower. The introduction of systemic fungicides has certainly contributed greatly to this aspect of farming and is widely used particularly in the storage of the higher value fruit crops and the more expensive vegetables; again exploiting the power of systemics to penetrate plant tissues and control both active and passive fungal infections.

Cost/benefit ratio

In an article in the ARC newsheet dated December 1974 it was reported that owing to the treatment of most onion seed stocks in the United

Kingdom with approximately £500-worth of benomyl, crop losses due to neck rot (*Botrytis allii*) were virtually eliminated. On a national basis this was estimated to have saved 33 per cent of the total crop valued at approximately £3m (Anon, 1975a). Cost effectiveness or benefit/cost ratios of this order are certainly exceptional and unique since most farming operations are concerned with ratios of the order of 3 to 4:1 or even less. Thus for the control of mildew a barley grower in Eastern England in 1974 might have spent approximately £5/ha for chemical and application costs in return for a possible 10 per cent increase in yield, and on the national average of 4 tonnes/ha and a selling price of £50/tonne this would provide a net return of £20/ha, a benefit/cost ratio of 4:1 and a net profit of £15/ha. Obviously a lower selling price for the barley or higher chemical costs would reduce this ratio even further, but within the limits of experimental error and allowing for the vast variation in prices that may occur these values are typical of the situation in low-priced crops where yield increases in excess of 10 per cent are frequently recorded (Little and Doodson, 1971).

By comparison much more money is spent on treating fruit crops such as grapes, a large proportion of which in France are treated with a mixed copper/sulphur schedule of six to eight sprays in one season which at 1974 prices cost approximately £50/ha. Based on a series of thirteen trials in Savoy over a 2-year period it was shown that such a treatment increased yields by approximately 50 per cent over the untreated control (M. Rickard, private communication). Applying this information to the national average production figures of 5 000 litres/ha quoted by FAO (1972) and assuming that most crops are treated in this fashion, this represents an increase in wine production of approximately 1 667 litres/ha. At an average possible price of 10p/litre to the grower this would give a benefit/cost ratio of £166/£50 = 3·3:1. As, however, the vine crop is more valuable than barley per unit area this represents a net profit of approximately £166 − £50 (£116)/ha to the vine grower compared with £20 − £5 (£15)/ha to the barley farmer. Thus the cost/benefit ratio is only a preliminary guide that needs to be studied carefully before any decision is taken regarding the use of any pesticide. As most of the newer systemics are said to be markedly non-phytotoxic when compared with some of the older and cheaper standard fungicides (Evans *et al.*, 1971), these calculations may be modified by improved yield or quality effects. In this case the financial return may be increased much more than the extra treatment cost, thus increasing the benefit/cost ratio and the net return to the grower.

It is usually fairly easy to establish the cost of a particular pesticide programme. What is difficult to predict is the financial outcome of the operation. This is particularly true of fungicide usage which is largely dependent on continually changing climatic conditions.

In order to avoid wastage of chemicals and effort, the study of disease forecasting systems is being encouraged by governments throughout the world, but most of the work to date has been on a short-term basis and it remains to be seen whether this can be extended to be of seasonal value to growers and chemical manufacturers alike.

As all newly introduced pesticides are of necessity more expensive than those previously available, it is perhaps surprising that systemic fungicides were accepted so freely in both agricultural and horticultural circles. Generally they have been used on previously intractable problems such as the control of loose smut of wheat (von Schmeling *et al.*, 1966); with novel methods of application such as the control of cereal mildew by seed dressing (Brooks, 1970) or for the replacement of less effective foliar treatments with a higher yielding spray programme (Evans *et al.*, 1971). In each case the increased value of the crop in terms of yield and quality exceeded that of the extra treatment costs, thus providing the grower with a greater net return/unit area of treated crop.

Biological activity and systemicity

As with all fungicide treatments the eventual performance of systemics is dependent on:

(*a*) the inherent biological activity of the product;
(*b*) correct timing of the fungicide application;
(*c*) accurate placement of the fungicide.

Correct placement implies knowing whether the fungus concerned is spread in seed, in soil or through the atmosphere and making maximum use of the compound's systemic properties. Systemicity generally leads to an improved distribution of the fungicide and to a better biological performance. However, the dosage must be adjusted with regard to crop safety, persistence and residues if systemic compounds are to be widely used in agriculture.

The subject of systemicity and the mechanisms of movement within higher plants has been reviewed by Crowdy (Chapter 5), but from the practical point of view the following characteristics are significant:

(*a*) The uptake and distribution of a systemic compound within a plant differ according to whether the compound is applied to the aerial or subterranean parts.
(*b*) The rate of movement is greater in herbaceous than in woody plants and even then is dependent on the physiological status of the tissues concerned (Shephard, 1971).
(*c*) Movement of most present-day systemics is unidirectionally upward following the transpiration stream, which in turn is affected by environmental conditions during and after treatment (Peterson and Edgington, 1971).
(*d*) The one exception to statement (*c*) concerns the newly introduced pyridine derivative pyroxychlor which appears to be translocated downwards from the foliage to the roots of several plants (Noveroske, 1975a; Hoitink and Schmitthenner, 1975).

Seed treatment

A seed dressing is defined as a substance applied to seed for the control of pest or disease problems of that species. Traditionally the objectives of fungicide dressings were to destroy pathogenic organisms on the seed, or to protect germinating seedlings from attack by soil-borne pathogens. The introduction of systemic fungicides has added two further possibilities.

(*a*) The control of pathogens situated within the seed and previously inaccessible to chemicals.
(*b*) The control of airborne diseases using the dressing as a reservoir of fungicide during the growth of the crop.

The current methods of treating seedborne diseases are summarised in Table 10.1. These were devised to deal with broad-spectrum non-systemic materials, largely based on a variety of organomercurials.

Table 10.1 Summary of seed dressing techniques

Formulations	Application methods	Examples of machinery
1. Adhesive dust	Batch or continuous flow processes	Simple drums, high speed dusters, e.g. Robinson and Plantector and in the row seed box treatments
2. Wet slurries	Dribble or spray applications	Panogen and 'Mistomatic' machines
3. Solutions	Dribble or spray treatments	

These techniques are equally applicable to systemic products and when used as sterilants of seed as well as protectants during germination, seed dressings provide a most economic and convenient method of control. Used in this manner the amount of chemical required is small and the operation is easily mechanised. When prolonged systemic protection is required the dilution effect of growth and the eventual breakdown of the active ingredient demand that larger amounts of chemical have to be applied to the seed. This can present considerable problems of formulation, which is already restricted by the physical nature of the active compound and by the kind of machinery to be used in dressing the seed. Thus dusting machines demand a solid adhesive product while the 'wet' seed dressers use either solutions or slurries. The dressing process is further limited by the retentive capacity of the seed itself, particularly when it has to be stored with unusually high rates of systemic dressings.

As the seed is handled roughly during the planting operation dressing may be lost but sufficient chemical must remain near the seed if it is to be taken up by the growing seedling. As roots are organs of absorption it is reasonable to suggest that this is the main method of entry of systemic dressings. When dealing with large dicotyledonous seeds however it can be shown that cotyledons are also absorptive organs.

Some *Phaseolus* beans were soaked for 24 h in distilled water. After stripping the testa from a sample, excess water was removed with dry filter paper and the naked seed dressed with 0·25 per cent by weight of carboxin. A sample of intact beans was then treated in a similar manner applying the fungicide to the testa. A third series of seedlings were grown in this manner devoid of chemical treatment. The beans were then grown in a humid chamber but not in direct contact with water. Within 14 days plants were produced which were inoculated with the bean rust pathogen *Uromyces phaseoli* and disease assessment made 10 days later still. The results are shown in Table 10.2.

Table 10.2 Per cent disease control and phytotoxicity index after various seed treatments

Treatments	Mean % disease control†	Mean phytotoxicity index
1. 0·25% w/w carboxin applied to testa	11	2·0
2. 0·25% w/w carboxin applied to naked seed	100	2·3
3. Untreated seedlings	0	0·3

† Bean rust pathogen *U. phaseoli* used as assay organism

The barrier to absorption set up by the testa is obvious from the small amount of protection afforded in treatment 1, while the excellent performance of treatment 2 reflects the good absorptive nature of the cotyledons. Phytotoxicity as recorded in this experiment consisted of the typical marginal necrosis described by Snel and Edgington (1970).

The success of the use of systemic fungicides as seed dressings is now clearly established and the oxathiin, carboxin, is widely used for the control of the deep-seated loose smut diseases of wheat and barley (Kuiper, 1968) caused by *Ustilago tritici* and *Ustilago nuda* respectively, diseases which were previously only controlled by the hot water treatment of seed. The first practical control of an airborne disease by seed treatment was that of barley mildew (*Erysiphe graminis*) with the pyrimidine, ethirimol (Brooks, 1970).

Soil treatment

Until the recent introduction of systemic products, fungicides were only applied to soils for

(*a*) The protection of germinating seed crops from soilborne organisms.
(*b*) The eradication of soilborne parasites prior to planting a new crop.

The highly selective character of present-day systemic fungicides does however enable crops to be treated *in situ* for the control of soilborne organisms and for the control of airborne pathogens following root uptake. The methods of application currently used for treating soil with fungicides depend primarily on the physical characteristics of the products and degree of sophistication of the machinery used. Gases, such as the general sterilant methyl bromide, are injected at set points from pressurised containers and similar 'point' treatments of liquids that liberate gases such as carbon disulphide and formaldehyde are also made. Solids, however, require thorough incorporation in soil. Liquids or suspensions of active ingredient in water, including most of the current systemics are frequently used as overall drenches. Thus benomyl and thiabendazole drenches are under test on a variety of crops for the control of typical soilborne pathogens such as *Verticillium* wilt and *Phymatotrichum* rots of cotton (Ranney, 1971; Lyda and Burnett, 1970). This technique is also recommended for a number of systemic products such as thiophanate-methyl, benomyl and ethirimol used against cucumber mildew, an airborne disease of major importance in glasshouse crops (Aelbers, 1970).

A novel modification of the drenching technique has been developed for use in Japan of the fungicide kitazin against rice blast (*Pyricularia oryzae*). Here the product is applied to the irrigation water as granules containing 17 per cent a.i. and since the active ingredient has only a limited degree of solubility (500 ppm in water) such granules were persistent for three weeks under field conditions (Yoshinaga, 1969).

As with any other chemicals applied to soil, systemic fungicides are potentially subject to inactivation. This may be due to the process of physical adsorption with a consequent lack of availability and reduced biological activity. Chemical and biological interference may also occur, leading to a breakdown of the active ingredient (see Chap. 8). The rate of breakdown is obviously of considerable importance in determining the fungicidal efficiency and residue status of treated soils. Uptake by plants is greatly affected by a variety of factors which are as yet not fully understood. Hock *et al.* (1970b) found that the uptake of benomyl by the roots of elm seedlings was much less in plants grown in potting compost than with those grown in sand cultures. Whether this is a matter of root structure or availability of fungicide was not specified. Similarly the uptake of substances dependent on water solubility might well be affected by drought, and their use correspondingly restricted to situations where irrigation or any other watering procedure is readily available. The complicated nature of this adsorptive phenomenon and its significance to the uptake of the pyrimidine fungicides by higher plants was discussed in some detail by Graham-Bryce and Coutts (1971). They showed that in this instance the process of adsorption was largely reversible if sufficient time was allowed. They also found that the distribution and consequent uptake from localised soil treatments was affected by various structural and textural characteristics of soils.

The activity of certain benzimidazole fungicides has even been influenced by formulating the active ingredient with certain surfactants (Pitblado and Edgington, 1971) although the benefit of such an effect has not yet been established in practice. Stipes and Oderwald (1971) and Biehn and Dimond (1970) investigated the effect of benomyl on Dutch elm disease when injected as a suspension of active ingredient into the feeding root zone of young elm trees with and without added surfactant. The former concluded that Tween 20 improved the performance of benomyl used in this manner.

Leaf and stem treatments

As airborne epidemics develop much more rapidly than seed or soil-borne diseases they must be tackled with urgency by means of spray or dust treatments of the aerial parts of the crop. The introduction of systemic products as seed or soil treatments provides an alternative approach for the control of certain airborne organisms, but most fungicides are still applied by spraying or dusting techniques. The field of application technology has received considerable attention from both physicists and biologists alike and was reviewed by Evans (1968). Table 10.3 is taken from this review and summarises the variety of techniques most commonly employed for the application of fungicides to the aerial parts of plants.

Table 10.3 Summary of most commonly employed commercial techniques of applying foliar fungicides (Evans, 1968)†

Growth form of crop	**Technique of interest**		
	High-volume‡	Low-volume§	Dusting
Ground or field crop	Hydraulic sprayer with horizontal spray gear	Aircraft with conventional spray bar and nozzles or rotary atomisers	Single or multi-outlet row crop dusters, or aircraft dusting operations
Bush or orchard plantation	Hydraulic sprayers with vertical-spray bars or hand lances	Airblast machines and occasionally aircraft, e.g. for control of downy mildew of vines	Power-operated dusters
Closed-canopy plantations	Hydraulic sprayers with hand lances	Airblast machines from below, and aircraft from above the canopy	Power dusters from below canopy, and dust applications from aircraft

† Taken from *Plant Diseases and their Chemical Control* (1968) by E. Evans, published by Blackwell Scientific Publications, Oxford

‡ High volume – any spray volume which leads to the coalescence of spray droplets and a consequent 'drip pattern' on sprayed foliage

§ Low volume – spray volumes where coalescence of spray droplets does not occur

In order to prevent the development of disease symptoms, a series of treatments is usually applied according to a pre-determined schedule designed to cover all potentially infective periods. Thus the efficiency of any treatment will depend on the persistence of the active ingredient between one treatment and the next and the distribution of the ingredient during these infective periods. Non-systemic products often have to withstand severe weathering as surface deposits while systemics are generally much less dependent on weather because they are able to penetrate host tissues. Thus a 21-day schedule of benomyl on apples gave a level of control of scab (*Venturia inaequalis*) equal to a 14-day schedule of the non-systemic captan standard (Evans *et al.*, 1971).

Another major factor governing the efficiency of any spray deposit is the distribution of the active product on or in the host plant. Final distribution of non-systemics is largely dependent on the method of application used and the redistribution of the deposits as solutions, suspensions or vapours on the surface of the crop (Evans, 1968). Tissue penetration and internal movement of the fungicide is implicit in the term systemicity and hence systemic distribution is usually superior to that of non-systemics. Morgan (1952) showed how a given quantity of a copper fungicide gave vastly different biological results dependent on the distribution pattern of that deposit. This effect is even more dramatic when the procedure is repeated with a systemic product. Commencing with a spray concentration of 0·025 per cent dinocap and benomyl respectively a series of five suspensions were prepared by repeated 1 : 1 dilutions of the previously higher concentration. By means of a micropipette droplets of approximately 0·05 ml in volume were placed on young cucumber foliage and allowed to dry prior to incubation for a period of 14 days in an atmosphere heavily loaded with spores of the mildew pathogen *Sphaerotheca fuliginea.* By placing equal amounts of active ingredient/leaf as 1, 2, 4, 8 or 16 droplets of the appropriate concentration, protectant action was then estimated as a percentage of the leaf area free from disease. The mean data recorded in this manner are summarised in Table 10.4.

Table 10.4 The effect of distribution of the activity of dinocap and benomyl deposits (mg of a.i. applied/leaf)

Concentration of applied droplets, %	**Number of drops/leaf**	**% disease control**	
		Dinocap	Benomyl
0·025	1	5	20
0·0125	2	10	40
0·0067	4	20	100
0·0033	8	55	100
0·0017	16	100	100

Since the final pattern of results of both compounds gave almost 100 per cent control it follows that the amount of active ingredient was not limiting the activity of either dinocap or benomyl and that the overall result was largely dependent on the distribution pattern. The superiority of the benomyl treatment was obvious throughout and its apoplastic movement in the transpiration stream clearly shown in the wedge-shaped pattern of control associated with single drops of benomyl, particularly when placed adjacent to main veins.

Ultimately the efficiency of any fungicide depends on the amount of active compound that arrives at the site of action. The active compound may not be the substance supplied as the formulated product, as shown for benomyl by Clemons and Sisler (1969) and for thiophanate-methyl by Selling *et al.* (1970).

As long ago as 1956 Gray was able to show that the uptake of streptomycin by leaves and flowers could be increased by the addition of glycerol to the spray liquid. This effect is generally attributed to the humectant properties of the glycerol with a corresponding delay in the rate of drying and greater uptake of the surface deposit. Surfactants have traditionally been added to improve biological efficiency of most fungicides in two ways: (*a*) by enhancing the wetting properties of sprays and possibly by improving distribution and (*b*) by increasing the solubility of the active ingredient as shown by the effect of Tween 20 on triarimol (Brown and Hall, 1971). These improvements apply equally well to both systemic and non-systemic products. Adhesive supplements were added to benomyl by Epstein (1969), who showed that improved control of apple scab was achieved by so doing. Thus although generally some improvement in disease control has been achieved by the addition of various supplements, the overall superiority of the systemics is largely dependent on better distribution and a more regulated persistence largely dependent on the applied dosage. The uptake, distribution, accumulation and persistence of toxicant by a given tissue is eventually responsible for its biological performance, but the level of toxicant which will control one pathogen may not be sufficient for another. This was clearly demonstrated by Frick (1971), who showed that while apple scab was well controlled by systemically moved benomyl in half leaf tests, this level of toxicant was insufficient to control the apple mildew pathogen (*Podosphaera leucotricha*).

While foliar spray treatments are convenient for the control of airborne pathogens of either annuals or perennials this is not a convenient technique for tackling systemically distributed pathogens of woody tissue such as Dutch elm disease (caused by *Ceratocystis ulmi*) or silver leaf disease of stone fruit (caused by *Stereum purpureum*). Stem treatments with or without wounding are most commonly used on large perennials (Schreiber, 1970) but various other techniques have been applied to a variety of crops for experimental purposes (Wain and Carter, 1967; North, 1962). In practice stem injection has been used for the correction of mineral deficiencies, particularly that of calcium-induced iron deficiency in top fruit and citrus. Tangential holes are bored into the trunks and solid or encapsulated doses of active ingredients implanted. Adhesive bands enclosing absorbent substrates containing the active ingredients are also used, but all such implantation techniques involve the use of locally high concentrations of chemical or else large reservoirs of dilute solutions. From the practical point of view the latter technique is cumbersome while the use of concentrated preparations tends to cause local necrosis and eventual failure to take up the toxicant from solution (see also pp. 312–318).

Post-harvest treatments

Harvest in this sense implies that part of the crop which is to be marketed as:

(*a*) Raw material for manufacturing or processing.
(*b*) Food for human or animal consumption.
(*c*) Propagating material for crop maintenance purposes.

As systemic fungicides generally are very active there has been a tendency of late to use them in preference to the traditional non-systemics but perhaps their greatest asset is their ability to penetrate host tissues, enabling them to eradicate many of the latent infections which might have occurred prior to harvest. Late sprays and post-harvest dips of benomyl greatly reduced the incidence of *Gloeosporium* fruit rot in stored apples (Pleisch *et al.*, 1971) and many boll rots of cotton (Ranney, 1971) at harvest. Dip treatments of benomyl at harvest also reduced the incidence of various vegetable rots in storage (Derbyshire and Crisp, 1971).

The category designated as propagating material includes seed pieces for potato growing, bulbs or ornamental cuttings, all of which have been treated with broad-spectrum fungicides such as benomyl in order to rid the planting material of diseases which might have affected the crop. Hide *et al.* (1969) used a 10 per cent benomyl dust for the treatment of potatoes and readily eradicated several potential pathogens from the seed tubers. Various benzimidazole fungicides have similarly been tested as preplanting treatment of bulbs with the same object in mind (Rooy, 1969).

Although the details of these processes would vary from crop to crop the basic principle of application remains, as the dust, sprays and dip treatments are aimed at attaining a uniform distribution of active ingredient on the potentially infected surface in order to destroy the pathogen *in situ*, and possibly in the case of propagating material to protect the crop during its establishment period.

Timing systemic fungicide treatments

The overall efficiency of any fungicide treatment is eventually dependent on accurately timing the application in a manner dictated by the biology of the pathogen. With non-systemic products the effect of timing can be extremely critical (Evans *et al.*, 1965) but the advent of systemics has in some instances made this requirement much less important due to tissue penetration giving improved distribution and weathering characteristics and to the curative nature of most present-day systemic products.

On the principle that a pathogen is most easily controlled when its inoculum potential is minimal it is generally agreed that treatment should be most effective when (*a*) the pathogen population is lowest, (*b*) the host least susceptible to the fungus and (*c*) the environmental conditions least congenial to infection. This entails treating seed with chemical dressings for the control of seedborne organisms and treating soils for the control of soilborne pathogens prior to planting (although this is not usually possible with perennial crops). It is more difficult to apply this principle to airborne diseases generally but some success has been achieved with post-harvest treatments of apple orchards for the control of scab with both non-systemic (Burchill and Hutton, 1965) and systemic fungicides (Darpoux, 1970). The objective here is to destroy the pathogen while the inoculum potential is declining naturally. Unfortunately however airborne pathogens tend to be extremely mobile and re-infection from outside the treated area could well nullify the effect of such treatments.

Accurate disease forecasting is now widespread for a variety of crops throughout the world, but the application of the chemical itself is often more of a practical difficulty than the decision to spray. For this reason there is a tendency to follow routine schedules throughout the normally infection season and it would seem that 'negative forecasting' could prove to be economically most desirable (Visser *et al.*, 1962). Negative forecasting is here taken to imply forecasts of when a given treatment could safely be omitted from the standard schedule of either systemic or non-systemic products. The actual interval recommended between successive treatments depends on the growth rate of the crop as well as the dosage applied on each occasion. The growth rate and the physiological status of the host tissues may affect the susceptibility of certain crops to specific pathogens and this may complicate the issue even further. However, the internal transfer of systemics means that previously unsprayed tissues may receive toxicant and theoretically this should improve its biological performance *vis-à-vis* non-systemics. Growth in this sense may simply imply expansion of a single organ such as a leaf or fruit which had previously received some spray deposit; more commonly however new structures such as leaves are produced between treatments and in order to gain a specific advantage in this case a systemic product would have to be redistributed from older sprayed organs to the newly developed structures. The first of these effects can frequently be seen in half-leaf tests on rapidly growing foliage on young laboratory-grown crops used for fungicide screening tests. However there is as yet no evidence to suggest that any of the present systemic fungicides are redistributed to the terminal buds in the manner of the aminotriazole and TBA herbicides which tend to accumulate in these areas, and until such products are available, the advantages of systemic products in controlling such pathogens as apple mildew (caused by *Podosphaera leucotricha*) will be minimal. Tissue penetration is essential to curative action of any fungicides as well as being the prelude to systemic movement and as all present-day systemics are curative in the early stages of infection their use should be less dependent on absolute precision timing than on their equivalent non-systemic standards. This generalisation seems to be particularly true of diseases that are essentially internal parasites but care should be taken when applying this philosophy to the superficial powdery mildew pathogens where even non-systemic products may have a considerable curative effect on the pathogen at any time (Evans *et al.*, 1970).

Root pathogens are particularly inaccessible to foliar treatments and although not widely used to date it is here that the unique downward movement of pyroxychlor might show its superiority (Noveroske, 1975a) although similar control could well be achieved with soil drench treatments of this and other fungicides having a similar spectrum of activity such as prothiocarb (Bastiansen *et al.*, 1974).

Future strategy for chemotherapy

One of the main characteristics of the chemotherapeutic approach to plant disease control to date has been the highly effective character of most treatments irrespective of the method of application used. This is largely attributed to the very good distribution achieved by systemic compounds and to the excellent crop safety record of these products due to their specially selective character. Although largely responsible for their biological success, this same specificity may account for the difficulties encountered in practice, namely those associated with fungicide resistance and changes in the parasitic balance of a given environment following the widespread use of these compounds (Williams and Ayanaha, 1975). Although these problems have been widely reported in the scientific literature over the last seven years the total area where effective control has been lost on a practical basis is still limited. Even so it would seem desirable to evolve a strategy that might delay or prevent the onset of resistance as soon as possible and to correct the selectivity problems wherever they occur (see Chapter 9).

In attempting to develop a logical approach to this objective it is pertinent to record that there is still no definite evidence to prove that resistance to any of the known systemics arose as a result of a mutagenesis, and Hastie's (1970) suggestion that its development was primarily a selective process still applies. On this basis it follows that resistance was the result of contact between the variable wild fungus population and sublethal dosages of the active ingredient. Any attempt to reduce the probability of resistance developing in the field should bear this factor in mind. Due to differing rates of uptake, dilution in the internal environment, breakdown of active ingredient within the host and surface losses due to weathering, etc., it is extremely difficult to control the amount of fungicide present in the environment of the fungus other than for a short period immediately following application. Based on this assumption it follows that maximum hindrance to the development of resistance would only occur if fungicidal action was confined to an immediate post-application period when high rates of fungicides could be assured. This implies using these substances in a curative fashion whenever possible and either limiting or avoiding their use as protectants.

In practical terms 'curative treatments' may be defined as applications made at any time after infection has become established. For economic reasons this should occur prior to the mass expression of disease symptoms. The use of oxathiins for the control of loose smut diseases on cereals is a classic example of such a philosophy, as is the use of benomyl dips for the control of latent fruit rot infections on apples, citrus and other crops. The curative approach is perhaps less practical when dealing with foliar sprays for the control of airborne diseases, since it needs a good forecasting system to establish when infection actually occurs. In this context therefore it might be more logical to use the curative approach when tackling the source of primary inoculum as a post-harvest or pre-season spray of perennial crops. In annual crops it may be used as a dip or dressing of all vegetative planting stocks where the uptake of systemic substances may prove extremely beneficial in cleaning-up the seed. The use of systemic fungicides for the control of soil-borne diseases is a classic example of this curative philosophy, since the product would only be used to control known soil infections. Up to the present, practical resistance problems have mainly developed where,

(*a*) seed or soil applications have been made for the control of airborne diseases sometimes after infection;
(*b*) repeated spray treatments have been made to a wild fungus population in one season under good reproductive conditions.

To pursue these recommendations successfully it is essential that research in this field of crop protection should continue with the objective of:

(*a*) discovering new and novel systemic fungicides with different spectra of activity and different modes of action;
(*b*) monitoring the onset and progress of resistance at all times and checking for cross resistance;
(*c*) establishing the mode of action of all known fungicides in order to forecast the possibility of cross resistance.

Chapter 11.I

Results in practice – I. Cereals and grasses

D. H. Brooks and N. G. Buckley

Introduction

The main world cereal crops are wheat, maize (corn), rice and barley. Cereals such as sorghum, millet, rye, oats and buckwheat are grown on a smaller scale, and are not covered in this review. Although sophisticated fungicides have been used for many years in Japan to protect the rice crop, it is only since about 1970 that the systemic fungicides have begun to achieve widespread acceptance in other cereal crops. However, with the increase in value of these crops, increased incidence and recognition of foliar diseases, and the large number of fungicides now available, the treatment of wheat and barley has become common in many countries. In 1974, well over half the barley crop in England was treated with systemic fungicides. Initially seed-borne diseases and powdery mildew were the main targets, but more recently mixtures of fungicides have become available which control a broader spectrum of diseases.

Barley and wheat

Powdery mildew

The extent to which fungicides for control of mildew have been developed depends, amongst other things, on the regularity with which mildew occurs, its severity and the cash return to the farmer of the cereal crop.

Erysiphe graminis DC is far less dependent on weather conditions than most cereal pathogens, and it is found wherever suitable hosts are grown. Its severity depends on such factors as summer temperature, rainfall and humidity, the presence of bridging crops overlapping with the main crops, and fertiliser usage. It is probably the most important disease attacking barley in West and East Europe, particularly in the UK, Denmark, West and East Germany, parts of Holland and Belgium, Northern France and Hungary. Outside Europe, it is sometimes important in New Zealand, West Australia, Canada and parts of the USA (particularly the eastern and western seaboards and south of the Great Lakes). Mildew on wheat has so far proved to be of lesser importance than on barley. It occurs wherever wheat is grown, particularly in dense rank growth following heavy nitrogen fertiliser application, producing a microclimate favourable to disease development. On wheat, mildew is possibly of most importance in eastern European countries such as Hungary and Yugoslavia where the susceptible cultivar Bezostaja is widely grown. Similarly in the United States certain of the soft red wheat varieties have been restricted in their use because of susceptibility to the disease. Oats are severely attacked by mildew in many countries, but are not widely enough grown to have attracted much attention from those concerned in developing fungicides.

Because of the low returns to the farmer in the Americas and Australasia, most attention to fungicide development has been centred in West Europe on barley (and to a lesser extent wheat) and East Europe on wheat (and to a lesser extent barley), and commercial usage is almost entirely limited to these areas.

Soil application

The attraction of this method is that the soil acts as a reservoir for the chemical, and the plant continues to absorb and translocate the fungicide over relatively long periods. The period of control obtained in this way depends on the length of time the chemical remains available in the region of root absorption, and this depends, amongst other things, on the physical properties of the molecule. Although promising experimental results have been obtained with benomyl, triarimol, triadimefon, ethirimol and thiophanate methyl, only ethirimol is used commercially in this way, and most of the available results have been obtained with this compound. Since the cereal crop is one of the few situations where systemic properties have been fully exploited, some discussion of this usage is appropriate here.

Method of application Numerous methods of application have been used, including seed dressing, pelleted seed, liquid sprayed into the furrow, broadcast granules and combine-drilled granules. With ethirimol, the chemical leaches only very slowly in most soil types, and broadcast granules have given variable and generally poor disease control. The other methods have all been used successfully. Where the chemical is placed apart from the seed, the degree of disease control is affected by the evenness of application; for example, where fertiliser granules coated with ethirimol were combine-drilled in an accurate experimental drill, good results were obtained, but when a range of farmer drills were used (most drills dispense fertiliser very unevenly) variable results were obtained. Always the most consistent results have been obtained with ethirimol used as a seed dressing.

One of the major problems with seed dressing is that many of the commercial seed-treating machines are not capable, without modification, of handling the relatively large quantity of chemical required (600–1 000 g formulation/100 kg) at the same time as the standard fungicide/

insecticide dressing (usually 200 g/100 kg); it has therefore been necessary to carry out costly conversions to machinery.

Using converted machinery, very good adhesion of both compounds to the seed is obtained.

Factors affecting activity The duration and rate of uptake, and rate of translocation of a soil-applied systemic fungicide depend on a large number of factors, including those which affect foliage-applied systemic fungicides (transpiration rate, etc.; see Chapter 5).

The chemical can only be taken up if it is in an available form in the root region. The mobility of the chemical depends on adsorption, and the degree to which it is reversible, and on structural and textural features of the soil (Graham-Bryce and Coutts, 1971). With ethirimol, adsorption increases with decreasing pH and increasing organic matter content; for this reason ethirimol is not very effective as a seed treatment in acid peat soils. In light sandy soils with little tendency to form aggregates, the chemical leaches in a roughly conical pattern beneath the seed. In heavier-textured soils which are usually well aggregated leaching is much reduced (Graham-Bryce and Coutts, 1971). In practice, excess leaching does not seem to be a factor limiting uptake in most situations.

Free water in the soil is necessary for uptake of fungicides. In pot experiments with ethirimol incorporated into the soil, uptake did not cease until wilting occurred. In the field, however, the plant may get its water from deep in the soil whilst the surface soil containing the fungicide is dry. In such conditions uptake will cease. This does not appear to limit uptake in most areas, in most seasons, but during periods of drought poor disease control has sometimes been noted: uptake of chemical appears to be resumed when the drought is broken.

In practice in the field, uptake of ethirimol by the barley plant appears to cease whilst there is still available chemical in the surface layers; probably the roots remaining in the treated surface layers of the soil are no longer capable of uptake.

Spray application

The economic return from mildew control does not normally justify more than one spray application. In most of Europe, mildew generally begins to appear in the spring barley crop towards the end of tillering, although in some years it can appear earlier (particularly in the vicinity of winter barley) or later. With the xylem-translocated fungicides currently available, leaves formed after the time of spraying receive little or no direct chemical protection. The degree of mildew control achieved, therefore, depends not only on the inherent activity, rate of application and persistence of the fungicide, but also, most important, on the timing of application and the inoculum pressure on the crop. When applied as sprays, the systemic fungicides are superior to non-systemic fungicides such as drazoxolon and quinomethionate in their limited movement within the leaves to which they are applied, and in their powerful eradicant activity.

The fungicides have normally been applied by ground equipment in 200–400 l water/ha. Aerial application, including ultra-low-volume application, has also been used with some success.

Results on spring barley

The duration and degree of control of mildew achieved with ethirimol seed dressing depends, not only on the factors already outlined, but also on such factors as inoculum pressure and, obviously, rate of application. There is, however, a relatively flat dosage–response curve, possibly because root distribution is one of the main factors limiting uptake. Mildew usually appears at the same time on treated and untreated crops, but the speed of development is greatly reduced by treatment. When high rates of application are used (675–2 000 g/100 kg) almost complete control of disease can be expected for 15 weeks or more (Brooks, 1970; Little and Doodson, 1971). At lower (commercial) rates (370–500 g/100 kg) the degree of control is somewhat less although the persistence does not appear to be altered appreciably. The precise time of breakdown of control is difficult to pinpoint since environmental factors and increased host resistance frequently cause the rate of mildew spread to decline in later (post-flowering) growth stages. In a series of trials carried out in 1968 (Brooks, 1970) seed treatment apparently kept the disease in check throughout the life of the crop; there was, however, little mildew spread on to the upper leaves even in untreated plots.

In 1969, when the disease continued unchecked, mildew started to build up in the treated plots during the grain-filling stages. Probably, therefore, under commercial conditions the effectiveness of ethirimol drops off very sharply after flowering. The degree to which the crop becomes infected at this stage depends on environmental conditions and the intensity of inoculum; re-invasion certainly occurs more rapidly on small plots surrounded by untreated areas than in large treated blocks. In Northern France, crops have sometimes become heavily infected in the late growth stages in spite of excellent control at flowering. The average mildew control at growth stage 10·5 is usually about 90 per cent.

On spring barley the persistence of ethirimol seems to be unaffected by sowing date; there is no evidence that early drilled crops lose their protection before those drilled later. One complicating factor is that the degree of control achieved with ethirimol depends not only on the load of chemical on the seed, but also on the density of sowing of the seed. Consequently, if very low seed rates are used, disease control will be less good than where very high seed rates are used. Within the normal range of rates used in any one country, however, the differences are probably not significant.

Yield responses from seed treatment in experimental plots have varied from slight decreases to increases of up to 80 per cent, with not always a very close correlation with the amount of mildew. The average yield increase from a very large number of field trials over several years in England was greater than 10 per cent.

Of the sprays, the most widely tested and used material is tridemorph. In over 200 trials carried out in Germany in the period 1962–8, an average increase in yield of 10 per cent was obtained with an application rate of 0·5–0·6 kg a.i./ha; similar results have been obtained in England and many other countries (Anon, 1971a). From 1969 to 1972 an average 7 per cent increase was obtained from 426 trials (5–11 per cent range) (Anon, 1974). Mildew is normally controlled for 4 to 5 weeks following application, and the effect of this delay on further epidemic build-up is frequently seen until harvest. In general the yield response has been well correlated with mildew control (Kradel *et al.*, 1969) but on some occasions yield increases have been obtained at very low mildew levels. When tridemorph is applied to an established infection it rapidly kills the mildew, and the crop sometimes appears to regain its green colour, giving what has been called a cosmetic effect.

Ethirimol is also widely used commercially as a foliar spray at rates ranging from about 0·25 to 5 kg/ha. Average yield response for a series of trials carried out in England and Germany in 1970 and 1971 varied from 5 to 12 per cent.

Other fungicides to have been widely used commercially include chloraniformethan, benomyl, thiophanate methyl and chlorquinox.

The discontinued fungicide triarimol was notable for its extremely low rate of application (30–75 g/ha): the new analogue EL222 is similar in its level of activity. Triadimefon and fluotrimazole are proving very promising in field trials.

Thus a large number of new systemic compounds have given good control of mildew on spring barley. All of these compounds have a similar systemic action when applied as sprays, and all have eradicant as well as protectant properties. Since timing of application has a very marked effect on the performance of sprays (see below) it is almost impossible to make valid comparisons between sets of data on individual compounds obtained in separate trials, and very few trials have been carried out in which all the compounds have been compared with one another under similar conditions. Even in these trials, the results are not always very useful, because the rates and formulations used have not always been those recommended commercially. Usually tridemorph, ethirimol, EL222 and triadimefon have been slightly more effective in controlling mildew than thiophanate methyl, chloraniformethan and chlorquinox, although the results are by no means consistent. In terms of yield the broad-spectrum compounds have sometimes given bigger yield response than expected from mildew control, presumably due to control of other pathogens. Almost always, spray applications have been inferior to ethirimol seed dressing (Mundy and Page, 1973).

From a comparison made of the result obtained with sprays at different times, and of sprays with seed dressings, it is obvious that mildew affects the barley plant in different ways depending on the time of attack. In large numbers of trials carried out on spring barley in 1969 and 1970, when mildew attack was relatively late, yield increases could almost always be accounted for in terms of increase in grain size, whether resulting from seed dressing or spray application (Brooks, 1971). In 1968, and more so in 1971, mildew attacked early and caused a decrease in tiller number, a marked reduction in root growth and a small reduction in straw length (Brooks, 1971). In these trials yield increases usually resulted from a combination of more tillers and plumper grain although in a few cases grain size was actually reduced, presumably because of increased competition between the tillers. Similar results were obtained by Jenkyn (1974).

Several workers have noticed that different varieties of barley respond to different extents to mildew control. For example, the variety Julia responded far more than was expected to control of mildew with ethirimol (Doodson and Saunders, 1969). With tridemorph it has been possible to divide both susceptible and resistant varieties into two groups, one of which responded well to treatment (even with little mildew), and one which did not respond well (Kradel *et al.*, 1969). Commercial experience in England has shown the same to be true for other fungicides.

Time of spray application Much work has been carried out on the timing of spray application, mainly by the commercial companies. The time has sometimes been expressed as calendar dates, sometimes in terms of growth stage of the crop and sometimes in relation to mildew attack.

From trials carried out in 1968 and 1969, Rosser (1969) concluded that spraying should be carried out in May or June when the mean temperature had exceeded 20°C for seven days (i.e. conditions suitable for rapid epidemic development); the correct dates were approximately 5 June in 1968 and 18 June in 1969, and these represented relatively late growth stage in the crop. In 1971, attempts to determine the correct time of spraying in relation to 'Rosser Periods' were unsuccessful since although mildew was early and severe, the first 'Rosser Period' did not occur until July (Evans and Hawkins, 1971; Gilmour, 1971). Gilmour suggested that a better forecasting system could be based on the accumulated day-degrees in excess of 15°C from 1 May. From trials carried out with tridemorph in 1970, Lester (1971) concluded that the optimum time for spray application in that year was at growth stage 8 (Feekes Scale – roughly mid-June).

The experience of the chemical companies has generally been that somewhat earlier sprays (GS 5–6) give optimum yield response (Anon, 1971c; Jung and Bedford, 1971; Rea and Brown, 1971; Brooks, 1970). For example, in 58 trials carried out in 1961–4 in Western Germany, sprays of tridemorph at GS 5–6 increased yield by an average 419 kg/ha, compared to 339 kg/ha at GS 8–9. A similar pattern was found in England in 1970, although in a series of trials in 1968 sprays at GS 6 were markedly superior to sprays at GS 4 (Anon, 1971a). Frequently, although the later sprays have given maximum disease control (when assessed at GS 10–11) earlier sprays have given greater yield response. For example, Evans and Hawkins (1971) found that in 1971 progressively later sprays gave increased mildew control up until early June, but that sprays in mid-May gave the biggest yield response.

It is clear from the data available that it would be quite wrong to standardise spray timing recommendations in terms of time of year or stage of growth of crop, since these will vary from area to area and from year to year. The new chemicals have powerful eradicant activity, and the optimum time to spray is probably at the beginning of the rapid phase of epidemic build-up, erring on the early side rather than the late. In Western Europe it is normally between GS 4 and 8, mid-May to mid-June. Where there is abundant inoculum coming into the field from adjacent infected crops, the spray could be made at the expected time of appearance of infection, bearing in mind the weather conditions needed (Rosser, 1969); where there is not an adjacent source, it would probably be better to wait until the appearance of mildew in the crop, although marginally lower yields might be expected.

The discrepancy between maximum yield and maximum mildew control makes it very difficult for the farmer to judge the correct timing of application, and it is here that he needs expert advice.

Results on winter barley

In Europe, winter barley is generally drilled in September or October. It may become infected with mildew immediately after emergence before cold weather prevents the rapid spread of mildew, or in spring when the temperature rises, or both. Even following a severe attack in the autumn, there is not always an appreciable amount of mildew in the spring, but the yield potential may already have been reduced (Brooks, 1971b). The problem of spray timing, and whether more than one spray is justified is therefore far more complex on winter barley than on spring barley.

Single sprays of tridemorph applied in October have caused increased tillering and root growth, and have greatly improved the vigour of crops (Lester, 1971). Yield response has, however, not always been obtained (Jung and Bedford, 1971).

Sprays of tridemorph (Kradel and Pommer, 1967; Jung and Bedford, 1971), triarimol (James *et al.*, 1971) and ethirimol applied in the spring (usually at GS 5–6) have also given marked disease control and have led to significant yield increases. For example, in two trials sprayed with tridemorph at GS 5–6 in 1970, excellent mildew control resulted for four weeks, and yield was increased by 15 per cent.

As would be expected from the result on spring barley, ethirimol as a seed dressing gives good control of mildew after emergence until weather prevents further spread of the disease. In Autumn 1969 almost complete control was achieved during this period in trials in Europe and Western Germany (Hall, 1971); in 1970, infection continued later in the autumn and disease started to appear in the treated plots (Hall, 1971; Yarham *et al.*, 1971). In both years mildew control was resumed in the spring and, particularly in 1970, the disease was largely checked until after ear emergence (Hall, 1971). Where comparisons were made, disease control achieved in the spring following seed dressing application was as good as that obtained with a spray applied in the spring with a hormone weed killer. As with the autumn spray of tridemorph, the seed dressing of ethirimol had a profound effect on the vigour of the crop. In one trial in 1970 (assessed in November) tiller no./plant was increased from 3·0 to 5·4 and tiller weight from 0·05 to 0·14 g; mean adventitious root length was increased from 1·5 mm to 77·5 mm, and the number of adventitious roots per plant from 0·1 to 1·8 (Finney and Hall, 1971). In 1969 mildewed plants suffered much more from winter kill and were much less vigorous in the spring than seed-treated ones; in 1970 there was less winter kill and the effects in spring were less striking.

The yield response of winter barley is difficult to interpret. In 1969, ethirimol applied as a spring spray gave an average yield increase of 6 per cent, but the seed dressing, which controlled autumn mildew and gave moderate control of mildew in the spring, increased yield by 12 per cent. The extra increase obtained from control of autumn mildew was associated with an increase (16 per cent) in the number of tillers at harvest. In 1970 there was an equally marked increase of tillering. In some trials this resulted in large yield increases. In other trials the treated plants produced fewer grains per ear, and less plump grain, and this largely offset the effect of increased tiller numbers. In one trial where there was a 15 per cent yield increase, tiller number was increased by 37 per cent but 1 000-grain weight was decreased by 7 per cent and the number of grains per ear by 12 per cent (Finney and Hall, 1971). The average yield response varied from 20 per cent increase (early, severe attack in autumn) to 5 per cent (late, mild attack in autumn) (Hall, 1971). A comparison of the two sets of trials suggests that in a year where autumn attack is followed by winter kill and perhaps by adverse growing conditions in the spring (e.g. the drought of 1970 or low soil fertility) the increase in tiller number will result in a direct yield benefit. In a year with little winter kill and good growing conditions, the crop attacked by mildew in autumn can compensate to such an extent that the increased tiller number resulting from chemical treatment can result in less productivity per tiller. These results were obtained with ethirimol seed dressing, but a similar pattern of results has been obtained with sprays of tridemorph (Jung and Bedford, 1971). Clearly more work is necessary to establish the best method of controlling mildew on winter barley.

Because of the recognition of ethirimol-tolerant strains of *E. graminis*, particularly in treated crops, it was recommended that commercial seed treatment of winter barley with ethirimol should be avoided. There is then a longer break between successive treated crops during which time the tolerant strains can decline in the population (Wolfe *et al.*, 1973). There is evidence that this approach has been successful, and that ethirimol-tolerant strains are not becoming dominant (Shephard *et al.*, 1975).

Effect on other diseases of barley

Barley is attacked by a number of other foliar diseases, one of the most important of which in Western Europe is brown rust (*Puccinia hordei*). The benzimidazoles, oxycarboxin, benodanil (BAS 3170F), triforine and tridemorph all have some activity against brown rust, but since the attacks of brown rust usually occur much later than the optimum time for mildew sprays, it is not normally possible to exploit this activity. There has, in fact, been speculation that the control of mildew could lead to an increase in brown rust, particularly with specific compounds like ethirimol. This does not, however, seem to be the case.

In three winter barley trials in 1969–70, ethirimol seed dressing apparently significantly reduced brown rust attack (Brooks, 1971) and similar observations were made in 1971 (Yarham *et al.*, 1971). In spring barley trials in 1970, in three out of the 14 trials infected with brown rust there was an apparent slight increase on ethirimol-treated plots (Brooks, 1971). Where the differences are relatively small, it is difficult to judge the reliability of the results because of the difficulty of accurately assessing brown rust in the presence of mildew on the untreated plots (it is easy to underestimate brown rust on mildewed plants). In trials carried out by the National Institute of Agricultural Botany in 1971 (Little and Doodson, 1971; Little, 1971 – personal communication), taking the trials as a whole, there was a slight tendency for brown rust to be worse on ethirimol-treated crops. On individual trials, there was an apparently large increase in brown rust on seven of the sixty-four variety/site combinations, but an apparent decrease in two. These differences were not always connected with the control of mildew. For example, on the highly mildew-resistant variety Feronia, there was a large increase in brown rust at one site, but no increase at the remaining seven sites. Thus this change in level of brown rust was not related to the increased clean leaf presented by the control of mildew.

There is thus little evidence that the control of mildew will lead to an increase in the levels of brown rust. Nevertheless, rust can limit yield potential gained by controlling mildew, and an integrated approach using rust-resistant varieties should be adopted.

Results on wheat

Curiously, the first two fungicides to achieve prominence, ethirimol and tridemorph, are less active against wheat powdery mildew than against barley powdery mildew, at least under field conditions. In the case of tridemorph, leaf uptake and systemic movement is less efficient in wheat than barley (Anon, 1971a) and also it is phytotoxic on some cultivars. Nevertheless good yield increases have sometimes been obtained.

In soft and hard spring wheat, ethirimol has given useful increases in yield in France. On winter wheat, the seed dressing is not usually persistent enough, but where mildew has attacked very early, increases in yield of 10–20 per cent have been observed in Hungary and the US (Arkansas).

As a foliar spray, ethirimol has given promising results in Europe and the US. In the US, up to 80 per cent disease control from a single spray has been obtained on soft red winter wheat in the Southern states where mildew frequently attacks at GS 4–5: however, good yield increases have not always been realised because of later attacks of glume blotch or rust. Less good disease control has generally been obtained in north-eastern US where mildew attacks the flag leaf and ear (GS 9–10·5), but yield increases have generally been greater there.

Yield results with tridemorph on winter wheat have also been disappointing and less than would have been expected from equivalent mildew levels on spring barley; the size of the response has varied with variety. For example, in a trial in Germany in 1970, the yield of Cappelle was decreased by a programme of sprays but that of Hanno was increased by over 20 per cent (Anon, 1971a).

Benomyl is active on wheat both as a spray (Jenkins, 1973; Carroll and Webb, 1974; Allison *et al.*, 1975b) and a seed/soil treatment (Johnston, 1970), although as a seed dressing it is usually less effective than ethirimol (Johnson, 1972; Skorda, 1975) and it has not been exploited in this outlet.

As sprays, benomyl, carbendazim and thiophanate methyl are sold commercially in Europe, usually in mixtures with maneb or other dithiocarbamates, at rates of 0·28 kg/ha. These combinations have proven no more effective for control of mildew than ethirimol/captafol mixtures or tridemorph.

Chloraniformethan and triforine have given results on wheat roughly comparable with those of tridemorph. The new compound triadimefon is, however, reported to be more effective as a seed dressing at 200 g/100 kg seed than tridemorph as a spray, although some delay and reduction in emergence has occurred.

More work is needed to establish the correct timing for application of mildewicides to wheat, but most workers stress the importance of protecting the flag leaf and ear: hence applications are likely to be later than those on spring barley where protection of the foliage is the objective.

Rusts

Wheat and barley are attacked by a number of rust diseases. The rusts are far more dependent than mildew on suitable weather conditions, and in a highly variable climate like that of West Europe rust intensity varies greatly from year to year. Thus on winter wheat yellow (stripe) rust (*Puccinia striiformis* West.) can cause losses of 20 per cent or more (Batts and Elliott, 1952) but is sometimes quite insignificant. Brown rust of barley (*P. hordei* Otth.) is common in Europe, but usually occurs only late in the season, so losses are not normally very great. Black (stem) rust (*P. graminis* Pers.) and leaf (brown) rust of wheat (*P. recondita* Rob. ex Desm.) are important in some areas in some years. In the United States, severe rust epidemics occur more frequently, and black rust and leaf rust are particularly damaging: in some years almost complete losses can occur locally (Rowell, 1968a).

Chemical control

The compounds most widely evaluated for the control of rust are oxycarboxin (von Schmeling and Kulka, 1966) and RH-124 (4-n-butyl-1,2,4-triazole) (von Meyer *et al.*, 1970). Both chemicals are freely translocated within the xylem, and both can be applied either as spray or soil treatment. Oxycarboxin is active against all the rusts, but curiously, RH-124 is active only against leaf rust (von Meyer *et al.*, 1970). Other chemicals being widely evaluated against rust include the benzimidazole derivatives, EL222, benodanil, pyracarbolid, triforine and triadimefon.

Seed treatments

Following seed treatment with carboxin at 2 630 g/100 kg, almost complete control of yellow rust on wheat was obtained for 31 days, and after a further 30 days (when the plants were in 'boot') only 5 per cent of the foliage was infected, compared to 25 per cent on untreated plants (Powelson and Shaner, 1966). There was no impairment of germination at that high rate, but no yield measurements were made.

At the much lower rate of 600 g/100 kg, Rowell (1967) noted a reduction in germination of about 20 per cent, and partial disease control of leaf rust lasting for about 60 days. Application of a further spray to the seed-treated plots provided control as good as that obtained with two sprays. In further trials (Rowell, 1968b), applications of granules combine-drilled with the seed to give 1·12 kg a.i./ha followed by two sprays at 1·68 kg/ha again illustrated the importance of control of early and mid-season stages of rust, although in this trial there was a relatively late attack of rust. Poorer results were achieved by Hagborg (1970) when using granules drilled between the seed rows. Combinations of seed treatment and sprays have also been used with some success against black rust and leaf rust of wheat in India (Pathak and Joshi, 1970).

RH-124 appears to be more active from root uptake than carboxin (von Meyer *et al.*, 1970), giving complete control after 14 days compared to no control with carboxin, both at the rate of 1 ppm in the soil.

Thus, at least where rust attacks early in the life of the crop, seed dressings provide some useful protection. In Europe, where attack normally comes later, it is doubtful whether useful control could be obtained with the chemicals currently available.

Spray application

Numbers of trials have been reported where single or multiple spray programmes have been evaluated for the control of various rusts, but it is very difficult to make a valid comparison between them because of the wide range of rates and conditions used. Furthermore, interpretation of yield data is complicated by the fact that artificial inoculation was frequently used.

In the United States, single sprays of oxycarboxin have not, in general, given sufficient control, due partly to the short half-life of the chemical giving insufficient persistence in early sprays, and insufficient eradicant activity in late sprays (Rowell, 1967). Nevertheless significant yield increases have sometimes been obtained, particularly with late sprays. For example, Powelson and Beaver (1970) obtained a significant increase from a single spray of 2·2 kg/ha against yellow rust on wheat applied just before flag-leaf emergence. In Europe, where inoculum pressure is often less severe, single sprays of oxycarboxin are used successfully commercially (Judge *et al.*, 1975).

RH-124 has been widely evaluated and results show that on leaf rust it is as effective or more effective than oxycarboxin but at lower rates (Rowell, 1969; Reed and Chambers, 1973; Line *et al.*, 1974). Control until late into the season has been obtained on spring wheat from single sprays of 0·11 and 0·45 kg/ha (Buchenau, 1970). A single spray of 1·12 kg/ha was at least as effective as three sprays of oxycarboxin each at 2·24 kg/ha. RH-124 is inactive against other rusts.

Against yellow rust in Europe, benodanil (BAS 3170F) at 1·5 kg/ha applied once gave an average increase of 32 per cent in two trials on wheat when sprayed at growth stage 8. Later application (GS 9·5–10) yielded 12 per cent less, indicating the importance of correct spray timing. On winter barley an increase of 19 per cent was obtained with a single spray at 1·0 kg/ha at growth stage 10·1. After five applications, complete control of yellow rust was obtained giving a yield increase of 29 per cent (Frost *et al.*, 1973). In another series of trials on winter wheat in England, Mundy (1973) showed benodanil to be more effective than triforine. On barley, benodanil at 1·0 kg plus tridemorph 0·53 kg/ha as a tank mix gave excellent results and has further enhanced control (May *et al.*, 1975). In the United States the compound has proved equally effective (Line *et al.*, 1974; Powelson, 1974).

In three trials on brown rust of spring barley, benodanil at 1·5 kg/ha gave good control when applied before, or at the onset of rust infection. The average yield increase over the three trials was 11 per cent. The action of benodanil against brown rust in barley seems to be protectant with little or no eradicant activity (Frost *et al.*, 1973).

Tridemorph at 0.53 kg/ha plus metiram 2·25 kg/ha is currently recommended for the control of yellow rust on barley in Europe. This combination is not as effective as benodanil (Mundy, 1973).

EL279 (α-2-Chlorophenyl)-α-cyclohexyl-5-pyrimidine methanol is also active against rust but has not yet been widely evaluated. In Washington State single and double sprays were inferior to RH-124 for control of leaf rust. Triadimefon performed similarly in the same test (Line *et al.*, 1974). Both compounds were effective against yellow rust.

Benomyl (2 kg/ha) was only moderately effective against stem and leaf rust on wheat (Rowell, 1970; Buchenau, 1970) and against brown rust of barley (at 1·7 kg) (Evans *et al.*, 1969). Thiophanate methyl was almost inactive (Soper, 1975).

Triforine gives good eradicant and protectant activity against leaf rust (Schicke and Veen, 1969; Ebenebe *et al.*, 1971) but has not been widely field-evaluated: some control of brown rust of barley and yellow rust of wheat and rye has been noted in field plots (Schicke *et al.*, 1971).

Discussion

Thus, although there are several systemic compounds active against the various rusts of wheat and barley, the difficulty of exploiting this in commercial practice hinges on the difficulty of timing the spray correctly in relation to epidemic build-up, particularly since rust frequently attacks in the later growth stages.

In conditions suitable for rust development, there is at first a slow build-up of disease while inoculum is limiting; in the later stages there is abundant incoulum and the disease builds up rapidly on individual plants (van der Plank, 1963). With the older fungicides, control of the later stage alone was almost impossible, and repeated sprays were necessary. With the new systemic fungicides which have more powerful eradicant activity, later applications alone can be effective. The earlier rust attacks a plant, the greater the damage done (Batts and Elliott, 1952); ideally, therefore, a spray should be applied as soon as possible after rust appears (Rapilly, 1970). In Europe, promising results have been obtained using this approach (Pommer and Kradel, 1969), although a second spray is necessary during a prolonged attack. In the United States, where rust can cause premature death of the crop before the grain starts to fill, a spray applied to protect the crop during the last 30 days might be essential (Rowell, 1968a). In such severe conditions, more than one application is economic (Hagborg, 1970, 1971) and this can be a programme of sprays, or seed treatment followed by spray application (Rowell, 1968b). The seed treatment, if used widely, could be important in restricting inoculum and thus prolonging the slow stage of disease build-up. Its use also makes the timing of subsequent sprays less critical (Rowell, 1968b).

Septoria diseases

The common septoria diseases of cereals are leaf and glume blotch caused by *Septoria nodorum* (Berk.) Berk. and leaf blotch caused by *S. tritici* Rob. ex Desm. *Septoria avenae* f. sp. *triticea* T. Johnson is also reported as common in wheat in Canada and the north central United States. Generally septoria is less important on barley and for this reason most recent studies on the effect of fungicides against the disease have been largely confined to wheat. Septoria diseases are widespread and recent surveys indicate that they are a major problem on wheat (Shipton *et al.*, 1971; King, 1973; Jenkins, 1973). They are prevalent in the western and coastal parts of the British Isles and Europe in areas of relatively high rainfall. Likewise the disease is frequently severe in the eastern region of the United States particularly on soft red winter wheats. *Septoria nodorum* has also been reported to be highly damaging in Argentina and Brazil (Ceñoz, 1961) and of importance in 50 per cent of the wheat belt in Western Australia.

Although there have been relatively few detailed studies to establish losses attributable to septoria there is no lack of evidence that the disease can be economically significant (Jenkins, 1973; Shipton *et al.*, 1971). Estimates of yield losses have been very varied. In Australia, Shipton (1968) recorded yield reductions up to 29 per cent with natural infections of both diseases, whilst in Switzerland, Bronniman (1968) showed losses of up to 65 per cent on small plots under conditions of artificial inoculation and from 10·5 to 27·6 per cent for natural infection. In Israel, Ellal and Dinoor (1973) found that treatment of spring wheat against *S. tritici* increased yield up to 23 per cent, and in Wales, Jenkins and Morgan (1969) obtained yield increases of about 35 per cent in plots of winter wheat sprayed frequently with zineb to control both septoria diseases. In USA, Hosford and Busch (1974) obtained yield increases of about 15 per cent with several applications of mancozeb.

Chemical control

A number of systemic fungicides are effective against septoria. The development of the benzimidazole derivatives has opened up new possibilities for the control of septoria disease. Many of these fungicides are now sold commercially in Europe, frequently as mixtures with other fungicides to supplement activity. Thus, carbenzadim, benomyl and thiophanate methyl are being developed and applied as foliar sprays either alone or with maneb or mancozeb. Tridemorph has also some activity against septoria (Anon, 1974) and has been tested in combination with either carbenzadim or a dithiocarbamate compound.

Spray application

Of the two most common septoria species attacking wheat, it is generally considered that glume blotch causes the most damage at the later growth stages of the crop. The disease is most noticeable on the ears, causing shrivelled grain, but the disease also affects the leaves and stems. In contrast, *S. tritica* damages the leaf and stems but is rarely seen on the ear and is more important early in the season. Cooke and Jones (1970) found that following artificial infection of the heads of spring wheat, yield losses were greater with *S. nodorum* than with *S. tritici.*, 19·5 per cent and 9·1 per cent respectively. If follows therefore that greater emphasis has been placed on glume blotch control.

In laboratory tests, a number of fungicides, including the benzimidazole derivatives, dithiocarbamates and captafol, prove very effective against *Septoria* spp.; they do, however, appear less effective when tested in the field. The reason for this discrepancy is unclear but it may be associated with the fungicides' inability to reach infection sites. Levels of septoria infection on leaves also prove very difficult to assess with accuracy since the disease symptoms are often barely distinguishable from natural scenescence.

In England, following a total of nine spray applications, benomyl alone at 1·12 kg/ha gave moderately good control of the disease on the ears in field-grown winter wheat. Spraying was found to increase the persistence of green leaf tissue to a degree corresponding with the effectiveness of disease control, and benomyl prolonged flag leaf growth by about 10 days more than in unsprayed plots. Grain yield was increased 43 per cent by benomyl, accountable for in terms of 1 000-grain weights (Melville and Jemmett, 1971).

Whilst the early trials against septoria were carried out with benomyl alone, more recently benomyl has mainly been applied with maneb or other dithiocarbamates as a tank mix. This served to broaden the spectrum of activity, particularly against rusts which are unaffected by benomyl. In France, trials on winter wheat and barley with the benomyl/maneb mixture applied at the end of tillering and half-earing yielded increases of 9 per cent on winter wheat and more than 15 per cent on winter barley (at rates of 0·3 kg/ha benomyl plus 2 kg/ha maneb) (Thiollière *et al.*, 1972). From seventeen trials in 1972 in England on a wide range of varieties, benomyl/dithiocarbamate gave average yield increases of 10·5 per cent and there were indications that the effects of the mixture were more than just additive (Anon, 1974).

More recently a formulated mixture of carbendazim (10 per cent) and maneb (64 per cent) applied at 2·8 kg/ha has proved more effective than tank mixes (Anon, 1974). In Germany, twelve trials against glume blotch and other ear diseases have also demonstrated the efficacy of carbendazim and indicate that control measures for glume blotch can be taken until the end of flowering (Hampel and Locher, 1973). Good results were also obtained in England in 1975 (Allison *et al.*, 1975b).

Thiophanate methyl has given similar results to benomyl and carbendazim (Locher and Hampel, 1973). In France, trials on soft and hard wheat, and winter and spring barley showed the mixture of thiophanate methyl (0·7 kg/ha) and maneb (2 kg/ha) to be effective against glume blotch (Douchet *et al.*, 1972).

A mixture of tridemorph at 0·52 kg/ha and metiram (zineb/polyethylene thiuram disulphide complex) at 1·6 kg/ha has shown effects on glume and leaf blotch giving 22 per cent and 48 per cent control respectively which resulted in a 13 per cent yield increase (Frost and Brown, 1973). Tridemorph in combination with carbendazim is also reported as very effective. Ethirimol in mixtures with captafol gives excellent control of septoria and mildew in Europe. Thus a number of systemic compounds are active against *Septoria* spp. of wheat and barley. They are capable of delaying the spread of the disease in the crop, as a result of which the flag leaf, in particular, remains green for a longer period so that the grain-filling period is lengthened. The effect of application also appears greater in varieties most susceptible to septoria (Spiertz, 1973). So far, results have indicated little success in directly correlating disease control with yield increases.

Other foliar diseases

Some systemic fungicides are now used commercially against *Rhyncosporium secalis*, causal agent of leaf blotch of barley. Tridemorph, alone and in mixture with carbendazim, has given marked reduction in disease levels, the mixture giving a greater degree of control. Yield increases of up to 21 per cent have been obtained (Frost *et al.*, 1973; Phillips *et al.*, 1975). The activity of tridemorph against this disease may represent an added advantage from the use of the material for treatment of mildew in barley. The normal rate of application is 0·52 kg/ha of tridemorph plus 0·125 kg/ha of carbendazim.

Observations late in the growing season have indicated that tridemorph and the benzimidazole-dithiocarbamate mixtures suppress or control a range of potential grain spoilage organisms such as *Cladosporium* spp. and *Alternaria* spp. – components of the sooty mould complex.

Seed and soil-borne diseases

Loose smuts

Since the loose smuts of barley (*Ustilago nuda* (Jens.) Rostr.) and wheat (*U. tritici* (Pers.) Rostr.) are borne internally in the seed, control is not possible with non-systemic fungicides. The most usual method hitherto has been the use of 'hot' water steep which eradicates the fungus without causing major damage to the seed. This method is, however, costly and difficult to control, and the advent of suitable systemic fungicides has greatly increased the ease and efficiency of disease control.

Successful control of *U. nuda* with carboxin was first reported by von Schmeling *et al.* (1966). This has since been confirmed and expanded by numerous workers in several countries (early results were collated by Moseman, 1968). The rates used vary very widely within the range 25 g to 600 g a.i./100 kg seed. On barley, rates of 100 g/100 kg and above almost always gave complete disease control. At lower rates, small numbers of smutted ears were found, but even at 25 g/100 kg infection was reduced from 293 to 2 smutted plants per 55 000 seeds (Maude and Shuring, 1969).

Extremely good results have also been obtained with carboxin on *U. tritici* but here higher rates of chemical are required to give complete control (Jones *et al.*, 1968; Maude and Shuring, 1969). Some workers have found almost complete control at rates above 100 g/100 kg (e.g. Hansing, 1967); other workers have failed to get complete control even at rates as high as 375 g/100 kg (Maude and Shuring, 1969; Kuiper,

1968). Nevertheless, at the rate of 110 g/100 kg, carboxin gives excellent control on both wheat and barley in commercial usage in England (Macer *et al.*, 1969). Because of its relatively narrow spectrum of disease control, carboxin is normally used in mixtures with other fungicides such as copper oxyquinolate (with which it has some synergistic effect) (Richard and Vallier, 1969), mercury (Macer, 1969), maneb or thiram. In the US, carboxin is registered with the EPA for use in mixtures with thiram at a rate of 90–140 g carboxin/100 kg of seed.

Carboxin appears to be without phytotoxicity except at very high rates (Reinbergs and Edgington, 1968); furthermore, in some conditions it appears to be directly beneficial to the plant (von Schmeling and Clark, 1970). Yield increases are not normally obtained unless very high levels of disease are present; Stoker and Dewey (1970) however found yield differences at smut levels as low as 5 per cent.

Pyracarbolid (closely related chemically to carboxin) has a similar range of activity to carboxin, but is slightly more potent (Stingl *et al.*, 1970; Jank and Grossman, 1971). For example, in field trials against *U. nuda*, pyracarbolid reduced infection from 13·3 to 0·1 per cent, compared to 1·9 per cent with carboxin when both compounds were applied at 25 g a.i./100 kg (Stingl *et al.*, 1970). BAS 3191 (2,5-dimethylfuran-3-carboxylic acid anilide) gives excellent control of the smuts, at least comparable with that obtained with carboxin (Pommer, 1970; Ventura *et al.*, 1970). It has, however, been replaced by BAS 3270 (2,5-dimethylfuran-3-carboxylic acid cyclohexylamide) which has a similar spectrum of activity but is less phytotoxic (Pommer *et al.*, 1971).

Newer, more potent analogues of carboxin are now being developed.

Thiabendazole gives only moderate control of loose smut. For example, an application rate of 200 g a.i./100 kg only reduced infection from 16·0 to 6·3 per cent, and there was slight reduction in germination at this high rate (Darpoux *et al.*, 1968).

Benomyl can be used to obtain complete control of loose smut on wheat and barley, but on barley at rates much higher than those needed with carboxin. At very low inoculum levels. Manking and Wood (1969) obtained good control on barley at 62·5 g/100 kg, but in other trials at higher inoculum levels, 250 g was inferior to 94 g of carboxin (Manking and Wood 1970). Benomyl appears to be more active against wheat smut than against barley smut, complete control against the former being obtained at 100 g/100 kg (Ventura *et al.*, 1970). Benomyl at 227 g/100 kg seed in mixture with thiram is used commercially in England, and at this rate gives almost complete disease control (Allison *et al.*, 1975). Thiophanate methyl is identical in behaviour to benomyl.

Covered smuts, bunts

The main diseases in this group are wheat bunt (*Tilletia caries* (DC.) Tul. and *T. foetida* (Wallr.) Liro), dwarf bunt (*T. contraversa* Kühn), covered smut of barley (*Ustilago hordei* (Pers.) Lagerh.), loose smut of oats (*U. avenae* (Pers.) Rostr.) and covered smut of oats (*U. kolleri* Wille).

These diseases are either carried superficially on the seed, or are soil-borne; in general, they have been very well controlled by non-systemic chemicals such as mercury and maneb. Many of the new systemic fungicides also give good or excellent control.

Against covered smut on oats, carboxin gave almost complete control at 150 g/100 kg (Wallace, 1968) and 90 g/100 kg (Kuiper, 1968); Stingl *et al.* (1970) found complete control at all rates down to 40 g and Ventura *et al.* (1968) down to 30 g. Very similar results were obtained against covered smut of barley (Wallace, 1968; Ventura *et al.*, 1968). Carboxin is somewhat less active against wheat bunt, incomplete control being obtained within the range 150–225 g/100 kg (Wallace, 1968; Ventura *et al.*, 1968; Kuiper, 1968). Against soil-borne *T. foetida* carboxin was considerably inferior to HCB (hexachlorobenzene) on all except HCB-resistant strains (Kuiper, 1968), and was ineffective against dwarf bunt when used as a seed dressing, although it has given some control when used as a spray at 6·72 kg/ha (Hoffmann 1969, 1971a).

Pyracarbolid has been reported to be slightly more active than carboxin against covered smut of oats (complete control at 10 g/100 kg) (Stingl *et al.*, 1970). Results are not available against the other diseases.

Thiabendazole gave complete control of seed-borne bunt at 120 g/100 kg (Hoffmann, 1970), and good control at rates as low as 20 g/100 kg (Darpoux *et al.*, 1968). Similar results have been obtained against covered smuts of oats and barley. Like carboxin, thiabendazole is not very effective against soil-borne bunt, but it has given some control of dwarf bunt at high rates (250–480 g/100 kg) (Hoffmann, 1971b; Dickens and Oshima, 1971). At high rates, slight yield reductions have been observed in spite of good control of dwarf bunt (Dewey *et al.*, 1974).

Benomyl is highly effective against seed-borne bunts and smuts, giving complete control at 100 g/100 kg on wheat, barley and oats (Ventura *et al.*, 1970; Manking and Wood, 1969). Benomyl (227 g/100 kg) + thiram is used commercially in England, and a carbendazim/maneb mixture has also given excellent disease control (Allison *et al.*, 1975). Benomyl is relatively ineffective against soil-borne bunt and dwarf bunt (Hoffmann, 1970). Where thiophanate methyl has been compared with benomyl results have been approximately equal.

Flag smut

Flag smut of wheat (*Urocystis agropyri* (Preuss) Schroet.) is mainly soil-borne and has not been very well controlled by standard fungicides. Of the new systemic fungicides, oxycarboxin gives the best control. Almost complete eradication was obtained at 30 g/100 kg: carboxin and pyracarbolid were also extremely effective (control at 75 and 50 g/100 kg respectively): benomyl was only moderately effective and thiabendazole was ineffective (Line, 1970). A similar pattern of results has since been found by several workers.

Seedling blights

The diseases in this group are seed- or soil-borne and are mainly well controlled by organomercurial or non-mercurial seed dressings such as maneb. Whilst many authors have noted the effect of systemic fungicides on field emergence, few have studied the effect of the chemicals on specific diseases.

Snow mould, Fusarium nivale Fr. None of the new fungicides is as effective as the standard compounds. Benomyl, thiophanate methyl, thiabendazole, BAS 3191, BAS 3270 and carboxin all have some activity at 100 g/100 kg (Ventura *et al.*, 1970; Darpoux *et al.*, 1968; Pommer *et al.*, 1971). Benomyl is used commercially at 100 g/100 kg as a seed dressing and at 200 g/ha as a spray for the control of snow mould in Finland.

Brown foot rot. Fusarium culmorum (Sacc.) Sn. and H. and *F. graminearum* (Schwabe) Sn. and H. Where wheat seed internally infected was used, thiabendazole gave complete control at 160 g/100 kg whilst maneb and mercury were relatively ineffective (Darpoux *et al.*, 1968).

Leaf stripe. Helminthosporium gramineum Rabenh. In trials on spring and winter barley, triarimol, BAS 3191 and BAS 3270 gave slight control, but benomyl and thiophanate methyl were ineffective. All treatments were inferior to maneb and mercury (Ventura *et al.*, 1970; Pommer *et al.*, 1971). Poor results were also obtained with thiabendazole (Darpoux *et al.*, 1968). Good control of leaf stripe has been obtained in the field with carboxin applied at a rate of 125 g/100 kg (Kingsland 1970) and 150 g/100 kg (Macer *et al.*, 1969). However, carboxin alone does not usually give adequate control: good control has been obtained in mixture with thiram (Kline, 1972).

Glume blotch. Septoria nodorum Berk. Benomyl (150 g) and thiophanate methyl 100 g per 100 kg reduced infection of seedlings, but were inferior to mercury; oxycarboxin and BAS 3191 were ineffective (Ventura *et al.*, 1970). Thiabendazole, however, was found to have good activity, comparable with that of maneb and mercury (Darpoux *et al.*, 1968). The activity of the bendimidazoles is increased by mixing with maneb or thiram.

Discussion

The great advantage of the systemic fungicides as seed dressings is that they provide control of pathogens borne within the seed (the loose smuts); these are inaccessible to non-systemic fungicides. None of the new chemicals has a sufficiently broad spectrum to control the wide range of diseases currently controlled by the organomercurials or certain non-mercurial dressings like maneb, thiram or copper oxyquinolate. It is therefore necessary to use mixtures of systemic and non-systemic fungicides: because of the relatively high cost of these mixtures their use has largely been confined to treatment with stocks known to be heavily infected with loose smut, or when the crop is being produced for seed. However, the newer analogues of carboxin currently being tested might be much cheaper to produce, and this could allow more widespread use than has been possible with carboxin.

Eyespot

Cercosporella herpotrichoides Fron occurs wherever wheat and barley are grown. It reduces yield by directly reducing the amount of carbohydrate going to the grain, and by causing lodging. Benomyl and thiophanate methyl are now used commercially for the control of eyespot; in addition the growth regulant chlormequat has been widely used to shorten and stiffen straw, and so make it less prone to lodging. When used together, benomyl and chlormequat have an additive effect on yield particularly after a high dose of nitrogen fertiliser has been applied (120 kg N/ha) (Fehrmann and Schroedter, 1972).

Many workers have investigated the effectiveness of benomyl in controlling eyespot in glasshouse experiments (Defosse, 1970; Lemaire *et al.*, 1970; Fehrmann, 1970). In these experiments, excellent control was achieved by spray, seed dressing and soil applications. Even when applied 30–50 days after inoculation, progress of the disease was halted, indicating a very powerful eradicant activity (Defosse, 1970; Fehrmann, 1970). Protective sprays were only effective if applied not more than 10 days before infection. The effectiveness of the benzimidazole compounds has been confirmed in the field, where the eradicant effect has again been established, allowing good control over widely different spray times. In the United States, in artificially inoculated plots of winter wheat, benomyl spray at 5·6 kg/ha in November, March or April gave almost complete disease control and significantly increased yields (Powelson and Cook, 1969). Other experiments in the United States have indicated that spring applications of benomyl were generally more effective at lower rates than fall applications (Huber and Mulanax, 1972; Bruehl and Cunfer, 1972). Similarly, in Germany, autumn application was relatively ineffective, but sprays of 240 g/ha in the spring were very effective, a spray in May reducing infection from 98 to 7–10 per cent when assessed 6 weeks before harvest (Fehrmann, 1970). Trials with carbendazim in Germany and England have further confirmed the most favourable time for application is at a period between the end of tillering to the formation of the first node (Hampel and Locher, 1973; Taylor and Waterhouse, 1975). In severely infected artificially inoculated plots in England, a spray at GS 9 (Feekes Scale) reduced infection by 95 per cent and increased yields on average from 6 to 8 per cent on moderately resistant varieties to 43 per cent on very susceptible varieties (Doodson and Saunders, 1969).

Equally striking results have been obtained in naturally infected wheat in the field using low rates of application. In England, sprays of 1·12 kg/ha at GS 3–4 gave excellent control, but slightly increased sharp eyespot (*Rhizoctonia solani* Kuhn). A seed treatment at 280 g a.i./ha gave poor control of eyespot (Prew, personal communication). In New Zealand, a single spray of 0·28 kg/ha applied at GS 7–8 reduced lodging from 36 per cent to almost nil, and increased yield by 46 per cent (Witchalls and Close, 1971). The efficiency of benomyl spray depends on correct timing in relation to the main infection period, and Rapilly (1970) has attempted to forecast this, using the method devised by Ponchet (1959) based on the mean daily temperature. Benomyl has also proved active against eyespot of rye (Rawlinson and Davies, 1971) and barley (Davies, 1970).

Less work has been reported on other chemicals, but thiabendazole was found to be almost as active as benomyl when applied as a seed dressing (Lemaire *et al.*, 1970). Thiophanate methyl in mixtures with maneb has also been found to be effective as a spray (Baker, 1975).

Rice

Rice is attacked by a number of fungal diseases, the most important of which is blast (*Pyricularia oryzae* Cav.), and this is the only disease on which extensive results with systemic fungicides have been published. Although blast is important in virtually every rice-growing country, it is largely in Japan that progress with fungicidal control has been made.

In the early 1960s, phenylmercury acetate and the antibiotic blasticidin S were widely used for control of blast; both compounds have now been replaced because of their high toxicity to mammals. Blasticidin is absorbed into plants, but is not effective as a systemic fungicide because of its rapid degradation in the plant (Misato *et al.*, 1959). More recently a far safer antibiotic, kasugamycin, has been used for blast control. This compound, obtained from the culture filtrate of *Streptomyces kasugaensis*, is very active when sprayed at 20–30 ppm, and there is a wide margin of safety from phytotoxicity. The chemical is very effective in preventing secondary spread and sporulation, but has little effect until after penetration (Okamoto, 1968). It is therefore rather more effective as an eradicant than as a protectant fungicide. It is absorbed into the leaves and roots and translocated readily into the upper parts of the plant, where further lesion enlargement is prevented (Ishiyama *et al.*, 1965a, b; 1967). In practice, however, it is used only as a foliar spray since excessively high quantities are required for application through the paddy water (Kozaka, 1969).

More recently a number of closely relatea organophosphorus fungicides have been developed in Japan, including kitazin, kitazin P, Hinozan, inazin and cohnen. All of these compounds give effective control of blast in sprays of 400–500 ppm, but above this there is a danger of phytotoxicity. The chemicals have eradicant as well as protectant properties, but are best applied not later than 2–3 days after infection. By far the most readily systemic of these compounds is kitazin P (Kozaka,

1969), and this compound is freely translocated from roots to leaves.

In practice kitazin P can be applied in a programme of 2–3 sprays at 450 ppm or it can be applied to paddy water at a rate of 5–10 kg/ha (Yoshinaga, 1969). The granules appear to become active about 3 days after application, to reach their maximum effectiveness after 5 to 7 days but to retain marked activity for at least 30 days (Kozaka, 1969). For leaf blight control, granule application at the time blast was observed, or 1 week earlier or later, gave equivalent disease control. With neck and node blast, best yield results were achieved with an application about 1 week before heading, although later applications gave slightly better disease control. Even at the lowest rate of application, granules were markedly more effective than a programme of three sprays (Yoshinaga, 1969).

A newly introduced organo-sulphur systemic fungicide, iso-prothiolane, shows promise for control of rice blast in both foliar and paddy water applications. It has both eradicant and protectant properties. Rates of application are 0·5 kg/ha as a foliar spray; and 1–5 kg a.i./ha as a granular, depending on the stage of the crop and on disease intensity. Best results are obtained when granule application is made 10–14 days before leaf blight infection or 10–20 days before heading for control of ear blight.

Another new fungicide, tricyclozole, has excellent systemic properties and is being tested in many countries as a foliar spray, a granule and a drench applied to the nursery boxes. Tricyclozole gives very long residual control by all methods of application (Yamaguchi, 1974) and is probably the most promising of the newer compounds.

Benomyl has marked activity against blast, although this does not appear to have been widely exploited. In trials in Louisiana, two sprays of benomyl each at 560 g/ha gave control of neck blast equivalent to that obtained, with phenylmercury acetate, and superior to blasticidin S (Lindberg, 1969). Thiophanate methyl has subsequently been shown to have similar activity to benomyl against this disease (Lindberg, 1974). In the Philippines, two sprays of benomyl at 200 g/ha were markedly superior to Kasumin for control of neck rot (Kasumin is reported to be relatively ineffective in the tropics) (Anon, 1970).

Benomyl has also been used as a seed treatment to control seedling blast. In box tests, benomyl at 400 g/100 kg of seed gave almost complete control during the seedling stage, whilst untreated plants were almost killed. Soil treatment with benomyl also gave good control of leaf and neck blast (superior to Kitazin) but only at rates unlikely to be economic (Anon, 1970).

Carboxin in a formulated mixture with thiram has been reported to increase stand counts of rice infected with several seed-borne fungi other than blast (Rush *et al.*, 1974).

The benzimidazole fungicides and thiophanate methyl have also controlled stem rot caused by *Sclerotium oryzae*, and benomyl has reduced the incidence of narrow brown leaf spot (*Cercospora oryzae*) following two applications at mid-boot and heading stages (Rush and Lindberg, 1974).

Maize

Most maize diseases are not amenable to control with the available systemic fungicides, and work has largely been restricted to trials against seed- and soil-born diseases. Moderate control of head smut *Sphacelotheca reiliana* (Kuhn) Clint. has been obtained in naturally infested field plots at a rate of 250 g/ka (Simpson and Fenwick, 1971) and in seedlings transplanted from artificially inoculated soil in boxes to the field at 400 g/100 kg (Jooste, 1969).

Carboxin is also reported to have some activity against southern leaf blight (*Helminthosporium maydis*, Nisikada and Miyake) when used as a seed dressing (Futrell, 1972).

Benomyl, thiophanate methyl and carboxin when applied as seed treatments or soil drenches have given up to 75 per cent control of loose smut (*Ustilago maydis* (DC) Corda) in field trials on seedlings injected with sporidial suspensions (Enyinnia and Halisky, 1974). Benomyl and thiabendazole have also shown some activity in arresting *Fusarium moniliforme* Sheldon infection in corn seed.

Grasses and turf

Grass and turf diseases are important throughout the world. The majority of the research work has been performed in the United States, however, where systemic fungicides are used extensively in the golf course and home owner outlets.

Since 1970 benomyl and thiabendizole have been applied commercially for the control of turf diseases, particularly on golf course tees and greens and for disease control in seed production crops. Prior to 1970 many turf grass diseases were successfully controlled with organo-mercury fungicides and thiram. These had the advantage of a wide spectrum of control. The current systemic fungicides, however, whilst narrower in their spectrum of activity, have provided excellent, relatively long-term control of those disease for which they are specific.

Dollar spot

Dollar spot caused by *Sclerotinia homoeocarpa* F. T. Bennett is common in bentgrasses and fescues in the northern United States and the British Isles. In the warmer regions of the United States bermudagrass (*Cynodon dactylon* (L.) Pers.), Bahiagrass (*Paspalum notatum* Flugga) and zoysiagrasses (*Zoysia* spp.) can be severely affected with the disease (Freeman, 1967).

Excellent control of dollar spot has been achieved with benomyl, thiophanate methyl and thiophanate ethyl. All three compounds appear to have the same level of activity (Larsen *et al.*, 1974; Bowyer and Shurtleff, 1973) when applied at 0·15 g/m^2 in 20 l of water as preventive treatments. A higher rate of 0·30 g/m^2 is required for control when the disease is established. Excellent disease suppression also occurs when thiabendazole is applied; however, some phytoxicity has been reported on bentgrass (*Agrostis palustris* L.) at 0·37/m^2 when turf was sprayed at weekly intervals (Harrison, 1970).

The systemic fungicides have largely replaced the cadmium-based fungicides for control of dollar spot. They need only be applied at relatively long (10- to 14-day) intervals, and in some instances spray intervals of up to 3 weeks suffice (Brown *et al.*, 1970). This contrasts with cadmium-based compounds which have to be applied at 3- to 4-day intervals as a curative measure. The benzimidazoles have also shown control 8 weeks after application (Bowyer and Shurtleff, 1973). As yet, tolerance of the fungus to the systemic fungicides is not a problem, although some strains tolerant to the benzimidazoles are known (Warren *et al.*, 1974b).

Dollar spot usually occurs in conditions of low fertility; benomyl is, therefore, frequently applied in a granular fertiliser formulation. Treatment starts in the late spring and benomyl is applied at 0·15 g/m^2 and repeated at 2- to 3-week intervals.

Brown patch

Brown patch, caused by *Rhizoctonia solani* Kuhn, is one of the most important turf diseases and is most active when there are periods of high relative humidity and high night temperatures. For this reason the disease is particularly prevalent in the southern United States and it is usually severe on St Augustine grass (*Stenotaphrum secondatum* (Walt.) Kuntze) and bentgrass (*Agrostis* spp.). Bermudagrass and bahiagrass are somewhat less susceptible but nevertheless prone to attack whilst Kentucky bluegrass (*Poa pratensis* L.) is seldom damaged.

The disease is characterised by a dark blue ring at the edges of diseased areas, which may range from a few inches to several feet in diameter. The severity of attack tends to increase in situations where there is excess nitrogen available in conjunction with high soil moisture. Rhizoctonia brown patch control has been best achieved with benomyl and thiabendazole by following a programme of preventive application starting the day following night air temperatures of not less than 20°C (Goldberg and Cole, 1972). Subsequent applications are made at 7- to 10-day intervals and more frequently if high night temperatures persist. Rates of application are similar to those used for dollar spot control.

Pythium or cottony blight

Pythium or cottony blight mainly caused by *Pythium ultimum* Trow and *Pythium aphanidermatum* (Edson) Fitzpatrick is primarily a disease of cool-season grasses. It is most destructive; and bentgrasses, fescues and Kentucky bluegrass are particularly susceptible. Bermudagrass is only slightly affected, and the disease does not usually occur in other southern grass species, although during winter months in the southern United States Pythium blight can be a problem on golf greens and tees overseeded with ryegrass.

Pythium blight is capable of destroying both seedling and mature grass stands and has proved difficult to control with fungicides. Until the 1970s mercury fungicides and chloroneb (1,4-dichloro-2,5-dimethoxybenzene) were applied in preventive programmes during periods of weather favourable for disease development.

More recently pyroxychlor has given promising results. This compound is narrow in its spectrum of activity, limited mainly to phycomycetous fungi. It is of particular interest in that foliar application of pyroxychlor can result in disease control by basipetal translocation.

Pyroxychlor has given good control of *Pythium* species on *Agrostis palustris* and *A. tenuis* at rates as low as 0·3 g/m^2 sprayed to wet (Cole *et al.*, 1974), and at 0·6 g and 1·2 g excellent residual control has been achieved for 14 days after treatment. Control with pyroxychlor is generally superior to chloroneb provided pyroxychlor sprays are irrigated with 1–2 cm of water soon after an early morning application (Cole *et al.*, 1974).

Rusts

There are numerous rust fungi that attack turf grasses but they rarely occur at a frequency or intensity to be of major importance on mowed lawns. Rusts can be a problem, however, on crops grown for seed production, hay or pastures. The 'Merion' variety of Kentucky bluegrass, certain fine leaved perennial rye grasses and bermudagrass, are highly susceptible to rust.

Presumably because of their lesser importance in turf, there are relatively few reports relating to control of rust diseases specifically in grasses. Oxycarboxin has been reported to control stripe rust (*Puccinia striiformis*) (Hardison, 1971; Murdoch *et al.*, 1973). It has been shown to be effective against leaf rust (*Puccinia brachypodia* var. *poaenemoralis* (Otth.) Cummins and H. C. Green) and stem rust (*P. graminis* Pers.) on bluegrass, *P. zoysia* Diet on zoysiagrasses and against crown rust (*P. coronata* Cda.) on Italian rye grasses at rates of 0·45 g/m^2.

Amongst the more recent systemic compounds triadimefon has shown excellent control of stripe rust (*P. striiformis*) on Kentucky bluegrass. Control occurred in the field after a single application of triadimefon at rates ranging from 0·78 to 2·35 kg/ha (Hardison, 1975). In the same test Hardison showed BAS 3170F to give excellent control of stripe rust at 1·12–2·2 kg/ha after three applications. Both compounds performed better than the commercial standard. a nickel sulphate/maneb mixture.

Stripe smut

Ustilago striiformis (West) Niesel and flag smut *Urocystis agropyri* (Preuss) Schrot, are widely distributed and can be of great importance in seed production. These smuts are particularly destructive on several cold-season grasses causing reduction in yields, and seed of low viability.

Stripe and flag smut have been controlled by soil drenches of benomyl and thiabendazole. Significant reduction was achieved with benomyl at 0·6 and 1·2 g in 3 l of water/m^2. Thiabendazole is slightly less active than benomyl but has given control at 0·9 g/m^2 (Muse, 1970). Thiophanate methyl is also active against stripe smut (commercial rate 0·8–1·6 g/m^2) but has proved less effective in controlling flag smut (Jackson and Fenstermacher, 1974). Thiophanate ethyl is inactive against both diseases (Hardison, 1972a). Satisfactory control of stripe smut has also been demonstrated with benomyl at a low-volume application of 0·04 l/m^2 containing 0·9 g of active ingredient (Cole *et al.*, 1972). BAS 3191F, in a single application to soil, reduced stripe smut by up to 95 per cent at 160 mg/600 ml of soil. 40 mg/600 ml of soil controlled flag smut and eradicant activity against the fungus was indicated (Hardison, 1972b).

Oxycarboxin is active against smuts of grasses (Hardison, 1971), but triforine was proved inactive in one test (Smiley, 1974).

Fusarium blights

Fusarium roseum f. sp. *cerealis* (Cke) and *F. tricinctum* f. sp. *poae* (Pk. P) Snyder and Hansen occur widely on bluegrass, bents and fescues. Benomyl adequately controls these blights at rates ranging from 0·9 to 1·0 g/m^2. Thiabendazole has proven somewhat less effective (Fulton *et al.*, 1971). Thiophanate methyl also successfully suppresses the disease and is used from 0·07 to 0·15 g in 0·2 l of water/m^2 followed immediately by 2–3 cm of irrigation water. Thiophanate methyl has consistently proved more effective than thiophanate ethyl (Goldberg *et al.*, 1973; Palmer, 1973; Smiley and Harrison, 1974).

Fusarium patch

Fusarium nivale (Fr.) Snyder and Hansen is common on turfs in the northern United States and northern Europe when there are cool wet months during spring or autumn. The disease is controlled by benzimidazole fungicides (Dickens and Butler, 1974) which are inactive, however, on snow moulds caused by *Typhula* sp.

Systemic fungicides have also been reported to control a number of grass diseases of lesser economic significance than those already described. Thiophanate methyl has shown good activity against copper spot (*Gloeocercospora sorghi* Bain and Edgerton) which can cause serious problems of bent grasses. Benomyl has been reported to give prolonged control of powdery mildew in orchardgrass (*Dactylis glomerata* L.) under greenhouse conditions (Leath and Berg, 1972). On ryegrass seed, benomyl applied at 5·6 kg/ha has given 90 per cent suppression of apothecia of *Gloeotinia temulenta*, causal agent of blind seed disease (McGee, 1972), and thiabendazole has significantly reduced a leaf spot caused by *Cercospora fusimaculans* of St Augustine grass applied as a soil drench at 0·3 g/m^2.

Amongst the most important diseases that are not satisfactorily controlled by available systemic fungicides are those incited by *Helminthosporium* spp. The benzimidazoles have proven inactive against these diseases (Conway and Rich, 1970) but it is of interest to note, however, that benomyl combined with maneb performed significantly better against helminthosporium leaf spot and crown rust than maneb alone (Warren *et al.*, 1974a).

Results in practice – II. Glasshouse crops

D. M. Spencer

Introduction

The glasshouse crops industry exists almost entirely in the northern hemisphere. Indeed, it is mainly confined to the temperate areas of Western Europe and North America with less extensive areas in Japan and Eastern Europe, particularly Romania, Bulgaria and the USSR. Consequently the range of crops with which the glasshouse industry is concerned is somewhat limited. In Europe, for example, tomatoes, cucumbers and lettuces constitute the bulk of the industry with flowers and other decorative plants making up most of the remainder.

Holland has a larger glasshouse acreage than any other country (Table 11.II.1) and of this the major part is devoted to vegetable production. The same is true for the British Isles where flower production accounts for about 30 per cent of the glasshouse acreage. In Denmark and Germany on the other hand a relatively larger proportion, though still less than half of the glasshouse space, is occupied by decorative crops. In the USA three-quarters of the glasshouse acreage is used for flower crops.

These proportions have changed in recent years with an increase in the acreage of ornamental crops.

Italy and Japan have relatively small areas of traditional glasshouses,

Table 11.II.1 Area and function of glasshouses in major glasshouse-crop-producing countries

		% Acreage	
	Acreage	**Vegetables**	**Flowers**
Holland	18 456	70	30
USA	6 570	25	75
France	2 865	63	27
Germany	5 808	37	63
British Isles:			
(*a*) England	4 521	70	30
(*b*) Scotland	265	†	†
(*c*) N. Ireland	54	†	†
(*d*) Guernsey	1 144	80	20
(*e*) Jersey	112	†	†
(*f*) Eire	450	†	†

† Figures not available; mainly vegetables

but both countries have rapidly increasing areas of crops grown under plastic structures. It is against this background of an established range of crops with a corresponding range of diseases that the practical application of systemic fungicides must be considered.

Crops in glasshouses are usually grown without rotation and thus tend to suffer from diseases which can very quickly become epidemic in one house or on one holding. Within the protection of the glasshouse, however, the grower does have the facility for a measure of control over the environment in which his crop is growing and hence, by careful manipulation of this control, the experienced grower can minimise the losses caused by disease. For example, by the judicious use of his ventilators the astute tomato-grower will so control the relative humidity of the atmosphere within his crop as to reduce the likelihood of spores of leaf-invading pathogens such as *Botrytis cinerea* and *Cladosporium fulvum* (*Fulvia fulva*) being able to germinate and to invade their host. Even with the wilt organism *Fusarium oxysporum*, the inoculum of which may be present in the deeper soil layers where steam or chemical soil partial-sterilants cannot penetrate, the experienced grower can accommodate the infection that may develop and can grow his crop with the minimum of loss through wilt. In considering the use of chemical crop-protectants, such as systemic fungicides, in the glasshouse it is therefore important also to take account of the non-chemical methods of disease control available to the grower and as far as possible to integrate the various methods of disease control.

The 1975 edition of the Agricultural Chemicals Approval Scheme (Ministry of Agriculture, 1975) which, together with its supplements, lists the 820 proprietary products officially approved for crop protection in horticulture and agriculture in the United Kingdom now mentions twelve systemic fungicides. Among these, benomyl, carbendazim and thiophanate methyl depend for their activity on the carbendazim moiety (methyl benzimidazol-2-yl carbamate). Therefore, for the purposes of this chapter these three compounds will be referred to as the carbendazim fungicides.

Disease control in the glasshouse

Special requirements

The two most important glasshouse vegetable crops, tomatoes and cucumbers, each have a long cropping season. From the time that the first fruit is harvested to the time that the plants are pulled out may be as long as 8 months and for some of this time fruit may be picked almost daily. Thus, any compound will be unacceptable for use on such crops unless it can be shown that the compound is either non-toxic to those who eat the fruit or that such toxicity is lost very soon after the compound has been applied. Benomyl, for example, was cleared in 1970 for use in the United Kingdom on all edible crops, with no time interval between application and harvest. Similarly there is no restriction on the time between application of dimethirimol and the picking of cucumber fruit. Other carbendazim fungicides can be used up to and during harvest, and the newer systemic fungicides recommended for use on these glasshouse vegetables also have a tolerably short interval between application and harvest.

Because of the short interval between successive harvests those protectant fungicides, which have to be applied as sprays at frequent intervals to protect new growth and which leave noticeable deposits on the fruit, will be unacceptable. Furthermore, glasshouse crops are grown intensively and therefore bulky machinery cannot be used without the danger of damaging tall-growing crops. It follows, therefore, that a material which has to be applied at short intervals by hand spraying will be less attractive to the grower than one which can be applied automatically and at longer intervals.

Nevertheless the manufacturers of all the systemic fungicides suitable for use as drenches, to control foliar diseases, give recommendations for spray application as well as drenches. Indeed, drench application is not recommended for woody subjects. The full potential of systemic fungicides is only really used when they are applied to the roots and it seems that at this stage root application is not fully satisfactory for many of the currently available systemic fungicides. Those which can be applied as drenches need to be used only monthly, whereas sprays are applied fortnightly. Furthermore, when applied to the soil and taken up by the roots these compounds will not cause blemishing deposits on the outside of the crop. For these reasons systemic fungicides have been long-awaited by the glasshouse industry and it is, perhaps, not surprising that growers have shown great enterprise in making full and enthusiastic use of benomyl, the first wide-spectrum systemic fungicide to become available to them.

In the past the glasshouse industry has been obliged to make use of crop-protectant compounds primarily developed for other agricultural applications. Dimethirimol is a notable exception in that it was developed specifically for use against powdery mildews of cucurbits and therefore had direct applicability to the glasshouse cucumber crop.

Use of selected systemic fungicides

Tomato

Grey mould (*Botrytis cinerea*) This organism can cause death of the whole plant if lesions girdle the lower stem. It also causes loss through reduced yield and reduced value of 'ghost-spotted' fruits. Primary infections usually occur on leaves and flowers (Smith, 1970), and infection spreads to other parts either where infected flowers fall or through the distribution of airborne spores.

Though Bollen and Fuchs (1970) showed the organism to be very sensitive to benomyl none of a group of eight systemic fungicides was able to control ghost-spotting of the fruit (Smith, 1971a). Indeed, though an *in vitro* bioassay method, using *Fulvia fulva* and *B. cinerea* as test organisms indicated the presence of fungitoxic materials in the leaves, sepals and petals of plants treated with benomyl or with two thiophanate compounds, the method failed to demonstrate fungitoxicity in fruits (Smith and Spencer, 1971). Several compounds, applied as sprays or drenches, were able to inhibit the spread of stem and leaf lesions, suggesting that if they were to be applied as a protectant treatment to the entire crop in a glasshouse they could prevent the build-up of conidial inoculum and thus reduce ghost-spotting of the fruit.

Spray applications of benomyl and carbendazim 50 g/100 l at 14-day intervals or 500 ml per plant of a drench containing 25 g/100 l applied at monthly intervals are used for the control of *B. cinerea*. Thiophanate methyl suspensions used for spray and drench applications contain 100 g/100 l and 50 g/100 l respectively.

Strains of *B. cinerea* able to tolerate high doses of carbendazim fungicides are widespread in tomato-growing areas. Thus the use of these fungicides alone may be less reliable than heretofore, and the manufacturers draw attention to this on their product labels. Indeed, one manufacturer recommends the use of dichlofluanid before picking tomatoes from an unheated crop where control by carbendazim fungicides is poor.

Leaf mould (*Fulvia fulva*) This disease is associated with high temperatures and high humidity and can often be controlled by manipulation of the environment, where this is possible. The causal organism is sensitive to low concentrations of thiabendazole, benomyl, carbendazim and thiophanate methyl, but less so to triforine and has been used in tests *in vitro* for the presence of fungitoxic materials in treated plants (Smith and Spencer, 1971). In small-scale glasshouse tests, six systemic fungicides were found to control the disease for at least 6 weeks after a single drench treatment (Smith, 1971a).

The carbendazim fungicides are recommended for use as a spray for the control of *Fulvia fulva* infection in the same way as it is used for the control of grey mould (*B. cinerea*).

Stem rot (*Didymella lycopersici*) This disease is much worse in some years than in others and usually attacks first at soil level. The sunken dark brown lesions with barely visible black pycnidia may appear on other parts of the plant later in the season, often at the site of injury but not necessarily so, and usually cause the death of those parts of the plant above the infection.

The ED_{50} of benomyl against both *D. lycopersici* and *B. cinerea* is 1–5 μg/ml (Bollen and Fuchs, 1970) and the symptoms of stem rot (*D. lycopersici*) have much in common with those of grey mould (*B. cinerea*). Smith (1971a) found that benomyl applied as root drench or spray, though retarding the growth of stem lesions caused by *B. cinerea*, was not completely inhibitory and results obtained for control of *D. lycopersici* by benomyl have been variable. However, it has been found that carbendazim fungicides applied monthly as a whole-stem spray or as a stem-base drench can reduce the level of disease caused by *D. lycopersici.*

Wilt diseases (*Verticillium albo-atrum*, *Verticillium dahliae and Fusarium oxysporum*) These important diseases are soil-borne and therefore efficient soil sterilisation is essential where they are a problem. The losses they cause can, to some extent, be mitigated by careful control of the environment: with *Fusarium* wilt the temperature needs to be kept down and with *Verticillium* wilt, higher temperatures are required. Until the advent of systemic fungicides the growers' only alternative to adequate environmental control was to pull out all infected plants as they appeared.

Bollen and Fuchs (1970) found benomyl to be highly active against *V. albo-atrum* and *V. dahliae* but less active against *F. oxysporum*. Fuchs *et al.* (1970) found the last-named fungus to be inhibited by 1 μg/ml benomyl *in vitro* and when they supplied the compound as a daily drench to potted seedlings which had been inoculated with *F. oxysporum*, symptoms of wilt were suppressed. Weinke *et al.* (1969) showed thiabendazole to be active *in vitro* against *Verticillium* and *Fusarium* species and were able to detect some movement of the compound into roots of tomato plants after foliar applications. Muller *et al.* (1972) found that wilt symptoms in seedlings following stem inoculation with *F. oxysporum* and *V. albo-atrum* were restricted in soil treated with thiabendazole at 25 and 50 μg/g. Thiabendazole persisted in the seedlings for up to 7 days after root uptake and persisted in the soil for up to 20 weeks.

Soil drenches of benomyl applied before or after inoculation were found to control *Fusarium* wilt of potted tomato plants by Channon and Thomson (1973) and also by Thanassoulopoulos *et al.* (1970), who noted that spraying the leaves with an aqueous suspension had no effect on the disease. They also affirmed that the chemical had no apparent phytotoxic effect when applied as a soil drench at rates up to 10 000 ppm (a.i.) but in the writer's laboratory potted tomato plants treated with high levels of benomyl have shown leaf necrosis, particularly on younger leaves. In field-grown tomatoes Matta and Garibaldi (1970) found benomyl and thiophanate-methyl highly effective when applied as soil drenches in reducing *Verticillium* wilt, whilst Jones *et al.* (1971) achieved wilt control for over 4 months when benomyl was applied at the same time as the soil fumigant 1,2-dibromoethane.

Benomyl, used as a soil drench, at 500 ml per plant, of a suspension containing 25 g/100 l applied at 4-week intervals is said by the manufacturers to reduce the spread of *Fusarium* and *Verticillium* wilts.

Brown root rot (*Pyrenochaeta lycopersici*) This disease reduces plant vigour and crop production by destroying part of the root system. Normally the disease is kept under control by annual or twice-yearly soil sterilisation with steam or methyl bromide. Ebben (1971) found that benomyl applied at 4·5 g/m^2 as a whole-plot drench or applied dry and raked into the soil surface, reduced brown root rot and increased tomato yields in soils known to be heavily infested with *P. lycopersici*. Thiabendazole was phytotoxic at the concentration used (4·5 g/m^2) and though it brought about a decrease in brown root rot the yields from plots treated with this compound were less than those from untreated plots. The cost of materials for this treatment would be about £350 per acre (£870/ha) and, as Ebben (1971) points out, more than one treatment would be needed in the season to control brown root rot. In addition, benomyl is ineffective against some soil-borne pathogens and so the compound cannot be considered in any way as a substitute for effective soil sterilisation. It may be that post-planting drenches could be used to control brown root rot in patches where soil sterilisation had not been efficient.

Cucumber

Powdery mildew (*Sphaerotheca fuliginea*) Powdery mildew is the most important cucumber disease. It spreads very quickly in hot dry conditions and its control by standard protectant fungicides is complicated by the very rapid growth of the cucumber plant accompanied by a rapid proliferation of new and therefore unprotected leaves.

Dimethirimol is a systemic fungicide which is highly active against certain powdery mildews. It was shown by Elias *et al.* (1968) to be particularly effective and to give long-lasting protection against powdery mildews of cucurbits when applied to the soil. Bent (1970) found dimethirimol to be toxic to the spores of *S. fuliginea in vitro* and *in vivo* and was able to correlate the disease control provided by this compound with its direct fungitoxic action. The compound was widely used on the British cucumber crop in 1968 and gave spectacular control of powdery mildew. Since 1969 the material has been used widely in Holland but in both these countries its continued use has been complicated by the emergence of strains of the fungus which are tolerant to dimethirimol (see Chapter 9).

Benomyl and carbendazim are highly effective against cucumber powdery mildew and are widely used both as a spray at 250 ppm (0·5 kg/1 000 l) or soil drench at 200 ppm (0·4 kg/1 000 l, 500 ml/plant), each applied at monthly intervals. Thiophanate methyl is used at double these concentrations. The carbendazim compounds give good control of *B. cinerea* and the spray application has also been found to give a measure of control of *Didymella bryoniae* (*Mycosphaerella melonis*). Cole *et al.* (1970) found that soil amendment with benomyl at concentrations of 40 ppm gave complete control of powdery mildew of cucumber.

Other systemic fungicides provide control of *S. fuliginea.* Triforine was shown to be systemically active against cucumber powdery mildew when applied as a pre- or post-infectional treatment (Fuchs *et al.*, 1971). This compound is not recommended as a drench treatment by its manufacturers, but as a spray at 200 ppm it gives good control of powdery mildew and some control of stem rot (*D. bryoniae*).

The systemic fungicide pyrazophos was found by de Waard (1974) to have a greater specific protectant and curative action than any other organophosphorus compound he tested against cucumber powdery mildew. This compound is used as a spray at 90–300 ppm depending on the level of infection pressure. There is a waiting period of 3 days between last application and harvest.

Black root rot (*Phomopsis sclerotioides*) Where this disease is a problem, soil sterilisation is essential though root lesions have been found on plants from steamed plots after about 4 months (Ebben, 1970b). More powdery mildew was found in plants with black root rot than on those with healthy roots and the application of dimethirimol gave poorer control of mildew on plants with root rot.

Ebben and Last (1969) grew plants on plots infested and non-infested with *P. sclerotioides* and controlled both black root rot and powdery mildew (*S. fuliginea*) with benomyl drenches. Where no benomyl was applied yields were very much reduced by mildew on plots not infested with *P. sclerotioides* but were even lower in plots where root rot also occurred.

Cruger (1969) noted in his experiments on the control of cucumber mildew by benomyl and dimethirimol that the former compound applied as a drench at 500 mg a.i. per plant gave some control of *P. sclerotioides.*

Stem rot (*Didymella bryoniae*) In tests on 2-week-old seedlings growing in pots, Ebben (1971) found that benomyl, thiophanate and thiophanate methyl prevented the development of lesions when petioles were inoculated with mycelium from growing cultures of *M. melonis.* By contrast triarimol and thiabendazole did not prevent lesion development even when they were used at phytotoxic concentrations. In older cropping plants, benomyl sprays were only able to retard the development of *M. melonis* lesions. Triforine and the carbendazim fungicides all give some control of stem rot when used as sprays to control powdery mildew.

The use of systemic fungicides in sprays and drenches on cucumbers for the control of mildew, grey mould and stem rot has to be integrated with any scheme for the biological control of insects and mites.

Lettuce

Grey mould (*Botrytis cinerea*) *and downy mildew* (*Bremia lactucae*) These are the two most important diseases of lettuce: both are associated with adverse growing conditions, both occur first on lower leaves and can cause loss from unsaleable heads, when, for example, the disease appears on the heart leaves.

Sprays and dusts with thiram or zineb have given adequate protection and, as yet, there seems no real indication that systemic fungicides have any particular advantages. Carbendazim fungicides applied as a spray at 50–100 g/l are useful for the control of grey mould where tolerant strains of *B. cinerea* are not present. There have been reports of these compounds causing a slight check to growth when applied at high concentration as a spray or as a drench. The carbendazim fungicides are ineffective against *Bremia lactucae*, a member of the Phycomycetes, but recently there has been increasing interest in systemic fungicides with activity against this family of fungi (Noveroske, 1975a). Among these, prothiocarb has given protection, in experiments, when incorporated in the growing medium at 100 g (a.i.)/m^2 or sprayed fortnightly on the growing crop at 1 000 ppm, commencing 2 weeks after planting out.

Carnation

Fusarium wilt (*Fusarium oxysporum*) *and Verticillium wilt* (*Verticillium cinerescens*) These are both serious soil-borne diseases which can be introduced on planting material, and *F. oxysporum* can be distributed by air-borne spores. In recent years the wilt caused by *F. oxysporum* has become much the more common of the two in the United Kingdom. The usual recommendations of soil sterilisation and careful attention to hygiene have given far from satisfactory control of the diseases on some holdings, and therefore the introduction of systemic fungicides was particularly welcomed by carnation growers. They were quick to develop their own methods of using these compounds but their results have often been somewhat disappointing.

Benomyl has been widely used for the control of these diseases since it was demonstrated to give control of *V. cinerescens* infection in trials in the Lea Valley (Knight, 1971).

Garibaldi (1968) reported on the efficacy of thiabendazole and carboxin used as a drench of 0·5–1·0 g/m^2 every one or two weeks in controlling *Verticillium* wilt. He obtained only partial control with carboxin, which was phytotoxic and other workers have since found thiabendazole occasionally to cause phytotoxicity.

In her earlier experiments, Ebben (1970a), though achieving control of *V. cinerescens* infection in pot tests, found neither benomyl nor thiabendazole applied at 0·5 g and 1·0 g a.i. per plant to control the disease when applied, before planting, to plots infested with the disease organism. When she applied the compounds to a growing crop in which *Verticillium* wilt was spreading Ebben (1971) found the rate of spread to be greatly reduced in the succeeding weeks. Flower yields were reduced in the thiabendazole-treated plots. In more recent experiments Ebben and Spencer (1975) found control of *F. oxysporum* to be dependent on the level of inoculum of the pathogen in the soil and on the age of the plant when treated.

Work in France (Tramier and Antonini, 1971) showed that soaking cuttings in 5 000 ppm benomyl before planting them in soil infested with *V. cinerescens* delayed infection, but that better protection was obtained by applying 1 g/m^2 every 8–15 days in the irrigation water. In order to control established infection a first treatment with 5–10 g/m^2 was required followed by the fortnightly application of 1 g/m^2. *Fusarium oxysporum* is more resistant to benomyl than is *V. cinerescens* and therefore requires 2–3 g/m^2 applied fortnightly. These rates seem to be high and would lead to a high soil level of the fungicide when compared with the makers' recommendation for its use in the British Isles. Here a dose of 0·75 g/m^2 is repeated 2 weeks later and the grower is then recommended not to apply the fungicide again within 3 months of the second drench. However, though there have been occasional reports of slight checks to growth after applications of carbendazim fungicides it seems that the carnation crop can withstand high levels of these compounds without damage (Fletcher, 1971).

Rust diseases of carnation (*Uromyces dianthi* and *Puccinia arenariae* are controlled by sprays containing 300 ppm triforine, whilst Umgelter (1969) showed that carboxin would control the rusts of several ornamentals including carnation. Oxycarboxin is now recommended by its manufacturer for use at 750–1 500 ppm as a spray, whilst benodanil at 1 000 ppm has also been found to give good control when used as a spray at 10- to 14-day intervals (Evans, 1974).

Triforine (300 ppm) is also used in the control of *Alternaria* leaf spot (*Alternaria dianthi*). This disease was shown by Engelhard (1971) and by Manning and Papia (1972) to be more severe on benomyl-treated carnation plants than on untreated controls.

Rose

It is a feature of the systemic fungicides currently available that none of them is translocated to any great extent in woody tissues when applied to the roots. Therefore, whereas drench treatments can be applied to other crops, spraying at the present time is the only useful method of application of systemic fungicides to rose bushes. The compounds are thus used mainly as protectants though eradicant activity has been demonstrated for some of them.

Powdery mildew (*Sphaerotheca pannosa*), black spot (*Diplocarpon rosae*) and rust (*Phragmidium* spp.) are the three most important diseases of cultivated roses and, of these, powdery mildew is the more troublesome under glass. The carbendazim fungicides are able to control all three diseases in the glasshouse but they are recommended for use only on outdoor roses in some countries; elsewhere sprays of 250–500 ppm are used.

Triforine, shown by Fuchs *et al.* (1971) to be fungitoxic to a range of fungi *in vitro*, has effectively controlled black spot and rust as well as powdery mildew when used at 150–200 ppm (a.i.) as a spray at 7- to 14-day intervals (Schicke and Arndt, 1971). It is cleared for use in many countries and is recommended for spray application at 150–250 ppm.

Dodemorph has been shown to be effective in controlling powdery mildews in a wide range of ornamental plants (Kradel *et al.*, 1969) and used as a spray at 0·1 per cent a.i. at 10-day intervals controlled the disease on Hybrid Tea roses in the glasshouse (Frost and Pattisson, 1971).

More recently bupirimate (600 ppm) has been found to give excellent control of mildew on outdoor roses (Evans, 1975).

In trials of systemic fungicides for the control of black spot on roses, Kavanagh (1970) found triforine to give better results than benomyl, and thiabendazole gave the poorest results as measured by the extent of defoliation at the end of the season.

Of the compounds more specifically active against the rusts, oxycarboxin at 750–1 000 ppm, and benodanil at 500 ppm applied as sprays at 10- to 14-day intervals effectively control rose rust.

Chrysanthemum

Growers of year-round chrysanthemums and pot chrysanthemums in the glasshouse maintain crop protection schedules which generally keep diseases well under control. Occasionally black rust (*Puccinia chrysanthemi*) is troublesome, and benodanil (500 ppm) and oxycarboxin (400–600 ppm) are used in its control. Recently white rust (*P. horiana*) has become occasionally serious, and Grouet and Allaire (1973) found oxycarboxin and triforine to be effective in the control of this disease. These two compounds are now widely used for this purpose as fortnightly sprays at 750–1 000 ppm and 300 ppm respectively. At 300 ppm triforine also controls ray blight (*Didymella ligulicola*) and powdery mildew (*Oidium chrysanthemi*). *Didymella ligulicola* also causes losses in chrysanthemum cuttings as the disease known as Ascochyta decay. Judd and Walton (1973) found that the disease was almost completely controlled if the unrooted cuttings were dipped in benomyl suspension (500 ppm) before being set out in the mist bench for rooting.

The carbendazim fungicides (250–500 ppm) applied fortnightly as high volume sprays give good control of grey mould (*B. cinerea*), where tolerant strains of the fungus are not a problem. At this concentration they also control powdery mildew but more recently pyrazophos has found use against this disease as a spray (150 ppm) applied at 7- to 10-day intervals.

Other ornamentals

A wide range of ornamentals is produced in the glasshouse as bedding plants, pot plants or as cut flowers. Such plants include cyclamen, begonia, saintpaulia, poinsettia, hydrangea, pelargonium, kalanchoe and azalea. These species each suffer from specific diseases which commonly include damping off and foot rots, caused by members of the Phycomycetes, rust diseases and powdery mildews, grey mould (*B. cinerea*), and diseases caused by *Fusarium* and *Rhizoctonia* species.

None of the currently available systemic fungicides seems to give control of diseases caused by Phycomycetes. Powdery mildews and rusts of many ornamentals are well controlled by triforine as a spray at 150 ppm in water (Schicke and Veen, 1969) and this material as a spray at 100–300 ppm is now recommended in many countries for use against these diseases.

Thiabendazole and benomyl are both very effective against grey mould and powdery mildews, and the latter compound is registered in Europe and North America as a standard treatment as a spray or pot drench for the control of these diseases.

Thielaviopsis basicola on cyclamen (Stalder and Barben, 1970) and on poinsettia (Raabe and Hurliman, 1971) is controlled by carbendazim compounds. Pelargonium rust (*P. pelargonii-zonalis*) and several other rusts of flowering plants were shown by Umgelter (1969) to be controlled by carboxin, though oxycarboxin at 400–600 ppm is the material recommended by the manufacturer for use as a spray against this disease. Spencer (1974) showed benodanil (400 ppm) and pyracarbolid (500 ppm) to give good control of pelargonium rust if they were applied before infection had occurred and if the plants were sprayed fortnightly thereafter.

Bulbs and corms

Most bulbs and corms of commercially grown crops are used to produce stocks for forced flowers in the glasshouse. The most serious diseases to which such bulbs are subject are caused by species of *Fusarium* and *Botrytis*. Thiabendazole and the carbendazim fungicides are active against these fungi and are used routinely in their control.

Tulip fire (*Botrytis tulipae*) This serious disease has an underground phase in which infection is transferred from mother to daughter bulbs and is also carried-over from season to season by sclerotia remaining on bulbs and in the soil. Price (1970) showed that thiabendazole and benomyl were toxic to sclerotia of *B. tulipae* and that bulbs dipped in suspensions of the fungicides remained toxic to the pathogen for several weeks. Price and Briggs (1974) showed that the immersion of infected bulbs in 1 000 ppm benomyl or 600 ppm thiabendazole killed the sclerotia of *B. tulipae* and considerably lessened the amounts of infection which subsequently developed on dipped bulbs. They suggested that the amount of disease remaining in dipped bulbs would be so small as to obviate the need for foliar spraying.

Where spray applications are needed the carbendazim fungicides are used at 1·1 kg/ha in 200–300 l water (2·25 kg/ha thiophanate methyl) from the time the plants are 2·5 cm tall and at 14-day intervals thereafter.

Fusarium basal rots (*Fusarium oxysporum*) *Fusarium* rots of tulip, narcissus, iris, gladiolus, lily and freesia, grown for cut-flower production under glass, have each been treated successfully with thiabendazole and benomyl (Magie, 1970; Gould and Miller, 1970, 1971). Better results are obtained when bulbs are dipped in the fungicide suspension within 2 days of lifting, though with bulbs intended for forcing, the dip can be applied just before cold treatment and for protection during growth, bulbs and corms are dipped immediately before planting. Thiabendazole is recommended to be used for this purpose in the normal hot water treatment when bulbs are soaked for $3\frac{1}{2}$ hours at 45°C, but this compound, like the carbendazim fungicides, is often used as a 15–30 min. dip at 25°C. In all cases the suspension needs to be replenished with 0·5 kg fungicide per tonne of bulbs treated.

The benzimidazole fungicides have also been used to control *Penicillium* spp. in various bulbs, and Smith (1971b) found that benomyl, applied as a bulb dip or post-emergence foliar spray, could effectively decrease the incidence of primary leaf infection by *Stagonospora curtisii* (Narcissus leaf scorch), as well as subsequent spread and thus could increase bulb weight. This compound is now registered for use on ornamental crops, including bulbs in Europe and North America.

Mushroom

Though not widely grown in the glasshouse nowadays, the mushroom (*Agaricus bisporus*), being a crop of the protected environment, is included in this chapter. This crop has shown outstanding annual increase in output. In 1950 the world annual production of mushrooms was about 50 000 tonnes, whereas now the annual output is over 500 000 tonnes. Of the major producing countries, USA and France each produce about 110 000 tonnes annually, Taiwan 80 000 tonnes and Holland 40 000 tonnes. In the British Isles the annual output of about 60 000 tonnes, worth some £26 million, is now almost equal in value to that of the tomato crop at £26·1 million per annum.

The composted animal manure on which mushrooms are usually grown provides a selective medium for the growth of mushroom mycelium, though under adverse conditions other fungi can colonise the medium and compete directly with the mushroom for space and nutrients. In addition, the cultivated mushroom, like any other crop, is attacked by insect pests and suffers from diseases caused by viruses, bacteria and fungi. Of the latter, the diseases known as dry bubble, wet bubble and cobweb caused respectively by *Verticillium fungicola*, *Mycogone perniciosa* and *Dactylium dendroides* are the three most serious. Each of these seems to have been important at various times in the past but, at present, *V. fungicola* is most frequently encountered on mushroom farms and can cause serious crop losses wherever *A. bisporus* is grown (Cross and Jacobs, 1968).

Chemical control of the disease poses problems because both host and parasite are fungi. In the past, zineb was widely used but the variable results obtained with this and other compounds led Gandy (1971) to evaluate more recently available compounds including systemic fungicides for use against *V. fungicola.* She found that benomyl (1·0 g (a.i.)/m^2) applied to the casing, a layer of peat and chalk which must be placed on the compost to induce fruiting, gave a considerable degree of control without any apparent toxic effect on the crop. Since the disease develops very rapidly the compound must be applied before cropping begins, otherwise one or two flushes of mushrooms may be lost before the fungicide becomes fully effective (Gandy, 1972).

Snel and Fletcher (1971) showed that whereas the ED_{50} values of benomyl and thiabendazole against mycelial growth of *A. bisporus* were 48 and 100 ppm the ED_{50} values for both compounds against *V. fungicola, D. dendroides* and *M. perniciosa* were all less than 1 ppm. When the two fungicides were incorporated in the casing layer of mushrooms grown in plastic pots both gave complete control of *M. perniciosa* infection.

Further work by Gandy (1972) showed that thiophanate methyl and carbendazim are also active against *V. fungicola.* The carbendazim fungicides are therefore used at 1 g/m^2 as a casing incorporation or as a drench in 0·25–1·25 l/m^2 as the first watering to control the three major mushroom diseases, though where *V. fungicola* is particularly troublesome thiophanate methyl is recommended to be used at 2 g/m^2.

In view of their spectacular success in disease control the carbendazim fungicides rapidly achieved very widespread use in the mushroom industry and still give adequate control of *M. perniciosa* and *D. dendroides.* However, strains of *V. fungicola* tolerant to these fungicides quickly made their appearance (Gandy and Spencer, 1974) so that Gaze and Fletcher (1975) concluded that the failure of many growers to control *V. fungicola* with carbendazim fungicides is largely attributable to the presence on the farms of tolerant strains of this pathogen.

Methods of application in glasshouses

The systemic fungicides at present available for practical application to the growing crop move in the plant only in the direction of the transpiration stream. Accordingly, the grower can apply them to the soil and thereby make use of their systemic properties, always assuming that there is efficient root uptake, or he can apply them to the above-ground parts of the plant. In this case there will be little uptake and movement and, consequently, no protection of new growth formed after the treatment has been applied so, in effect, the compounds are used as orthodox protectant fungicides, though in some cases they can eradicate established infections.

There is no doubt that systemic fungicides have been welcomed by glasshouse growers who are now applying benomyl by almost any means at their disposal.

It is likely that growers of those crops in which it would be more satisfactory to apply systemic fungicides as sprays will have suitable equipment for this purpose and there is little need here to discuss the merits of the various types of spraying equipment. There are circumstances, however, in which fixed equipment in the glasshouse, used for example for overhead watering or for the application of foliar feed, can be used for the application of systemic fungicides. It is difficult to get uniform distribution from such equipment particularly in tall-growing crops and, on very large holdings, this is accentuated by the large volume of liquid needed to fill the system and to be purged from it at the end of treatment.

The use of low-level fixed irrigation lines to apply systemic fungicides as soil drenches is less costly in labour than high-volume spraying though at present some compounds are recommended only for spray application. Trickle irrigation, by which plants in pots or larger plants such as tomatoes and cucumbers are each watered from a single nozzle, has been used for the application of systemic fungicides to the soil. So too has the drip-watering system in which the resistance required to give a standard dosage to each plant is provided by lengths of small-bore plastic tubes on the ends of the distribution lines. The outlets in both of these systems are very liable to become blocked by dirt, pipe scale, hard-water scale or, when used for systemic fungicides, by settlement of improperly mixed wettable powder formulations.

Plants such as carnations, grown in beds, are watered by a modification of the low-level trickle system. In this a harness of lay-flat polythene tubes, punched with small holes are used to apply the water. This is a low-pressure system and thus, if used for the application of systemic fungicides, it must be borne in mind that a large amount of suspension will drain to the lower end of the system each time the supply is turned off.

None of these systems is ideal; all of them suffer to a greater or lesser extent from lack of uniformity in delivery. Some degree of uneven application of systemic fungicides may be acceptable but overdosage could lead to phytotoxicity and undesirably high residues. Blocked outlets on the other hand could mean that some plants were not treated, thus leading to the false impression that the compound had failed to control the disease or, more ominously, that a strain of the pathogen resistant to the compound had made its appearance.

It would be safer, where possible, to apply a measured dose of the compound to each plant. Dimethirimol is recommended to be applied directly to the base of the plant from a special lancet applicator which delivers 20 ml of the diluted liquid from a knapsack container. Benomyl is also recommended to be applied as a soil drench at 1 pint per plant which could well be provided by a suitable applicator supplied from the sprayer tank.

Tolerance

The development of tolerance of pests and diseases to the action of crop protectants is a problem of long standing. Insects quickly develop tolerance to insecticides and strains of disease organisms quickly appear to infect the resistant varieties of crop plants produced by the plant breeder. Therefore, when systemic fungicides became commercially available, it was feared that tolerant strains of disease organisms might not be long in making their appearance. This is more likely under the selection pressure exerted by the use of these specific compounds which effectively protect the whole of the plant and the whole of the crop.

Indeed strains of organisms tolerant to some of the systemic fungicides currently available appeared very quickly (Woodcock, 1971). Though this subject is dealt with at length elsewhere in this volume (Dekker, Chapter 9), it would seem appropriate to make observations on the problem, particularly in relation to the diseases of glasshouse crops.

It was suggested long ago for insecticides and has now been suggested for systemic fungicides that to ring the changes, by using different compounds in succession, would prolong the effective life of each compound (Anon, 1971c). There would seem to be more likelihood of success if a mixture of compounds with different modes of action were to be used in that the pathogen would be less likely to be able to circumvent more than one metabolic block at one time (Crowdy, 1970). Ebben and Spencer (1973) were able to achieve complete protection of cucumbers inoculated with benomyl-tolerant and dimethirimol-tolerant strains of the powdery mildew organism, *Sphaerotheca fuliginea*, by the use of the two fungicides either alternately or in admixture.

However, at the present time there is very little choice, for most diseases, of systemic fungicides with different modes of action. For the control of *Botrytis cinerea* in tomato for example the very active compounds are all carbendazim fungicides. Therefore, the manufacturers of some of these compounds recommend, on the labels of their products, that where poor control due to tolerance is established or suspected, other suitable protectant fungicides should be used. Indeed, there is at present considerable interest in the glasshouse industry in spraying programmes in which the use of systemic and protectant fungicides is integrated.

Any attempt to draw parallels between the development of resistance by insects to insecticides and the development of resistance by fungi to fungicides should take account of the fact that resistance has now been demonstrated for several fungi and that Hastie (1970) found benomyl to induce instability in diploids of *Aspergillus nidulans* and that Dassenoy and Meyer (1973) found benomyl to have a direct mutagenic effect on *Fusarium oxysporum.* Nevertheless, there is good evidence that tolerant strains of some pathogens were already in existence before the advent of systemic fungicides and have simply been selected by the applied treatment. This certainly seems to be true for *Verticillium fungicola* in mushroom (Wuest *et al.*, 1974).

The experience at the Glasshouse Crops Research Institute is of interest here. When tolerance of cucumber powdery mildew to dimethirimol had appeared it was decided to arrange for a period of 6 weeks in autumn 1970 when no cucurbit plants would be grown at the Institute. At the end of this time, in order to re-start work on the control of this disease, infected plants were introduced from another site some 100 km distant. The new strain was susceptible to dimethirimol and during 1971 there was no indication of the failure of the compound to control the disease.

Non-fungicidal effects of benzimidazole compounds

It is possible, experimentally, to produce phytotoxic symptoms such as stunting and leaf necrosis with drench applications of benzimidazole derivatives at high concentrations. There have been reports of other physiological effects of some of these compounds. Cole *et al.* (1970) found benomyl applied to the soil at levels above 20 ppm to bring about a progressive reduction in plant weights but at 10 ppm the compound caused an increase in fresh and dry weight of cucumber plants.

Lettuce growers, using benomyl for the control of *B. cinerea,* have noticed a more intense green coloration of the crop and plants which have a mature appearance when young. Such plants may be heavier than those not treated with the fungicide. Benomyl-treated celery has been found to be crisp and the green parts still fresh after winter storage whereas the untreated crop becomes yellow in store. Tomato plants which have been treated with benomyl, thiabendazole or thiophanate derivatives are usually a darker shade of green, particularly in the younger leaves, than are the untreated controls. Producers of chrysanthemum cuttings who have used benomyl to control *Verticillium albo-atrum* infection of stock plants have commented on the darker green of treated plants.

These responses, in which chlorophyll retention is prolonged, are all of the type which could be associated with the activity of cytokinins. Compounds with cytokinin-like activity are related to adenine (I) (Rothwell and Wright, 1967). The similarity in structure between this compound and methyl benzimidazol-2-ylcarbamate (II), which is the systemically fungitoxic moiety derived from benomyl, supports the possibility that the benomyl-derivative could have cytokinin-like activity.

Indeed, when detached wheat leaves were floated on benzimidazole solutions not only was breakdown of proteins and chlorophyll prevented but the leaves were resistant to rust infection (Person *et al.*, 1957; Samborski *et al.*, 1958).

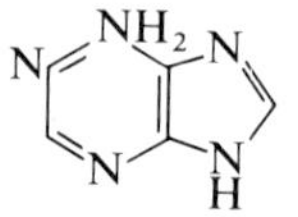

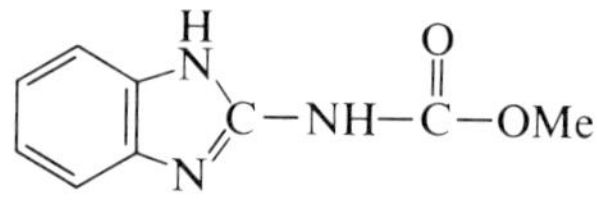

Biological control

An aspect of the use of systemic fungicides in the glasshouse which is commanding increasing attention concerns their integration into a programme which includes the use of predators and parasites to control pests.

The biological control of red spider mite by *Phytoseiulus persimilis* and white fly by *Encarsia formosa* on cucumber and tomato is now in commercial use, but for this method of pest control to be successful the balance between the pest and its natural enemy must be maintained and thus the use of chemicals can only be tolerated if they do not seriously upset this balance. The introduction of systemic fungicides has been of benefit to the development of biological control as a means of reducing the losses caused by pests of glasshouse crops. The chemicals used previously, such as sulphur, quinomethionate and dinocap, though effective against cucumber powdery mildew were also phytotoxic and acaricidal and could not be used in the presence of the red spider mite predator.

Dimethirimol was one of the first systemic fungicides to be used in conjunction with biological control and proved to be completely compatible. When cucumber powdery mildew became tolerant to this compound, benomyl used in a spray programme was found to be less satisfactory (Parr and Binns, 1970). This compound when applied as a spray was found to have a sterilising effect on female predators, severely reducing their fecundity and thus destroying the balance between the predator and its prey. However, when benomyl is applied as a drench to cucumber plants it controls powdery mildew and its effect on the biological control of red spider mite is much less serious.

The more recently available carbendazim fungicides seem to behave in a manner similar to that of benomyl. As high-volume sprays at 0·01–0·05 per cent concentrations they are strongly acaricidal and harmful to *Phytoseiulus* and would probably be better used as drenches (Parr, 1971).

Pyrazophos applied as a spray at 0·03 per cent is highly toxic to the red spider mite predator, but triforine as a spray (0·02 per cent) is much less harmful and can be used with comparative safety in an integrated control programme.

Present trends: future needs

In the years before the introduction of the first commercial systemic fungicides it was often said that such compounds were being sought so that they could be applied to the roots, to be taken up by the plant and thus protect all parts of it against fungal pathogens (Spencer, 1957). Though dimethirimol was always used as a drench treatment, the carbendazim fungicides are widely used in the glasshouse industry only as sprays and, indeed, the newer systemic fungicides such as triforine and pyrazophos are recommended for use only as sprays.

When used as protectant sprays the systemic properties of the compounds are not being utilised at all and it would seem that future development of systemic fungicides could be aimed at the use of materials which can be applied to the roots and be retained within the plant at a level well above that which is toxic to the pathogen.

Minimising the effects of subtoxic concentrations of fungicide, which might encourage the development of resistant strains by continued selection, is one of the suggestions often made to prolong the active life of systemic fungicides. Other suggestions, as summarised by Greenaway and Whatley (1975), include:

(*a*) using the least specific fungicides available
(*b*) using mixtures of fungicides
(*c*) monitoring fungi to provide an early warning of the development of resistance
(*d*) treating the minimum area necessary at the time, when optimum results will be obtained from treatments.

The glasshouse grower who uses systemic fungicides as sprays needs to ensure that the entire crop is covered with an adequate amount of the fungicide; this is difficult to achieve in practice. Furthermore, the new growth produced between spray applications will be inadequately protected and it is on such tissues that tolerant strains might appear and develop.

None of the systemic fungicides currently in use is active against Phycomycetes. This, in itself, is a drawback which must be remedied by the future introduction of compounds active against this group. At present, however, there seem to be indications of further hazards resulting from the use of selective compounds active against organisms other than Phycomycetes. The inference is that the systemic fungicides not only control disease organisms but also kill soil-borne organisms which are themselves antagonists of such Phycomycetes as *Phytophthora* and *Pythium*. These latter are thus liable to cause disease when systemic fungicides are used, though they may not have done so in the absence of the chemical. An example of this was demonstrated by Dekker (1971a), who found that aster seedlings were killed by *Phytophthora parasitica* in the presence of benomyl but were unharmed when benomyl was not used. It is interesting to speculate here on the relative potential for antagonism of soil-borne fungi as compared with bacteria and actinomycetes, which are much less affected by benomyl and which therefore may well continue to antagonise the Phycomycetes in the presence of the systemic fungicide.

It is likely that the selective effect of the systemic fungicides extends beyond the Phycomycetes and their antagonists. Indeed the soil microflora is likely to be considerably affected by drenches of the systemic compounds so that the applied compound will become subject to the action of a wide range of organisms, some of which will increase in number because of the presence of the compound. Smith and Worthing (1973) found that fungicide loss was less rapid from sterilised soil than from unsterilised soil. They also found that in 1973 there was a more rapid loss of fungicide from soil to which benomyl had been applied than had been the case in 1972. This was later shown to be due to the presence in the soil of a *Flavobacterium* which was capable of the rapid degradation of carbendazim fungicides (Smith and Worthing, 1975). If such organisms were not operating, the build-up of systemic fungicides in soils over a long period could well lead to problems of phytotoxicity, bearing in mind that the benzimidazole fungicides seem to be unaffected by steam sterilisation.

The requirements for future systemic fungicides can now be summarised to include compounds with different and preferably wider spectra of activity and different modes of action from those presently available. Compounds active against Phycomycetes will be valuable and, in view of the activity of present compounds against antagonistic fungi, there could conceivably be an increase in bacterial diseases, and it would be an advantage if materials with antibacterial properties were available. In order to be truly systemic the compounds should be translocated downwards as well as upwards so that effective methods of application specifically for glasshouses can be used. Compounds applied by spray or drench should move freely, even in woody tissues, and be retained within the plants at levels well above those required to control the disease, so that the development of tolerant strains can be avoided.

It must be borne in mind that, unlike the protectant fungicides which, in the main, stay on the surface, the systemic fungicides are intended to be retained within the tissues of the plant. Therefore the compounds themselves, as well as all possible degradation products within the plant, must be free from any danger to those who eat the crop.

Chapter 11.III

Results in practice – III. Vegetable crops

R. B. Maude

Introduction

The role of conventional fungicides in the treatment of vegetable diseases has been mainly that of protection rather than cure. Compounds were applied to seeds, soil and foliage to form local barriers against fungal invasion. Such treatments were most successful when used before the onset of infection and little disease control was obtained after fungi had become established in the plant tissues. This situation has been radically altered by the discovery and development of systemic fungicides which are mobile and fungitoxic within the plant system. Their use has made possible the therapy of established infections and the internal protection of the roots and leaves of plants from invasion by external fungal pathogens.

The potential of such compounds for the improved control of vegetable diseases has been assessed in practice by comparing their performance with that of the non-systemic fungicides. They have been applied in a variety of ways for the control of seed-borne, soil-bornc, leaf and post-harvest diseases of vegetables. The effectiveness of these treatments is discussed here.

Seed-borne diseases

Seed dressings with fungicidal dusts, whilst effective against superficial pathogens, have in the past given relatively little control of fungi which are deep in the seed tissues and these have been treated either by steeping the seeds in hot water (50°C for 25 min; Bant and Storey, 1952) or soaking them in water containing 0·2 per cent a.i. thiram for 24 h at 30°C (Maude *et al.*, 1969).

The latter method, which is effective against many internal seed pathogens, is a systemic treatment, involving the penetration of the seed by thiram. This only occurs, however, under the very special conditions of the treatment and requires that the seed shall become fully imbibed in the constant presence of 20–30 ppm thiram in a large volume of water. Because of the soaking and subsequent drying of seeds necessary in the thiram soak treatment, the method has proved of commercial use mainly against the pathogens of small, high-value seeds such as *Septoria apiicola* on celery and was not practicable for the treatment of bulky seeds such as peas and beans. For such seeds there was obviously a need for systemic fungicides which could be applied to the seeds as a dry powder or slurry and then be absorbed during the imbibition of the seeds in the soil.

The possibility that such fungicides might be found was opened up by the discovery by von Schmeling and Kulka (1966) that dry seed dressings of the systemic oxathiin fungicide, carboxin, controlled deep-seated infection by *Ustilago nuda* in barley seed.

The narrow spectrum of activity of the oxathiin compounds (effective mainly against Basidiomycete fungi; Edgington *et al*., 1966) severely limited their potential value for the treatment of seed-borne vegetable diseases, the majority of which are caused by Ascomycetes or Fungi Imperfecti. The situation was considerably changed, however, by the introduction of the benzimidazole compounds, particularly benomyl (Delp and Klopping, 1968) and thiabendazole (Weinke *et al*., 1969) which are toxic to many fungi (Bollen and Fuchs, 1970; Edgington *et al*., 1971).

It was shown by Maude and Kyle (1970a) that slurry (1·25 g a.i./kg seed) or dry (1·875 g a.i./kg seed) dressings of benomyl applied to pea seeds eliminated internal infection by *Ascochyta pisi*. Thiabendazole dressings were equally effective in this respect (Maude and Kyle, 1971) though carboxin dressings failed to control the disease. Benomyl and thiabendazole seed dressings also controlled *A. fabae* (leaf and pod spot) of field beans and *Colletotrichum lindemuthianum* (anthracnose) of French beans (Maude and Kyle, 1971). Recently, Kaiser *et al*. (1973) also found that benomyl and thiabendazole dusts (3 g a.i. fungicide/kg seed) gave more effective control of seed-borne *Ascochyta rabiei* of chick pea (*Cicer arietinum*) than certain other systemic fungicides including chloroneb, carboxin and oxycarboxin. This use of systemic fungicides represents the first practical and effective seed treatment for these internal diseases of the larger vegetable seeds. Smaller seeds can also be treated, and dry and slurry dressings of benomyl (1 g a.i./kg seed) effectively disinfected onion seeds of *Botrytis allii* which commonly occurs on imported seed samples, the main source of this serious post-harvest disease (i.e. neck rot) of stored bulb onions in Britain (Maude and Presly, 1973). Benlate T, a commercial seed dressing containing 30 per cent benomyl and 30 per cent thiram, is now marketed for the treatment of the seed-borne *Ascochyta* diseases of peas and field beans and *Botrytis* of onion seeds.

Although the benzimidazoles are designated 'broad-spectrum fungicides' there are certain seed-borne fungi against which they have little toxicity in culture, e.g. *Alternaria* and *Stemphylium* spp. (Edgington *et al*., 1971), *Phoma betae* (Bollen and Fuchs, 1970) and *Colletotrichum spinaceae* (Maude and Kyle, 1971).

In addition, the use of dry dressings of a systemic fungicide demonstrably toxic to a fungus in culture has sometimes failed to give effective control of the same pathogen on the seed. Thus dry dressings of benomyl failed to eliminate *Septoria apiicola* from celery seed or to rid cabbage seeds of *Phoma lingam* (Maude and Kyle, 1970b). Possibly in these cases lethal concentrations of the fungicide failed to reach the deeper seated mycelium of the fungi. When maximum penetration of fungicide was ensured by 24 h soaks at 30°C of celery seeds in 0·2 per cent benomyl (Maude and Kyle, 1970b) or cabbage seeds in 0·2 per cent benomyl or thiabendazole (Jacobsen and Williams, 1971) both seed stocks were completely disinfected. It is obvious that the search must continue for other systemic fungicides which are effective against these and other fungi and that each pathogen may require separate treatment. Other compounds are available but several, e.g. tridemorph, are rather phytotoxic as seed dressings, and although this compound inhibited the growth of *A. brassicicola* and *P. betae* in culture (Maude and Kyle, 1971) it depressed seed germination when applied at fungitoxic rates.

The selectivity of systemic fungicides is also illustrated in the treatment of tuber-borne diseases of seed potatoes. Here the latent fungi such as *Oospora pustulans* (skin spot), *Helminthosporium solani* (silver scurf) and *Rhizoctonia solani* (black scurf) are normally controlled by tuber dips in an organomercury fungicide soon after lifting (Western, 1971). Recently, Hide *et al.* (1969) demonstrated that tuber dips in thiabendazole or tuber dusting with benomyl effectively reduced the incidence of the first two fungi on seed tubers in store and on potatoes raised in the field from this seed material. These fungicides were not similarly effective, however, against the seed-borne Basidiomycete *R solani*, better control of which was obtained with dusts of carboxin or oxycarboxin.

Seed dusts with benomyl or thiabendazole (1–10 per cent a.i. in kaolin, applied at 5 g dust/kg seed) are now preferred to dips in thiabendazole for the control of skin spot and silver scurf on seed potato tubers (Hide and Hirst, 1971), and a prototype machine capable of applying systemic dust treatments to small quantities of potato tubers has been developed (Hirst, 1971). More thiabendazole, however, was taken up by seed tubers from dips in aqueous solutions and suspension of the fungicide than from dusts (Tisdale and Lord, 1973), and thiabendazole on the surface of tubers was not metabolised when these were stored for 3 months. In America, where tuber infection by *Verticillium albo-atrum* has been reported in certified seed stocks (Beckman *et al.*, 1969), 3-min dips in benomyl (12 g a.i./l) gave some control of the disease on the harvested crops (Biehn, 1970). Dust applications (10 g formulated product/kg seed) of benomyl (1 g a.i./kg seed) and thiophanate methyl (1 g a.i./kg seed) made to artificially infected seed tubers immediately before planting also substantially reduced tuber rot (*Fusarium caeruleum*) and wilt (*V. albo-atrum*) in the subsequent crop (Ayers, 1974). In contrast, however, benomyl (4·5 g of a 10 per cent a.i. dust/kg seed tubers), dusted on to gangrene (*Phoma exigua* var. *foveata*) infected tubers planted in *Phoma* free soil or on to healthy tubers planted in

Phoma infested soil, increased gangrene in the plants raised from them (Logan, 1974).

Thus where formerly thiram soak or organomercurial dip treatments were effective against a wide range of seed- and tuber-borne diseases of vegetables, the same latitude of use is not equally possible with the systemic compounds which will have to be carefully selected for use against specific pathogens. None the less, the simplicity of their application coupled with their effectiveness against deep-seated fungi makes their use as seed dressings in commercial practice almost certain.

Soil-borne diseases

The problems of treating field soil with conventional fungicides to control soil-borne fungi are considerable (Torgeson, 1967). The most effective and economic control has usually been achieved by localised treatments using heavy seed dressings or row or furrow applications. Under these circumstances the fungicides can sometimes provide protective barriers near the young plant.

While similar restrictions may also apply to the placement of systemic fungicides, their uptake by the seed tissues or by the roots of plants affords additional protection against invasion by soil-borne fungi.

In practice, both seed and soil applications have been used to protect vegetables from attack by these pathogens.

Seed and soil applications of systemic fungicides have given considerable control of *Rhizoctonia solani*, which infects the hypocotyls of seedling vegetables causing damping-off or wire-stem. Thus thiabendazole seed dressings (1·9 g a.i./kg seed) considerably reduced damping-off of peas due to *Rhizoctonia* in field soil (Crosier *et al.*, 1970), and seed slurries, seed steeps and soil drenches with benomyl prevented pea seedlings from becoming infected in artificially contaminated garden soil (Jhooty and Behar, 1970). In naturally infected field soil, however, drenches of benomyl (2·0 kg a.i./ha) or chloroneb (4·7 kg a.i./ha) were less effective than quintozene (1·5 kg a.i./ha) in controlling *Rhizoctonia* infection of carrot seedlings (Shlevin and Katan, 1975). Carboxin seed dressings more effectively reduced damping-off of peas than benomyl, and both oxathiin fungicides (carboxin and oxycarboxin) gave superior control of seedling losses in French beans due to *R. solani* (Beyries, 1969). In fact, benomyl slurries (5 g a.i./kg seed) were less effective than similar slurries of quintozene and neither fungicide gave the same measure of control as carboxin (5 g a.i./kg seed) which completely prevented seedling infection. Seed dressings of this fungicide also prevented seedling losses due to the fungus in *Phaseolus* beans (*P. mungo* var. *aureas*) and at application rates of 20 g a.i./kg seed gave almost complete control of *Rhizoctonia* infection of cabbage seedlings. Both carboxin and oxycarboxin, when dusted on to potato tubers (10 per cent dusts – 1 g a.i./kg seed), have given considerable control of stem canker infection due to *R. solani* (Edgington and Busch, 1967), and benomyl similarly applied proved more effective than broadcast applications of quintozene (28 kg PCNB/ha) (Davis *et al.*, 1971a) against the same disease. However, *R. solani* infection of lettuce (bottom rot) caused by contact of the outer older leaves with soil infested with the fungus was best controlled in field plantings by spraying the under surfaces of the outer leaves at soil level with three to four applications of benomyl 1·1 kg a.i./ha at 7-day intervals (Pieczarka and Lorbeer, 1974).

Where several fungi, e.g. *R. solani*, *Fusarium solani* and *Thielaviopsis basicola*, caused complex rotting and damping-off of beans in artificially infested glasshouse soil, Papavizas and Lewis (1975) demonstrated that these were best controlled by seed dressings prepared from mixtures of fungicides, e.g. carboxin, choroneb and thiophanate methyl.

It is apparent that the hypocotyls and stem bases of plants below soil level can be effectively protected from invasion by *R. solani* by soil applications but more particularly by seed applications of systemic fungicides. It has not been shown, however, whether the compounds are acting as external protectants by inhibiting the growth of the fungus in the soil or whether by their presence in the vascular system of seedlings and plants they are preventing fungal penetration of the hypocotyl and stem base tissues. Possibly a combination of both modes of action occurs. In one instance, however, ineffectual internal protection of the hypocotyl region may be responsible for the failure of a systemic fungicide to control field outbreaks of brown stem rot (*Cephalosporium gregatum*) on soybeans. Thus although soil drenches of thiabendazole (Gray and Sinclair, 1971) and benomyl (Thapliyal, 1970) in glasshouse tests restricted stem invasion by the fungus, furrow applications (1 or 2 g a.i./10 kg soil) of thiabendazole were not effective (Gray and Sinclair, 1971). Since thiabendazole was found to accumulate in the root and epicotyl but not in the hypocotyl tissues of the soybean the authors suggested that possibly insufficient fungicide was present in this region to prevent invasion by the fungus.

Seedling infection of onions by the onion smut fungus (*Urocystis cepulae*) occurs on the first leaves of plants at or slightly below soil level, only during emergence. Some protection during this susceptible period is afforded by heavy seed dressings of thiram applied at 50 per cent or more of the seed weight (Cruickshank and Jacks, 1953). Far more successful control of the disease has recently been reported by Harrow (1970) using seed dressings of benomyl (12·5–50 per cent a.i. applied w/w seed) which virtually eliminated smut in onion seedlings. Seed dressings with thiram (40–80 per cent a.i. applied w/w seed) were much less effective than this and dressings of carboxin and oxycarboxin proved too toxic to onion seeds for practical use.

A more prolonged period of protection is required against the onion white rot fungus (*Sclerotium cepivorum*) which may attack salad, bulb onions and garlic at any stage in the growth of these crops. In France, Lafon and Bugaret (1968) found that slurry dressings of benomyl (2·0 g a.i./kg cloves) applied to garlic cloves considerably reduced the incidence of white rot in heavily contaminated field soil and increased the numbers and weight of healthy plants at harvest. Recently, Ryan *et al.* (1971) also reported that larger numbers of Rijnsburger bulb onions survived to harvest if the sets from which they were raised were dressed with benomyl at rates ranging above and below 2·0 g a.i./kg set. They also stated that benomyl (62·5–500 g a.i. fungicide/kg seed) used on White Lisbon onion seed gave almost as effective control of onion white rot in artificially contaminated field soil as calomel used at 500–1 000 g a.i./kg seed. Recently, Entwistle and Munasinghe (1973 and 1974) obtained similar results with several methyl benzimidazol-2-ylcarbamate-containing compounds including benomyl, thiophanate methyl and carbendazim (50 per cent and 60 per cent a.i. formulations) applied to onion seeds with a 1 per cent methyl cellulose sticker at rates of 100–150 g a.i./kg seed. Thiabendazole used at similar rates was not as effective. Seed treatment with thiophanate methyl (125 g a.i./kg seed) is currently widely used in commercial practice in Britain in preference to calomel for control of white rot in salad (green) onions. In America, also, thiophanate methyl gave excellent control of white rot and an increased yield when tested in furrow (1·8 kg a.i./ha), and root dip applications (60 g a.i./l) on bulb onions transplanted into heavily infested field soil (Maloy and Machtmes, 1974).

Thus systemic fungicides have been applied to soil, seeds, seed tubers and bulbs to protect extremely local areas on the developing plants, e.g. hypocotyls, epicotyls or the fairly restricted root systems of onion, from infection. In these circumstances considerable control of several diseases has been achieved by their use. Where, however, a much wider area of plant root systems, e.g. stolons and tubers which develop from and are at some distance to the original seed potato, require protection, then systemic seed dressings have not been effective though some measure of control has been achieved by soil treatments with these compounds.

Dusts of carboxin and oxycarboxin (Edgington and Busch, 1967), dry dressings of benomyl (Davis and Callihan, 1970) and 3-min dips in benomyl or dusts of carboxin (Biehn, 1969) applied to seed potatoes failed to reduce the incidence of black scurf (*R. solani*) on tubers harvested from naturally or artificially contaminated field soil. It is presumed that effective control of infection on the developing tubers would only have occurred had sufficient systemic fungicide been supplied to them via the stolons from the treated tubers during the growth of the crop. This apparently was not the case. Slightly more effective control of tuber infection occurred when seed piece treatments with benomyl were followed by soil drenches with the same fungicide (Biehn, 1969) possibly because of the more effective placement of benomyl in a readily available form in relation to the developing stolons and tubers. Furrow applications of granules of (5 per cent a.i.) carboxin and oxycarboxin (4·5 kg a.i./ha), however, failed to reduce the incidence of *R. solani* on harvested tubers (Edgington and Busch, 1967). Similarly, seed tuber dressings of benomyl have given little reduction in the incidence of *Verticillium dahliae* (potato wilt) propagules in the soil (Corbett and Hide, 1971), but some control of vascular infection of harvested tubers was obtained where seed piece dips in benomyl were followed by soil drenches of the same fungicide applied 2 (7·3 kg/ha) and 9 (18 kg/ha) days after planting (Biehn, 1970). The rotovating of benomyl into the soil (22·4 kg a.i./ha) also effectively decreased the number of *Verticillium* propagules present in the soil at harvest and increased the yield of potatoes (Corbett and Hide, 1971).

Thus where fungi are dispersed through the soil the task of protecting fairly extensive plant root systems by means of systemic fungicides is practically difficult and economically may not be feasible.

The problem of adequately protecting the root system also arises in the use of conventional fungicides for the control of brassica clubroot (*Plasmodiophora brassicae*). Seed pelleting with calomel is unreliable and offers ineffective protection to brassica root systems in direct-drilled crops, as does similar treatment with benomyl when applied to cabbage seed at 2 kg a.i./kg seed (Buczacki, 1973). Practical, if not completely effective, control on the transplanted crop, however, is achieved by seed-bed treatments and the dipping of transplants in calomel. From the work of Jacobsen and Williams (1970) it is apparent that benomyl mixed with clubroot-infected pot soil prevented the roots of cabbage seedlings from becoming infected, apparently by preventing zoospore invasion of the root zones. Control of the disease on cauliflower seedlings was also achieved by van Assche and Vanachter (1970) with benomyl, thiophanate or thiophanate methyl mixed with pot soil. Findlayson and Campbell (1971) found, however, that soil type was critical in determining the effectiveness of benomyl. Thus in naturally infested peat soil benomyl (10 g a.i./l soil) failed to control clubroot on cauliflower seedlings although this was effectively treated by much smaller applications of calomel (2·5 g a.i./l soil). In mineral soil, however, benomyl was more effective by comparison at lower rates of application.

Thus the possibility exists of using systemic fungicides as alternatives to calomel for the treatment of seedlings in mineral soil prior to transplanting them into the field. Van Assche and Vanachter (1970) consider that such treatments may give sufficient initial protection to young plants to ensure that these give rise, when planted in the field, to a completely marketable if slightly infected crop. The fact remains that in the limited soil system of a pot there may be little to choose between the protection afforded to root systems by conventional and systemic compounds, and more effective use of the systemic action of benomyl for the control of clubroot might be made by using it as a soil drench treatment on a limited scale in the field as is suggested by Jacobsen and Williams (1970). These suggestions now can be reappraised, as recently various methods of application of systemic fungicides have been tested for the control of the disease in field soil.

Slurry applications containing 0·5 per cent thiophanate methyl or 1·0 per cent benomyl made to the roots of cauliflower plants before transplanting them into field soil gave as effective control of clubroot in the subsequent crop as similar applications of 9·6 per cent calomel (Anon, 1973b). In tests of a range of benzimidazole-derived systemic compounds, Buczacki *et al.* (1976) found that dipping the roots of cabbage transplants in thiophanate methyl (0·4–0·5 per cent a.i.) gave yield increases equivalent to those obtained with similar applications of 10 per cent calomel when the plants were raised to maturity in severely infested field soil. Dips in aqueous thiophanate methyl were as effective as methyl cellulose slurries with the same fungicide, and it is suggested that the addition of a sticker is unnecessary for systemic compounds which are readily held in suspension.

Drenching of individual cabbage plants in the field with 0·28 g a.i. benomyl in either 170 or 400 ml solutions at planting and with higher volumes 1 month later significantly reduced clubroot and gave a 900 per cent plant weight increase (Reyes *et al.*, 1974).

Dibble hole applications of thiophanate methyl and benomyl (0·05, 0·1 or 0·2 g a.i. in 500 ml suspension per plant) reduced the severity of infection and the number of affected plants and increased the weight of cabbage heads and crop yield (Lopatin *et al.*, 1973). Comparable results were obtained by Il'ina *et al.* (1974).

Although band applications of several benzimidazole-derived compounds substantially reduced clubroot of direct drilled cabbages and considerably increased their yield, it is considered that such a treatment (involving the use of 12 kg a.i. chemical/ha) would be prohibitively expensive (Buczacki *et al.*, 1976).

It is probable that the benzimidazole-derived systemic fungicides can be used most effectively as alternatives to calomel in the root dip treatment of brassica transplants for the control of clubroot.

Furrow applications of small amounts of benomyl or thiabendazole in conjunction with crop rotations are suggested by Papavizas and Lewis (1971) as potentially feasible measures for the control of black root rot of beans (*Thielaviopsis basicola*) which was considerably reduced in pot soil by drenches of these fungicides. In later tests (Papavizas *et al.*, 1972) seed treatments with benomyl and thiabendazole used singly and in various mixtures proved effective against the fungus and are now preferred to furrow applications for this purpose.

One important group of soil-borne diseases which cannot be controlled by commercially available systemic fungicides is that attributable to attack by Phycomycetes, which usually cause damping-off of seedlings or root rot of older plants. Thus soil applications of benomyl and thiabendazole failed to control root infection of peas by *Aphanomyces euteiches* (Papavizas and Lewis, 1970), and seed dressings of both fungicides were ineffective against the Phycomycete fungi causing pre-emergence losses in peas (Harper, 1968; Crosier *et al.*, 1970). Increases in emergence of peas and in their fresh weight at harvest were achieved, however, if the seeds were dressed with captan (to prevent seedling losses by Phycomycete fungi) and benomyl was added to the soil or the seeds to control hypocotyl infections by *Fusarium solani* var. *pisi* and *Rhizoctonia solani.* Similar results were obtained by Natti and Crosier (1971) who found that furrow sprays of benomyl, benomyl plus chloroneb or thiabendazole provided effective control of bean (*Phaseolus vulgaris*) root rot (*Fusarium solani* f. *phaseoli*) during the early part of the growing season. Dual-function dressings of captan plus benomyl were also used by Maude and Kyle (1971) on early-sown peas. The addition of captan to the mixture substantially improved the emergence of the peas and benomyl successfully eliminated seed-borne *Ascochyta pisi.* Seed dressing mixtures of captan + thiabendazole proved equally effective, as did Benlate T (a mixture of thiram with benomyl) which is now recommended for use as a seed dressing for the control of seed-borne *Ascochyta* and soil-borne pathogens (*Pythium*, *Fusarium* and *Rhizoctonia* spp.) of peas and field beans in Britain.

There is some evidence that mixtures of certain fungicides may have synergistic effects. Thus Richardson (1973) demonstrated that mixtures of chloroneb and thiram (1 : 1 ratio 2·5 g mixture/kg seed) applied to pea seeds gave more effective control of seed rot and damping-off (*Pythium ultimum*) than could be explained by their additive effects alone.

Leaf diseases

Historically, conventional fungicides probably achieved their greatest use as foliar sprays. In the vegetable crop the spraying of potato foliage for the control of blight (*Phytophthora infestans*) is an outstanding example of this. Some of the newer systemic fungicides show specific activity against *Phytophthora* spp. and other phycomycete fungi though none are yet commercially available. The majority of systemic fungicides, however, are effective against other leaf pathogens, in some cases by penetration and limited movement within the leaf tissues; in others by their direct protectant action. Additional uses, not possible with non-systemic compounds, are the control of leaf diseases by seed or soil applications. In concept, the treatment of root-invading fungi by foliar sprays of systemic fungicides, although highly desirable, may not be realistic, as movement of these chemicals occurs mainly in an upward direction in plant tissues. There is some cause for optimism for the future, however, since recently Knauss (1974). Noveroske (1975a) and Hoitink and Schmitthenner (1975) independently demonstrated probable basipetal transport of a new systemic compound, pyroxychlor, in plant tissues.

Improved control of disease by the use of sprays of a systemic fungicide was first reported by Froyd and Johnson (1966) who found that two high-volume sprays of thiabendazole (421 g a.i./ha) gave more effective control of leaf spot of sugar beet than five applications of much greater concentrations of captan. Later, Delp and Klopping (1968) reported the superiority of benomyl over maneb in controlling the same disease. Erwin (1970) suggested that the improved disease control obtained with thiabendazole may have been due to local systemic movement of the fungicide within the leaf tissues, and the same may apply to benomyl. Solel (1970a) demonstrated that thiabendazole, benomyl and thiophanate, when sprayed on to the lower surfaces of sugar beet leaves, prevented infection of the upper surfaces of the leaves inoculated with *Cercospora beticola* spores, indicating translaminar movement of the fungicides. Such an effect did not occur if the non-systemic fungicide Brestan 60 (fentin acetate with maneb) was used instead. Recently (1971a), Solel has shown that the vapour of benomyl but not that of thiabendazole is toxic to *C. beticola* spores. Possibly both modes of action contribute to the effectiveness of foliar sprays of this fungicide against leaf spot of sugar beet. Certainly, in field spray trials (Solel, 1970b) three benomyl sprays (421 g a.i./ha each) at 3-weekly intervals completely controlled the disease and were superior to similar thiabendazole or fentin acetate applications. Under conditions conducive to disease expression (i.e. successive applications of spores to plants and frequent sprinkler irrigation) Paulus *et al.* (1970b) also found that benomyl gave superior disease control to thiabendazole when both fungicides were sprayed at 18- to 22-day intervals. Recently, low-volume aerial sprays of benomyl (168 g a.i./ha) have given effective control of this disease (Solel, 1971b) in Israel and promising results were also obtained in America (Potter and Sneider, 1973) with benomyl and thiabendazole sprays applied in a similar fashion.

However, the appearance of virulent strains of *C. beticola* tolerant to benomyl in sugar beet fields in Greece (Georgopoulos and Dovas, 1973) and America (Ruppel and Scott, 1974) resulted in loss of disease control. Related benzimidazoles were equally ineffective against benomyl-tolerant strains of the fungus and it is probable that in the future alternating spray programmes of systemic and non-systemic fungicides will be required to maintain control of *Cercospora* in regions where tolerance has developed.

Weekly high-volume sprays of benomyl (80–560 g a.i./ha) gave more effective control of late (*Septoria apiicola*) and early (*Cercospora apii*) blights of celery than similar applications with the non-systemic fungicide maneb (Delp and Klopping, 1968). Benomyl sprays, however, proved less efficient against celery late blight if applied at fortnightly intervals. Similar results were obtained by Paulus *et al.* (1970a), who found that fortnightly sprays of benomyl (280 g a.i./ha) were no more effective than those of several non-systemic compounds when infection in the crop was severe. When sprays were applied as soon as the disease appeared in the crop more successful disease control was achieved with fortnightly sprays though thiabendazole proved slightly superior to benomyl. Ryan and Kavanagh (1971) reported, however, that four sprays of benomyl (450 g a.i./ha) at similar intervals gave more effective control of the disease under conditions of moderate and severe infection than a number of conventional fungicides. Benomyl in these trials was inferior only to the tin fungicides. Similar results were obtained by Maude and Shuring (1970) and Maude and Moule (1971a) who found benomyl to be superior to several non-systemic fungicides but slightly less effective than the tin fungicide Brestan 60 (fentin acetate with maneb) in reducing the severity of the disease on the foliage. Both fungicides (560 g a.i./ha), however, gave comparable reductions in petiole infection when sprayed at fortnightly intervals. Thiabendazole was less effective than benomyl. Lacy (1973) also found that benomyl was more effective than thiabendazole in reducing celery leaf infection and that 1·1 kg benomyl/ha sprayed at 14-day intervals gave as effective control of *Septoria* as several non-systemic fungicides applied every 7 days. No evidence was found by Maude and Moule (1971a) that either of these fungicides moved within the plant in sufficient concentrations to give any effect against the fungus other than that which could be ascribed to their direct protectant action.

Benomyl-tolerant strains of *S. apiicola* (Wilson, 1974) and *C. apii* (Berger, 1973) now have been reported, a fact which must influence the future choice of fungicides for the control of these diseases.

More effective control of ringspot (*Mycosphaerella brassicicola*) of Brussels sprouts was achieved by Welch *et al.* (1969) with four fortnightly benomyl sprays (1·7 kg a.i./ha) than with a similar number of sprays of dithiocarbamate, copper and other fungicides. Infection of the seed inflorescences of cabbages by *Alternaria brassicicola* (dark leaf spot) was not prevented, however, by seven weekly sprays of tridemorph (1·1 kg a.i./ha) although this fungicide was inhibitory at low concentrations to the growth of the fungus in agar culture (Maude and Moule, 1971b).

The timing of the application of sprays of systemic fungicides to coincide with a period of maximum susceptibility of crop to infection has sometimes given very effective control of disease. Thus *Sclerotinia sclerotiorum* (white mould of snapbean), which infects the flowers of beans, was controlled by Natti (1967) with one (full bloom) or two (full- and post-bloom) high-volume sprays of benomyl (0·8 or 1·7 kg a.i./ha) which effectively prevented foliage and pod infection for a further 22 days. Although similar sprays of thiabendazole (0·56, 1·1 and 2·2 kg a.i./ha) and dicloran (1·1 or 2·2 kg a.i./ha) reduced infection initially, neither maintained a high level of disease control. Pegg (1972) with two (full- and post-bloom) sprays of benomyl (2·2 kg a.i./ha) obtained similar results and found that not only was leaf and pod infection considerably reduced in the field but that the pods remained free of mould in store. Since none of the fungicides effectively controlled the disease when applied as single post-bloom sprays, the timing of such applications is clearly critical, and early spraying is presumably necessary to prevent the establishment of the fungus in the flower tissues.

Although a similar control of *Botrytis cinerea* on the pods of French beans at harvest with two spray applications (full- and post-bloom) of benomyl (0·56 kg a.i./ha on each occasion) was reported by King (1971), single full-bloom applications (0·56 kg a.i./ha) were also shown to be just as effective (King, 1971; Ryan *et al.*, 1971; Gabrielson *et al.*, 1971). In trials, sprays of thiophanate methyl and benomyl gave equal control of *Botrytis* and these fungicides were superior in this respect to thiabendazole and several non-systemic fungicides. A single high-volume spray of benomyl (0·56 kg a.i./ha) at full bloom is now recommended by the manufacturers of Benlate (Du Pont and Co.) for the control of this disease in the French bean crop in Britain.

In controlling both diseases on bean pods at harvest by applications made to the flowers the effect of benomyl when compared with that of thiabendazole appears to be far greater than that which can be ascribed to protectant action alone. Its additional effect of reducing anthracnose (*Colletotrichum lindemuthianum*) on the pods of French bean (King, 1971; Ryan *et al.*, 1971) is furthermore difficult to understand since this fungus normally causes pod infections by the transfer of spores from stem lesions. Possibly the effect is one of reduction of inoculum potential in this case.

Considerable control of bean rust (*Uromyces phaseoli*) has also been obtained with one or, at most, two sprays of systemic fungicides applied at high volume at the first sign of the disease. In America one spray (0·56 or 1·1 kg a.i./ha) (Baldwin, 1970) and in Tanzania two sprays (2·5 kg a.i./ha) (Hudson and Jaffer, 1970) of oxycarboxin gave excellent control of the disease. In both countries non-systemic fungicides failed to give effective control, though in Tanzania sprays of other systemic fungicides, triarimol, BAS 3050 F (mebenil) and HOE 6952, also considerably reduced infection. In Brazil, also, sprays of oxycarboxin and oxycarboxin plus carboxin gave effective control of bean rust (Iamamoto *et al.*, 1971).

The application of systemic fungicides to seeds can result in some protection of the cotyledons and first leaves of some vegetables from attack by airborne fungi. Von Schmeling and Kulka (1966) were the first to demonstrate this effect by showing that infection of the unifoliate and trifoliate leaves of beans by *Uromyces phaseoli* (bean rust) was largely prevented by dusting the seeds with carboxin or oxycarboxin. Natti (1970) reported that the incidence of *Erysiphe polygoni* (powdery mildew) on kidney bean cotyledons was reduced if the seeds from which they were grown were dressed with benomyl. These effects, however, possibly may be too transient to represent a practical control method under field conditions of infection, and a more lasting control of leaf pathogens has been obtained by incorporating systemic fungicides with field soil. Thus soil drenches of benomyl controlled *Erysiphe cichoracearum* (powdery mildew) on the foliage of muskmelons (Wensley and Huang, 1970; Netzer and Dishon, 1970). In this respect one benomyl drench (0·2–0·5 g a.i. in 100 ml/plant) was as effective as four fortnightly sprays of the same fungicide (800 g a.i./ha) (Netzer and Dishon, 1970). Similarly Derbyshire (Anon, 1971b) found that a single benomyl drench applied to the soil (0·25 or 0·35 g a.i./l per m^2) of nursery beds containing young celery plants infected with *Septoria apiicola* 5 days before transplanting them into field soil prevented the development of the disease in the main crop. The treatment was practised in some celery nurseries in 1971. Control of the disease on infected celery seedlings was also obtained by transplanting them into compost containing benomyl (4·2 g a.i./m^2) to a depth of 5 cm. When these plants were later grown to maturity they remained healthy and the disappearance of the disease was correlated with the natural loss of affected foliage associated with the complete protection afforded to the new foliage by the systemic uptake of benomyl from the soil and its presence in the plant tissues (Maude and Moule, unpublished results).

As well as controlling leaf-infecting fungi, drenches of systemic fungicides possibly may find a new role in the treatment of certain virus disorders of vegetables. Thus Tomlinson (1974) demonstrated that two applications of carbendazim (a total of 0·1 g a.i. in 200 ml water/plant) to the soil of pots containing young lettuce plants (infected with beet western yellows virus) completely prevented symptom expression in their leaves at all stages of plant growth in the glasshouse and in the field. Plants which received carbendazim were healthy and robust at maturity although their leaf tissues contained as much virus as those of the non-treated infected plants which were chlorotic and weak. Similar applications of carbendazim also suppressed symptoms of tobacco mosaic virus in the leaves of infected tobacco plants.

The majority of systemic fungicides although they control many of the leaf-infecting fungi have little effect against those phycomycete fungi which cause leaf diseases in a number of vegetable species. Recently, however, new systemic compounds have been produced which are toxic to this group of plant pathogens. These include, pyroxychlor, effective against *Phytophthora palmivora* (Knauss, 1974) and *P. parasitica* (Noveroske, 1975b); prothiocarb, partially systemic and effective against a number of genera of Phycomycetes (Bastiaansen *et al.*, 1974); and 2-cyano-n-(ethylaminocarbamyl)-2-(ethoxyimino)-acetamide (E. I. Du Pont de Nemours & Co.) and LS 73-1038 (May and Baker, Ltd). In laboratory tests pyroxychlor and 2-cyano-n-(ethylaminocarbamyl)-2-(ethoxyimino)-acetamide applied in aqueous suspension (at 0·025 and 0·1 per cent a.i.) to the roots of lettuce seedlings failed to prevent infection of the cotyledon leaves inoculated with spores of *Bremia lactucae* (downy mildew) (Crute *et al.*, 1976). Prothiocarb and LS 73-1038 when applied at similar rates, however, effectively inhibited fungal invasion of the seedling leaves. Foliar sprays of these two fungicides, at similar concentrations, also prevented leaf infection of seedlings in the laboratory and of older plants raised in the glasshouse.

Post-harvest diseases

Systemic fungicides already have a well-established use as post-harvest treatments for the control of several storage rots of bananas and citrus fruits (Eckert, 1969). Thus the dipping of bananas and citrus in suspensions of thiabendazole to prevent outbreaks of blue and green mould (*Penicillium italicum* and *P. digitatum*) during storage had been practised in America since 1966 where the treatment has been cleared for commercial use (Erwin, 1970). More recently benomyl dips have been used for the same purpose.

By comparison, the post-harvest treatment of vegetable produce with systemic fungicides for the control of storage rots has received little attention.

Recently, however, 2-min dips in, and the dusting of tubers with, benomyl or thiabendazole were shown by Leach (1970) to prevent the development of dry rot (*Fusarium roseum*) during a period of 5 months' storage of artificially inoculated potato tubers. Similar treatments with non-systemic fungicides were not successful. Ten-second dips in benomyl, thiabendazole and in the non-systemic fungicide dicloran (1·8 g a.i. of the fungicide/l) were used by McMillan (1970) to control outbreaks of watery soft rot (*Sclerotinia sclerotiorum*) which frequently occur during the long-distance transport of pole beans to market. Although the benzimidazole fungicides considerably reduced outbreaks of the disease, it was completely eliminated by the dicloran treatment which is therefore preferred for this use. Largely similar results were recently obtained by Spalding and Reeder (1974). Dipping or soaking (15 min) of carrot roots or celery butts in benomyl (1 g a.i./l) more effectively controlled rotting by *Centrospora acerina* (liquorice rot of carrots, crown rot of celery) during storage than did similar treatments in dicloran (0·5 g a.i./l) (Derbyshire and Crisp, 1971). So effective were both benomyl treatments that after 7 months' storage, carrot root rotting due to the fungus was reduced by two-thirds and the marketability of the roots was doubled. Likewise, decay of stored celery sticks by *C. acerina* was prevented for 12 weeks by the treatment. Since deterioration in stored carrot roots is often complete after 3–4 months and celery *Centrospora* may produce losses of about 30 per cent after 8 weeks' storage the extensions of the shelf-life of both carrots and celery by these treatments was of considerable practical importance. Further work in 1971–2 and 1972–3 by Crisp (1974) on hand and mechanically harvested carrot roots, dipped in benomyl (0·5 or 1·0 g a.i./l) and stored, indicated that this treatment also controlled storage rots caused by *Botrytis cinerea* and *Sclerotinia sclerotiorum*. A similar effect using benomyl dips (0·5 g a.i./l) is reported from America (Wells and Merwarth, 1973), and in Canada (Lockhart and Delbridge, 1974) benomyl and thiabendazole (0·7 g a.i./l) sprayed on to the roots of washed carrots gave more effective control of storage rots caused by *B. cinerea* and *Rhizoctonia solani* (crater rot) than did drenching the roots in sodium orthophenylphenate.

Botrytis and *Sclerotinia* rotting of polythene-wrapped celery sticks in store in Israel (Barkai-Golan and Aharoni, 1972) was prevented for 70 days by post-harvest dips in or sprays of thiabendazole (0·6 g a.i./l). These treatments in addition reduced leaflet infection caused by *Cercospora apii* and leaflet and petiole infection caused by *Septoria apiicola*. The use of benomyl dips has been approved in Britain (Anon, 1975a) for the post-harvest treatment of carrots and celery. The dipping of Dutch White cabbage heads in aqueous suspensions of benomyl (0·7 g a.i./l) and thiabendazole (0·8 g a.i./l) substantially reduced *Botrytis* soft rotting of the heads during 24 weeks in store but increased losses resulting from physiological disorders (Kear and Symons, 1973).

In one instance, sprays of fungicide mixtures of benomyl and maneb, and of benomyl and daconil, applied to a lettuce crop in the field substantially reduced soft rotting (caused by *B. cinerea* and *S. sclerotiorum*) of the plants in store (Barkai-Golan *et al.*, 1974).

It is apparent that in their effectiveness against the storage diseases mentioned here the benzimidazoles exhibit activity against some of the main genera of fungi which cause post-harvest decay and therefore are of considerable value for the treatment of mixed populations of spoilage fungi on stored vegetable produce.

Results in practice – IV. Fruit crops

R. J. W. Byrde and V. W. L. Jordan

Introduction

An account of the performance of systemic fungicides on fruit crops – particularly on top fruits – is somewhat a contradiction in terms. Fungicides which are systemic in some herbaceous plants, such as cucumbers or cereals often show little movement in top fruits, particularly when applied as foliage sprays. Nevertheless, these compounds are powerful fungicides in their own right; with even the very limited movement that can be achieved – perhaps aided by an increased tenacity (Doma *et al.*, 1971) – they have opened up new possibilities for the control of diseases which have hitherto proved intractable.

The poor translocation in top fruits is well illustrated by the work of Cavell *et al.* (1971). Autoradiographs of young apple seedlings treated at the root with ^{14}C-ethirimol show clearly the localisation of the compound or its breakdown products within the vascular tissue, whereas this compound is fully systemic in cereals. However, benomyl has occasionally been effective as a root application to young plants for the control of apple powdery mildew caused by *Podosphaera leucotricha*. Cimanowski *et al.* (1970) obtained control of the disease on 1-year-old rootstocks growing in soil, but failed to do so when they were growing in Perlite: clearly soil texture can be an important factor. Gilpatrick (1969) obtained good control on very small seedlings, but Hammett (1968) was unsuccessful with 1-year-old seedlings growing in John Innes compost. Lewis (1968) recorded the failure of soil applications of benomyl to control scab (caused by *Venturia inaequalis*) on apple trees 8–10 ft high, even by root injection, and Byrde *et al.* (1969) did not obtain a significant reduction of the disease by application of benomyl as a spray to the grass sward. Gilpatrick and Bourke (1973) demonstrated significant uptake of triforine from soil into apple seedlings.

When applied as a foliar spray to apples, the compounds do not appear to move from one leaf to another, at least in concentrations sufficient to control diseases, e.g. Ben-Aziz *et al.* (1974); moreover, Mercer (1971) reported little or no movement of thiophanate methyl even within a treated leaf, and Doma *et al.* (1971) found only triarimol capable of giving a marked zone of mildew protection on a leaf from a single deposit. Bupirimate has since been shown to move within the apple leaf, and uptake occurs from spray deposits on green stems, allowing some movement into young leaves (Finney *et al.*, 1975).

Translocation over greater distances has been obtained with special trunk injection methods. Thus Pinkas *et al.* (1973) detected thiabendazole in the top branches and leaves of apple trees following pressure injection, and Shabi *et al.* (1974) showed that pre-leaf-fall injection of pear trees with carbendazim or thiabendazole was followed by rapid translocation. In the following season scab (*V. pirina*) was controlled on leaves reached by the fungicide. Some phytotoxicity was, however, observed.

On cherries, there is a single report of effective control of powdery mildew (caused by *P. oxyacanthea*) following root application of benomyl (Gilpatrick, 1969), and the systemic activity of carbendazim in almond blossom has been demonstrated (Ramsdell and Ogawa, 1973).

Fruit tissue can be penetrated by systemic compounds: thus Ben-Arie (1975) detected benomyl and, to a lesser degree, thiabendazole within post-harvest-treated pears; stone fruit tissue is also penetrated to a limited depth by benomyl (Kable, 1971a, b; Ravetto and Ogawa, 1972; Phillips, 1975).

In strawberries, there is some evidence to suggest a more systemic distribution (see p. 287).

Systemic fungicides are generally applied to fruit trees by spraying from the ground, but aircraft spraying has given promising results (Ogawa *et al.*, 1972; Ramsdell *et al.*, 1975a) against some diseases, and post-harvest dips are particularly effective against storage rots. A recent report (Ogawa *et al.*, 1975) describes the use of overhead sprinklers, for brown rot control.

This review is not exhaustive: references to trial reports, such as those in *Fungicide–Nematicide Tests* or Annual Reports, have been included only when giving information apparently not yet described in scientific papers.

Temperate fruit

Top fruit

Apples

The most important diseases of apples are scab and powdery mildew (Table 11.IV.1). Although lime-sulphur was formerly used to control both diseases, the superior fruit quality and disease control obtained with organic fungicides led to the adoption of the latter for routine spraying during the 1950s. Where both diseases are severe, a separate fungicide is used for each, generally at 7 to 10-day intervals – e.g. captan, dithianon or dodine against scab, and dinocap or binapacryl against mildew. Since some of the new systemic fungicides are active against both fungi, to a varying degree, a single compound can again be used for disease control.

The control of apple scab by benomyl, first described by Delp and Klöpping (1968) is so well known that a comprehensive reference list

Table 11.IV.1 Control of apple scab (*Venturia inaequalis*) and powdery mildew (*Podosphaera leucotricha*)†

Disease	Compound	References
Scab	Benomyl	Delp and Klöpping (1968) and see text
	Carbendazim	Ellis *et al.* (1974); Swinburne *et al.* (1975)
	Thiabendazole	von Keyserlingk and Wihrheim (1971)
	Thiophanate methyl	Borecka *et al.* (1971); Cole *et al.* (1971); Doma *et al.* (1971); Douchet, Lhoste and Quere (1971); Formigoni *et al.* (1971); Ishii (1970)
	Triforine	Adlung and Drandarevski (1971); Schicke and Veen (1969)
Mildew	Benomyl	Bourdin *et al.* (1969); Burchill and Williamson (1971); Catling (1971); Cole *et al.* (1971); Covey (1971b); Coyier (1971); Delp and Klöpping (1968); Doma *et al.* (1971); Evans *et al.* (1971); Foschi *et al.* (1972); Butt *et al.* (1973)
	Carbendazim	Ellis *et al.* (1974)
	Pyrazophos	Hay (1971); Butt *et al.* (1973); Shabi (1975)
	Thiabendazole	von Keyserlingk and Wihrheim (1971)
	Thiophanate methyl	Aelbers (1971); Burchill and Williamson (1971); Cole *et al.* (1971); Doma *et al.* (1971); Douchet, Lhosteand Quere (1971); Formigoni *et al.* (1971); Foschi *et al.* (1972)
	Triforine	Adlung and Drandarevski (1971); Schicke and Veen (1969); Gilpatrick *et al.* (1972); Butt *et al.* (1973)

† See also Gilpatrick (1975)

is not given: many examples are cited by Hickey (1971a). The compound has often given good control even when used at extended intervals, e.g. up to 28 days (Burchill and Williamson, 1971; Byrde *et al.*, 1969; Catling, 1971; Delp and Klöpping, 1968; Evans *et al.*, 1971; Kavanagh and O'Kennedy, 1971; Moller *et al.*, 1969). Benomyl may owe its success partly to its activity as a curative fungicide (e.g. Catling, 1971; Kavanagh and O'Kennedy, 1971; Zehr, 1971a). Szkolnik (1969) considered that its post-infectional activity was intermediate between those of captan and dodine.

The compound is also suitable for use as an autumn (pre-leaf-fall) spray to prevent perithecial formation on fallen leaves (Borecka *et al.*, 1971; Borecki and Nowacka, 1969; Burchill and Swait, 1971; Connor and Heuberger, 1968; McIntosh, 1969; Vojvodic, 1970; Burchill, 1972; Latorre, 1972; Paisley and Heuberger, 1972), so bringing about a drastic reduction in inoculum the following season and so probably allowing a reduced spray programme. It has also been used in the spring to prevent ascospore discharge (Miller, 1970). Even when a normal spray programme is used, finishing in July, a residual effect on scab the following season has been observed (Byrde *et al.*, 1973), and such a programme has been shown to stop perithecial formation (Ross, 1973; Ross and Newbery, 1975).

Although almost always highly effective against scab, a few exceptions have been recorded and seem worthy of special mention. Thus Fridrich and Gilpatrick (1971) found unsatisfactory control of leaf scab (but good fruit scab control) in the Hudson Valley, and Palmiter (1970), in the same area, noted leaf infection earlier in the season: most of the lesions were abnormal and produced few if any spores. The addition of Tween 20 spreader improved the control of foliage infection. Klos *et al.* (1971), in Michigan, similarly recorded a high incidence of flower cluster leaf scab under severe infection conditions although, again, conidiophores and conidia were absent from the lesions, which soon stopped developing, (see also Foschi *et al.*, 1972). There is no reason to think that resistant races of *Venturia inaequalis* are responsible for the above results. The addition of oil to benomyl has given improved scab control (Wicks, 1973), but reduction in yield following such treatment has also been recorded (Spotts *et al.*, 1975).

Benomyl is also effective against powdery mildew, though not to the same degree: thus, it has not been satisfactory when used at extended intervals (e.g. Burchill and Williamson, 1971; Evans *et al.*, 1971). One of its attributes is its marked effect on primary mildew, both in 'cleaning up' blossom truss infections in the spring and in controlling overwintering bud infections, as recorded the following season (e.g. Catling, 1971). From greenhouse tests its curative action against mildew appears greater than that of dinocap and binapacryl (Frick, 1971; Sztejnberg *et al.*, 1975).

Use of benomyl as a routine spray for scab and mildew control has given rise to several side-effects. Thus, an increased level of fruit russet (by comparison with standard fungicides) has sometimes been encountered, e.g. on Golden and Red Delicious (Gilpatrick, 1970) and on Cox's Orange Pippin (Byrde, 1972; Kirby, 1971). A mixture with lead arsenate proved highly phytotoxic (Ross and Sanford, 1971). The compound sometimes appears to stimulate vegetative growth (e.g. Evans *et al.*, 1971), and yield increases, sometimes statistically significant, have been recorded (Byrde *et al.*, 1970; Evans *et al.*, 1971; Kirby, 1971). Although the basis of this effect is not clear, it may well reflect increased fruit set, and it is significant that the compound, or its breakdown products, has been detected by bioassay in extracts of floral parts including the stigma (Corke *et al.*, 1972). Carbendazim application can also result in increased fruit set (Ellis *et al.*, 1974).

Benomyl and related compounds, after breakdown to carbendazim, are also relatively persistent in the orchard, including the soil (see Chapter 8). Thus Ben-Aziz *et al.* (1974) detected carbendazim in pear leaves 7 months after spraying, and Byrde *et al.* (1973) detected carbendazim residues in apple bark after a similar period.

Carbendazim and thiophanate fungicides have generally given similar results to benomyl. Thiophanate methyl has been found to have excellent curative action against apple scab (Formigoni *et al.*, 1971). When applied as a routine spray thiabendazole, in contrast, has given erratic results, well illustrated in the reports edited by Hickey (1971a)

Table 11.IV.2 Control of apple diseases†

Causal fungus	Disease	Compound	References
Botryosphaeria ribis	White rot	Benomyl, thiophanate methyl	Gilpatrick (1975)
Botrytis cinerea	Grey mould rot	Benomyl, thiabendazole	Hardenburg (1974)
Gloeodes pomigena	Sooty blotch	Benomyl	Delp and Klöpping (1968)
Gloeosporium spp.	Bitter rot	Benomyl	Bompeix and Morgat (1969); Borecka *et al.* (1971); Catling (1971); Cole *et al.* (1971); Burchill and Edney (1972); Edney (1970)
		Thiabendazole	Bompeix and Morgat (1969); Bondoux (1967); Edney (1970); Pleisch *et al.* (1971); Pratella and Tonini (1968); Burchill and Edney (1972); Cole *et al.* (1973)
		Thiophanate methyl	Borecka *et al.* (1971); Cole *et al.* (1971); Formigoni *et al.* (1971); Ishii (1970)
Glomerella cingulata	Necrotic leaf blotch	Benomyl	Taylor (1971)
Gymnosporangium juniperi-virginianae	Cedar-apple rust	Triforine	Szkolnik (1974)
Microthyriella rubi	Fly speck	Benomyl	Delp and Klöpping (1968)
Nectria galligena	Canker	Benomyl	Byrde *et al.* (1973); Catling (1971); Corbin (1971); Swinburne *et al.* (1975)
		Carbendazim	Swinburne *et al.* (1975)
		Thiophanate methyl	Cole *et al.* (1971)
	Fruit rot	Benomyl	McDonnell (1970); Burchill and Edney (1972); Cole *et al.* (1973)
		Thiabendazole	Bondoux (1967)
Penicillium expansum	Blue mould	Benomyl	Cargo and Dewey (1970); Daines and Snee (1969); Spalding *et al.* (1969); Spalding and Hardenburg (1971); Hardenburg (1974); Roy (1975)
		Thiabendazole	Blanpied and Purnasiri (1968, 1970); Cargo and Dewey (1970); Daines and Snee (1969); Pierson (1966); Pleisch *et al.* (1971); Spalding *et al.* (1969); Spalding and Hardenburg (1971)
Pezicula (*Gloeosporium*) spp.	Canker IsoPropyl	Benomyl	Borecki *et al.* (1970)
Physalospora obtusa	Frog-eye leaf spot	Triarimol	Lade (1970)
	Black rot	Benomyl	Delp and Klöpping (1968)
Podosphaera leucotricha	Powdery mildew	See Table 11.IV.1.	
Sclerotinia fructigena	Brown rot	Benomyl	Cole *et al.* (1971)
		Thiophanate methyl	Cole *et al.* (1973); Burchill and Edney (1972)
S. laxa f. *mali*	Blossom wilt	Benomyl	Byrde and Melville (1971); Catling (1971)
Trichoseptoria fructigena	Trichoseptoria rot	Benomyl	Bompeix and Morgat (1969)
		Thiabendazole	Bompeix and Morgat (1969); Bondoux (1967)
Venturia inaequalis	Scab	See Table 11.IV.1.	

† See also Gilpatrick (1975)

although it has proved moderately effective as a spring spray to reduce ascospore inoculum (Miller, 1970).

Among systemic fungicides for apple mildew, pyrazophos has useful antisporulant properties (Sztejnberg *et al.*, 1975), but can cause damage (Butt *et al.*, 1973). Triforine is less active as an antisporulant, but is a good mildew fungicide. Sometimes its use is followed by greener foliage (e.g. Adlung and Drandarevski, 1971), but an early formulation sometimes caused increased fruit russet (Adlung and Drandarevski, 1971; Butt *et al.*, 1973).

Systemic fungicides have been useful against other apple diseases (Table 11.IV.2) and several have shown antisporulant activity against *Nectria galligena*, the apple canker pathogen. Thus Bennett (1971) showed a marked and prolonged inhibition of spore production in the orchard following benomyl treatment: thiabendazole was less effective. Cole *et al.* (1971) demonstrated the antisporulant activity of a canker paint based on thiophanate methyl. Benomyl (McDonnell, 1970) was also superior to thiabendazole (McDonnell, 1971) for Nectria fruit rot control, and Manners and Corke (1971) showed that Nectria spores were much more resistant to thiabendazole than to benomyl. Spore output from cankers due to *Gloeosporium perennans* was also affected more by benomyl than by thiabendazole (Corke *et al.*, 1971). Routine benomyl applications for scab and mildew control result in the accumu-

Table 11.IV.3 Control of pear diseases†

Causal fungus	Disease	Compound	References
Botrytis cinerea	Storage rots	Benomyl	Beattie and Outbred (1970); Coyier (1970); Ben-Arie and Guelfat-Reich (1973)
		Thiabendazole	Beattie and Outbred (1970); Ben-Arie and Guelfat-Reich (1973)
Entomosporium maculatum	Leaf spot	Benomyl	Kantzes and Weaver (1971)
Penicillium expansum	Fruit rot	Benomyl	Maas and MacSwan (1970); Spalding (1970); Ben-Arie and Guelfat-Reich (1973)
		Thiabendazole	Maas and MacSwan (1970); Scott and Roberts (1970); Spalding (1970); Ben-Arie and Guelfat-Reich (1973)
Venturia nashicola	Scab	Benomyl	Kishi (1971)
V. pirina	Scab	Benomyl	Bourdin *et al.* (1969); Kirby *et al.* (1970); Williamson and Burchill (1974)
		Thiophanate methyl	Aelbers (1971); Cole *et al.* (1971); Douchet *et al.* (1971); Formigoni *et al.* (1971); Ishii (1970); Williamson and Burchill (1974)
		Triforine	Schicke and Veen (1969)

† See also Gilpatrick (1975)

lation of toxic residues in the bark (Byrde *et al.*, 1973). Swinburne *et al.* (1975) convincingly demonstrated the superiority of carbendazim over benomyl and, particularly, thiabendazole, both in suppressing sporulation and controlling the disease.

The use of benomyl, thiabendazole and thiophanate methyl as sprays or post-harvest dips against Gloeosporium rots represents a greatly improved control measure for this serious disease.

Diseases not readily controlled by benomyl include Alternaria rot (Spalding, 1970), which is also not controlled by thiabendazole (Blanpied and Purnasiri, 1970), cedar apple rust (Delp and Klöpping, 1968) and collar rot caused by *Phytophthora cactorum* and *P. syringae.* A tendency for a disease caused by a *Phytophthora* sp. to increase when benomyl was applied has even been reported (van den Berg and Bollen, 1971).

Pears

The most important of the diseases of pears (Table 11.IV.3) is scab, which can cause severe fruit disfigurement and also persistent lesions on the wood. Benomyl and thiophanate-methyl have given successful control, and have post-infectional activity (Ostermann, 1974). By contrast, control by thiabendazole has proved erratic. Williamson and Burchill (1974) showed these compounds could prevent ascospore formation, but sometimes benomyl treatment stimulated conidial production from stem lesions. Benomyl increased fruit russeting (Kirby, 1971) and internal breakdown (Ben-Arie and Guelfat-Reich, 1975).

Stone fruit

The most important group of diseases are those caused by the brown rot fungi which attack the fruit both in the orchard and after harvest, and also the blossom trusses. Three species of *Monilinia* (*Sclerotinia*) are responsible, with different geographical distributions (Wormald, 1954; Byrde and Willetts, 1977). Although control by conventional fungicides is often unsatisfactory, the advent of benomyl and related compounds resulted in a wide measure of control (Table 11.IV.4). As well as acting as a prophylactic spray on living tissue, the compounds suppress sporulation of the fungus on infected peduncles and mummified fruits of peach (Kable, 1970, 1975). Instances of less effective brown rot control have been recorded (MacSwan, 1971a; Kable, 1975). The action of benomyl (and that of thiabendazole) as fruit dips has been found to be enhanced by higher water temperatures, (Daines, 1970; Smith, 1971), or the use of oil or wax formulations (e.g. Wells, 1972). Benomyl stimulated the leaf size of cherry seedlings (Cimanowski *et al.*, 1970). There is also evidence that a pre-blossom application can advance the flowering date of plums and somewhat increase the effective pollination period (Stott and Jefferies, 1972).

Although also controlling several other diseases of stone fruit (Table 11.IV.4), benomyl is ineffective against peach leaf curl (e.g. Corbin and

Kitchener, 1970) and Rhizopus rot of peach (e.g. Hickey, 1971b). Often the latter disease is controlled by the addition of dicloran to post-harvest dips (e.g. Wells, 1971). Triforine is another useful systemic fungicide for stone fruit, although damage to cherry has been reported (Gilpatrick, 1973).

Soft fruit

Bush and cane fruit

The principal diseases of black currant and gooseberry are grey mould, leaf spot and American gooseberry mildew. Control of these three diseases can be achieved by the use of three separate conventional fungicides; dichlofluanid is particularly valuable against grey mould, the dithiocarbamates have been extensively used against leaf spot, and quinomethionate has brought mildew under control. The advent of broad-spectrum systemic fungicides has opened the way to control of all three diseases by a single compound, and Table 11.IV.5 shows that benomyl and carbendazim can achieve this. On occasions thiabendazole (O'Riordain *et al.*, 1971) and thiophanate methyl (Borecki *et al.*, 1971) have proved somewhat inferior against *Pseudopeziza ribis*. O'Riordain (1973) reported poor control with benomyl, carbendazim and thiophanate methyl, whereas triforine gave better control than the standard mancozeb. Ingram (1971) found benomyl (without a wetting agent), pyrazophos and thiophanate were relatively ineffective against *Sphaerotheca mors-uvae* on black currant, whereas Yakovleva (1974) obtained good control with bemomyl, pyrazophos or thiophanate methyl. Clifford and Hislop (1971) showed that thiophanate and thiophanate methyl had a strong vapour action against this fungus *in vitro*.

Although less information is available on the control of raspberry diseases, of which grey mould, cane spot and spur blight are the most common, benomyl and thiophanate methyl have given good control (Burchill and Swait, 1975; Cole and Cox, 1973): however, control of spur blight with benomyl in Ireland has been erratic (O'Riordain *et al.*, 1971).

Benomyl has proved useful against three blueberry diseases, but it was relatively ineffective against powdery mildew on this host (caused by *Microsphaera penicillata* var. *vaccinii*) (Hilborn, 1969), and neither benomyl nor thiabendazole successfully eradicated *Botryosphaeria corticis* in dormant cutting wood (Beute and Milholland, 1971). Benomyl, triforine and thiophanate methyl were effective in controlling conidial infection by *Monilinia vaccinii-corymbosi*, but only triforine controlled the primary ascospore infection (Pepin and Ormrod, 1974).

Strawberries

The chief fungal diseases of strawberries are grey mould, powdery mildew, *Verticillium* wilt and red core (caused by the Phycomycete *Phytophthora fragariae*). Good control of grey mould can now be

Table 11.IV.4 Control of stone fruit diseases†

Host	Causal fungus	Disease	Compound	References
Almond	*Monilinia laxa*	Blossom blight	Benomyl, carbendazim	Ramsdell and Ogawa (1973)
Apricot	*Eutypa armeniacae*	Dieback	Benomyl	Moller and Carter (1969, 1971)
	Monilia (*Sclerotinia*) *fructicola*	Brown rot	Benomyl, thiabendazole	Beattie and Outbred (1970); Jones, Burton and Tennes (1973)
	M. laxa	Blossom blight	Thiophanate methyl	Ogawa *et al.* (1972)
	Monilia sp.	Blossom blight	Benomyl	Waffelaert (1969)
	Oidium sp.	Mildew	Benomyl	Waffelaert (1969)
Cherry	*Coccomyces hiemalis*	Leaf spot	Benomyl	Cimanowski *et al.* (1970); Delp and Klöpping (1968)
	M. fructicola	Brown rot	Benomyl, thiabendazole	Beattie and Outbred (1970)
			Triforine	Gilpatrick (1973); Jones, Burton and Tennes (1973); Jones (1975); Koffman and Kable (1975)
	M. fructicola and *M. laxa*	Blossom blight	Benomyl, carbendazim, thiophanate methyl	Ogawa *et al.* (1968a); Byrde and Willetts (1977)
	Podosphaera oxyacanthae	Mildew	Benomyl	Gilpatrick (1969)
Nectarine	*M. fructicola* and *M. laxa*	Blossom wilt and fruit rot	Benomyl	Wells and Gerdts (1971); Byrde and Willetts (1977)
Peach	*Cladosporium carpophilum*	Scab	Benomyl	Delp and Klöpping (1968)
	Cytospora (*Valsa*) *cincta*	Cytospora canker	Cyloheximide semicarbazone	Helton and Rohrbach (1967)
			Benomyl, thiabendazole	Springer (1970); Harder and Luepschen (1975)

	Fusicoccum amygdali	Canker	Benomyl, carbendazim, thiophanate methyl	Jailloux and Froidefond (1974)
	M. fructicola	Brown rot, blossom blight	Benomyl, thiabendazole, thiophanate methyl	See review by Byrde and Willetts (1977)
			Triforine	Gilpatrick (1973)
	Monilinia spp.	Brown rot	Thiophanate methyl	Ishii (1970)
	Sphaerotheca pannosa	Mildew	Benomyl	Bregadze (1972)
			Bupirimate	Finney *et al.* (1975)
	Stigmina carpophila	'Stigmina'	Benomyl	Corbin and Kitchener (1970)
	Tranzschelia discolor	Rust	Triforine	Decker and Buchanan (1974)
Plum	*Cytospora cincta*	Cytospora canker	Cycloheximide thiosemicarbazone	Helton and Kochan (1967)
	Dibotryon morbosum	Black knot	Benomyl, thiophanate methyl	Ritchie *et al.* (1975)
	M. fructicola	Brown rot	Triforine	Gilpatrick and Szkolnik (1972); Gilpatrick (1973); Jones, Burton and Tennes (1973)
	Monilinia spp.	Blossom blight, brown rot	Benomyl, thiabendazole, thiophanate methyl	Renaud (1968, 1970); Jones and Burton (1973); Jones, Burton and Tennes (1973)
	Tranzschelia prunispinosae	Rust	Benodanil	Jordan and Tarr (1975)

† See also Gilpatrick (1975)

Table 11.IV.5 Control of bush and cane fruit diseases

Host	Causal fungus	Disease	Compound	References
Black currant	*Botrytis cinerea*	Grey mould	Benomyl	Tapley *et al.* (1969); O'Riordain (1973); Phillips *et al.* (1973)
	Cronartium ribicola	Rust	Benomyl, thiophanate methyl	Borecki *et al.* (1971)
	Microsphaera grossulariae	European gooseberry mildew	Benomyl	Borecki *et al.* (1971)
	Pseudopeziza ribis	Leaf spot	Benomyl	Borecki *et al.* (1971); Gilchrist and Cole (1971); O'Riordain *et al.* (1971); Phillips *et al.* (1973)
			Carbendazim	Phillips *et al.* (1973)
			Thiabendazole	Ayers and Gunary (1971)
			Thiophanate methyl	Gilchrist and Cole (1971)
			Triforine	O'Riordain *et al.* (1971); O'Riordain (1973)
	Sphaerotheca mors-uvae	American gooseberry mildew	Benomyl	Gilchrist and Cole (1971); Phillips *et al.* (1973)
			Carbendazim	Phillips *et al.* (1973)
			Pyrazophos	Yakovleva (1974)
			Thiophanate methyl	Gilchrist and Cole (1971); Ingram (1971)
Gooseberry	*Botrytis cinerea*	Grey mould	Benomyl	Tapley *et al.* (1969); Phillips *et al.* (1973)
			Carbendazim	Phillips *et al.* (1973)
	Pseudopeziza ribis	Leaf spot	Benomyl	O'Riordain *et al.* (1971); Tapley *et al.* (1969); Woodward (1971); Phillips *et al.* (1973)
			Carbendazim	Phillips *et al.* (1973)
			Thiophanate methyl	Woodward (1971)
			Triforine	O'Riordain *et al.* (1971)
	Sphaerotheca mors-uvae	American gooseberry mildew	Benomyl	O'Riordain *et al.* (1971); Woodward (1971); Phillips *et al.* (1973)
			Carbendazim	Phillips *et al.* (1973)
			Pyrazophos, thiophanate methyl	Woodward (1971)
			Triforine	O'Riordain *et al.* (1971)

Raspberry	*Botrytis cinerea*	Grey mould (fruit)	Benomyl	Freeman and Pepin (1967); Cole and Cox (1973); Burchill and Swait (1975)
			Thiophanate methyl	Cole and Cox (1973)
		Grey mould (stem)	Thiabendazole	O'Riordain *et al.* (1971)
	Didymella applanata	Spur blight	Benomyl	Burchill and Swait (1975)
			Thiophanate methyl	Cole and Cox (1973)
	Elsinoe veneta	Cane spot	Benomyl, thiabendazole	O'Riordain *et al.* (1971); Cole and Cox (1973)
			Thiophanate methyl	Cole and Cox (1973)
	Nectria sp.	Cane death	Benomyl	O'Riordain *et al.* (1971)
	Sphaerotheca macularis	Powdery mildew	Benomyl	Burchill and Swait (1975)
Blackberry	Four pathogens	Storage rot	Benomyl, thiabendazole	MacSwan (1971b)
Blueberry	*Botrytis cinerea*	Grey mould	Benomyl	Delp and Klöpping (1968)
	Exobasidium vaccinii	Red leaf	Carboxin	Lockhart (1971)
	Gloeosporium minus and *Gloeocercospora inconspicuia*	Leaf spot	Benomyl	Milholland and Jones (1975)
	Monilinia vaccinii-corymbosi	Twig and blossom blight	Benomyl	Pepin and Ormrod (1968)
		Mummy berry	Benomyl, triforine	Pepin and Ormrod (1974); Ramsdell *et al.* (1975a, b)
			Thiophanate methyl	Pepin and Ormrod (1974)

Table 11.IV.6 Control of strawberry diseases†

Causal fungus	Disease	Compound	References
Botrytis cinerea	Grey mould	Benomyl	Freeman and Pepin (1967, 1968); Gilchrist and Cole (1971); Kavanagh and O'Callaghan (1969); Maas (1970, 1971); Müller (1970); Paulus *et al.* (1969); Tapley *et al.* (1969); Borecka *et al.* (1973); Jordan (1973a)
		Carbendazim	Jordan (1973a)
		Thiabendazole	Maas (1970); Jordan (1973a)
		Thiophanate methyl	Aelbers (1971); Formigoni *et al.* (1971); Gilchrist and Cole (1971); Borecka *et al.* (1973); Jordan (1973a)
Botrytis cinerea, Rhizoctonia sp.	Storage rots (fruit)	Benomyl, carbendazim, thiophanate methyl	Jordan (1973a)
	Storage rots (runners)	Benomyl	Maas and Scott (1971, 1973)
		Thiophanate methyl	Maas and Scott (1973)
Colletotrichum fragariae	Anthracnose	Benomyl	Howard (1971); Horn *et al.* (1972)
Dendrophoma, Diplocarpon, Mycosphaerella	Leaf spots	Benomyl	Pessanha *et al.* (1970)
		Thiabendazole	Howard and Albregts (1972)
		Thiophanate methyl	Paulus *et al.* (1973)
Sphaerotheca macularis	Powdery mildew	Benomyl	Freeman and Pepin (1968); Jordan (1971); Maas (1970); Scott *et al.* (1970)†; Kavanagh (1972)†
		Carbendazim	Kavanagh (1972)†; Jordan (1973a)
		Thiabendazole, thiophanate methyl	Jordan (1973a)
Verticillium dahliae	Wilt	Benomyl	Jordan (1972a, 1975); Lockhart *et al.* (1969); Talboys *et al.* (1975)
		Carbendazim	Jordan (1975)
		Thiabendazole	Jordan (1972a, 1975)
		Thiophanate methyl	Jordan (1972b, 1975)

† This reference describes a trial on greenhouse-grown plants

achieved by the use of dichlofluanid, but powdery mildew is more difficult to control with conventional fungicides where infection is severe, whilst control of the two soil-borne diseases has been impracticable. Now, however, as shown in Table 11.IV.6, these and other minor diseases have been controlled by benomyl, carbendazim, thiabendazole and thiophanate methyl, with the expected exception of red core. Control of *Botrytis* fruit rot has generally been of the same order as that given by dichlofluanid, as found in 1970 by O'Riordain *et al.* (1971), but in 1971 these authors found benomyl and thiophanate methyl inferior to dichlofluanid. Benomyl (Maas, 1970), carbendazim and thiophanate methyl (Jordan, 1973a) were reported as being more effective than thiabendazole against grey mould, and much more effective against mildew. Against this disease, pyrazophos and a substituted azepine were less effective than benomyl (Jordan, 1971).

Full systemic action is likely to be more readily achieved in the herbaceous strawberry plant than with top fruits; the systemic distribution of benomyl (Nicholson *et al.*, 1972), carbendazim (Siegel, 1973), thiabendazole and thiophanate methyl (Jordan, 1973b) has been demonstrated (in strawberry plants). It is probably significant that Howard (1971) found benomyl effective against *Colletotrichum fragariae* whereas twelve non-systemic fungicides were not. Kavanagh (1972) demonstrated control of powdery mildew in the glasshouse with a benomyl soil drench, and pre-planting dips with benomyl or thiophanate methyl (Paulus *et al.*, 1973) controlled Ramularia leaf spot.

Even so, phloem translocation does not seem to occur to any extent, as Lockhart *et al.* (1969) found benomyl ineffective as a foliar spray against *Verticillium* wilt although giving good control as a soil treatment in the planting hole. Using such a benomyl treatment, Jordan (1972a) could detect the pathogen within the tissues of a visually healthy plant, and later demonstrated (Jordan, 1973b) that although benomyl, thiabendazole and thiophanate methyl were fungistatic in their action against *Verticillium dahliae*, root dips and soil application provided sufficient fungitoxicant in plant tissues to reduce disease severity. Fungitoxicant was detected in tissues of daughter plants following an application of benomyl to mother plants (Nicholson *et al.*, 1972); this treatment followed by an application to young runners, with either benomyl or thiophanate methyl, prevented transmission of Verticillium wilt infection into runners.

Benomyl, when used as a pre-lifting spray, controlled storage decay of cold-stored runners (Maas and Scott, 1973) and increased survival, quality and first-year fruit yields.

Sub-tropical fruit

Table 11.IV.7 summarises reports on the control of citrus and grape diseases by systemic fungicides. The most important diseases of citrus

Table 11.IV.7 Control of diseases of sub-tropical fruits

Host	Causal fungus	Disease	Compound	References
Citrus (mainly orange)	*Cercospora* sp.	Pink pitting	Benomyl	Suit and Ducharme (1971)
	Deuterophoma tracheiphila	'Mal secco'	Benomyl	Elia (1969); Salerno and Somma (1971); Solel *et al.* (1972)
			Carboxin, cycloheximide semicarbazone, oxycarboxin, thiabenazole	Solel *et al.* (1972)
	Diaporthe medusaea	Melanose	Benomyl	Timmer (1974)
	Diplodia natalensis	Stem-end rot	Benomyl	Brown (1968); Timmer (1973)
			Thiabendazole, thiophanate methyl	Timmer (1973)
	Elsinoe fawcettii	Scab	Benomyl	Hearn and Childs (1969); Hearn *et al.* (1971); Phelps (1969)
	Guignardia citricarpa	Black spot	Benomyl	McOnie *et al.* (1969)
	Mycosphaerella sp.	Greasy spot	Benomyl	Phelps (1969); Whiteside (1973)
	Penicillium digitatum	Penicillium rots†	Benomyl	Baron and Cayuela (1969); Brown (1968); Delp and Klöpping (1968); Gutter (1969a‡, 1969b, 1970); Harding (1968); McCornack and Brown (1970); Tugwell and Wicks (1969); Timmer (1973)
			Thiabendazole	Baron and Cayuela (1968); Cuillé and Bur-Ravault (1969)‡; Gutter (1969a)‡; Harding (1967‡, 1968); McCornack and Brown (1970); Seberry and Baldwin (1968); Tugwell and Wicks (1969); Eckert and Kölbezen (1971)†; Timmer (1973); Gutter *et al.* (1974)

			Thiophanate methyl	Ishii (1970); Timmer (1973)
	Unknown	Fruit rot	Benomyl, thiabendazole	Smoot and Melvin (1970)
	Unknown	Die back	Benomyl	Hearn and Fenton (1970)
Grape	*Botrytis cinerea*	Grey mould	Benomyl	Deitrich and Brechbuhler (1970); Frazao (1970); Thiollière (1970); Kuprashvili and Chekheidze (1973); Parle and Dodanis (1973); Hunter and Whiteley (1976)
			Carbendazim	Kuprashvili and Chekheidze (1973)
			Thiophanate methyl	Formigoni *et al.* (1971); Kuprashvili and Chekheidze (1973)
	Dibotryon morbosum	Black knot	Benomyl, thiophanate methyl	Ritchie *et al.* (1975)
	Elsinoe ampelina	Anthracnose	Benomyl	Hopkins (1974)
	Guignardia bidwellii	Black rot	Benomyl	Delp and Klöpping (1968); Klos *et al.* (1968); Welch (1968)
	Isariopsis clavispora	Leaf spot	Benomyl	Hopkins (1974)
	Melanconium fuligineum	Bitter rot	Benomyl	Delp and Klöpping (1968); Welch (1968)
	Phomopsis viticola	Dead arm	Benomyl	Gartel (1974)
	Plasmopara viticola	Downy mildew	Benomyl	Malenin and Kobokhidze (1973)
	Pseudopeziza tracheiphila	Red fire	Benomyl	Kirch (1972)
	Sphaerotheca pannosa	Mildew	Benomyl	Bregadze (1972)
			Bupirimate	Finney *et al.* (1975)
	Tranzschelia discolor	Rust	Triforine	Decker and Buchanan (1974)
	Uncinula necator	Powdery mildew	Benomyl	Bottalico (1969); Braun (1971); Melenin and Kobokhidze (1973)
			Bupirimate	Finney *et al.* (1975)
			Pyrazophos	Jovanovic (1974)
			Thiophanate methyl	Formigoni *et al.* (1971)

† Generally treated by post-harvest dips
‡ Includes use in wax formulation

are the post-harvest rots due to *Penicillium digitatum* ('green mould') and *P. italicum* ('blue mould'), which for convenience are covered together in the table. Previous control measures involved the use of diphenyl wraps, or post-harvest treatment with *o*-phenyl phenol or 2-aminobutane, but these have been superseded by benomyl, thiabendazole and thiophanate methyl. Of these, there are indications that the former is the more active (Baron and Cayuela, 1969; Brown, 1968; Gutter, 1970).

In addition to post-harvest dips, pre-harvest sprays with these compounds have effectively controlled post-harvest decays (Timmer, 1973), and two post-bloom applications of benomyl controlled melanose on Texas grapefruit (Timmer, 1974).

Many diseases of less economic importance are also controlled by these compounds. Thiabendazole has been found to be toxic *in vitro* to *Diplodia natalensis*, *Phomopsis citri* and *Penicillium digitatum* (Brown *et al.*, 1967) and has suppressed growth and sporulation *in vivo* (Eckert and Kolbezen, 1971). Whiteside (1970) demonstrated the antisporulant activity of this compound and of benomyl against the perithecial stage of *Mycosphaerella* sp. which causes 'greasy spot' disease.

The systemic properties of benomyl (Elia, 1969; Salerno and Somma, 1971), cycloheximide semicarbazone, oxycarboxin and thiabendazole (Solel *et al.*, 1972) have proved useful in the control of 'mal secco' by soil application to orange seedlings and lemon plants.

Of grape diseases the most serious are powdery mildew, *Botrytis* fruit rot and downy mildew (caused by *Plasmopara viticola*). Conventionally, a different fungicide is used for each disease – sulphur or dinocap for powdery mildew, dichlofluanid for *Botrytis*, and a copper or dithiocarbamate fungicide for downy mildew. There is now the prospect of controlling powdery mildew with benomyl, pyrazophos or thiophanate fungicides; however, although many reports have shown these systemics to be inactive against downy mildew, Malenin and Kobokhidze (1973) have reported good control of *P. viticola* with benomyl. Thiophanate methyl (Formigoni *et al.*, 1971) and benomyl can often be used without affecting fermentation processes, although Hunter and Whiteley (1976) found that fermentation of fruit treated with benomyl was delayed by 7 days.

Tropical fruit

Disease control under conditions of high temperature and rainfall is difficult to achieve with conventional fungicides, and under such conditions systemic compounds seem likely to have particular advantages (Byrde, 1970) despite their high cost. The latter is probably a major factor in their current extensive use as dips rather than as sprays.

The most important banana pathogens are *Mycosphaerella musicola* (*Cercospora musarum*) which causes 'Sigatoka' leaf spot. *Gloeosporium*

Table 11.IV.8 Control of diseases of tropical fruits

Host	Causal fungus	Disease	Compound	References
Banana	*Colletotrichum musae*, *Fusarium* spp.	Crown rot	Benomyl, thiabendazole, carbendazim, 1,5-cyanpentyl-carbamoyl (methoxy carbomyl-amino)-benzimidazole	Griffee and Burden (1973); Griffee and Pinegar (1974)
	Fusarium roseum	*Fusarium* fruit rot	Benomyl†, thiabendazole†	Shillingford (1970)
	Gloeosporium musarum	Anthracnose	Benomyl†	Allen (1970); Burden (1969); Frossard (1969, 1970a); Long (1970); Ogawa *et al.* (1968b, 1969); Shillingford (1970); Tsai *et al.* (1968); Frossard *et al.* (1973)
			Thiabendazole†	Allen (1970); Beaudoin *et al.* (1969); Burden (1969); Frossard (1969); Long (1970); Scott and Roberts (1967); Shillingford (1970); Swarts *et al.* (1969); Frossard *et al.* (1973)
			Thiophanate methyl	Ishii (1970); Frossard *et al.* (1973)
	Mycosphaerella fijiensis	Black leaf streak	Benomyl	Long (1971); Firman (1972)
	M. musicola (*Cercospora musarum*)	Sigatoka leaf spot	Benomyl	Brun and Melin (1970); Melin (1970); Stover (1969); Griffee and Burden (1973); Melin (1973)
			Carbendazim	Melin (1973)
			Tridemorph	Hampel and Pommer (1971)
	Nigrospora sphaerica	'Squirter' disease	Benomyl†, benomyl and related compounds	Allen (1970); Griffee and Burden (1974)
			Thiabendazole†	Allen (1970); Rippon *et al.* (1970)
			Benomyl	Temkin-Gorodeiski and Peled (1971)
	Five fungi	Stem-end rot complex	Thiabendazole	Weinke *et al.* (1969); Temkin-Gorodeiski and Peled (1971)
Pineapple	*Ceratocystis* (*Thielaviopsis*) *paradoxa*	Fruit rot	Benomyl†	Frossard (1970b); Ogawa *et al.* (1969)
			Thiabendazole†	Frossard (1968, 1970b)
		Butt rot	Benomyl	Rohrbach and Apt (1971)
Mango	*Colletotrichum gloeosporoides*	Anthracnose	Benomyl†	Jacobs *et al.* (1973)
			Benomyl	McMillan (1973)
	Oidium sp.	Powdery mildew	Benomyl	McMillan (1973)

† As post-harvest dip

musarum which causes anthracnose (stem-end rot, black end), and *Fusarium oxysporum* f. *cubense* which causes the 'Panama' wilt disease. No report appears to exist of the attempted control of the latter disease with a systemic fungicide. Copper and/or oil sprays are conventionally used against Sigatoka disease, but anthracnose is difficult to control with fungicides (Wardlaw, 1961).

Table 11.IV.8 summarises the control to date of banana and pineapple diseases, and thiabendazole and benomyl have clearly undergone extensive testing. There is some evidence that, of the two, benomyl tends to be the more effective against banana anthracnose (e.g. Burden, 1969; Long, 1970). Thiabendazole was found by Beaudoin *et al.* (1969) to be effective against this disease provided it was applied within three days of inoculation. The performance of benomyl against both *M. musicola* (Brun and Melin, 1970) and *M. fijiensis* (Long, 1971) was greatly enhanced by the addition of oil. Stover (1969), in a detailed examination of the mode of action of benomyl against *M. musicola*, showed that it reduced germination, sporulation and spot formation, had translaminar activity and could be translocated from roots to leaf, whereas Baron (1971) implied downward movement from banana leaves. There was also some evidence of an increased host resistance following its use. Similar results were obtained with tridemorph, which is transported acropetally (Hampel and Pommer, 1971); it was less effective against *M. musicola in vitro*, but gave good control after application to banana leaves.

Conclusion

Even though not fully systemic in most of the fruit crops described above, these fungicides represent a marked advance in chemical disease control. Benomyl and related fungicides have controlled a wide range of fruit diseases, while triforine shows particular promise against powdery mildews. Thiabendazole, although not always quite as effective as benomyl, is particularly favoured for post-harvest use, at least partly because of long experience of its use as an anthelminthic in clinical and veterinary practice.

Resistance to these new fungicides by fruit pathogens has been reported, and may well develop further with increasing usage. On soft fruit, insensitive isolates of *Botrytis cinerea* have been recorded on raspberry and strawberry (Jarvis and Hargreaves, 1973), and on grape (Schuepp, 1974). The poor control of black currant leaf spot obtained with benomyl, carbendazim and thiophanate methyl in 1972 and 1973 (O'Riordain, 1973), fungicides previously very effective, may indicate the development of insensitivity in this pathogen. Resistance to benomyl has been reported in *Venturia inaequalis* (Wicks, 1974; Iida, 1975), *Monilinia fructicola* (Whan, 1976), *Penicillium digitatum* (Harding, 1972; Smoot and Brown, 1974) and *P. italicum* (Harding, 1972; Muirhead, 1974). Kundert (1975) considered that the emergence of resistant strains might be responsible for the decreasing effect of benzimidazole fungicides against *Podosphaera leucotricha.* (See also Chapter 9).

There remains a need for compounds capable of full systemic activity in woody plants: the ultimate feasibility of this is encouraged by the demonstration by Helton and Williams (1970) of the translocation in apple trees of the antibiotic derivative cycloheximide (actidione) semicarbazone.

The Phycomycetes, represented among fruit pathogens by *Phytophthora cactorum, P. syringae, P. fragariae, Plasmopara viticola* and *Rhizopus* spp., also present a particular challenge. It is nevertheless clear that many fruit diseases have been better controlled than hitherto by the use of the systemic fungicides now available.

Chapter 11.V

Results in practice – V. Forestry

Eugene B. Smalley

Introduction

The practical uses of new systemic fungicides in forestry have been quite limited. With the exception of Dutch elm disease (DED) major research efforts have been directed towards disease problems occurring in forest nurseries, in young stands or in Christmas tree plantations. These activities have mainly involved attempts to control foliage diseases, which for the most part are partially controlled with non-systemic fungicides. The use of suitably active systemic fungicides could drastically improve these controls while reducing the high cost of repeated applications. They might also replace some chemicals, such as the organic mercurials, which have worked very well in the past, but are now either banned or are being phased-out because of possible ecological hazards.

The practical prevention, eradication or therapy of fungus diseases of forest trees with systemic chemicals involves special problems and possibly certain advantages over similar efforts with crop plants. Since the host plant or its products are rarely eaten by humans, chemicals can be utilised which possess longer residuals than would be tolerable in crops destined for human consumption. The long-lived host-plant in the landscape is often of very high value to the owner, and funds to be expended towards control efforts might be far greater than would be feasible for crop production. In these cases it may be aesthetically of extreme importance to the owner (municipality, arborist, park manager or forester) to keep the individual tree free of disease and obvious defects. In the forest and woodland, the reverse might be true. Thus, except in the case of the intensively managed Christmas tree plantation or in the forest nursery, the amount of money that could be expended on chemical control practices might be quite low per individual tree. In the forest, except in rare situations, the main endemic fungal parasites (e.g. foliage parasites canker producers, etc.) would not normally be controlled by chemical means. Chemical control efforts might occasionally be feasible, however, when whole stands were threatened with destruction.

For these reasons, major research efforts and funds have been expended to discover 'chemotherapeutants' to control such diseases as chestnut blight (*Endothia parasitica*), white pine blister rust (*Cronartium ribicola*), oak wilt (*Ceratocystis fagacearum*) and Dutch elm disease (*Ceratocystis ulmi*). Early historical details concerning these efforts are reviewed by Wain and Carter (see Chapter 2). The present discussion covers the recent literature concerning research towards development of useful systemic fungicides for the control of forest tree diseases. Chemicals for control of tree diseases such as the powdery mildews and other problems which also affect fruit crops have not been included. Major emphasis is given to Dutch elm disease investigations where progress towards practical control has been made recently with the various benzimidazole fungicides. Much of the research, unfortunately, has been reported in rather tentative form (e.g. Information Reports, Annual Reports, Fungicide–Nematicide tests, conference proceedings, etc.). The merits of many of these individual reports are often difficult to assess with accuracy. Emphasis throughout will be given to reports of controlled studies published as scientific papers in referred journals where conditions, limitations, or reality of the experimental design can be considered in comparing conflicting results. Reports of experimental control cited in this report should not be confused with the actual recommendation of specific uses of legally registered systemic fungicides, since only a few of the systemic fungicides have as yet achieved such status in the United States and elsewhere. Tables 11.V.1 to 11.V.5 summarise the major forest tree diseases reported to be controlled by systemic fungicides. Reports of poor control or failure are not included in the tables; these are discussed and the results contrasted with those obtained with non-systemic fungicides.

Root diseases

The majority of the important root diseases of forest trees such as those caused by *Armillariella mellea*, *Fomes annosus* or the various species of *Phytophthora* have not been controlled with systemic fungicides. In fact, there are few reports of studies with systemic fungicides to control root diseases of forest trees. One notable exception to this is the attempt to control the Cylindrocladium root rot (*Cylindrocladium scoparium* and *C. floridanum*) (Table 11.V.1). This disease, which is particularly severe on species of *Pinus*, *Abies*, *Pseudotsuga*, *Juglans* and *Liriodendron*, has in the past been prevented in forest nurseries by expensive soil fumigation techniques using various formulations with methyl bromide (Cordell and Skilling, 1975). Affeltranger (1969), in comparing benomyl (methyl 1-(butylcarbamoyl)benzimidazol-2-ylcarbamate) (0·5 lb a.i./100 gal at 2 pints/sq. ft) with soil fumigation, found that the benomyl treatment was about equivalent to the application of 300 lb/acre of MC-33 (67 per cent methyl bromide + 33 per cent chloropicrin) as measured by pure culture analysis of surviving *C. scoparium* propagules in treated soils. MC-33 at 600 lb/acre was the most effective treatment in his trial.

More recently Berbee and Hadi (1977) using higher dosages found that soil treatment with benomyl protected white pine seedlings from

Table 11.V.1 Control of root diseases of forest trees

Host	Causal fungus	Disease	Compound	References
Red pine	*Cylindrocladium scoparium*	Root rot	Benomyl	Berbee and Hadi, 1976
White pine	*Cylindrocladium scoparium*	Root rot	Benomyl	Berbee and Hadi, 1976
White pine	*Cylindrocladium scoparium*	Root rot	Benomyl	Berbee and Hadi, 1976
Norway spruce	*Cylindrocladium scoparium*	Root rot	Benomyl	Berbee and Hadi, 1976
Silver buttonwood	*Cylindrocladium scoparium*	Root rot	Benomyl	Miller and McRitchie, 1973
Miscellaneous woody ornamentals	*Cylindrocladium* spp.	Root rot	Benomyl	Self and Suberatz, 1974a; Horst and Hoitink, 1968

the stem canker phase of this disease. Bioassays of pines growing in the treated soil demonstrated the mobility of benomyl (as methyl benzimidazole-2-ylcarbamate (carbendazim)) from roots to branches and needles. These authors reported that over a 2-year period benomyl root dips (8 g a.i./l) provided better control in transplanted red pine, white pine, Norway spruce and white spruce than did conventional soil fumigation procedures.

Horst and Hoitink (1968) reported complete control of Cylindrocladium blight in rooted azalea cuttings with soil drenches of benomyl. Self and Suberatz (1974a, b) and Suberatz and Self (1974) also reported excellent control of Cylindrocladium root rot with benomyl in cutting beds of various woody ornamentals. However, they indicated that such treatments often favoured the subsequent development of *Pestalotia* and *Alternaria* rots. Good control of root rot disease in potted silver buttonwood trees (*Conocarpus erectus* var. *sericeus*) in South Florida using protective foliar sprays of benomyl (0·57 g a.i./l) was reported by Miller and McRitchie (1973). Soil treatments also provided moderate control of the cutting bed phase of the disease; cuttings soaked in benomyl suspensions, however, were not protected.

Foliage diseases

Many of the major foliage diseases of trees are particularly severe in Christmas tree plantations or forest nurseries. Such situations lend themselves particularly well to chemical control procedures involving soil applications or foliar sprays. Table 11.V.2 summarises the major foliage diseases of forest trees reported to be controlled by systemic fungicides.

Several of the major foliar diseases of conifers have been controlled in the nursery by appropriate systemic fungicide treatment. Nicholls (1973) found that cycloheximide (β-(2-(3,5-dimethyl-2-oxocyclohexyl)-2-hydroxyethyl)glutarimide) and benomyl (along with maneb (manganese ethylenebisdithiocarbamate) and chloroneb (1,4-dichloro-2,5-dimethoxybenzene)) were effective in controlling *Lophodermium pinastri* on 2-0 red pine seedlings in nursery beds. He observed that cycloheximide and benomyl were more effective in protecting the foliage than in eradication of the fungus after infection had occurred. In their studies using naturally infected 5- to 7-year-old plantation-grown Scotch pine trees in northern Ohio, Powell and Leben (1972), although reducing the disease severity with thiophanate methyl (1,2-di-(3-methoxycarbonyl-2-thioureido) benzene), thiabendazole (2-(thiazol-4-yl)benzimidazole) and triarimol (2,4-dichloro)-α-pyrimidin-5-yl-benzhydrol) sprays, failed to obtain improved control over the registered maneb fungicide. In studies in South Carolina, Witcher *et al.* (1975) reported that their results with benomyl were slightly less effective than those obtained with maneb.

Table 11.V.2 Control of foliage diseases of forest trees

Host	Causal fungus	Disease	Compound	References
Red pine	*Lophodermium pinastri*	Needle cast	Cycloheximide	Nicholls, 1973
Scotch pine	*Scirrhia acicola*	Brown spot	Benomyl	Willis, 1971, 1972
	Lophodermium pinastri	Needle cast	Thiophanate methyl	Powell and Leben, 1972; Witcher *et al.*, 1975
			Thiabendazole	
			Triarimol	
Long leaf pine	*Scirrhia acicola*	Brown spot	Cycloheximide	Snow *et al.*, 1964
			Benomyl	Kais, 1975
Austrian pine	*Dothistroma pini*	Dothistroma Needle blight	Benomyl	Epstein, 1970
Monterey pine	*Dothistroma pini*	Dothistroma Needle blight	Benomyl	Gilmore and Vanner, 1971, 1972
Jack pine	*Various fungi*	Storage moulds	Thiophanate methyl	Carlson, 1972
			Benomyl	
Blue spruce	*Rhizosphaera kalkhoffi*	Needle cast	Benomyl	Skilling and Waddell, 1975
Grand fir	*Pucciniastrum goeppertianium*	Needle rust	Oxycarboxin	Davison, 1969
Douglas fir	*Rhabdocline pseudotsugae*	Needle cast	Cycloheximide	Weir, 1963
	Phaeocryptopus gaeumanni	Swiss needle cast	Benomyl	Vanner and Hood, 1974
			Triforine	
Arizona cypress	*Phomopsis juniperovora*	Phomopsis blight	Benomyl	Smyly and Filer, 1973
	Cercospora sequoiae	Arizona cypress blight		Gill, 1973
Oriental arborvitae	*Cercospora sequoiae*	Arborvitae blight	Benomyl	Gill, 1973; Witcher *et al.*, 1973
Eastern red cedar	*Phomopsis juniperovora*	Phomopsis blight	Thiophanate methyl	Otta, 1973, 1974
			Benomyl	Otta, 1973, 1974; Witcher *et al.*, 1973
Juniperus spp.	*Phomopsis juniperovora*	Phomopsis blight	Benomyl	Gill, 1974; Morrison, 1972
Giant Sequoia	*Botrytis cinerea*	Storage rot	Benomyl	Smith *et al.*, 1973
Balsam poplar	*Septoria musiva*	Leaf spot	Benomyl	Droron and Kusch, 1974
	Marssonina populi	Leaf spot		
Big-tooth aspen	*Marssonina populi*	Marssonina blight	Benomyl	Dockinger, 1975
Cottonwood	*Melampsora medusae*	Leaf rust	Benomyl	Filer, 1975
Ohio buckeye	*Guignardia aesculi*	Leaf blotch	Benomyl	Neely, 1975
Sycamore	*Gnomonia platani*	Anthracnose	Benomyl	Hock, 1971; Himelick and Ricketts, 1974; Neely, 1975
			Triarimol	Hock, 1974; Hock and Berry, 1975
			Thiophanate methyl	Hock, 1971; Hock and Berry, 1975
			Thiabendazole	Hock, 1971; Himelick and Ricketts, 1974
Black walnut	*Gnomonia leptostyla*	Anthracnose	Triarimol	Berry and Hock, 1975

In early studies with the systemic antibiotic cycloheximide semicarbazone, Snow *et al.* (1964) reported that one spray application gave significant control of the Brown Spot disease (*Scirrhia acicola*) on small long-leaf pine seedlings. However, no residual action was evident 3 to 6 months after treatment, since developing needles were not protected. In these trials at the concentrations used, Bordeaux mixture was more effective than cycloheximide. Willis (1971 and 1972) reported excellent control of this disease with a single application of benomyl in spray trials on heavily infected Scotch pine in a Kansas plantation. Two applications of Bordeaux mixture, however, gave superior results. Kais (1975) reported control of *Scirrhia acicola* on long-leaf pine seedlings with benomyl soil drenches proving almost equal to Bordeaux mixture sprays. Foliar applications of benomyl, however, were ineffective.

Control of Dothistroma needle blight of pines (*Dothistroma pini*) has been achieved in seedling beds, shelter belts and Christmas tree plantings using Bordeaux mixture applied twice in the growing season (Peterson, 1975a). Gilmour and Vanner (1972) reported significant control of the disease in *Pinus radiata* seedlings with benomyl sprays. However, cuprous oxide fungicides in various formulations yielded significantly better control. In tests for control of this disease on 15-year-old Austrian pines, Epstein (1970) reported that benomyl with polyvinyl chloride added provided control equal to results obtained with copper fungicides. Other foliar diseases of conifers have also been controlled by systemic fungicides. As early as 1963, Weir (1963) reported temporary control of *Rhabdocline pseudotsugae* on Douglas fir with foliar application of cycloheximide. Protection was not prolonged as might have been expected with systemic action, and phytotoxicity resulted in heavy loss of needles. The Swiss needle cast of Douglas fir (*Phaeocryptopus gaeumanni*) was controlled equally well in potted 2-year-old seedlings by 2 sprays of benomyl or triforine (1,4-di(2,2,2-trichloro-1-formamido-ethyl)piperazine) (Hood, 1973; Vanner and Hood, 1974). Results were significantly better than from copper fungicides used in the same trial.

Skilling and Waddell (1975) reported excellent control of Rhizosphaera needle cast (*Rhizosphaera kalkhoffii*) in a blue spruce Christmas tree plantation with two spray applications of either benomyl, cycloheximide or chlorothalonil. The authors considered benomyl slightly less effective than Bordeaux mixture but suggested that it would be an adequate commercial substitute. *Pucciniastrum goeppertianium*, the cause of needle rust of *Abies grandis*, was prevented in Christmas tree plantings in western Washington by foliar applications of oxycarboxin (2,3-dihydro-6-methyl-5-phenylcarbamoyl-1,4-oxathiin-4,4 dioxide) Davison, 1969). Ferbam in these trials was less effective.

In forest nurseries, Phomopsis blight (*Phomopsis juniperovora*) can become a serious and destructive disease to members of the Cupressaceae. According to Peterson (1975), epidemics in nurseries in the Great Plains (US) have resulted in total losses in seedling and transplant beds. In the past this disease has been controlled by weekly applications of phenyl mercury fungicides. Benomyl sprays have been reported to control Phomopsis blight on Arizona cypress (Smyly and Filer, 1973), *Juniperus* spp. (Morrison, 1972; Gill, 1974), eastern red cedar (Otta, 1973 and 1974; Witcher *et al.*, 1973) and arborvitae (Gill, 1973). Otta (1974) reported that thiophanate methyl sprays were similarly effective on eastern red cedar. He suggested that both chemicals might be satisfactory replacements for the phenyl mercury fungicides. Peterson and Davis (1973), however, reported poor control of Phomopsis blight of juniper with benomyl sprays and no control with triforine. Although benomyl is now registered in the US for controlling this disease, Peterson (1975) suggests that control may not be entirely effective under conditions of high infection.

Cercospora blight (*Cercospora sequoiae*) causes a serious needle blight of giant sequoia and red cedar seedlings in the nursery. It can also be very damaging to Arizona cypress Christmas tree plantations and to ornamental plantings of various other members of the Cupressaceae (Hodges, 1975). Periodic sprays with Bordeaux mixture have usually controlled this disease satisfactorily. Gill (1973), however, reported control of this disease on field-grown oriental arborvitae and Arizona cypress by benomyl sprays applied at 2- or 4-week intervals. Witcher *et al.* (1973) also reported good control of this disease on red cedar with eight benomyl applications in three years.

Conifer nursery stock in cold storage may occasionally be seriously damaged by various storage moulds. Carlson (1972) and Carlson and Huber (1974) reported increased planting survival of Jack pine seedlings when benomyl or thiophanate methyl sprays were applied to lifted seedlings before storage. Captan, however, was equally or slightly more effective. Botrytis storage rot (*Botrytis cinerea*) of giant sequoia was controlled by benomyl applied either as a pre-lifting spray or as a post-lifting dip (Smith *et al.*, 1973).

Several important foliage diseases of hardwoods in the nursery have been controlled by systemic fungicides. Dockinger (1975) found impressive control of Marssonina blight (*Marssonina populi*) of bigtooth aspen with benomyl foliar sprays; Berry and Hock (1975) obtained 'promising' control of black walnut anthracnose (*Marssonina juglandis*) with triarimol, and Filer (1975) achieved control of Melampsora rust (*Melampsora medusae*) of cottonwood seedlings with a single benomyl foliar application. Drorun and Kusch (1974) reported control of balsam poplar leaf spots (*Septoria musiva* and *Marssonina populi*) on 15-ft trees sprayed by back-pack mist-blower with benomyl.

Hock (1971) reported excellent control of sycamore anthracnose (*Gnomonia platani*) with thiophanate methyl and triarimol. Benomyl and thiabendazole were less effective. Later, Hock and Berry (1975) reported control of sycamore anthracnose with benomyl, thiophanate methyl and triarimol on nursery seedlings. On large trees sprayed with triforine, thiophanate methyl and triarimol Himelick and Neely (1972) obtained poor results as compared with conventional organomercury sprays. In their trials on large nursery trees, Davis and Peterson (1973) failed to control sycamore anthracnose with mist-blown sprays of triforine. Himelick and Ricketts (1974) reported that foliar sprays with benomyl, thiabendazole, triforine and thiophanate methyl, while reducing leaf infections, were not practically effective in controlling the disease. In other trials the same year, however, these authors reported significant reductions in the amount of leaf infections with thiabendazole sprays. Neely (1975) reported that the resulting control of the shoot blight stage of sycamore anthracnose by soil treatments with benomyl was as effective as foliar sprays with Bordeaux mixture or phenyl mercuric acetate. In these same studies he also reported almost complete control of leaf blotch of Ohio buckeye (*Guignardia aesculi*) with similar soil applications of benomyl.

Stem diseases and trunk cankers

Table 11.V.3 summarises the various reported controls of stem diseases and trunk cankers of forest trees with systemic fungicides. Fusiform rust (*Cronartium fusiforme*), the most serious nursery disease of loblolly pine in the United States, has been controlled by foliar sprays of carboxin (2,3-dihydro-6-methyl-5-phenylcarbamoyl-1,4-oxathiin) and oxycarboxin (Rowan, 1972). In these trials, however, ferbam was equally effective in control and benomyl was less effective. In greenhouse trials, Hare (1973) partially controlled the disease on inoculated slash pine seedlings with soil drenches of benomyl, oxycarboxin and triforine. Growth of the active galls was also restricted by benomyl and oxycarboxin. The author suggested that it might be possible to use these systemics to eradicate galls on pines in seed orchards and in ornamental plantings.

In early work on conifer rusts, Hinds and Peterson (1966) reported that direct canker application of cycloheximide at 200 ppm temporarily reduced canker fruiting of western gall rust (*Peridermium harknessii*), but no evidence of systemic action was observed. This work was preceded by a series of investigations on the control of white pine blister rust (*Cronartium ribicola*) with antibiotics. Beginning about 1957, Moss (1957, 1958, 1961) reported that cycloheximide killed 83 per cent of the blister rust cankers on pole-sized western white pine if, prior to treatment, the cankers had been excised by cutting back from the inside margin of surface discoloration to expose the rust. Lemin *et al.* (1960) established the systemic nature of cycloheximide, and reported that this antibiotic is absorbed by, persists in and is translocated upward in pole-size western white pine. They found that it persists for 2 years in trunk bark treated by the 'basal stem' method and that it can be recovered in the needles in such trees. Tritium labelling studies confirmed the translocation and persistence of this compound in basal stem treated eastern white pine (Lemin and Thomas, 1961).

Table 11.V.3 Control of stem diseases and trunk cankers of forest trees

Host	Causal fungus	Disease	Compound	References
Lodge pole pine	*Endocronartium harknessii*	Western gall rust	Cycloheximide	Hinds and Peterson, 1966
White pine	*Cronartium ribicola*	Blister rust	Cycloheximide	Moss, 1957, 1958; Lemin *et al.*, 1969; Lemin and Thomas 1961; Phelps and Weber, 1966, 1967, 1969, 1970
Slash pine	*Cronartium fusiforme*	Fusiform rust	Benomyl Oxycarboxin Carboxin Triforine	Hare, 1973
Loblolly pine	*Cronartium fusiforme*	Fusiform rust	Carboxin Triforine	Rowan, 1972
Portuguese cypress	*Monochaetia unicornis*	Graft failure	Benomyl	Gibson and Howard, 1969
Arizona cypress	*Monochaetia unicornis*	Monochaetia canker	Benomyl	Witcher *et al.*, 1973
Aspen	*Hypoxylon pruinatum*	Hypoxylon canker	Cycloheximide	Brown and Silverborg, 1970
Poplar hybrids	*Septoria* sp. *Marssonina* sp.	Cutting bed cankers	Benomyl Thiophanate methyl	Carlson, 1973
American chestnut	*Endothia parasitica*	Chestnut blight	Benomyl Carbendazim-HCl	Anagnostakis and Jaynes, 1971 Jaynes and Van Alfen, 1974
Sycamore	*Ceratocystis fimbriata*	Canker stain	Benomyl	Davis and Peterson, 1975

Investigators in other parts of the United States obtained less dramatic results. Powers and Stegall (1965) in studies in North Carolina found no measurable differences in canker development between cycloheximide-treated cankers and untreated or fuel-oil-treated controls. Similar results were reported by Leaphart and Wicher (1968). Working with eastern white pine in the Lake States, Phelps and Weber (1966, 1967, 1969) reported that cycloheximide suppressed canker growth and activity when applied directly to cankers as a basal stem treatment on scarified cankers, but found that it had no effect on other cankers above the treated area. They found that, to be successful, treatments had initially to kill the host tissue. In summarising, Phelps and Weber (1970) found that cycloheximide gave best results by direct chemical applications to scarified cankers. It did not suppress other cankers above the treated cankers although upward translocations of the chemical could be demonstrated. Apparently dose levels in the upper tree parts were not high enough to control the disease. They indicated that none of the formulations tested killed the blister rust mycelium without first killing the host tissue. Reports of control of white pine blister rust with the newer systemic fungicides have so far not been forthcoming.

In other conifer canker diseases, Witcher *et al.* (1973) reported the control of Monochaetia canker (*Monochaetia unicornis*) of Arizona cypress seedlings by eight foliar applications of benomyl in 3 years. In Kenya, Gibson and Howard (1969) controlled Monochaetia-incited graft failure of Portuguese cypress. This was achieved with benomyl sprays applied to stock plants 24 h before grafting, which suggested systemic action in preventing graft failure.

A few stem or canker diseases of hardwoods have also been controlled by systemic fungicides. Direct canker applications of cycloheximide in fuel oil arrested Hypoxylon canker (*Hypoxylon pruinatum*) activity for 5 years (Brown and Silverborg, 1970). Tree survival increased with increased concentrations of the chemical up to 200 ppm. However, Hinds and Peterson (1966) failed to control this disease with cycloheximide sprays to lower trunk cankers. Carlson (1973) reported excellent control of cutting bed cankers of poplar hybrids (var Brooks No. 6) caused by *Septoria* and *Marssonina* spp. with spray applications of benomyl and thiophanate methyl. Davis and Peterson (1975) prevented the growth of the sycamore canker strain pathogen (*Ceratocystis fimbriata* f. *platani* as well as *Verticillium dahliae* and *Cytospora cincta*) by incorporation of benomyl at 50 ppm into an asphalt-base tree-wound dressing. The wound dressing had no toxic effect on callus formation.

Anagnostakis and Jaynes (1971) reported that benomyl applied as a soil drench at the rate of 175 or 375 lb a.i./acre to 2-year-old American chestnut trees limited the growth of *Endothia parasitica* cankers. In later studies Jaynes and Van Alfen (1974) found that trunk injection of 6-year-old chestnuts with methyl benzimidazol-2-ylcarbamate hydrochloride (carbendazim-HCl) by pressure techniques resulted in significant reductions in canker growth both when treated before or after inoculation. However, total remission was only observed on one of the treated trees 9 months after treatment.

Vascular wilt diseases (other than Dutch elm disease)

Although the majority of the studies investigating the control of vascular wilt diseases of forest trees with systemic fungicides has been on Dutch elm disease, a few have concerned other diseases (Table 11.V.4). Following the early work by Beckman (1959) on elms, Geary and Kuntz (1962) reported an 82 per cent disease reduction in oak trees inoculated with *Ceratocystis fagacearum* after topical bark applications of the plant growth regulator 2,3,5-triiodobenzoic acid. Venn *et al.* (1968) found that oak wilt symptoms rarely progress in trees treated with another growth regulator, sodium-2,3,6-trichlorophenyl acetate (TCPA) injected into the tree trunks *via* Mauget capsules (J. J. Mauget Co., Burbank, California) (Heffernan, 1968). Nair *et al.* (1969) found that high doses of the cytokinin (6-benzylaminopurine) also suppressed or arrested wilt development in inoculated oak seedlings, but allowed widespread distribution of the pathogen in the xylem. In more recent investigations, Gregory *et al.* (1971, 1972) obtained excellent protection in inoculated red oaks trunk-injected with either 4 or 40 mg carbendazim-HCl. Similar treatment of diseased seedlings also resulted in some remission of symptoms. Jones *et al.* (1973) reported that injecting solubilised benomyl (carbendazim-HCl) could also effectively prevent oak wilt in inoculated mature red oaks. These authors (Gregory and Jones, 1974) also found that 'sand-grown' red oak seedlings in the greenhouse drenched with benomyl failed to develop symptoms following inoculation, and the pathogen could not be isolated from treated seedlings.

Stipes (1969b) reported that thiabendazole and benomyl applied to the phloem and xylem tissues of mimosa translocated upward in such seedlings. Stipes *et al.* (1971) and Phipps and Stipes (1972), in preliminary reports, observed that symptom frequency of mimosa wilt was reduced in 'sand-grown' mimosa seedlings treated with thiabendazole or benomyl drenches and inoculated with *Fusarium oxysporum* f. sp. *perniciosium*. In 1972, Attabhanya and Holcomb (1972) reported that benomyl-soil or sand-culture drenches at 50 to 200 ppm gave complete prevention of mimosa wilt and partial therapy. In their most recent report, Phipps and Stipes (1975) found that benomyl was more effective than thiabendazole in reducing the symptoms of mimosa wilt, but a concentration of 500 μg/ml was required for disease prevention. Both compounds failed to eradicate the parasite from inoculated roots. Very recently Attabhanya and Holcomb (1976) reported that benomyl gave complete control of mimosa wilt at concentrations from 50 to 4 000 μg a.i./g soil. Concentrations from 400 to 4 000 μg a.i./g soil, however, induced mild chlorosis, stunt and non-lethal defoliation. Thiabendazole failed to control the disease up to 200 μg a.i./g soil, but above 400 μg

Table 11.V.4 Control of vascular wilt diseases (other than Dutch elm disease)

Host	Causal fungus	Disease	Compound	References
Red oak	*Ceratocystis fagacearum*	Oak wilt	2,3,5-Triiodobenzoic acid	Geary and Kuntz, 1962
			Sodium 2,3,6-trichlorophenylacetate	Venn *et al.*, 1968
			6-Benzyl amino purine	Nair *et al.*, 1969
			Benomyl	Jones *et al.*, 1973; Gregory *et al.*, 1972; Gregory and Jones, 1974
Maple	*Verticillium dahliae*	Verticillium wilt	Carbendazim-HCl	Sinclair, 1975
Oaks (various)	*Cephalosporium diospyri*	Oak decline Cephalosporium wilt	Benomyl	Van Arsdel *et al.*, 1972; Van Arsdel and Bush, 1972
Mimosa	*Fusarium oxysporum* f. sp. *perniciosum*	Mimosa wilt	Benomyl	Stipes, 1969b, Stipes *et al.*, 1971; Phipps and Stipes, 1972, 1975; Attabhanya and Holcomb, 1972, 1976
			Thiabendazole	Stipes, 1969; Stipes *et al.*, 1971; Phipps and Stipes, 1972, 1975

it yielded 100 per cent control although accompanied by severe stunting. Thiophanate methyl plus ethazole (5-ethoxy-3-(trichloromethyl)-1,2,4-thiadiazole) gave 100 per cent control at 100 and 200 μg a.i./g, but was phytotoxic.

Although extensive investigations on the control of Verticillium wilt of such important crops as cotton have been carried out (Erwin, 1976), few studies with systemic fungicides have been reported with this parasite on forest trees in spite of the widespread occurrence of the diseases. Sinclair (1975) in preliminary greenhouse and nursery studies reported that carbendazim-HCl or benomyl applied to the soil prevented foliar symptoms in potted infected *Acer platanoides* trees and *A. rubrum* nursery trees. In oak decline, a problem related to the vascular wilts caused by *Cepholosporium diospyri*, Van Arsdel *et al.* (1972a, b) reported successful treatment of infected oaks with benomyl dissolved in dimethyl formamide, and applied in various ways. Treatments, however, were apparently quite phytotoxic.

Dutch elm disease

The Pesticide Regulation Division, United States Environmental Protection Agency (EPA), in March 1972 gave its approval to E. I. du Pont de Nemours and Co. for label changes on its systemic fungicide benomyl to allow its use as an aid in controlling Dutch elm disease (Anon, 1973c). The approved label suggested that benomyl suspensions be applied to elm trees either as mist-blown foliar spray or by trunk injections using Mauget-type trunk injector needles (Heffernan, 1968) with non-pressurised 65 ml capacity cups. In 1973, this label was further amended to allow the alternative use of hydraulic spray equipment as well as improved techniques for trunk injection using pressure devices (Anon, 1975). Highly accurate figures are not available concerning the practical commercial success of these Dutch elm disease control efforts in the US during the past 4 years, and rather conflicting opinions have been forthcoming concerning the usefulness of benomyl in Dutch elm disease control. In 1974, to obtain a more accurate picture of benomyl use and practice in Wisconsin, a questionnaire on this subject was mailed to 550 practising arborists in Wisconsin by the Cooperative Extension Service. With about 20 per cent of the arborists responding, of 2 735 elms treated therapeutically with benomyl in 1972 and 1973, the arborists reported that 1 859 survived with apparent remission through 1974, or about 68 per cent (Smalley, 1976a). Such levels of control on the average were economically useful and about the same as those obtained in large-scale experimental trials in municipal situations (Smalley *et al.*, 1973).

The present discussion on Dutch elm disease covers the major methods and techniques of systemic fungicide application, and the various new systemic fungicides and their various formulations now in late development stages or which have recently been incorporated into

Table 11.V.5 Control of Dutch elm disease

Application method	Compound	Experimental situation	References
Soil injections	Thiabendazole	Small, nursery-grown American elms	Stipes, 1973
	Thiabendazole	Mature American elms, urban situation	Epstein, 1976
	Thiophanate methyl	Mature American elms, urban situation	Epstein, 1976
	Benomyl	Container-grown American elms, greenhouse	Stipes, 1969a; Hock *et al.*, 1970
	Benomyl	Small, nursery-grown American elms	Stipes, 1973
	Benomyl	Large, nursery-grown American elms	Smalley, 1971; Biehn and Dimond, 1971; Biehn, 1972b, 1973; Neely, 1974
	Benomyl	Mature American elms, urban situation	Smalley *et al.*, 1973
Foliar sprays	Benomyl	Large, nursery-grown American elms	Biehn, 1972b; Neely, 1974
	Benomyl	Mature American elms, urban situation	Hart, 1972; Smalley *et al.*, 1973
	Carbendazim-HCl	Large, nursery-grown American elms	Neely, 1974
Bark applications	Sodium 2,3,5,6-tetrachlorobenzoate	Large, nursery-grown American elms	Beckman, 1959
	Sodium 2,3,6-trichlorophenyl acetate (TCPA)	Large, nursery-grown American elms	Smalley, 1962; Brener and Beckman, 1968
Trunk injection	TCPA	Large, nursery-grown American elms	Smalley, 1962
	TCPA	Mature American elms, urban situation	Smalley, 1967
	Benomyl	Mature American elms, urban situation	Smalley *et al.*, 1973; Holmes, 1974; Van Alfen and Walton, 1974
	Carbendaxim-(lactic acid)	Mature American elms, urban situations	Neely *et al.*, 1973
	Carbendazim-(lactic acid)	Mature English elms, urban situations	Gibbs and Clifford, 1974; Gibbs and Dickinson, 1975
	Carbendazim-HCl	Medium-sized nursery-grown American elms	King and Campana, 1973
	Carbendazim-HCl	Large, nursery-grown elms, *U.* × *Hollandica* '*Belgica*'	Gibbs and Clifford, 1974; Gibbs and Dickinson, 1975
	Carbendazim-HCl	Mature American elms, urban and forest situations	Gregory and Jones, 1975; Campana *et al.*, 1973; Prasad, 1974a, 1974b; Campana, 1975

Table 11.V.5—*Continued*

Application method	Compound	Experimental situation	References
Trunk injection	Carbendazim-$H_2PO_2 \cdot H_2O$ Thiabendazole $H_2PO_2 \cdot H_2O$ M & B 21914	Young, nursery-grown elms *U.* × *Hollandica* *'Vegeta'*	Gibbs and Dickinson, 1975
Root injection	Carbendazim-H_3PO_4	Mature American elms, urban and forest situations	Kondo and Huntley, 1973; Kondo *et al.*, 1973

municipal practice for protection or therapy of elms against Dutch elm disease (Table 11.V.5).

Soil applications

In the spring of 1967, the Du Pont Company made available to Dutch elm disease investigators small quantities of a new systemic fungicide coded No. 1991 and later named Benlate. Company representatives suggested that for *C. ulmi* control it might most usefully be applied to the soil for root uptake. Most early studies in the greenhouse and nursery made use of this application route. Merck's similar systemic fungicide, thiabendazole, also became available at this time and was included in some of the early trials. In a short report without detail, Stipes (1969) reported that *C. ulmi* could not be isolated from stem-inoculated 2-year-old greenhouse elms growing in soil–sand mixture drenched with 1 000 or 4 000 ppm of benomyl. Thiabendazole at the same concentrations proved to be less active. Hock *et al.* (1970a) observed that benomyl (or its active derivative) was readily absorbed by potted elm seedlings growing in sand, and fungicidal activity could be detected by bioassay procedures in all parts of the treated plants. Forty-three per cent of the untreated plants inoculated with *C. ulmi* developed symptoms compared to 1·7 per cent in the treated trees. They also observed that *C. ulmi* could not be isolated from treated seedlings.

Using large inoculated nursery elms, Smalley (1971) reported that trees treated in 1967 at the rate of 328 kg benomyl/hectare in the root zone developed 2 per cent foliar symptoms as contrasted with 42 per cent in controls, and none of the treated trees died the following season. Thiabendazole applied at 183 kg/ha in the same study failed to affect the course of the disease. In 1968 trials, inoculated elms treated with 308 kg benomyl/ha developed 31 per cent foliar symptoms *v.* 72 per cent in controls, but even at dose levels as high as 412 kg a.i./ha thiabendazole failed to reduce foliar symptoms. In a large replicated dose–response study begun in 1969, Smalley (1971) reported that foliar symptoms were not reduced until dose levels reached 224 kg benomyl/ha. At 336 kg benomyl/ha (the highest level in the study), inoculated treated trees developed 21 per cent foliar symptoms *v.* 54 per cent in controls. In similar trials in 1969, Biehn and Dimond (1971) reported that benomyl applied at 291 kg a.i./ha resulted in 9 per cent foliar symptoms in inoculated nursery elms *v.* 45 per cent in controls. Using a higher dose in 1970 (454 kg a.i./ha), their treated elms developed less than 3 per cent foliar symptoms *v.* 83 per cent in controls. Elms in these treatment plots inoculated again in 1971 and 1972 developed, respectively, 3·1 per cent and 6·2 per cent foliar symptoms compared to 70 per cent and 92 per cent in new controls. Biehn (1973) concluded that a single benomyl soil application could provide protection against DED in nursery elms for at least three consecutive seasons.

In studies in Illinois begun in 1970 and carried through 1973, 34 per cent of the large inoculated nursery elms treated with benomyl at dose levels of 180 kg a.i./ha died *v.* 63·6 per cent of the controls (Neely, 1974). At higher doses (360 kg a.i./ha), 6·5 per cent of the treated trees died *v.* 51 per cent of controls during the three seasons of the study. Holmes (1975) also reported successful prevention of DED with benomyl soil applications, but failed to obtain second-year control.

A few of the reported investigations on soil treatment with systemic fungicides have been less successful. Stipes and Weinke (1972) obtained only slight reduction in disease in small nursery elms treated at very high dose levels (up to 4 000 kg a.i./ha) of benomyl. Somewhat better results were obtained with thiabendazole (20 per cent foliar symptoms *v.* 80 per cent in controls). In larger nursery elms, Stipes (1972a) also failed to control the disease with dose levels as high as 20 000 kg thiabendazole/hectare. The following year, however, using small nursery elms, Stipes (1973) reported 4, 23 and 76 per cent foliar symptoms in inoculated elms treated respectively with 1 962 kg/ha benomyl, thiabendazole or water. The addition of 1·0 per cent Tween 80 reduced the effectiveness of both materials. In his studies, Stipes inoculated trees in the lower trunk. In studies yielding poor or no control, treatments were made at or shortly before symptoms appeared. His successful results were obtained as preventive soil applications (Stipes, 1972b). Hock and Schreiber (1971) also reported that benomyl failed to reduce symptoms in spite of soil applications of 343 kg a.i./ha. These authors also inoculated their test trees in the lower trunk shortly after treatment (7 day). It seemed likely in both cases that the trials failed simply because of the massive infections resulting from lower trunk inoculation which allowed fungus colonisation of the active xylem prior to any uptake and distribution of benomyl (carbendazim) in fungicidal concentrations. The expectation of bringing dead or dying branches back to life seems a little unrealistic, even with the 'ideal' systemic fungicide. Simulation of natural branch inoculation by the vector *Scolytis multistratus* might have been more realistic.

Only a few studies involving soil applications have utilised mature elms in urban situations. Smalley *et al.* (1973) reported that therapeutic treatment of early stage *C. ulmi* infections in municipal elms by pressure soil applications of benomyl were surprisingly effective and two-thirds of the treated elms survived with complete disease remission through the season following treatment. Holmes (1974) treated sixty-one large street and park elms preventively with benomyl soil applications, while leaving fifty other similar elms untreated as controls. Although similar numbers of new natural infections developed in each group, symptoms were milder (5 per cent foliar symptoms) in the treated than in the untreated (45 to 85 per cent foliar symptoms).

All of the successful benomyl soil application studies have indicated that to obtain high levels of prevention or therapy dose levels on the order of 400 kg a.i./ha or more are required. Soil types, rainfall and other factors, of course, will account for variations in this requirement. Attempts to obtain satisfactory control at reduced doses by the addition of surfactants (Stipes, 1973) or by treating only restricted portions of the root area (Smalley, 1976a) have not been very successful. Recently, Epstein (1977) succeeded in reducing the incidence of new infection significantly by soil application of benomyl, thiabendazole and thiophanate methyl using restricted site 'feeder stations' and 'feeder collars'. In these techniques, the systemic fungicide was applied periodically to specially prepared locations, spaced randomly in the root zone (feeder stations) or applied in evacuated sand-filled areas at the base of the trunk (feeder collars). Newly developing feeder roots in these locations then contacted and absorbed the fungicide.

Foliar and branch sprays

Biehn (1972b) reported that foliar sprays of benomyl (8 g/1) plus Biofilm provided substantial protection of large nursery elms against infection by *C. ulmi*. He reported that 'acidified' Benlate (carbendazim-HCl) sprays were not as effective. Stipes (1972a, b) failed to revive small basally inoculated nursery elms treated with single foliar sprays of benomyl, thiabendazole or thiophanate methyl at the time most elms were exhibiting incipient or moderate wilt symptoms.

Hart (1972) found that of seventy-one mature elm trees sprayed protectively with benomyl (applied at the rate of 4 lb benomyl/100 gal water by mist blower) in late May, 1970, only one tree developed Dutch elm disease and none of forty-four trees treated in 1971 became diseased. In 1970, 14 per cent of his untreated controls became infected with *C. ulmi* and 11 per cent in 1971.

Smalley *et al.* (1973) reported that bioassay of xylem segments from benomyl-sprayed greenhouse elms indicated high antifungal activity from all spray concentrations greater than 1·1 g/l. Various additives did not greatly improve this systemic antifungal activity. Of 763 mature elms sprayed protectively with mist-blown sprays at 4·8 g/l in 1971 and 1972, four elms developed new Dutch elm disease (0·52 per cent), while in 838 adjacent unsprayed controls, twenty-four elms became infected (2·9 per cent). He concluded that although bioassays of branches from selected municipally sprayed elms in these trials indicated erratic distribution of active material in the xylem, practical protection on a mass basis was achieved. He suggested that large hydraulic sprayers which can achieve better wetting of the small branches (the actual site of uptake), rather than mist blowers, would be the better choice of equipment for municipal use. Neeley (1974) using large nursery elms also achieved moderate therapy and prevention of DED with benomyl foliar sprays applied alone or in various formulation (acidified, in oil or as carbendazim-HCl).

Trunk and root injections

Beckman (1958) reduced the incidence of Dutch elm disease with bark applications of the plant growth regulator 2,3,5,6-tetrachlorobenzoic acid. He suggested that inhibition of spring wood development might account for the observed alterations in susceptibility (Beckman and Howard, 1957; Beckman, 1958, 1959). In field trials in Wisconsin Smalley (1962) confirmed his results, but reported that severe foliar malformations which developed following treatment obviously precluded the use of the chemical in the practical prevention of Dutch elm disease. Smalley (1962) also reported that another plant growth regulator sodium-2,3,6-trichlorophenylacetate (TCPA) gave nearly complete protection from Dutch elm disease when injected at low concentrations (5 to 12 ppm) and high volumes into large nursery elms prior to inoculation. Foliar malformations were not observed at those dose levels giving optimal protection. Topical bark applications of this material yielded only slight protective effects. Time of trunk injection was apparently quite important, and optimum protection was not achieved in the spring until trees had reached the full leaf stage. Bark applications, however, induced maximum effects when applied at bud break and prior to full leaf. Brener and Beckman (1968) reported similar effects.

Using the marginally effective bark application techniques, Brener and Beckman (1968) reported that TCPA treatment, while failing to control DED to a commercially accepted degree, did enhance tylose development following inoculation with *C. ulmi*. They suggested that TCPA might condition the xylem parenchyma by altering plasticity and permeability of cell membranes or the metabolic poise of cell protoplasts so that they could respond more rapidly to infection. Whether such conclusions could also be drawn when TCPA was applied by the more effective trunk injection route of application was not determined. The absence of observed foliar malformations in elms injected with high-volume, low concentrations of TCPA (Smalley, 1962) which Brener and Beckman (1968) associated with disease control might suggest the existence of other mechanisms.

No effective practical device for trunk injection of large volumes of low TCPA concentrations into mature urban elms was available in the early 1960s which might have been used to test these nursery tree control effects in municipal situations. Since preliminary studies with large elms suggested that a moderate degree of control could also be achieved by injection of small quantities of TCPA concentrates (0·6 ml of 1·5 lb a.i./gal) by Mauget injection tubes placed at 5-inch intervals around the tree trunk (Heffernan, 1968), a very large municipal trial for Dutch elm disease was carried out in Milwaukee in 1966 in cooperation with the Bureau of Forestry. TCPA was injected preventively into 62 593 elms in the City of Milwaukee, while 38 871 other elms in the same areas remained untreated and served as controls (Smalley, 1967). In the untreated areas the incidence of DED in the season of treatment climbed from 4·4 per cent (in 1965) to 16·9 per cent, while in the TCPA-treated areas incidence rose to only 6·6 per cent. These trials, while commercially effective, failed to duplicate the higher levels of control observed in the nursery elms treated with the high-volume, low-concentration trunk injections. Widespread second- and third-year bark-splitting at the sites of concentrate application precluded the recommendation of this method as a practical control procedure.

Although the major early studies with the new systemic benzimidazole fungicides made use of the soil route of application, a few attempts were made to prevent the disease by trunk applications. Smalley (1971) in 1967 failed to prevent the disease by spring trunk implantations of a benomyl slurry (0·3 g a.i./cm dia) in large nursery elms. Hock and Schreiber (1971) preventively injected benomyl at various concentrations suspended in 50 per cent ethanol (1 l/tree) directly into the base of the trunk of large nursery elms. Such treatments also failed to control the disease. Topical bark applications of carbendazim-HCl and thiabendazole in moist cheese-cloth 'bandages' likewise failed to prevent Dutch elm disease in small inoculated nursery elms (Stiples, 1972a).

In spite of these earlier failures with trunk implantation techniques and other problems of insolubility of benzimidazoles. Smalley *et al.* (1973) reported that protective treatment of 185 healthy municipal elms in 1971 with benomyl suspensions applied with Mauget trunk injectors reduced the incidence of new infection from 16·4 per cent in controls to 6·5 per cent in treated trees. In larger trials in 1972 (688 treated *v.* 619 controls), the untreated healthy elms developed over three times as many new infections as did the treated trees.

Using similar procedures in extensive therapy trials on naturally infected urban elms from 1969 to 1971, Smalley *et al.* (1973) reported moderate to highly successful disease remission in benomyl-treated trees having less than 5 per cent crown wilt at the time of treatment. These studies indicated that the most effective period for treatment in Wisconsin was after 23 June when new bark-beetle-induced branch infections were beginning to appear as foliar wilt. Early, rapidly developing infections such as occur with root graft transmission or recurrent disease from previous year infection were clearly not controlled by this method of trunk injection. In therapy trials using mature inoculated municipal elms treated in this manner three times during the year of inoculation and again the following spring, 83 per cent remained healthy while only 20 per cent of the controls survived.

The Mauget injector system introduced the fungicide directly into the large spring-wood vessels of the current season's xylem through small (5 mm dia) aluminium tubes attached to a 65-ml plastic cup. With this system, moderate systemic fungitoxic activity could be detected with bioassays for more than 1 week following treatment of large urban elms with benomyl suspensions (Smalley *et al.*, 1973). For use in mass-protection of healthy elms it had the advantage over other systems (to be discussed later) of being inexpensive while at the same time having the capability of injecting moderate to large volumes of fluid through very small holes (which healed rapidly). The system had several serious disadvantages which made it quite unpopular with the practising arborist. The success of the injections required very careful placement of the aluminium tubes in the xylem to make contact with the ring of active vessels, and good uptake was totally dependent upon the weather conditions. As a result, the time period required to complete an injection might be quite long (24 to 48 h). During this long injection period the cups attached to the tree could become an attractive target for vandalism and required costly supervision during injection. For maximum protective effects such treatments needed to be repeated several times during the growing season (Smalley *et al.*, 1973). Increased doses without retreatment could, however, be achieved by multiple-ring capsule applications.

In spite of usefulness of this technique, more soluble formulations would markedly improve the effectiveness of the systemic treatments by allowing the injection of larger volumes to achieve higher doses and better distribution. Towards this goal McWain and Gregory (1971) reported that benomyl, in varying amounts, could be solubilised in several solvents including: 91 per cent glacial acetic acid (24 g benomyl/l), 88 per cent formic acid (150 g benomyl/l), 85 per cent hot lactic acid (190 g/l) and 42·5 per cent hot lactic acid (50 g/l). They concluded that because of its solvent properties and the fact that the solubilised benomyl could be 'infinitely' diluted with water, the lactic acid preparation seemed to be the most promising product for tree injection. Smalley *et al.* (1973) confirmed the fact that it possessed the highest level of fungitoxic activity of all the formulations tested over a wide range of concentrations in his trials. They reported, however, that the lactic acid formulations were extremely phytotoxic to elm leaves even when solutions were diluted from the 50 g benomyl/l of 85 per cent lactic acid to 0·5 g/l. Gibbs and Clifford (1974) also reported that benomyl dissolved in 2 per cent lactic acid resulted in phytotoxicity to the xylem, the vascular cambium and the foliage.

Stipes (1969) attempted without success to improve benomyl and thiabendazole uptake by applications in 90 to 100 per cent dimethyl sulphoxide (DMSO) in topical bark application. Busch and Hall (1971) reported increased solubility of benomyl using Tween 20. However, Hoch and Schreiber (1971) tested twenty-three different combinations of adjuvants and benomyl added to elm seedling potting mixes without obtaining improved uptake. Smalley *et al.* (1973) reported that with agitation the addition of Tergitol NPX (10 ml/l), Tween 80 (10 ml/l) and Tween 20 (10 ml/l) to benomyl suspension increased fungitoxic activity over benomyl alone. However, no increases in activity were observed when the suspensions were allowed to stand for 72 h.

Following the early work of Staron *et al.* (1966), Buchenauer and Erwin (1971) prepared HCl solutions of thiabendazole and benomyl for foliar spray control of Verticillium wilt of cotton. Modifications of their procedure in preparation of carbendazim-HCl solutions for trunk injections have been reported by Smalley *et al.* (1973) and McWain and Gregory (1973). Kondo *et al.* (1973) reported similar techniques for preparation of the water-soluble sulphate, phosphate, nitrate and 'sodio' salts of carbendazim. They reported improved *in vitro* fungitoxicity with carbendazim-nitrate, phosphate and sulphate over benomyl or carbendazim-HCl. Similar results were reported by Janutolo and Stipes (1975). Gibbs and Clifford (1974) prepared soluble carbendazim-HCl stabilised with HCl, carbendazim-nitrate stabilised with lactic acid and carbendazim-bisulphate stabilised with $KHSO_4$. Following comparative studies they selected the carbendazim-HCl preparation for their field trials because of its low order of phytotoxicity. Gibbs and Dickinson (1975) in later studies utilised the hypophosphite salt of carbendazim which could be prepared at 22 per cent *v.* only 0·8 per cent for the phosphoric acid salt as prepared by Kondo *et al.* (1973).

The reports by Jones and Gregory (1971), Gregory *et al.* (1971) and McWain and Gregory (1971) concerning the successful high-pressure trunk injections of elms with large volumes of 'solubilised' benomyl stimulated new interest and resulted in a large amount of new chemical control research on DED. Their original apparatus consisted of an injection head attached by a small hydraulic jack with a belt to hold the head on the tree. A neoprene gasket sealed the head to the exposed sapwood surface. The rest of the system consisted of a fluid supply-hose with valve at the injector head, a solution reservoir, a pressure regulator and a supply of compressed gas. For attachment a $1\frac{3}{8}$-inch diameter hole is cut in the back to the cambium. In the centre of the exposed wood a half-inch wedge-shaped hole is cut two or three annual rings deep. Then the apparatus was connected to one of several injection sites per tree and fluid injected at 96 psi. At the time, the authors estimate that the cost of each unit would be less than $200.00. This device has apparently only been used mainly in research studies and was not marketed.

In 1975 Gregory and Jones (1975) reported an improved version of their apparatus which they described as being more convenient and efficient. The major change involved a larger solution reservoir and totally different injector head. This head was attached to a $\frac{7}{16}$-inch diameter hole drilled into the sapwood through a larger 1-inch diameter bark hole and held in place by four nails. This system bypassed the functional xylem by injecting the fluid into the underlying sapwood. Nitrogen under pressure was used as the propellent.

Himelick (1972) described a somewhat different 'high pressure' system for injecting chemicals into trees. This system made use of a hollow coarse-threaded lag bolt ($\frac{5}{8}$-inch diameter) attachment screw. To prepare the injector, the lag bolt head was cut off, a $\frac{3}{16}$-inch diameter hole drilled through the centre and a $\frac{3}{4}$-inch square nut welded on to the head end. A $\frac{1}{2} \times 2$ inch pipe nipple was then attached to the nut and welded in place. These attachment screws were placed in pre-drilled holes 8 to 10 inches apart around the trunk. A gasoline-powered hydraulic sprayer forced the injection fluid through a manifold pressure-hose system to the attachment screws at 100 psi. Himelick indicated that compressed air or nitrogen tanks could also be used as pressure sources. His injection procedure as in the 1975 Gregory and Jones model bypassed the functioning xylem to inject into the older sapwood. Himelick (1972) reported that pressure injection of fluids shortened the injection time over so-called gravity devices from days to minutes and allowed constant supervision of the equipment. Solutions applied at a convenient trunk height moved both upward into the branches and downward into the roots. This system in principle did not differ radically from that described by Gregory *et al.* (1971) and Jones and Gregory (1975) except for the more efficient injection head and the elimination of some gadgetry. Although the system has not been commercially available *in toto*, it can be readily constructed from hardware and plumbing supplies available to most municipalities and aborists. As a result this method or modifications thereof have been used in many places in the US for the commercial applications of benomyl for DED control.

Kondo (1972) described an apparatus for root injection of water-soluble chemicals. Aside from the direct application of solutions to freshly severed roots at several points around the tree, the system did not differ remarkably in plumbing design from the methods of Gregory *et al.* (1971) and Jones and Gregory (1975). Kondo (1972) claimed that the device provided uniform, high-volume coverage throughout the vascular system of the tree. Using this technique Kondo and Huntley (1973) reported that the average cost of root injections in an eighty-tree study was $92.68 per elm tree.

A simple pressure injection device was first marketed in 1972 by the Elm Research Institute (Smalley *et al.*, 1973; Van Alfen and Walton, 1974). The unit consisted of a hand-pump-operated pressure tank (modified garden sprayer) and a plastic manifold system containing a number of injection outlets. Trunk attachment in early models was made in $\frac{1}{2}$-inch holes 6 to 10 inches apart around the trunk with modified galvanised pipe nipples. Later designs used smaller injection holes. The pipe threads carefully placed in the trunk adequately sealed the injection of fluids at 30 to 50 psi. This device with its extreme simplicity, low cost, and highly effective operation has been widely sold to or duplicated by arborists in the US since 1972. A similar device called the Tree Saver has been marketed in England by Tree Conservation Ltd. It made use of specially designed 'injection lances' which held a reservoir of fluid under pressure directly over the current season's xylem to overcome the objection of bypassing the functional xylem vessels. The lance is held in place with a coarse threaded lag bolt similar to the one used by Himelick (1972). A modification of this device was used by Gibbs and Dickinson (1975) in their trunk injection studies.

Research workers using the various plumbing devices for pressure trunk injection of large volumes of solubilised systemic fungicides into elms have tended to concentrate their efforts on accumulation of bioassay data of fungal toxicity present in treated trees. Published efficacy data relative to prevention and therapy of Dutch elm disease have been slower to accumulate. Few attempts have been published on correlations between fungitoxic elm tissue and actual control of DED.

Jones and Gregory (1971) and Gregory *et al.* (1971) with their early pressure injection system found that it was not possible to pressure-inject very much benomyl in suspension into dormant trees (elms, oak and maples) in November. However, they reported that 'quantities' of solubilised benomyl could be injected with subsequent distribution within the tree.

Smalley *et al.* (1973) reported that a low order of fungitoxicity could be detected in small elm branches 4 weeks after trunk injections of benomyl solubilised in HCl-acetone (carbendazim-HCl) applied with Mauget injectors. Using the ERI pressure system, moderate antifungal activity was detected in small branches after trunk injection of 7 litres of a 1 000 ppm benomyl suspension. Activity, however, was greatest in the lower trunk, and distribution into the upper branches, as reported by Van Alfen and Walton (1974), was erratic. The greatest fungitoxic activity in the small branches was obtained by pressure trunk injection (ERI system) of HCl-acetone solubilised benomyl. In one case, 25 g solubilised benomyl was injected into a 12·2 m elm in less than 30 minutes. Uptake in larger elms was even more rapid. Bioassays indicated high antifungal activity throughout the tree's functioning xylem.

In a 2-year study with 100 mature elm trees, Prasad and Travnick (1973b) reported that carbendazim-HCl and to some extent benomyl suspensions could be injected under pressure into elms with patterns of distribution dependent upon time and physiology of the tree. Using essentially the pressure-injection technique described by Himelick (1972), they were able to inject 10 litres of carbendazim-HCl (1 000 to 16 000 ppm) in about 90 minutes into trees with 7- to 10-inch diameter at breast height (dbh) with subsequent excellent coverage (as indicated by bioassay). Actual efficacy data against Dutch elm disease was not presented.

Kondo *et al.* (1973) using the root injection system with twelve large healthy elms achieved 70 to 100 per cent coverage when injecting large volumes of carbendazim-H_3PO_4 at concentrations from 250 ppm to 1 000 ppm. They indicated that two treated elms inoculated with *C. ulmi* developed foliar symptoms in inoculated branches, but that *C. ulmi* could not be recovered from such branches.

Gregory *et al.* (1973) using carbendazim-HCl solutions in their pressure-injection system reported that injection of 24 g carbendazim-HCl/litre (325 ml/inch dbh) provided nearly complete protection from Dutch elm disease. Using similar equipment, Campana *et al.* (1973, 1975) treated naturally infected elms having visible symptoms with carbendazim-HCl and reported that treated trees had significantly less disease increase than non-treated controls in the year of treatment. However, at the end of the next season only 35 per cent of the treated trees remained alive. None of the controls survived.

Neely *et al.* (1973) pressure-injected twenty-six large naturally infected elms in Evanston, Illinois, with benomyl solubilised in lactic acid. The chemical was injected at 3 000 ppm at $\frac{1}{2}$ gal/inch diameter. Each tree had a single site infection and less than 20 per cent foliar wilt at the time of treatment. Twenty-one trees with <15 per cent wilt survived through the season of treatment with no advance in foliar wilt. By the end of the season following treatment, however, only seven trees (28 per cent) remained with no increase in symptoms.

Kondo and Huntley (1973) treated thirty-seven large healthy elms by root injection with carbendazim-H_3PO_4. Results were assessed by bioassay of the crown twigs for fungitoxicity. Coverage was variable and based on tree size and volumes applied: they postulate the dose levels required for various degrees of coverage. In eighty diseased elms treated in the same manner, all remained alive at leaf fall, and all but one leafed out normally the following spring. Four of ten untreated infected controls died during the same period. Further reports on the final outcome of this study have not apparently been published.

King and Campana (1974) pressure-injected carbendazim-HCl prophylactically and therapeutically into medium-sized nursery elms. Two of the trees receiving prophylactic treatment developed symptoms. Therapeutic treatment apparently resulted in significant reduction in foliar symptoms, although details concerning the actual numbers of trees used, etc., were not presented. In protective studies on medium-sized nursery elms treated by pressure injection with carbendazim-HCl only five of eighteen injected trees became infected while controls were severely diseased (Gibbs and Clifford, 1974). Their therapy treatments in these studies held foliar symptoms to <5 per cent in treated trees as compared to 76 per cent in untreated elms. The authors observed a decline in fungitoxicity within treated trees over time which they suggested may indicate that treatments gradually lose protective effects with the passage of time. Van Alfen and Walton (1974) reported that medium-size nursery elms pressure-injected (ERI) with carbendazim-HCl were effectively protected from DED. Treated trees developed 8 per cent crown wilt *v.* 56 per cent in controls 6 weeks after treatment. Critical second-year results were not reported.

Prasad (1974a) found that pressurised trunk injection of carbendazim-HCl in diseased landmark and historic trees in Canada suppressed DED and prolonged the life expectancy, with trees remaining in good health for at least 2 years after treatment. Costs of treatment varied from $80 to $475. In another study, field elms of moderate size were pressure-injected with carbendazim-HCl and inoculated 19 days later (Prasad, 1974b). None of the treated elms developed the disease while 85 per cent of the controls became severely infected. The author estimated the cost of treatment at $25/tree. Second-year results were not reported. Bioassays in these trials suggested that the fungitoxicant tended to concentrate in the leaves with the passage of time, while activity in the xylem declined.

Most recently, Gibbs and Dickinson (1975) and Clifford *et al.* (1976) reported that pressure-injected 'carbendazim' salts (hydrochloride and hypophosphite) gave superior results to thiabendazole hypophosphite in controlling DED. They experienced difficulty, however, in achieving good results in trees over 25 m tall. Their comprehensive studies utilised artificially inoculated or naturally infected European elms (*Ulmus × hollandica* 'Belgica', *U. × hollandica* 'vegeta', *U. carpinifolia* var. *sarniensis*, *U. carpinifolia* var. Wheatley, and *U. procera*).

Thus, the therapeutic and protective results obtained by the pressure-trunk injection of various water-soluble carbendazim salts appear extremely promising for the practical control of Dutch elm disease. These materials, given legal registration and approval, seem to offer the public much improved control methods over previously registered benomyl formulations. Although appropriate dose levels for various sized trees do not appear to be well worked out, the general low-order phytotoxicity of the benzimidazoles allows a considerable margin of safety between control level and overdose.

Several serious problems remain to be resolved, which in the long run will determine the success or failure of these materials in practice. Clearly, the carbendazim salts (as well as the benzimidazoles generally) tended to be fungistatic at low doses. As a result, various authors have observed initial remission of Dutch elm disease following treatment, but with the fungus remaining alive in the infected branches (King and Campana, 1974; Campana *et al.*, 1973). Recurrence of disease in apparently arrested cases a year or two later has been observed (Neely *et al.*, 1973; Campana, 1975). Combined with this, the more soluble carbendazim salts apparently move rapidly into the leaves following treatment and are partially lost to the tree at leaf fall (Prasad, 1973b, 1974; Prasad and Travnick, 1973b, 1974). The reduced activity in the branches in the second season may thus not be sufficient to prevent further pathogen development without repeat treatments. For really permanent therapy, the fungistatic compound in the branches must remain sufficiently high through the second season to allow complete healing of all infected areas in the tree. The fungus, once walled-off by healthy annual rings, has only a limited capability to penetrate into the new functioning xylem (Banfield, 1968).

It is also clear that strains of *C. ulmi* are naturally present which are highly tolerant to carbendazim salts (Schreiber and Townsend, 1976; Brasier and Gibbs, 1975). Schreiber and Townsend (1976) found that six to twenty-nine randomly selected isolates of *C. ulmi* from various parts of the US were capable of growing on 1 000 ppm of carbendazim-HCl incorporated into culture media. They suggested that the wide variability in tolerance to carbendazim of the isolates indicates multiple gene control for this character. The occurrence of such natural strains may explain observed remission failure of treatments in the season following treatment, or control levels less than expected in properly dosed trees as indicated by bioassay of fungitoxic levels in the branches.

For the near future, several new experimental materials now in developmental stages have appeared quite promising in preliminary studies for control of Dutch elm disease (Gibbs and Dickinson, 1975; Smalley, 1976a, b). Of these, two materials, thiabendazole hypophosphite and M & B 21914 (1-methoxycarbonyl-3-(dimethylaminoacetamidophenyl)-thiourea HCl) ('Rhodia') have yielded therapeutic results equal to or better than the carbendazim salts. Both compounds are water-soluble, can be pressure-trunk-injected in large volumes. Their close relationship to carbendazim, however, may not solve the problem of overcoming the emergence of resistant strains of the pathogen. Materials with similar biological properties are still needed whose mechanisms of action differ radically from the benzimidazoles.

In 1975, Du Pont received an experimental label from EPA to test on a limited commercial scale the water-soluble benomyl derivative carbendazim-HCl in a formulation called Lignasan. The Elm Research Institute, Harrisville, New Hampshire, at the same time and with the assistance of Du Pont also received from EPA an experimental label for testing another benomyl derivative carbendazim-H_3PO_4 called BLP. The outcome of these parallel studies is not yet complete, but one or the other of these water-soluble derivatives will probably be cleared for commercial use by EPA in 1976*. Attempts have thus far not been made to secure label clearance for soil applications of benomyl in spite of the fact that many promising studies have shown efficacy via this route of application. The spectre of ecological damage from the high benomyl doses necessary to achieve high level elm protection has no doubt influenced the lack of corporate interest in this route of application (Wright and Stringer, 1973; Stringer and Wright, 1973; Stringer and Lyons, 1974; Prasad and Moody, 1974; Black and Neely, 1975).

* Lignasan B. L. P. (carbendazim-H_3PO_4) at a low and probably unrealistic dose was cleared for commercial use by EPA on 11 May 1976.

References

Abeles, F. B. (1973) *Ethylene in plant biology*. New York and London: Academic Press, 302 pp.

Abeles, F. B., Bosshart, R. P., Forrence, L. E. and Habig, W. H. (1971) Preparation and purification of glucanase and chitinase from bean leaves. *Pl. Physiol.* **47,** 129.

Abiko, K., Kishi, K. and Yoshioka, A. (1975) Emergence of plantvax-resistant strains of chrysanthemum rust fungus *Puccinia horiana. Ann. Phytop. Soc. Japan* **41,** 100 (abs.).

Acosta, A. and Livingston, J. E. (1955) Effects of calcium sulfamate and sodium sulfanilate on small grains and on stem rust development. *Phytopathology* **45,** 503.

Adlung, K. G. and Drandarevski, C. A. (1971) The evaluation of 'CELA W524', a systemic fungicide, for the control of powdery mildew and apple scab. *Proc. 6th Br. Insectic. Fungic. Conf.* **2,** 577.

Aelbers, E. (1970) Thiophanate and thiophanate methyl, two new fungicides with systemic action. *Proc. 22nd int. Symposium for Phytopharmacy and Phytiatry, Ghent* (May 1970).

Aelbers, E. (1971) Thiophanate and thiophanate methyl, two new fungicides with systemic action. *Mededelingen Fakulteit Landb, wetensch, Gent* **36,** 126.

Affeltrauger, C. E. (1969) An evaluation of a fumigation–fungicide test for control of Cylindrocladium root rot in a North Carolina State Forest Nursery. USDA, USFA, SA, SX, PF, Div. FPC, Asheville, N.C. *Rpt No. 70-1-1.*

Aharonson, N. and Kafkafi, U. (1975) Absorption, mobility and persistence of thiabendazole and methyl 2-benzimidazolecarbamate in soils. *J. Agr. Food Chem.* **23,** 720.

Aikman, D. P. and Anderson, W. P. (1971) A quantitative investigation of a peristaltic model for phloem translocation. *Ann. Bot.* **35,** 761.

Akai, S. (1955) Chemotherapeutic application of some compounds to rice plants and outbreak of *Helminthosporium* leaf spot. *Univ. Kyoto Forsch. PflKrankh.* **5,** 45.

Akai, S. and Kato, H. (1962) Internal application of pentachlorophenol to rice plants and the outbreak of *Helminthosporium* leaf spot. *Univ. Kyto Forsch. PflKrankh.* **7,** 35.

Albersheim, P., Jones, T. M. and English, P. D. (1969) Biochemistry of the cell wall in relation to infective processes. *Ann. Rev. Phytopathol.* **7,** 171.

Albert, J. J. and Groves, A. B. (1966) Eradicant action of certain fungicides and fungicide combinations on *Venturia inaequalis. Phytopathology* **56,** 583.

Alcorn, S. M. and Ark, P. A. (1956) Movement of certain antibiotics in cuttings of *Pyracantha* and carnation. *Appl. Microbiol.* **4,** 126.

Alfen, N. K. van and Walton, G. S. (1974) Pressure injection of benomyl and methyl-2-benzimidazole carbamate hydrochloride for control of Dutch elm disease. *Phytopathology* **64,** 1231.

Allen, N. S. (1974) Endoplasmic filaments generate the motive force for rotational streaming in *Nitella. J. Cell Biol.* **63,** 270.

Allen, P. M. (1969) Ph.D. Thesis, University of Illinois, Urbana.

Allen, P. M. and Gottlich, D. (1970) Mechanism of action of the fungicide thiabendazole, 2-(4′-thiazolyl) benzimidazole. *Appl. Microbiol.* **20,** 919.

Allen, R. N. (1970) Control of black-end and squirter diseases in bananas with benzimidazole and salicylanilide compounds. *Aust. J. exp. Agric. Anim. Husb.* **10,** 490. (*Chem. Abs.* **74,** 12092x, 1971.)

Allison, D. A., Peake, R. and Tomalin, J. (1975a) Control of seed-borne diseases of winter wheat with benomyl and carbendazim-dithiocarbamate mixtures. *Proc. 8th Br. Insectic. Fungic. Conf.* **1**, p. 197.

Allison, D. A., Peake, R. and Tomalin, J. (1975b) Trials in the UK to evaluate benomyl, carbendazim and dithiocarbamate mixtures for yield response and control of Septoria, eyespot and mildew in winter wheat. *Proc. 8th Br. Insectic. Fungic. Conf.* **1**, p. 305.

Anagnostakis, S. L. and Jaynes, R. A. (1971) Inhibition of *Endothia parasitica* by benomyl in field-grown American chestnuts. *Pl. Dis. Reptr* **55**, 199.

van Andel, O. M. (1958) Investigations on plant chemotherapy. II. Influence of amino-acids on the relation plant-pathogen. *Tijdschr. PlZiekt.* **64**, 307.

van Andel, O. M. (1962a) Fluorophenylalanine as a systemic fungicide. *Nature, Lond.* **194**, 790.

van Andel, O. M. (1962b) Growth regulating effects of amino acids and dithiocarbamic acid derivatives and their possible relation with chemotherapeutic activity. *Phytopath. Z.* **45**, 66.

van Andel, O. M. (1966a). Mode of action of L-threo-β-phenylserine as a chemotherapeutant of cucumber scab. *Nature, Lond.* **211**, 326.

van Andel, O. M. (1966b) Amino acids and plant diseases. *A. Rev. Phytopath.* **4**, 349.

van Andel, O. M. (1968) Shifts in disease resistance induced by growth regulators. *Neth. J. Pl. Pathol, Suppl. 1* **74**, 113.

Andersen, A. S. and Rowell, J. B. (1960) Half-life effectiveness against stem rust of systemic chemicals in wheat seedlings. *Phytopathology* **50**, 627.

Andersen, A. S. and Rowell, J. B. (1962) Duration of protective activity in wheat seedlings of various compounds against stem rust. *Phytopathology* **52**, 909.

Anderson, H. W. and Gottlieb, D. (1952) Plant disease control with antibiotics. *Econ. Bot.* **6**, 294.

Anderson, H. W. and Nienow, I. (1947) Effect of streptomycin on higher plants. *Phytopathology* **37**, 1.

Anderson, W. P. and Reilley, E. J. (1968) A study of the exudation of excised maize roots after removal of the epidermis and outer cortex. *J. exp. Bot.* **19**, 19.

Anon (1970) Systemic fungicides against blast. *The IRRI Reporter* **6**, 1.

Anon (1971a) *Technical manual concerning the activity of tridemorph.* Badische Anilin und Soda-fabrik.

Anon (1971b) Benomyl drench effective against celery leaf spot. *Grower* **76**, 13.

Anon (1971c) Ring the changes with chemicals. *Grower* **75**, 1333.

Anon (1972) *FAO Production Year Book*, published by the Food and Agriculture Organisation.

Anon (1973a) *Ann. Rept East Malling Research Stn* p. 147.

Anon (1973b) Cauliflower, winter – Control of clubroot by fungicidal dips. *Rep. Sts. exp. Stn Jersey*, 1973, p. 20.

Anon (1973c) *Annotated Index of Registered Fungicides and Nematicides.* U.S. Environmental Protection Agency. Benomyl I-B-02-00.06.

Anon (1974) *Delsan M and Benlate for winter wheat. The technical background to the recommendations.* E. I. du Pont de Nemours (UK) Ltd.

Anon (1975a) *Approved products and their uses for farmers and growers, 1975 list.* London, Minist. Agric. Fish. Fd.

Anon (1975b) *Du Pont Agrichemicals Label Manual 1975.* E. I. du Pont de Nemours and Co. (Inc.) Biochemicals Department. Wilmington, Del., 19898. 96 pp.

Ark, P. A. and Dekker, J. (1958) Uptake of antibiotic GS1 by seeds and plants. *Phytopathology* **48**, *391.*

Ark, P. A. and Thompson, J. P. (1958) Prevention of antibiotic injury with Na-K chlorophyllin. *Pl. Dis. Reptr* **42**, 1203.

Arny, D. C. and Leben, C. (1954) Vapor action of certain mercury seed treatment materials. *Phytopathology* **44**, 380.

van Arsdel, E. P. and Bush, D. L. (1972) Uptake of dimethylformamide solutions of benomyl in sycamore, live oak and port oak. *Phytopathology* **62**, 807.

van Arsdel, E. P., Lyda, S. D. and Jones, T. W. (1972) Treatment of Cephalosporium wilt disease with a benomyl fungicide. *Phytopathology* **62**, 807.

Asakawa, M., Misato, T. and Fukunaga, K. (1963) Studies on the prevention of the phytotoxicity of blasticidin S. *Pesticide and Technique (Tokyo)* **8,** 24.

Ashida, J. (1965) Adaptation of fungi to metal toxicants. *Phytopathology* **3,** 153.

van Asshe, C. and Vanachter, A. (1970) Systemic fungicides to control fungal diseases in vegetables. *Parasitica* **26,** 117.

Attabhanya, A. and Holcomb, G. E. (1972) Benomyl controls Fusarium wilt and is translocated in mimosa seedlings. *Phytopathology* **62,** 744 (abs.).

Attabhanya, A. and Holcomb, G. E. (1976) Control of fusarium wilt of mimosa with systemic fungicides. *Pl. Dis. Reptr* **60,** 56.

Ayers, G. W. (1974) Potato seed treatment for control of *Verticillium* wilt and *Fusarium* seed piece decay. *Can. Pl. Dis. Surv.* **54,** 74.

Ayers, M. R. L. W. and Gunary, D. (1971) Thiabendazole – broad spectrum fungicide – field experience of its performance on a range of crops in the United Kingdom. *Proc. 6th Br. Insectic. Fungic. Conf.* **2,** 341.

Aytoun, R. S. C. (1956) The effects of griseofulvin on certain phytopathogenic fungi. *Ann. Bot., Lond. (N.S.)* **20,** 297.

Baker, C. J. (1975) A survey of foot rot diseases of wheat in Germany and the effect of systemic fungicides on the fungus spectrum. *Proc. 8th Br. Insectic. Fungic. Conf.* **2,** p. 405.

Baker, E. A. (1971) Chemical and physical characteristics of cuticular membranes. *Ecology of leaf-surface micro-organisms* (eds T. F. Preece and C. H. Dickinson), p. 55. New York and London: Academic Press.

Baldwin, R. E. (1970) Control of snapbean rust with systemic and non-systemic fungicides. *Phytopathology* **60,** 1013.

Banfield, W. M. (1968) Dutch elm disease recurrence and recovery in American Elm. *Phytopathol. Z.* **62,** 21.

Bant, J. H. and Storey, I. F. (1952) Hot-water treatment of celery seed in Lancashire. *Pl. Path.* **1,** 81.

Barkai-Golan, R. and Aharoni, N. (1972) Post-harvest thiabendazole treatments to prevent spoilage in stored celery. *Israel J. agric. Res.* **22,** 157.

Barkai-Golan, R., Aharoni, N., Dubitski, E. and Karadavid, R. (1974) Control of storage diseases of lettuce by fungicidal field sprays. *Res. Summ. Agric. Res. Org., Bet Dagan,* 1971–3, p. 117.

Baron, J. M. Q. and Cayuela, J. C. (1969) Trials on the effect of fungicide 1.991 on the preservation of Valencia oranges and comparative studies on the effect of other fungicides. (In Spanish.) *Ann. Inst. nac. Invest. agron.* **18,** 285. (*Rev. Pl. Path.* **49,** 2861, 1970.)

Baron, M. (1971). Dosage, migration et distribution d'un fungicide systemic (benomyl) dans les feuilles de bananier. *Fruits d'outre-mer* **26,** 643.

Bartels-Schooley, J. and MacNeill, B. H. (1971) A comparison of the modes of action of three benzimidazoles. *Phytopathology* **61,** 816.

Bartnicki-Garcia, S. and Lippman, E. (1972) Inhibition of *Mucor rouxii* by polyoxin D: Effects on chitin synthetase and morphological development. *J. Gen. Microbiol.* **71,** 301.

Bastiaansen, M. G., Pieroh, E. A. and Aelbers, E. (1974) Prothiocarb, a new fungicide to control *Phytophthora fragariae* in strawberries and *Pythium ultimum* in flowerbulbs. *Meded. Faculteit Landbouwwetenschappen Gent* **39,** 1019.

Bateman, D. F. and Millar, R. L. (1966) Pectic enzymes and tissue degradation. *Ann. Rev. Phytopathol.* **4,** 119.

Bates, A. N., Spencer, D. M. and Wain, R. L. (1963) Investigations on fungicides. VII. The antifungal activity of certain hydroxy nitroalkanes and related compounds. *Ann. appl. Biol.* **51,** 153.

Bates, J. A. R. (1975) Private communication.

Batts, C. C. V. and Elliott, C. S. (1952) Indications of effect of yellow rust on yield of wheat. *Pl. Path.* **1,** 130.

Baude, F. J., Gardiner, J. A. and Han, J. C. Y. (1973) Characterization of residues on plants following foliar spray applications of benomyl. *J. Agr. Food Chem.* **21,** 1084.

Baude, J. F., Peace, H. L. and Holt, R. F. (1974) Fate of benomyl on field soil and turf. *J. Agr. Food Chem.* **22,** 413.

Beattie, B. B. and Outbred, N. L. (1970) Benzimidazole derivatives as post-harvest fungicides to control rotting of pears, cherries and apricots. *Aust. J. exp. Agric. Anim. Husb.* **10,** 65. (*Rev. Pl. Path.* **50,** 1301, 1971.)

Beaudoin, C., Champion, J. and Mallessard, R. (1969) Essais de traitements des Bananes au thiabendazole. *Fruits d'outre-mer* **24,** 89. (*Rev. appl. Mycol.* **48,** 2497, 1969.)

Bebbington, R. M., Brooks, D. H., Geoghegan, M. J. and Snell, B. K. (1969) Ethirimol: a new systemic fungicide for the control of cereal powdery mildews. *Chem. Ind.* 1512.

Beckman, C. H. (1958) Growth inhibition as a mechanism in Dutch elm disease therapy. *Phytopathology* **48,** 172.

Beckman, C. H. (1959) Dutch elm disease control with polychlorobenzoic acid. *Phytopathology* **49,** 227.

Beckman, C. H. (1964) Host responses to vascular infection. *Ann. Rev. Phytopathol.* **2,** 231.

Beckman, C. H. (1969a) The mechanics of gel formation by swelling of simulated plant cell wall membranes and perforation plates of banana root vessels. *Phytopathology* **59,** 837.

Beckman, C. H. (1969b) Plasticizing of walls and gel induction in banana root vessels, infected with *Fusarium oxysporum. Phytopathology* **59,** 1477.

Beckman, C. H., Elgersma, D. M. and McHardy, W. E. (1972) The localization of fusarial infections in the vascular tissue of single-dominant-gene resistant tomatoes. *Phytopathology* **62,** 1265.

Beckman, C. H., Halmos, S. and Mace, M. E. (1962) The interaction of host, pathogen and soil temperature in relation to susceptibility to Fusarium wilt of banana. *Phytopathology* **52,** 134.

Beckman, C. H. and Howard, F. L. (1957) Chemical inhibition of springwood development in relation to infection and symptoms of Dutch elm disease. *Phytopathology* **47,** 3.

Beckman, C. H., Kuntz, J. E., Riker, A. J. and Berbee, J. G. (1953) Host responses associated with the development of oak wilt. *Phytopathology* **43,** 448.

Beckman, C. H., Mace, M. E., Halmos, S. and McGahan, M. W. (1961) Physical barriers associated with resistance in *Fusarium* wilt of bananas. *Phytopathology* **51,** 507.

Beckman, C. H., Mueller, W. C. and Mace, M. E. (1974) The stabilization of artificial and natural cell wall membranes by phenolic infusion and its relation to wilt disease resistance. *Phytopathology* **64,** 1214.

Beckman, C. H., Stessel, G. J. and Howard, F. L. (1969) *Verticillium* spp. and associated fungi from certified potato seed tubers. *Pl. Dis. Reptr* **53,** 771.

Beckman, C. H. and Zaroogian, G. E. (1967) Origin and composition of vascular gel in infected banana roots. *Phytopathology* **57,** 11.

Ben-Arie, R. (1975). Benzimidazole penetration, distribution and persistence in post-harvest-treated pears. *Phytopathology* **65,** 1 187.

Ben-Arie, R. and Guelfat-Reich, S. (1973). Preharvest and postharvest applications of benzimidazole for control of storage decay of pears. *Hortscience* **8,** 181.

Ben-Arie, R. and Guelfat-Reich, S. (1975). Stimulation of internal breakdown, respiration and ethylene evolution of Spadona pears with benzimidazole fungicides. *Can. J. Pl. Sci.* **55,** 593.

Ben-Aziz, A. and Aharonson, N. (1974) Dynamics of uptake, translocation and disappearance of thiabendazole and methyl-2-benzimidazolecarbamate in pepper and tomato plants. *Pestic. Biochem. Physiol.* **4,** 120.

Ben-Aziz, A., Shabi, E. and Aharonson, N. (1974). Persistence and distribution of three benzimidazole fungicides in pear trees sprayed for the control of *Venturia pirina. Phytoparasitica* **2,** 117.

Bennett, M. (1962) An approach to the chemotherapy of silver leaf disease (*Stereum purpureum* (Fr.) Fr.) of plum trees. *Ann. appl. Biol.* **50,** 515.

Bennett, M. (1971) Effect of fungicides on inoculum potential of apple canker disease (*Nectria galligena*). *Proc. 6th Br. Insectic. Fungic. Conf.* **1,** 98.

Bent, K. J. (1967) Vapour action of fungicides against powdery mildews. *Ann. appl. Biol.* **60,** 251.

Bent, K. J. (1970) Fungitoxic action of dimethirimol and ethirimol. *Ann. appl. Biol.* **66,** 103.

Bent, K. J., Cole, A. M., Turner, J. A. W. and Woolner, M. (1971) Resistance of cucumber powdery mildew to dimethirimol. *Proc. 6th Br. Insectic. Fungic. Conf.* **1,** 274.

Ben-Yephet, Y., Henis, Y. and Dinoor, A. (1974) Genetic studies on tolerance of carboxin and benomyl at the asexual phase of *Ustilago hordei. Phytopathology* **64,** 51.

Ben-Yephet, Y., Dinoor, A. and Henis, Y. (1975) The physiological basis of carboxin sensitivity and tolerance in *Ustilago hordei. Phytopathology* **65,** 936.

Berbee, J. G. and Hadi, S. (1977) Control of Cylindrocladium root rot and stem canker of conifers with benomyl. *Phytopathology* **67** (in press).

Berg, G. A. van den and Bollen, G. J. (1971) Effect of benomyl on incidence of wilting of *Callistephus chinensis* caused by *Phytophthora cryptogea. Acta bot. neerl.* **20,** 256.

Berger, R. D. (1973) Disease progress of *Cercospora apii* resistant to benomyl, *Pl. Dis. Reptr* **57,** 837.

Bergmann, H., Lyr, H., Kluge, E. and Ritter, G. (1975) Untersuchungen zur Wirkungsweise von Tridemorph. *Systemfungizide* (p. 183), Intern. Symp. Reinhardsbrunn. (eds H. Lyr and C. Polter) Berlin: Akademie-Verlag.

Berry, F. H. and Hock, W. K. (1975) Walnut anthracnose in *Forest Nursery Diseases in the United States.* (eds G. S. Peterson and R. S. Smith) *USDA Forest Service Ag. Handbook* **470,** 92.

Betgem, J. (1960) Veldproeven met het 3-phenyl derivaat van 5-amino-1-bis(dimethylamido)-fosforyl-triazole-1,2,4. *Meded. Landb. Hogesch. Opzoekstns. Gent* **25,** 1227.

Beute, M. K. and Milholland, R. D. (1971) Eradication of *Botryosphaeria corticis* from blueberry propagation wood. *Pl. Dis. Reptr* **54,** 122.

Beyries, A. (1969) Effectiveness of some systemic fungicides against *Corticium solani* by seed treatment of bean and radish. *Phytiat. Phytopharm.* **18,** 107.

Bhate, D. S., Lavate, W. V. and Acharya, S. P. (1964) Isolation, crystallisation and physicochemical properties of neopentaene, a new antifungal antibiotic. *Hindustan Antibiot. Bull.* **6,** 153.

Biehn, W. L. (1969) Evaluation of seed and soil treatment for control of *Rhizoctonia* scurf and *Verticillium* wilt of potato. *Pl. Dis. Reptr* **53,** 425.

Biehn, W. L. (1970) Evaluation of seed treatments for control of seed-borne *Verticillium* wilt of potato. *Pl. Dis. Reptr* **54,** 254.

Biehn, W. L. (1972a) Specificity of *p*-fluorophenylalanine for inhibiting sporulation of *Ceratocystis ulmi. Phytopathology* **62,** 493.

Biehn, W. L. (1972b) Dutch elm disease (*Ceratocystis ulmi*). *Fungicide and Nematicide Tests* **28,** 124.

Biehn, W. L. (1973) Long-term protective action of benomyl soil treatment against Dutch elm disease. *Pl. Dis. Reptr* **57,** 35.

Biehn, W. L. and Dimond, A. E. (1971) Prophylactic action of benomyl against Dutch elm disease. *Pl. Dis. Reptr* **55,** 179.

Biehn, W. L. and Dimond, A. E. (1969) Reduction of tomato *Fusarium* wilt symptoms by 1-(butylcarbamoyl)-2-benzimidazole carbamic acid, methyl ester. *Phytopathology* **59,** 397.

Biehn, W. L. and Dimond, A. E. (1970) Protective action of benomyl against Dutch elm disease. *Phytopathology* **60,** 571 (abs.).

Black, W. M. and Neely, D. (1975) Effect of soil-injected benomyl on resident earthworm populations. *Pestic. Sci.* **6,** 543.

Blanpied, G. D. and Purnasiri, A. (1968) Thiabendazole control of *Penicillium* rot of McIntosh apples. *Pl. Dis. Reptr* **52,** 867.

Blanpied, G. D. and Purnasiri, A. (1970) Thiabendazole control of post-harvest decay. *HortScience* **5,** 476. (*Rev. Pl. Path.* **50,** 3033, 1971.)

Bollen, G. J. (1971) Resistance to benomyl and some chemically related compounds in strains of *Penicillium* species. *Neth. J. Pl. Pathol.* **77,** 187.

Bollen, G. J. (1972) A comparison of the *in vitro* antifungal spectrum of thiophanates and benomyl. *Neth. J. Pl. Pathol.* **78,** 55.

Bollen, G. J. and Fuchs, A. (1970) On the specificity of the *in vitro* and *in vivo* antifungal activity of benomyl. *Neth. J. Pl. Path.* **76,** 299.

Bollen, G. J. and Scholten, G. (1971) Acquired resistance to benomyl and some other systemic fungicides in a strain of *Botrytis cinerea* in cyclamen. *Neth. J. Pl. Path.* **77,** 83.

Bollen, G. J. and van Zaayen, A. (1975) Resistance to benzimidazole fungicides in pathogenic strains of *Verticillium fungicola. Neth. J. Pl. Path.* **81,** 157.

Bolley, H. L. (1891) Potato scab and possibilities of prevention. *Bull. N. Dakota Agric. Exp. Sta. no. 4,* p. 1.

Bolley, H. L. (1906) Tree feeding. *Rep. N. Dakota Agric. Exp. Sta. no. 17,* p. 104.

Bompeix, G. and Morgat, F. (1969) Lutte chimique contre les pourritures des pommes en conservation: efficacité du benomyl et du thiabendazole. *C. r. hebd. Seanc. Acad. Agric. Fr.* **55,** 776. (*Rev. Pl. Path.* **49,** 515, 1970.)

Bondoux, P. (1967) L'eau chaude et le thiabendazole en traitement curatif contre les parasites lenticellaires des Pommes. *C. r. hebd. Séanc. Acad. Agric. Fr.* **53,** 1314. (*Rev. appl. Mycol.* **47,** 2205, 1968.)

Bonner, J. (1961) On the mechanics of auxin-induced growth. *Plant Growth Regulation,* p. 307. Ames: Iowa State Univ. Press.

Booth, A., Moorby, J., Davies, C. R., Jones, H. and Wareing, P. F. (1962) Effects of indolyl-3-acetic acid on the movement of nutrients within plants. *Nature, Lond.* **194,** 204.

Borecka, H., Borecki, Z. and Millikan, D. F. (1971) Control of apple scab, bitter rot and sawfly in Poland with the use of some newer fungicides. *Pl. Dis. Reptr* **55,** 828.

Borecka, H., Borecki, Z. and Millikan, D. F. (1973) Recent investigation on the control of gray mold of strawberries in Poland. *Pl. Dis. Reptr* **57,** 31.

Borecki, Z., Masternak, H., Puchala, Z. and Millikan, D. F. (1970) Preliminary observations on Pezicula canker control in Poland. *Pl. Dis. Reptr* **54,** 640.

Borecki, Z., Millikan, D. F., Puchala, Z. and Bystydzienska, K. (1971) Effectiveness of benomyl and other chemicals for the control of leaf diseases on black currant in Poland. *Pl. Dis. Reptr* **55,** 932.

Borecki, Z. and Nowacka, H. (1969) Control of apple scab (*Venturia inaequalis*) with modern fungicides. *Biul. Inst. Ochr. Rósl, Poznán* **44,** 195. (*Rev. Pl. Path.* **49,** 1537j, 1970.)

Borum, D. E. and Sinclair, J. B. (1967) Systemic activity of 2,3-dihydro-5-carboxanilide-6-methyl-1,4-oxathiin against *Rhizoctonia solani* in cotton seedlings. *Phytopathology* **57,** 805.

van den Bos, B. G. (1960) Investigations on pesticidal phosphorus compounds. III. The structure of the reaction product of 3-amino-5-phenyl-1,2,4-triazole and bis(dimethylamido) phosphoryl chloride (WP155). *Rec. Trav. chim. Pays-Bas* **79,** 1129.

van den Bos, B. G., Koopmans, M. J. and Huisman, H. O. (1960) Investigations on pesticidal phosphorus compounds. I. Fungicides, insecticides and acaricides derived from 3-amino-1,2,4-triazole. *Rec. Trav. chim. Pays-Bas* **79,** 807.

van den Bos, B. G., Schipperheyn, A. and van Deuren, F. W. (1966) Investigations on pesticidal phosphorus compounds. V. The structure of the products of the reaction of 3-alkyl-5-anilino-1,2,4-triazoles with bis(dimethylamido) phosphoryl chloride. *Rec. Trav. chim. Pays-Bas* **85,** 429.

van den Bos, B. G., Shoot, C. J., Koopmans, M. J. and Meltzer, J. (1961) Investigations on pesticidal phosphorus compounds. IV. N-bis(dimethylamido) phosphoryl heterocycli. *Rec. Trav. chim. Pays-Bas* **80,** 1040.

van den Bossche, H. (1974) Biochemical effects of miconazole on fungi – I. Effects on the uptake and/or utilization of purines, pyrimidines, nucleosides, amino acids and glucose by *Candida albicans. Biochem. Pharmacol.* **23,** 887.

van den Bossche, H., Willemsens, G. and van Cutsem, J. M. (1975) The action of miconazole on the growth of *Candida albicans. Sabouraudia* **13,** 63.

Bottalico, A. (1969) Preliminary trials with benomyl against powdery mildew on grapevine in Apulia. (In Italian.) *Atti Giorn. fitopatol., 1969,* 433 (*Rev. Pl. Path.* **49,** 2144, 1970.)

Bouchereau, P. and Atkins, J. G. Jr. (1950) Studies on the action of copper seed treatments for water-planted rice. *Phytopathology* **40,** 3.

Boudru, M. (1935) La maladie de l'orme en Belgique. *Bull. Soc. chim. Belg.* **42,** 508.

Bourdin, J., Douchet, J. P. and Thollière, J. (1969) Efficacité du benomyl sur tavelures et oidium des arbres fruitiers à pepins. *Phytiat.-Phytopharm.* **18,** 15. (*Rev. Pl. Path.* **49,** 1422, 1970.)

Bowen, M. R. and Wareing, P. F. (1971) Further investigations into hormone-directed transport systems. *Planta* **99,** 120.

Bowers, B., Gevin, G. and Cabib, E. (1974) Effect of polyoxin D on chitin synthesis and septum formation in *Saccharomyces cerevisiae. J. Bacteriol.* **119,** 564.

Bowyer, T. H. and Shurtleff, M. C. (1973) *Fungicide and Nematicide Tests* **29,** 113.

Brasier, C. M. and Gibbs, J. N. (1975) MBC tolerance in aggressive and non-aggressive isolates of *Ceratocystis ulmi. Ann. appl. Biol.* **81,** 231.

Bratt, H., Daniel, J. W. and Monks, I. H. (1972) Metabolism of the systemic fungicide, dimethirimol by rats and dogs. *Food Cosmet. Toxicol.* **10,** 489.

Braun, A. J. (1971) Grape: powdery mildew (*Uncinula necator*). *Fungicide and Nematicide Tests* **26,** 56.

Braun, A. J. and Pringle, R. B. (1959) Pathogen factors in the physiology of disease. Toxins and other metabolites. *Plant Pathology: Problems and Progress 1908–1958* (ed. C. S. Holton), p. 88. Madison: Univ. Wisconsin Press.

Bregadze, A. G. (1972) Benlate in the control of powdery mildew of peach. *Zashch. Rast., Mosk.* **17,** 33. (*Rev. Pl. Path.* **51,** 4171.)

Brener, W. D. and Beckman, C. H. (1968) A mechanism of enhanced resistance to *Ceratocystis ulmi* in American elms treated with sodium trichlorophenyl-acetate. *Phytopathology* **58,** 555.

Brian, P. W. (1952a) Systemic fungicides. *Pl. Prot. Overseas Rev.* **3,** 5.

Brian, P. W. (1952b) Antibiotics as systemic fungicides and bactericides. *Ann. appl. Biol.* **39,** 434.

Brian, P. W. (1954) The use of antibiotics for control of plant diseases caused by bacteria and fungi. *J. appl. Bact.* **17,** 142.

Brian, P. W. (1960) Griseofulvin. *Trans. Br. mycol. Soc.* **43,** 1.

Brian, P. W., Wright, J. M., Stubbs, J. and Way, A. M. (1951) Uptake of antibiotic metabolites of soil micro-organisms by plants. *Nature, Lond.* **167,** 347.

Briggs, D. E., Waring, R. H. and Hackett, A. M. (1974) The metabolism of carboxin in growing barley. *Pestic. Science* **5,** 599.

Briggs, G. E. and Robertson, R. N. (1957) Apparent free space. *A. Rev. Pl. Physiol.* **8,** 11.

Bronniman, A. (1968) Zur Kenntris von *Septoria nodorum* Berk. dem Erreger der spelzenbraune und einer Blattdune des Weizens. *Phytopath. Z.* **61,** 101.

Brooks, D. H. (1970) Powdery mildew of barley and its control. *Outlook on Agriculture* **6,** 122.

Brooks, D. H. (1971) Observation on the effects of mildew on growth of spring and winter barley. *Ann. appl. Biol.* **70,** 149.

Brooks, F. T. and Bailey, M. A. (1919) Silver leaf disease (including observations upon the injection of trees with antiseptics). *J. Pomol.* **1,** 81.

Brooks, F. T. and Brenchley, G. H. (1931) Further injection experiments in relation to *Stereum purpureum. New Phytol.* **30,** 128.

Brooks, F. T. and Storey, H. H. (1923) Silver leaf disease. *J. Pomol.* **3,** 117.

Brown, D. and Woodcock, D. (1975a) Structure–activity relationships in fungitoxicants related to triforine and chloraniformethan. I. Synthesis and fungitoxicity to *E. graminis* on barley. *Pestic. Sci.* **6,** 347.

Brown, D. and Woodcock, D. (1975b) Structure–activity relationships in fungitoxicants related to triforine and chloraniformethan. II. Redistribution on barley and marrow seedlings. *Pestic. Sci.* **6,** 363.

Brown, D. and Woodcock, D. (1975c) Structure–activity relationships in fungitoxicants related to triforine and chloramiformethan. III. A quantitative Hansch analysis. *Pestic. Sci.* **6,** 371.

Brown, D. H. and Silverborg, S. B. (1970) Effect of cycloheximide on Hypoxylon cankers of aspen. *Pl. Dis. Reptr* **54,** 462.

Brown, G. E. (1968) Experimental fungicides applied pre-harvest for control of post-harvest decay in Florida citrus fruit. *Pl. Dis. Reptr* **52,** 844.

Brown, G. E., McCornack, A. A. and Smoot, J. J. (1967) Thiabendazole as a post-harvest fungicide for Florida citrus fruit. *Pl. Dis. Reptr* **51,** 95.

Brown, G. E., Massie, L. B. and Cole, H. (1970) *Fungicide and Nematicide Tests* **26,** 120.

Brown, I. F. (1970) Biological activity of a new broad spectrum systemic fungicide. *Proc. 7th Int. Congr. Plant Protection* p. 206 (abs.).

Brown, I. F. and Hall, H. R. (1971) Studies on the activity of triarimol (EL-273) against certain powdery mildew fungi. *Phytopathology* **61,** 886 (abs.).

Brown, I. F. (Jr), Whaley, J. W., Taylor, H. M. and van Heynigen, E. M. (1967) The structure–activity relationships of a new group of pyridine alkane and carbinol fungicides. *Phytopathology* **57,** 805 (abs.).

Brown, J. G. and Boyle, A. M. (1944) Effect of penicillin on a plant pathogen. *Phytopathology* **34,** 760.

Bruchhausen, V. von and Stiasni, M. (1973) Transport of the systemic fungicide Cela W 524 (triforine) in barley plants. II. Uptake and metabolism. *Pestic. Science* **4,** 767.

Bruehl, G. W. and Cunfer, B. (1972) Control of Cercosporella foot rot of wheat by benomyl. *Pl. Dis. Reptr* **56,** 20.

Brun, J. and Melin, P. (1970) Utilization of benomyl mixed with oil for the treatment against cercosporiose of bananas. *Summaries of Papers. VIIth Int. Cong. Plant Prot., Paris,* p. 231.

Büchel, K-H., Meiser, W., Krämer, W. and Grewe, F. (1975) Chemistry and properties of fluotrimazole and related azole fungicides. *Proc. 8th int. Plant Prot. Cong. Moscow* **3,** 111.

Bucheneau, G. W. (1970) *Fungicide and Nematicide Tests* **26,** 104.

Buchenauer, H. (1975) Various response of cell membranes of plant and fungal cells to different systemic fungicides. *Proc. 8th intern. Plant Prot. Cong. Moscow,* **3,** 94.

Bucheanauer, H. (1976) Hemmung der Ergosterinbiosynthese in *Ustilago avenae* durch Triadimefon und Fluotrimazol. *Z. Pfl. Krankh. u. Pfl. Schutz* **83,** 363.

Buchenauer, H., Edgington, L. V. and Grossmann, F. (1973) Photochemical transformation of thiophanate-methyl and thiophanate to alkyl benzimidazole-2-yl carbamates. *Pestic. Science* **4,** 343.

Buchenauer, H. and Erwin, D. C. (1971) Control of *Verticillium* wilt of cotton by spraying foliage with benomyl and thiabendazole solubilized with hydrochloric acid. *Phytopathology* **61,** 433.

Buchenauer, H. and Grossmann, F. (1969) Einfluss von Morphaktinen auf verschiedene Tomatenkrankheiten. *Dtsch. Bot. Ges., Neue Folge* **3,** 149.

Buczacki, S. T. (1973). An appraisal of systemic fungicides for the control of clubroot of brassicae. *Proc. 7th Br. Insectic. Fungic. Conf.* p. 573.

Buczacki, S. T., Susan, E., Cadd, Jean G., Ockendon and White, J. G. (1976) *Ann. appl. Biol.* **84,** 51. Field evaluation of systemic fungicides and derivatives of dithiocarbamic acid for the control of clubroot.

Burchill, R. T. (1972) Comparison of fungicides for suppressing ascospore production by *Venturia inaequalis* (Cke.) Wint. *Pl. Path.* **21,** 19.

Burchill, R. T. and Edney, K. L. (1972). An assessment of some new treatments for the control of rotting of stored apples. *Ann. appl. Biol.* **72,** 249.

Burchill, R. T. and Hutton, K. E. (1965) The suppression of ascospore production to facilitate the control of apple scab (*Venturia inaequalis*) (Cke.) (Wint.) *Ann. appl. Biol.* **56,** 285.

Burchill, R. T. and Swait, A. A. J. (1971) Apple scab (*Venturia inaequalis*). Epidemiology: suppression of ascospores by post-harvest sprays. *Rep. E. Malling Res. Stn, 1970,* p. 106.

Burchill, R. T. and Swait, A. A. J. (1975) A comparison of fungicides for the control of raspberry spur blight. *Ann. appl. Biol.* **81,** 341.

Burchill, R. T. and Williamson, C. J. (1971) Comparison of some new fungicides for the control of scab and powdery mildew of apple. *Pl. Path.* **20,** 173.

Burden, O. J. (1969) Control of ripe fruit rots of bananas by the use of post-harvest fungicide dips. *Aust. J. exp. Agric. Anim. Husb.* **9,** 655. (*Rev. Pl. Path.* **49,** 1736, 1970.)

Busch, L. V. and Hall, R. (1971) Verticillium wilt of chrysanthemum: suppression of fungal colonization of leaves and the prevention of wilt symptom development by foliar applications of benomyl. *Can. J. Bot.* **49,** 1987.

Butler, G. W. (1953) Ion uptake by young wheat plants. II. *Physiologia Pl.* **6,** 617.

Butt, D. J., Kirby, A. H. M. and Williamson, Cynthia J. (1973) Fungitoxic and phytotoxic effects of fungicides controlling powdery mildew on apple. *Ann. appl. Biol.* **75,** 217.

Byrde, R. J. W. (1956) The varietal resistance of fruits to brown rot. I. Infection experiments with *Sclerotinia frustigena* Aderh. and Ruhl. on certain dessert, culinary and cider varieties of apples. *J. hort. Sci.* **31,** 188.

Byrde, R. J. W. (1957) The varietal resistance of fruits to brown rot. II. The nature of resistance in some varieties of cider apples. *J. hort. Sci.* **32,** 227.

Byrde, R. J. W. (1959) A note on the control of brown rot of apples by griseofulvin. *Pl. Path.* **8,** 90.

Byrde, R. J. W. (1963) Natural inhibitors of fungal enzymes and toxins in disease resistance. *Conn. Agric. Exp. Sta. Bull.* **663,** 31.

Byrde, R. J. W. (1969) Systemic fungicides, including recent developments in the agricultural sphere. *Proc. 5th Br. Insectic. Fungic. Conf.* **3,** 675.

Byrde, R. J. W. (1970) The new systemic fungicides and their potential use in the tropics. *Trop. Sci.* **12,** 105.

Byrde, R. J. W. (1972) Apple diseases (with particular reference to mildew and canker control). *Proc. 6th Br. Insectic. Fungic. Conf.* **3,** 648.

Byrde, R. J. W. and Ainsworth, G. C. (1958) The chemotherapy of fungal diseases. *Symp. Soc. gen. Microbiol.* **8,** 309.

Byrde, R. J. W., Crowdy, S. H. and Woodcock, D. (1953) Studies on systemic fungicides. III. The activity of certain chlorine-substituted β-naphthols and naphthyloxy-*n*-aliphatic carboxylic acids as systemic fungicides. *Ann. appl. Biol.* **40,** 152.

Byrde, R. J. W., Harper, C. W. and Holgate, M. E. (1970) Spraying trials against apple mildew and apple scab at Long Ashton, 1968. *Rep. agric. hort. Res. Stn Univ. Bristol, 1968,* p. 146.

Byrde, R. J. W., Harper, C. W. and Holgate, M. E. (1969) Spraying trials against apple mildew and apple scab at Long Ashton, 1969. *Rep. agric. hort. Res. Stn Univ. Bristol, 1969,* p. 189.

Byrde, R. J. W., Harper, C. W., Holgate, M. E. and Hunter, T. (1973) Residual effects of fungicidal spray programmes on apple trees. *Pestic. Sci.* **4,** 327.

Byrde, R. J. W. and Melville, S. C. (1971) Chemical control of blossom wilt of apple. *Pl. Path.* **20,** 48.

Byrde, R. J. W. and Willetts, H. J. (1977) *The brown rot fungi of fruit: their biology and control.* Pp. 171. Oxford: Pergamon Press.

Calderbank, A. (1971) Metabolism and mode of action of dimethirimol and ethirimol. *Acta Phytopathol. Acad. Scient. Hungar.* **6,** 355.

Campana, R. J. (1975) Evaluation of pressure-injected, solubilized benomyl as a therapy for American elms, naturally inoculated by native elm bark beetles with the Dutch elm disease fungus. *Adden. to Final Report of 1972 Injections* (see Campana *et al.*, 1973).

Campana, R. J., Gregory, G. and Jones, T. (1973) Arrest of symptom development in Dutch elm disease with pressure injection of solubilized benomyl. *Proc. 2nd Internat. Cong. Plant Path.* abs. No. 0197.

Campos, A. and Borlaug, N. E. (1956) Wetting and penetrating agents combined with fungicides as protectants and eradicants for wheat stem rust. *Phytopathology* **46,** 8.

Canny, M. J. (1973) *Phloem translocation.* Cambridge University Press.

Cargo, C. A. and Dewey, D. H. (1970) Thiabendazole and benomyl for the control of postharvest decay of apples. *HortScience,* **5,** 259.

Carlson, L. W. (1972) Jack pine: storage molds. *Fungicide and Nematicide Tests 1972* **28,** 127.

Carlson, L. W. (1973) Poplar cutting bed canker (hybrid poplar) (*Septoria* sp. and *Marssonina* sp.). *Fungicide and Nematicide Tests* **29,** 124.

Carlson, L. W. and Huber, R. (1974) Jack pine: storage molds. *Fungicide and Nematicide Tests* **30,** 128.

Caroselli, N. E. and Howard, F. L. (1942) Response of diseased maple trees to chemotherapy and fertilization. *Phytopathology* **32,** 21.

Carroll, R. B. and Webb, F. J. (1974) *Fungicide and Nematicide Tests* **30,** 104.

Carter, G. A. (1968). Studies on systemic fungicides. Ph. D. thesis, University of London.

Carter, G. A., Chamberlain, K. and Wain, R. L. (1973) Investigations on fungicides. XV. The fungitoxicity and systemic antifungal activity of *N*-(2,2,2-trichloro-l-methoxyethyl) formamide and related compounds. *Ann. appl. Biol.* **75,** 49.

Carter, G. A., Summers, L. A. and Wain, R. L. (1972) Investigations on fungicides. XIV. Fungitoxicity of *N*-(2,2,2-trichloro-1-methoxyethyl) formamide and related compounds. *Ann. appl. Biol.* **70,** 233.

Carter, G. A. and Wain, R. L. (1964) Investigations on fungicides. IX. The fungitoxicity, phytotoxicity and systemic fungicidal activity of some inorganic salts. *Ann. appl. Biol.* **53,** 291.

Catling, W. S. (1971) The control of powdery mildew, canker and other diseases of apple with benomyl. *Proc. 6th Br. Insectic. Fungic. Conf.* **1,** 103.

Cavell, B. D., Hemingway, R. J. and Teal, G. (1971) Some aspects of the metabolism and translocation of the pyrimidine fungicides. *Proc. 6th Br. Insectic. Fungic. Conf.* **2,** 431.

Cenóz, H. P. (1961) Effect of Septoria diseases on wheat. *Aust. J. expt. Agric. Anim. Husb.* **8,** 89.

Channon, A. G. and Thomson, M. C. (1973) The effect of benomyl on the infection of tomatoes by *Fusarium oxysporum* f. sp. *lycopersici* and *Botrytis cinerea. Ann. appl. Biol.* **75,** 31.

Chapman, R. A. (1951) Relation of specific chemotherapeutants to the infection court. *Phytopathology* **41,** 6.

Charles, A. (1953) Uptake of dyes into cut leaves. *Nature, Lond.* **171,** 435.

Chiba, M. and Doornbos, F. (1974) Instability of benomyl in various conditions. *Bull. Environm. Contam. Toxicol.* **11,** 273.

Chin, W. T., Stone, G. M. and Smith, A. E. (1969) The fate of carboxin in soil, plants and animals. *Proc. 5th Br. Insectic. Fungic. Conf.* p. 322.

Chin, W. T., Stone, G. M. and Smith, A. E. (1970a) Metabolism of carboxin (Vitavax) by barley and wheat plants. *J. Agric. Food Chem.* **18,** 709.

Chin, W. T., Stone, G. M. and Smith, A. E. (1970b). Degradation of carboxin (Vitavax) in water and soil. *J. Agric. Food Chem.* **18,** 731.

Chin, W. T., Stone, G. M. and Smith, A. E. (1971) The fate of carboxin in soil, plants and animals. *Pesticide terminal residues* (ed. A. S. Tahori) p. 271. London: Butterworths.

Cimanowski, J., Masternak, A. and Millikan, D. F. (1970) Effectiveness of benomyl for controlling apple powdery mildew and cherry leaf spot in Poland. *Pl. Dis. Reptr* **54,** 81.

Clark, E. M., Backman, P. A. and Rodriguez-Kabana, R. (1974) *Cercospora* and *Cercosporidium* tolerance to benomyl and related fungicides in Alabama peanut fields. *Phytopathology* **64,** 1476.

Clarkson, D. T. (1974) *Ion transport and cell structure in plants.* New York: McGraw-Hill.

Cleland, R. (1963a) Independence of effects of auxin on cell wall methylation and elongation. *Pl. Physiol.* **38,** 12.

Cleland, R. (1963b) The occurrence of auxin-induced pectin methylation in plant tissue. *Pl. Physiol.* **38,** 738.

Clemons, G. P. and Sisler, H. D. (1969) Formation of a fungitoxic derivative from benlate. *Phytopathology* **59,** 705.

Clemons, G. P. and Sisler, H. D. (1971) Localization of the site of action of a fungitoxic benomyl derivative. *Pesticide Biochem. and Physiol.* **1,** 32.

Clifford, D. R., Fieldgate, D. M., Watkins, D. A. M. and Woodcock, D. (1971) The fungicidal activities of nitrophenols. *Proc. 2nd Int. Congress of Pestic. Chem. Tel Aviv* **5,** 281.

Clifford, D. R., Gendle, P. and Fribbins, E. A. (1976) Formulations for the control of Dutch elm disease. *Pestic. Sci.* **7,** 91.

Clifford, D. R. and Hislop, E. C. (1971) The activities of some systemic fungicides against powdery mildew fungi. *Proc. 6th Br. Insectic. Fungic. Conf.* **2,** 438.

Clifford, D. R., Hislop, E. C. and Holgate, M. (1970) Some factors affecting the activities of dinitrophenol fungicides. *Pestic. Sci.* **1,** 18.

Cohick, A. D. (1973) Bay Dam 18654, a new broad spectrum fungicide. *2nd Int. Cong. Plant Pathology,* 0116 (abs.).

Cole, H., Boyle, J. S. and Smith, C. B. (1970) Effect of benomyl and certain cucumber viruses on growth, powdery mildew and element accumulation by cucumber plants in the greenhouse. *Pl. Dis. Reptr* **54,** 141.

Cole, H., Goldberg, C. W. and Massie, L. B. (1972) *Fungicide and Nematicide Tests* **28,** 121.

Cole, H., Taylor, B. and Dutch, J. (1968) Evidence of differing tolerances to fungicides among isolates of *Sclerotinia homoeocarpa. Phytopathology* **58,** 683.

Cole, H., Warren, C. G. and Sanders, P. (1974) *Fungicide and Nematicide Tests* **30,** 124.

Cole, M. (1958) Oxidation products of leuco-anthocyanins as inhibitors of fungal polygalacturonase in rotting apple fruit. *Nature, Lond.* **181,** 1596.

Cole, R. J. and Cox, T. W. (1973) Control of raspberry diseases with thiophanate-methyl. *Proc. 7th Br. Insectic. Fungic. Conf.* **2,** 483.

Cole, R. J., Cox, T. W. and Soper, D. (1973) Control of storage diseases of apples and pears with thiophanate-methyl. *Proc. 7th Br. Insectic. Fungic. Conf.* **2,** 623.

Cole, R. J., Gilchrist, A. J. and Soper, D. (1971) The control of diseases of apples and pears in the United Kingdom with thiophanate methyl. *Proc. 6th Br. Insectic. Fungic. Conf.* **1,** 118.

Collander, R. (1954) The permeability of *Nitella* cells to non-electrolytes. *Physiologia Pl.* **7,** 420.

Colquhoun, T. T. (1940) Effect of manganese on powdery mildew of wheat. *J. Aust. Inst. agric. Sci.* **6,** 54.

Connor, S. R. and Heuberger, J. W. (1968) Apple scab. V. Effect of late season application of fungicides on prevention of perithecial developments by *Venturia inaequalis. Pl. Dis. Reptr* **52,** 654.

Conway, W. S. and Rich, A. (1970) *Fungicide and Nematicide Tests.* **26,** 121.

Cooke, B. M. and Jones, D. Gareth (1970) A field inoculation method for *Septoria tritici* and *Septoria nodorum. Pl. Path.* **19,** 72.

Cooper, D., Banthorpe, D. V. and Wilkie, D. (1967) Modified ribosomes conferring resistance to cycloheximide in mutants of *Saccharomyces cerevisiae. J. Mol. Biol.* **26,** 347.

Corbett, D. C. M. and Hide, G. A. (1971) Chemical control of *Verticillium dahliae* and *Heterodera rostochiensis* on potatoes. *Proc. 6th Br. Insectic. Fungic. Conf.,* p. 258.

Corbin, J. B. (1971) Benomyl beneficial for European canker control. *Orchard N.Z.* **44,** 55, 57. (*Rev. Pl. Path.* **50,** 3944, 1971).

Corbin, J. B. and Kitchener, R. A. (1970) Peach: brown rot (*Monilinia fructicola*); scab (*Venturia carpophila*); stigmina (*Stigmina carpophila*). *Fungicide and Nematicide Tests* **25,** 43.

Corcoran, J. W. and Hahn, F. E. (1975) *Antibiotics,* Vol. III. Mechanism of action of antimicrobial and antitumour agents. Berlin: Springer-Verlag.

Cordell, C. E. and Skilling, D. D. (1975) Cylindrocladium root rot in forest nursery diseases in the United States. (eds G. W. Peterson and R. S. Smith) *USDA Forest Service Agr. Handbook* **470,** 23.

Corden, M. E. and Dimond, A. E. (1957) Relation of plant-growth regulating properties of naphthalene aliphatic acids, to their activity in inducing disease resistance. *Phytopathology* **47,** 518.

Corden, M. E. and Dimond, A. E. (1959) The effect of growth-regulating substances on disease resistance and plant growth. *Phytopathology* **49,** 68.

Corden, M. E. and Edgington, L. V. (1960) A calcium requirement for growth-regulator-induced resistance to *Fusarium* wilt of tomato. *Phytopathology* **50,** 625.

Corke, A. T. K., Goldfinch, S. and Hunter, T. (1971) Bitter rot (caused by *Gloeosporium* spp.). *Rep. agric. hort. Res. Stn Bristol, 1970,* p. 115.

Corke, A. T. K., Hunter, T. and Goldfinch, S. (1972) The presence of benomyl (and/or breakdown products) in Cox flowers. *Rep. agric. hort. Res. Stn Bristol, 1971,* p. 147.

Covey, R. P. (1971a) The effect of methionine on the development of apple powdery mildew. *Phytopathology* **61,** 346.

Covey, R. P. (1971b) Orchard evaluation of two new fungicides for the control of apple powdery mildew. *Pl. Dis. Reptr* **55,** 514.

Coyier, D. L. (1970) Control of storage decay in 'd'Anjou' pear fruit by preharvest applications of benomyl. *Pl. Dis. Reptr* **54,** 647.

Coyier, D. L. (1971) Control of powdery mildew on apples with various fungicides as influenced by seasonal temperature. *Pl. Dis. Reptr* **55,** 263.

Crafts, A. S. (1933) The use of arsenical compounds in the control of deep-rooted perennial weeds. *Hilgardia* **7,** 361.

Crafts, A. S. (1961) *Translocation in plants.* New York: Holt, Rinehart and Winston.

Crafts, A. S. and Crisp, C. E. (1971) *Phloem transport in plants.* San Francisco: Freeman and Co.

Craker, L. E. (1971) Ethylene production from ozone-injured plants. *Environ. Pollut.* **1,** 299.

Cremlyn, R. J. W. (1961) Systemic fungicides. *J. Sci. Fd Agric.* **12,** 805.

Cremlyn, R. J. W. (1973) Systemic fungicides – a widening choice for agriculture. *Internat. Pest Control* **15,** 8.

Crisp, A. F. (1974) Control of wastage in cooled storage carrots. *Acta Hortic.* **2,** 389.

Crosier, W. F., Nash, G. T. and Crosier, D. C. (1970) Differential reaction of *Pythium* sp. and five isolates of *Rhizoctonia solani* to fungicides on pea seeds. *Pl. Dis. Reptr* **54,** 349.

Cross, M. J. and Jacobs, L. (1968) Some observations on the biology of spores of *Verticillium malthousei. Mushr. Sci.* **7,** 239.

Crosse, R., McWilliams, R. and Rhodes, A. (1964) Some relations between chemical structure and antifungal effects of griseofulvin analogues. *J. gen. Microbiol.* **34,** 51.

Crosse, R., McWilliams, R., Rhodes, A. and Dunn, A. T. (1960) Antifungal action of streptomycin–copper chelate against *Phytophthora infestans* on tomato (*Lycopersicon esculentum*). *Ann. appl. Biol.* **48,** 270.

Crowdy, S. H. (1959) Uptake and translocation of organic chemicals by higher plants. *Plant Pathology, Problems and Progress, 1908–1958* (ed. C. S. Holton), p. 231. Madison: University of Wisconsin Press.

Crowdy, S. H. (1970) Factors influencing the activity of systemic fungicides. *Proc. 7th Int. Congr. Pl. Prot* (Paris), 217.

Crowdy, S. H. (1972) Chapter 5 in *Systemic Fungicides* (ed. R. W. Marsh). London: Longman.

Crowdy, S. H. and Davies, M. E. (1952) Studies on systemic fungicides. II. Behaviour of groups of reported chemotherapeutants. *Phytopathology* **42,** 127.

Crowdy, S. H., Gardner, D., Grove, J. F. and Pramer, D. (1955) The translocation of antibiotics in higher plants. I. Isolation of griseofulvin and chloramphenicol from plant tissue. *J. exp. Bot.* **6,** 371.

Crowdy, S. H., Green, A. P., Grove, J. F., McCloskey, P. and Morrison, A. (1959a) The translocation of antibiotics in higher plants. 3. The estimation of griseofulvin relatives in plant tissues. *Biochem. J.* **72,** 230.

Crowdy, S. H., Grove, J. F., Hemming, H. G. and Robinson, K. C. (1956) The translocation of antibiotics in higher plants. II. The movement of griseofulvin in broad bean and tomato. *J. exp. Bot.* **7,** 42.

Crowdy, S. H., Grove, J. F. and McCloskey, P. (1959b) The translocation of antibiotics in higher plants. 4 Systemic fungicidal activity and chemical structure in griseofulvin relatives. *Biochem. J.* **72,** 241.

Crowdy, S. H. and Pramer, D. (1955) Movement of antibiotics in higher plants. *Chemy. Ind.* 160.

Crowdy, S. H. and Rudd Jones, D. (1956) Partition of sulphonamides in plant roots: a factor in their translocation. *Nature, Lond.* **178,** 1165.

Crowdy, S. H., Rudd Jones, D. and Witt, A. V. (1958) The translocation of sulphonamides in higher plants. II. *J. exp. Bot.* **9,** 206.

Crowdy, S. H. and Tanton, T. W. (1970) Water pathways in higher plants. I. *J. exp. Bot.* **21,** 102.

Crowdy, S. H. and Wain, R. L. (1950) Aryloxyaliphatic acids as systemic fungicides. *Nature, Lond.* **165,** 937.

Crowdy, S. H. and Wain, R. L. (1951) Studies on systemic fungicides. I. Fungicidal properties of the aryloxyalkylcarboxylic acids. *Ann. appl. Biol.* **38,** 318.

Cruger, G. (1969) Anwendung systemischer fungizide zur gurkenmehltaube kämpfung in giessverfahren. *Mitt. biol. Bund Anst. Ld-U. Forstw.* **132,** 216.

Cruickshank, I. A. M. (1963) Phytoalexins. *A. Rev. Phytopath.* **1,** 351.

Cruickshank, I. A. M. (1966) Defence mechanisms in plants. *Wld. Rev. Pest Control* **5,** 161.

Cruickshank, I. A. M. and Jacks, H. (1953) Seed disinfection. IX. Control of onion smut *Urocystis cepulae* Frost. *N.Z. J. Sci. and Tech. A* **35**, 390.

Crute, I. R., Wolfman, S. A. and Davis, A. A. (1977) A laboratory method of screening fungicides for systemic activity against *Bremia lactucae. Ann. appl. Biol.* **85,** 147.

Cuillé, J. and Bur-Ravault, L. (1969) Traitements des oranges contre les *Penicillium* avec des formules a base de thiabendazole. *Fruits d'outre-mer* **24,** 421. (*Rev. Pl. Path.* **49,** 759, 1970.)

van Cutsem, J. M. and Thienpont, D. (1972) Miconazole, a broad-spectrum antimycotic agent with antibacterial activity. *Chemotherapy* **17,** 392.

de Rooy – see under R.

de Waarde – see under W.

Daines, R. H. (1970) Effects of fungicide dip treatments and dip temperatures on post-harvest decay of peaches. *Pl. Dis. Reptr* **54,** 764.

Daines, R. H. and Snee, R. D. (1969) Control of blue mould of apples in storage. *Phytopathology* **59,** 792.

D'Armini, M. (1953) Esperienze di endoprevenzione condotte con carbonato di litio contro l'oidio del tabacco. *Tabacco, Roma* **57,** 319.

Darpoux, H. (1970) Actions du triarimol sur les tavelures du poirier et du pommier *Societé de Phytiatrie-Phytopharmacie Séance de 20.1.70.*

Darpoux, H., Catelot, Mme and Gorse, Mlle (1958a) Étude preliminaire sur l'action systèmique et endothèrapique du sel de manganese de la 2-pyridinethione-l-oxyde. *Phytiat.-Phytopharm.* **7,** 107.

Darpoux, H., Halmos, E. and Leblanc, R. (1958b) Étude de l'action systémique de diverses substances la plupart antibiotiques. *Ann. Épiphyties* **9,** 387.

Darpoux, H., Ventura, E. and Cassini, R. (1968) Activité fongicide du thiabendazole vis-à-vis des principaux champignons parasites des semences de céréales. *Phytiat.-Phytopharm.* **17,** 219.

Dassenoy, B. and Meyer, J. A. (1973) Mutagenic effect of benomyl on *Fusarium oxysporum. Mutation Res.* **21,** 119.

Daum, W., Scheinpflug, H., Frohberger, P-E. and Grewe, F. (1971) ω-Cyanoalkylcarbamylbenzimidazoles. Brit. Pat. 1228108.

Davidse, L. C. (1973) Antimitotic activity of methyl benzimidazol-2-yl carbamate (MBC) in *Aspergillus nidulans. Pestic. Biochem. Physiol.* **3,** 317.

Davidse, L. C. (1975a) Mode of action of methyl benzimidazole-2-yl carbamate (MBC) and some biochemical aspects of acquired resistance against this fungicide in *Aspergillus nidulans. Systemfungizide* (eds. H. Lyr and C. Polter) p. 137. Intern. Symp. Reinhardsbrunn. Berlin: Akademie-Verlag.

Davidse, L. C. (1975b) Antimitotic activity of methyl benzimidazole-2-yl carbamate in fungi and its binding to cellular protein. In *Microtubules and microtubule inhibitors* (eds Borgers, M. and de Brabander, M.) (p. 483). Amsterdam: North-Holland Publ. Co.

Davidse, L. C. (1976) Metabolic conversion of methyl benzimidazol-2-yl carbamate (MBC) in *Aspergillus nidulans*. *Pest. Biochem. Physiol.* (in press).

Davies, J. M. L. (1970) Studies on variation of *Cercosporella herpotrichoides* Fron. Ph.D. Thesis, University of Wales.

Davis, D. (1952) Chemotherapeutic activity may be independent of fungitoxicity. *Phytopathology* **42,** 6.

Davis, D., Becker, H. J. and Rogers, E. F. (1959) The chemotherapy of wheat and bean rust diseases with sydnones. *Phytopathology* **49,** 821.

Davis, D., Chaiet, L., Rothrock, J. W., Deak, J., Halmos, S. and Barger, J. D. (1960) Chemotherapy of cereal rusts with a new antibiotic. *Phytopathology* **50,** 841.

Davis, D. and Dimond, A. E. (1952) Altering resistance to disease with synthetic organic chemicals. *Phytopathology* **42,** 563.

Davis, D. and Dimond, A. E. (1953) Inducing disease resistance with plant growth-regulators. *Phytopathology* **43,** 137.

Davis, D. and Dimond, A. E. (1956) Site of disease resistance induced by plant growth regulators in tomato. *Phytopathology* **46,** 551.

Davis, D., Lo, C. P. and Dimond, A. E. (1954) Chemotherapeutic activity of unsubstituted heterocyclic compounds. *Phytopathology* **44,** 680.

Davis, J. R. and Callihan, R. H. (1970) Evaluation of several fungicides in Idaho for control of *Rhizoctonia* on potato. *Phytopathology* **60,** 1533.

Davis, J. R., Groskopp, M. D. and Callihan, R. H. (1971a) Seed and soil treatments for control of *Rhizoctonia* on stems and stolons of potato. *Pl. Dis. Reptr* **55,** 550.

Davis, M. E., Weeden, S. M. and Martin, T. J. (1971) Studies of powdery mildew control in spring barley using chloraniformethan. *Proc. 6th. Br. Insectic. Fungic. Conf.* **1,** p. 42.

Davis, S. H. and Peterson, J. L. (1973) Sycamore anthracnose (*Gnomonia platani*). *Fungicide and Nematicide Tests* **29,** 126.

Davis, S. H., Jr. and Peterson, J. L. (1975) A tree wound dressing to prevent spread of the *Ceratocystis* causing canker stain of the plane tree. *Pl. Dis. Reptr* **57,** 28.

Davison, A. D. (1969) Grand Fir needle rust (*Pucciniastrum goeppertianum*). *Fungicide and Nematicide Tests* **25,** 106.

Dawes, E. A. and Large, P. J. (1968) Class II reactions: synthesis of small molecules, *Biochemistry of bacterial growth* (eds J. Mandelstam and K. McQuillen), p. 203. New York: John Wiley and Sons.

Day, L. H. (1928). Pear blight control in California. *California Agric. Ext. Service Circ.* **20,** 1.

Decallonne, J. R., Genot, M. and Meyer, J. A. (1975) Effects of benomyl, carbendazim and thiophanates on respiration and oxidative phosphorylation of *Fusarium oxysporum* and *Saccharomyces cerevisiae*. *Pestic. Sci.* **6,** 113.

Decker, P. and Buchanan, D. W. (1974) The effect of rust (*Tranzschelia discolor* (F. Chl.) Trans. & Litr.) on tree performance and fruit yield of 'Early Amber' peach and 'Sungold' nectarine. *Proc. Fla St. hort. Soc.* **86,** 333. (*Rev. Pl. Path.* **54,** 4990, 1975.)

Deese, D. and Stahmann, M. E. (1960) Role of pectic enzymes in susceptibility and resistance to *Fusarium* and *Verticillium* wilts of plants *Phytopathology* **50,** 633.

Defosse, L. (1970) Essais de lutte chimique contre le piétin-verse, *Cercosporella herpotrichoides* Fron, avec un fongicide systemique. *Parasitica* **26,** 4.

Dekhuijzen, H. M. (1964) The systemic action of dimethyldithiocarbamates on cucumber scab caused by *Cladosporium cucumerinum* and the conversion of these compounds by plants. *Neth. J. Pl. Path.* **70,** Suppl. 1.

Dekhuijzen, H. M. and Dekker, J. (1971) Mechanism of resistance of *Cladosporium cucumerinum* against 6-azauracil and 6-azauridine monophosphate. *Pesticide Biochem. and Physiol.* **1,** 11.

Dekker, J. (1955) Internal seed disinfection by an antibiotic from *Streptomyces rimosus*. *Nature, Lond.* **175,** 689.

Dekker, J. (1957) Internal seed disinfection of peas infected by *Ascochyta pisi* by means of the antibiotics rimocidin and pimaricin and some aspects of the parasitism of this fungus. *Tijdschr. PlZiekt.* **63,** 65.

Dekker, J. (1961a) Systemic activity of procaine hydrochloride on powdery mildew. *Tijdschr. PlZiekt.* **67,** 25.

Dekker, J. (1961b) Mode of action of procaine hydrochloride and some derivatives on powdery mildew. *Meded. LandbHoogesch. OpzoekStns. Gent* **26,** 1378.

Dekker, J. (1962) Systemic control of powdery mildew by 6-azauracil and some other purine and pyrimidine derivatives. *Meded. LandbHoogesch. OpzoekStns. Gent* **27,** 1214.

Dekker, J. (1963a) Antibiotics in the control of plant diseases. *A. Rev. Microbiol.* **17,** 243.

Dekker, J. (1963b) Effect of kinetin on powdery mildew. *Nature, Lond.* **197,** 1027.

Dekker, J. (1967) Conversion of 6-azauracil in sensitive and resistant strains of *Cladosporium cucumerinum. Mechanism of action of fungicides and antibiotics* (ed. W. Girbardt), p. 333. Berlin: Akademie-Verlag.

Dekker, J. (1968a) Acquired resistance to fungicides. *World Rev. Pest Control* **8,** 79.

Dekker, J. (1968b) The development of resistance in *Cladosporium cucumerinum* against 6-azauracil, a chemotherapeutant of cucumber scab, and its relation to biosynthesis of RNA-precursors. *Neth. J. Pl. Path.* **74,** Suppl. 1, 127.

Dekker, J. (1969a) L-Methionine-induced inhibition of powdery mildew and its reversal by folic acid. *Neth. J. Pl. Path.* **75,** 182.

Dekker, J. (1969b) Antibiotics. *Fungicides. Vol. II* (ed. D. C. Torgeson), p. 579. New York: Academic Press.

Dekker, J. (1971a) Selective action of systemic fungicides and development of resistance. *Proceedings 2nd International Congress of Pesticide Chemistry IUPAC, Tel-Aviv, 1971*, p. 305.

Dekker, J. (1971b) Selective action of fungicides and development of resistance in fungi to fungicides. *Proc. 6th Br. Insectic. Fungic. Conf.* **3,** 715.

Dekker, J. (1973) Selectivity and resistance against systemic fungicides. *EPPO Bull. No. 10*, 47.

Dekker, J. (1977) Effect of fungicides on nucleic acid synthesis and nuclear function. In *Antifungal Compounds*, vol. 2. (eds Sisler, H. F. and Siegel, M. R.). Marcel Dekker, Inc. (in press).

Dekker, J., van Andel, O. M. and Kaars Sijpesteijn, A. (1958) Internal seed disinfection with pyridine-2-thiol-N-oxide and a derivative. *Nature, Lond.* **181,** 1017.

Dekker, J. and van der Hoek-Scheuer, R. G. (1964) A microscopic study of the wheat-powdery mildew relationship after application of the systemic compounds procaine, griseofulvin and 6-azauracil. *Neth. J. Pl. Path.* **70,** 142.

Dekker, J. and Oort, A. J. P. (1964) Mode of action of 6-azauracil against powdery mildew. *Phytopathology* **54,** 815.

Dekker, J. and Roosje, G. S. (1968) Effect of 6-azauracil against apple powdery mildew and apple scab. *Neth. J. Pl. Path.* **74,** 219.

Delp, C. J. and Klöpping, H. L. (1968) Performance attributes of a new fungicide and mite ovicide candidate. *Pl. Dis. Reptr* **52,** 95.

Derbyshire, D. M. and Crisp, A. F. (1971) Vegetable storage diseases in East Anglia. *Proc. 6th Br. Insectic. Fungic. Conf.*, p. 167.

Desai, B. G., Geypens, M. and Ramanujam, K. (1975) Evaluation of seventeen fungicides in inhibiting growth of *Pythium* species isolates *in vitro. Mededel. Faculteit Landbouwwetenschappen Gent* **40,** 683.

Deverall, B. J. (1964) Substances produced by pathogenic organisms that induce symptoms of disease in higher plants. *14th Symp. Soc. gen. Microbiol.*, p. 165.

Deverall, B. J. and Wood, R. K. S. (1961) Chocolate spot of beans (*Vicia fabae* L.) – interaction between phenolase of host and pectic enzymes of the pathogen. *Ann. appl. Biol.* **49,** 473.

Dewey, W. G. and Albrechtsen, R. S. (1974) Effect of thiabendazole seed treatment on the incidence of dwarf bunt and in the yield of winter wheat. *Pl. Dis. Reptr* **58,** 743.

Dicken, L. E. and Butler, J. D. (1974). *Fungicide and Nematicide Tests* **30,** 116.

Dickins, L. E. and Oshima, N. (1971) Chemotherapeutants evaluated for control of dwarf bunt of wheat. *Pl. Dis. Reptr* **55**(7), 613.

Dickson, J. G. (1959) Chemical control of cereal rusts. *Bot. Rev.* **25**, 486.

Dietrich, J. V. and Brechbuhler, C. (1970) Control of *Botrytis cinerea* Pers. in vineyards. *Summaries of Papers, VIIth Int. Cong. Plant Prot., Paris*, p. 253.

Dimond, A. E. (1947) Oxyquinoline benzoate aids suppression of symptoms of Dutch elm disease. *Phytopathology* **37**, 848.

Dimond, A. E. (1953) Progress in plant chemotherapy. *Agric. Chem.* **8**, 34.

Dimond, A. E. (1959) Plant chemotherapy. *Plant Pathology – Problems and Progress, 1908–1958*, (ed. C. S. Holton), p. 221. Wisconsin: University Press.

Dimond, A. E. (1962) Objectives in plant chemotherapy. *Phytopathology* **52**, 1115.

Dimond, A. E. (1963a) The modes of action of chemotherapeutic agents in plants. *Conn. Agric. Exp. Sta. Bull. no. 663*, p. 62.

Dimond, A. E. (1963b) The selective control of plant pathogens. *Wld Rev. Pest Control* **2**, 7.

Dimond, A. E. (1965) Natural models for plant chemotherapy. *Advances in Pest Control Research* (ed. R. L. Metcalf) **6**, 128. New York: Wiley Interscience Publishers Inc.

Dimond, A. E. (1970) Biophysics and biochemistry of the vascular wilt syndrome. *Ann. Rev. Phytopathol.* **8**, 301.

Dimond, A. E. and Chapman, R. A. (1951) Chemotherapeutic properties of two compounds against *Fusarium* wilts. *Phytopathology* **41**, 11.

Dimond, A. E. and Davis, D. (1952) 2-carboxymethylmercaptobenzothiazole salts as chemotherapeutants for plant diseases. *Phytopathology* **42**, 7.

Dimond, A. E. and Davis, D. (1953) The chemotherapeutic activity of benzothiazole and related compounds for *Fusarium* wilt of tomato. *Phytopathology* **43**, 43.

Dimond, A. E., Davis, D., Chapman, R. A. and Stoddard, E. M. (1952) Plant chemotherapy as evaluated by the *Fusarium* wilt assay on tomatoes. *Conn. Agric. Exp. Sta. Bull.* no. 557, 1.

Dimond, A. E. and Horsfall, J. G. (1959) Plant chemotherapy. *A. Rev. Pl. Physiol.* **10**, 257.

Dimond, A. E., Plumb, G. H., Stoddard, E. M. and Horsfall, J. G. (1949) An evaluation of chemotherapy and vector control by insecticides for combating Dutch elm disease. *Conn. Agric. Exp. Sta. Bull.* no. 531, 1.

Dimond, A. E. and Waggoner, P. E. (1953) On the nature and role of vivotoxins in plant disease. *Phytopathology* **43**, 229.

Dixon, G. R. and Pegg, G. F. (1969) Hyphal lysis and tylose formation in tomato cultivars infected by *Verticillium albo-atrum*. *Trans. Br. mycol. Soc.* **53**, 109.

Dockinger, L. S. (1975) Marssonina blight of big tooth aspen in forest nursery diseases in the United States. (eds G. W. Peterson and R. S. Smith) *USDA Forest Service Ag. Handbook* **470**, 95.

Doma, S., Clifford, D. R. and Byrde, R. J. W. (1971) Experiments with some of the newer systemic fungicides on apple and marrow. *Pestic. Sci.* **2**, 197.

Doodson, J. K. and Saunders, P. J. W. (1969) Observations on the effects of some systemic fungicides applied to cereals in trials at the NIAB. *Proc. 5th Br. Insect. Fungic. Conf.*, **1**, p. 1.

Douch, P. G. C. (1973) The metabolism of benomyl fungicide in mammals. *Xenobiotica* **3**, 367.

Douch, P. G. C. (1974) The metabolism of some thioureidobenzene fungicides in mice and sheep. *Xenobiotica* **4**, 457.

Douchet, J. P., Lhoste, J. and Quere, G. (1971) Thiophanate et methylthiophanate, nouveaux fongicides systèmiques actifs sur la tavelure et l'oidium des arbres fruitiers à pepins. *Meded. Fac. Landbouw. RijksUniversiteit Gent* **36**, 135.

Douchet, J. P., Quere, G. and Lhost, J. (1971) Action fongicide de l'association methyl thiophanate and manèbe dans la lutte contre les maladies des organes aériens des céréales. *Phytiatrie-Phytopharmacie*, No. Special. 37.

Drorun, J. A. and Kusch, D. S. (1974) Balsam poplar leaf spots (*Septoria musiva, Marssonina populi*). *Fungicide and Nematicide Tests* **30**, 127.

Eaton, F. M. (1930) The effect of boron on powdery mildew and spot blotch of barley. *Phytopathology* **20,** 967.

Ebben, M. H. (1970a) Tomato wilt caused by *Verticillium* spp. *A. Rep. Glasshouse Crops Res. Inst.* (1969), p. 119.

Ebben, M. H. (1970b) Cucumber diseases. *A. Rep. Glasshouse Crops Res. Inst.* (1969), p. 120.

Ebben, M. H. (1971) Cucumber stem rot caused by *Mycosphaerella melonis. A. Rep. Glasshouse Crops Res. Inst.* (1970), p. 137.

Ebben, M. H. and Last, F. T. (1969) Benomyl applied to soil for the control of some pathogens of tomato and cucumber. *Proc. 5th Br. Insectic. Fungic. Conf.*, p. 315.

Ebben, M. H. and Spencer, D. M. (1973) The integrated use of two systemic fungicides for the control of cucumber powdery mildew. *Proc. 7th Br. Insectic. Fungic. Conf.*, p. 211.

Ebben, M. H. and Spencer, D. M. (1975) *Fusarium* wilt of carnation. *A. Rep. Glasshouse Crops Res. Inst.* (1974), p. 107.

Ebenebe, C., Fehrmann, H. and Grossman, F. (1971) Effect of a new systemic fungicide, piperazin-1, 4-diyl-bis-(1-(2,2,2,-trichloroethyl) formamide) against wheat leaf rust. *Pl. Dis. Reptr* **55,** 691.

Eble, T. E., Bergy, M. E., Large, C. M., Herr, R. R. and Jackson, W. G. (1959) Isolation, purification and properties of streptovicins A and B. *Antibiot. Ann.* p. 555.

Eckert, J. W. (1969) Chemical treatments for control of post-harvest diseases. *Wld Rev. Pest Control* **8,** 138.

Eckert, J. W. and Kolbezen, M. J. (1971) Chemical treatments for the control of post-harvest diseases of citrus fruits. *Proc. 6th Br. Insectic. Fungic. Conf.* **3,** 683.

Edgington, L. V. (1963) A chemical that retards development of Dutch elm disease. *Phytopathology* **53,** 349.

Edgington, L. V. and Barron, G. L. (1967) Fungitoxic spectrum of oxathiin compounds. *Phytopathology* **57,** 1256.

Edgington, L. V., Buchenauer, H. and Grossmann, F. (1973) Bioassay and transcuticular movement of systemic fungicides. *Pestic. Sci.* **4,** 747.

Edgington, L. V. and Busch, L. V. (1967) Control of *Rhizoctonia* stem canker in potato. *Can. Pl. Dis. Surv.* **47,** 28.

Edginton, L. V., Corden, M. E. and Dimond, A. E. (1961) The role of pectic substances in chemically induced resistance to *Fusarium* wilt of tomato. *Phytopathology* **51,** 179.

Edgington, L. V. and Dimond, A. E. (1959) Correlation between the wilt resistance of plants and the nature of their pectic substances. *Phytopathology* **49,** 538.

Edgington, L. V. and Dimond, A. E. (1964) The effect of adsorption of organic cations to plant tissue on their use as systemic fungicides. *Phytopathology* **54,** 1193.

Edgington, L. V., Khew, K. L. and Barron, G. L. (1971) Fungitoxic spectrum of benzimidazole compounds. *Phytopathology* **61,** 42.

Edgington, L. V. and Walker, J. C. (1958) Influence of calcium and boron nutrition on development of *Fusarium* wilt of tomato. *Phytopathology* **48,** 324.

Edgington, L. V., Walton, G. S. and Miller, P. M. (1966) Fungicide selective for Basidiomycetes. *Science N.Y.* **153,** 307.

Edney, K. L. (1970) Some experiments with thiabendazole and benomyl as postharvest treatments for the control of storage rots of apples. *Pl. Path.* **19,** 189.

Elgersma, D. M. (1967) Factors determining resistance of elms to *Ceratocystis ulmi*. *Phytopathology* **57,** 641.

Elgersma, D. M. (1970) Length and diameter of xylem vessels as factors in resistance of elms to *Ceratocystis ulmi*. *Neth. J. Pl. Pathol.* **76,** 179.

Elia, M. (1969) Brief preliminary notes on endotherapeutic trials against citrus 'mal secco'. *Inftore fitopatol.* **19,** 403. (*Rev. Pl. Path.* **49,** 2482, 1970.)

Elias, R. S., Shephard, M. C., Snell, B. K. and Stubbs, J. (1968) 5-*n*-Butyl-2-dimethylamino-4-hydroxy-6-methylpyrimidine: a systemic fungicide. *Nature, Lond.* **219,** 1160.

Ellal, G. and Dinoor, A. (1973) Chemical control of Septoria leaf blotch in wheat. *Proc. 7th Br. Insectic. Fungic. Conf.* **1,** 193.

Ellis, C. A. V., Merryweather, J. W. and Phillips, W. H. (1974) Field evaluation of carbendazim (BAS 364F) as a fungicide on apples. *Proc. 7th Br. Insectic. Fungic. Conf.* **2,** 665.

El-Nakeeb, M. A. and Lampen, J. O. (1964a) Complexing griseofulvin by nucleic acids of fungi and its relation to griseofulvin sensitivity. *Biochem. J.* **92,** 59P.

El-Nakeeb, M. A. and Lampen, J. O. (1964b) Binding of tritiated griseofulvin by ribosomes of *Microsporum gypseum. Bacteriol. Proc.* **15,** 7.

El-Nakeeb, M. A., McLellan, W. L., Jr. and Lampen, J. O. (1965a) Antibiotic action of griseofulvin on dermatophytes. *J. Bacteriol.* **89,** 557.

El-Nakeeb, M. A., Moustafa, A. and Lampen, J. O. (1965b) Uptake of griseofulvin by the sensitive dermatophyte, *Microsporum gypseum. J. Bacteriol.* **89,** 564.

El-Zayat, M. M., Lukens, R. J., Dimond, A. E. and Horsfall, J. G. (1968) Systemic action of nitrophenols against powdery mildew. *Phytopathology* **58,** 434.

Endo, A., Kakiki, K. and Misato, T. (1970) Mechanism of action of the antifungal agent polyoxin D. *J. Bacteriol.* **104,** 189.

Endo, A. and Misato, T. (1969) Polyoxin D, a competitive inhibitor of UDP-N-acetylglucosamine: chitin N-acetylglucosaminyltransferase in *Neurospora crassa. Biochem. Biophys. Res. Comm.* **37,** 718.

Endo, K., Mori, Y., Kakiki, K. and Tomomasa, M. (1970) Absorption, translocation and metabolic fate of radioactive Inegin in rice plant. *Nippon Nogei Kagaku Kaishi,* **44,** 356. *Chem. Abstr.* **74,** 12100.

Engelhard, A. W. (1969) Preventive and residual fungicidal activity of three benzimidazole compounds and zinc ion + maneb against *Diplocarpon rosae* on two rose cultivars. *Pl. Dis. Reptr* **53,** 537.

Engelhard, A. W. (1971) Benlate, a systemic fungicide for cut flowers. *Florida Flower Grower,* **8,** 3.

Engelhardt, G., Wallnöfer, P. R. and Plapp, R. (1973) Purification and properties of an aryl acylamidase of *Bacillus sphaericus,* catalyzing the hydrolysis of various phenylamide herbicides and fungicides. *Appl. Microbiol.* **26,** 709.

Ennis, H. L. and Lubin, M. (1964) Cycloheximide: aspects of inhibition of protein synthesis in mammalian cells. *Science* **146,** 1474.

Entwistle, A. R. and Munasinghe, H. L. (1973) Recent studies on the fungicidal control of white rot disease of salad onions. *Proc. 7th Br. Insectic. Fungic. Conf.,* p. 581.

Entwistle, A. R. and Munasinghe, H. L. (1974) White rot disease of salad onions. *Rep. natn. Veg. Res. Sta. for 1974,* p. 110.

Enyinnia, T. and Halisky, P. M. (1974) Control of *Ustilago maydis* with systemic fungicides. *Phytopathology* **64,** 580.

Epstein, A. (1977) Soil application of systemic fungicides for suppression of Dutch elm disease infections. *Phytopathology* **67,** (in press).

Epstein, A. H. (1969) Enhancing efficacy of benomyl to control *Venturia inaequalis. Phytopathology* **59,** 1291 (abs.).

Epstein, A. H. (1970) Fungicidal control of Dothistroma needle blight of Austrian pine. *Pl. Dis. Reptr* **54,** 679.

Epstein, E. (1955) Passive permeation and active transport of ions into plant roots. *Pl. Physiol. Lancaster* **30,** 529.

Erwin, D. C. (1970) Progress in the development of systemic fungitoxic chemicals for control of plant diseases. *Pl. Prot. Bull. F.A.O.* **18,** 73.

Erwin, D. C. (1973) Systemic fungicides: Disease control, translocation and mode of action. *A. Rev. Phytopath.* **11,** 389.

Erwin, D. C. (1976) *Control of vascular pathogens by antifungal* compounds I. *Discovery, Development and Uses.* Chap. 10 (in press). New York: Marcel Dekker Press.

Erwin, D. C., Hee, H. and Sims, J. J. (1968a) The systemic effect of 2-benzimidazolecarbamic acid, methyl en *Verticillium* wilt of cotton. *Phytopathology* **58,** 528.

Erwin, D. C., Sims, J. J. and Partridge, J. (1968b) Evidence for the systemic fungitoxic activity of 2-(4′-thiazoly)-benzimidazole in the control of *Verticillium* wilt of cotton. *Phytopathology* **58,** 860.

Erwin, D. C., Wang, M. C. and Sims, J. J. (1970) Translocation of 2-(4-thiazolyl)benzimidazole in cotton. *Phytopathology* **60,** 1291.

Esau, K. (1965) *Plant anatomy*. 2nd ed. Wiley: New York.

Esau, K., Engelman, E. M. and Bisalputra, T. (1963) What are translocation strands? *Planta* **50**, 617.

Eschrich, W. (1967) Bidirektionelle Translocation in Siebröhren. *Planta* **73**, 37.

Eschrich, W. (1970) Biochemistry and fine structure of phloem in relation to transport. *A. Rev. Pl. Physiol.* **21**, 193.

Evans, E. (1968) *Plant diseases and their chemical control*. Oxford: Blackwells Scientific Publications.

Evans, E. (1971a) Systemic fungicides in practice. *Pest. Sci.* **2**, 192.

Evans, E. (1971b) Problems and progress in the use of systemic fungicides. *Proc. 6th Br. Insectic. Fungic. Conf.* **3**, 758.

Evans, E. (1973). Practical aspects of using systemic fungicides. *EPPO Bull. No. 10*, 59.

Evans, E., Couzens, B. J. and Griffiths, W. (1965) Timing experiments on the control of potato blight with copper fungicides in the United Kingdom. *Wld Rev. Pest Control* **4**, 84.

Evans, E. and Hawkins, J. H. (1971) Timing of fungicidal sprays for control of mildew on spring barley. *Proc. 6th Br. Insectic. Fungic. Conf.*, p. 33.

Evans, E., Marshall, J., Couzens, B. J. and Runham, R. L. (1970) The curative activity of non-ionic surface active agents against some powdery mildew diseases. *Ann. appl. Biol.* **65**, 473.

Evans, E., Marshall, J. and Hammond, C. (1971) An evaluation of benomyl as a scab/mildew fungicide in the United Kingdom. *Pl Path.* **20**, 1.

Evans, E., Rickard, M. and Whitehead, R. (1969) Effect of benomyl on some diseases of spring barley. *Proc. 5th Br. Insectic. Fungic. Conf.*, p. 610.

Evans, E. and Saggers, D. T. (1962) A new systemic rust therapeutant. *Nature, Lond.* **195**, 619.

Evans, E. J. (1975) *Min. of Ag. East Midlands Region, Results of Hort. Expts.* 1975, p. 83.

Evans, S. G. (1974) Chemical control of carnation rust. *Min. of Ag. Devel. Rep. S.E. Region, Hortic. Crops* (1974), p. 8.

Fargher, R. G., Galloway, L. D. and Probert, M. E. (1930) The inhibitory action of certain substances on the growth of mould fungi. *Mem. Shirley Inst.* **9**, 37.

Farrell, G. M., Preece, T. F. and Wren, M. J. (1969) Effects of infection by *Phytophthora infestans* (Mont) deBary on the stomata of potato leaves. *Ann. appl. Biol.* **63**, 265.

Fawcett, C. H. and Spencer, D. M. (1969) Natural antifungal compounds. *Fungicides Vol. 2*, (ed. D. C. Torgeson), p. 637. New York and London: Academic Press.

Fawcett, C. H. and Spencer, D. M. (1970) Plant chemotherapy with natural products. *A. Rev. Phytopath.* **8**, 403.

Fawcett, C. H., Spencer, D. M. and Wain, R. L. (1955) Investigations on fungicides. I. Fungicidal and systemic fungicidal activity in certain aryloxyalkanecarboxylic acids. *Ann. appl. Biol.* **43**, 553.

Fawcett, C. H., Spencer, D. M. and Wain, R. L. (1957) Investigations on fungicides. II. Aryloxy- and arylthio-alkanecarboxylic acids and their activity as fungicides and systemic fungicides. *Ann. appl. Biol.* **45**, 158.

Fawcett, C. H., Spencer, D. M. and Wain, R. L. (1958) Investigations on fungicides. IV. (Aryloxythio)trichloromethanes. *Ann. appl. Biol.* **46**, 651.

Fawcett, C. H., Spencer, D. M. and Wain, R. L. (1969) The isolation and properties of a fungicidal compound present in the seedlings of *Vicia faba*. *Neth. J. Pl. Path.* **75**, 72.

Fawcett, C. H., Spencer, D. M., Wain, R. L., Fallis, A. G., Jones, Sir E. R. H., LeQuan, M., Page, C. B., Thaller, V., Shubrook, D. C. and Whitham, P. M. (1968) Natural acetylenes. Part XXVIII. An antifungal acetylenic furanoid keto-ester (wyerone) from shoots of broad bean (*Vicia faba* L.) Fam. *Papilionaceae J. chem. Soc.* 2455.

Fawcett, C. H., Spencer, D. M., Wain, R. L., Jones, Sir E. R. H., LeQuan, M., Page, C. B. and Thaller, V. (1965) An antifungal acetylenic keto-ester from a plant of the *Papilionaceae* family. *Chem. Commun.* **1**, 422.

Fehrmann, H. F. (1970) Bekämpfung der Halmbruchkrankheit des Weizers mit Benomyl. *Nachrichtenbl. Deutsch. Pflanzenschutz* (Stuttgart) **22,** 136.

Fehrmann, H. and Schroedter, H. (1972) Ecological investigation on the epidemiology of *Cercosporella herpotrichoides*. IV. Elaboration of a practicable method for the control of the eye-spot disease in wheat with systemic fungicides. *Phytopath Z.* **74,** 161.

Feldman, A. W. and Caroselli, N. E. (1951) Soil and tree acidity in relation to susceptibility to Dutch elm disease. *Phytopathology* **41,** 12.

Feldman, A. W., Caroselli, N. E. and Howard, F. L. (1950) Physiology of toxin production by *Ceratostomella ulmi. Phytopathology* **40,** 341.

Fensom, D. S. (1957) The bioelectrical potentials of plants and their functional significance. I. *Can. J. Bot.* **35,** 573.

Fensom, D. S. (1959) The boelectrical potentials of plants and their functional significance. III. *Can. J. Bot.* **37,** 1003.

Fensom, D. S., Clattenburg, R., Chung, T., Lee, D. R. and Arnold, D. C. (1968) Moving particles in intact sieve tubes of *Heracleum mantegazzianum. Nature, Lond.* **219,** 531.

Fensom, D. S. and Williams, E. J. (1974) On Allen's suggestion for long-distance translocation in phloem of plants. *Nature, Lond.* **250,** 490.

Ferguson, I. B. and Clarkson, D. T. (1975) Ion transport and endodermal suberization in the roots of *Zea mays. New Phytol.* **75,** 69.

Fest, C. and Schmidt, K.-J. (1973) The chemistry of organophosphorus pesticides. New York: Springer Verlag.

Filer, T. H., Jr. (1975) Melampsora rust of cottonwood in forest nursery diseases in the United States. (eds G. W. Peterson and R. S. Smith) *USDA Forest Service Agr. Handbook* **470,** 99.

Findlayson, D. G. and Campbell, C. J. (1971) Fungicides for preventing clubroot of cauliflower in loam and peat soils. *Can. Pl. Dis. Surv.* **51,** 122.

Finney, J. R. and Hall, D. W. (1971) The effect of an autumn attack of mildew (*Erysiphe graminis* DC.) on the growth and development of winter barley. *Pl. Path.* **21,** 73.

Finney, J. R., Farrell, G. M. and Bent, K. J. (1975) Bupirimate—a new fungicide for the control of powdery mildews on apples and other crops. *Proc. 8th Br. Insectic. Fungic. Conf.* **2,** 667.

Firman, I. D. (1972) Black leaf streak of Bananas in Fiji. *Ann. appl. Biol.* **70,** 19.

Fisher, D. J. (1974) A note on the mode of action of the systemic fungicide tridemorph. *Pestic. Sci.* **5,** 219.

Fleeker, J. R., Lacy, H. M., Schultz, I. R. and Houkom, E. C. (1974) Persistence and metabolism of thiophanate-methyl in soil. *J. Agr. Food Chem.* **22,** 592.

Fletcher, J. T. (1971) Systemic fungicides – Progress and promise. *Proc. 13th East. Reg. Grow. Conf. Oaklands 1971*, p. 40.

Ford, J. H., Klomparens, W. and Hammer, C. L. (1958) Cycloheximide (Acti-dione) and its agricultural uses. *Pl. Dis. Reptr* **42,** 680.

Formigoni, A., Castagna, G. and Ciocca, P. (1971) Two years of trials in Italy on 1,2-bis-(3-ethoxy-carbonyl-2-thioureido) benzene, a new systemic fungicide. *Meded. Fac. Landbouw. RijksUniversiteit Gent* **36,** 96.

Forsyth, F. R. (1957) Effect of ions of certain metals on the development of stem rust in the wheat plant. *Nature, Lond.* **179,** 217.

Forsyth, F. R. and Peturson, B. (1959a) Chemical control of cereal rusts. IV. The influence of nickel compounds on wheat, oat and sunflower rusts in the greenhouse. *Phytopathology* **49,** 1.

Forsyth, F. R. and Peturson, B. (1959b) Control of leaf rust of wheat with inorganic nickel. *Pl. Dis. Reptr* **43,** 5.

Forsyth, F. R. and Peturson, B. (1960) Control of leaf and stem rust of wheat by zineb and inorganic nickel salts. *Pl. Dis. Reptr* **44,** 208.

Foschi, S., Rapparini, G. and Bencivelli, A. (1972) New results on scab and powdery mildew control on apples. *Informatore Agrario* **28,** 8489. (*Rev. Pl. Path.* **53,** 3540, 1974).

Foy, C. L. (1961) Absorption, distribution and metabolism of 2,2-dichloropropionic acid II. *Pl. Physiol. Lancaster* **36,** 698.

Frank, A. (1971) Studies on the metabolism of 2-(2-furyl)benzimidazole in certain mammals. *Acta Pharmacol. Toxicol.* **29,** suppl. 2, 1.

Franke, W. (1967) Mechanisms of foliar penetration of solutions. *A. Rev. Pl. Physiol.* **18,** 281.

Franklin, T. J. and Snow, G. A. (1971) *Biochemistry of antimicrobial action.* Chapman and Hall Ltd., London.

Frazao, A. (1970) Chemical control of *Botrytis cinerea* in vineyards. *Summaries of Papers, VIIth Int. Cong. Plant Prot., Paris,* p. 258.

Freeman, J. A. and Pepin, H. S. (1967) A systemic fungicide (Fungicide 1991) for the control of gray mould and powdery mildew in strawberries and raspberries. *Can. Pl. Dis. Surv.* **47,** 104.

Freeman, J. A. and Pepin, H. S. (1968) A comparison of two systemic fungicides with non-systemic for control of fruit rot and powdery mildew of strawberries. *Can. Pl. Dis. Surv.* **48,** 120.

Freeman, T. E. (1967) Diseases of southern turfgrasses. *Florida Agr. Exp. Sta. Tech. Bul.* **713,** 31 pp.

Frick, E. L. (1971) Fungicides. *Rep. E. Malling Res. Stn, 1970,* p. 132.

Fridrich, J. H. and Gilpatrick, J. D. (1971) Apple: scab (*Venturia inaequalis*). *Fungicide and Nematicide Tests* **26,** 10.

Frohberger, P.-E. (1975) New approaches to the control of cereal diseases with triadimefon (MEB 6447) *Proc. 8th int. Plant Prot. Cong. Moscow,* **3,** 247.

Fron, G. (1936) La maladie de l'orme. *C. r. hebd. Séanc. Acad. Agric. Fr.* **22,** 1081.

Frossard, P. (1968) Essais de désinfection des pedoncules d'Ananas contre le *Thielaviopsis paradoxa. Fruits d'outre-mer* **23,** 207.

Frossard, P. (1969) Action du thiabendazole et du benlate sur l'anthracnose des Bananes et son champignon pathogène: *Colletotrichum musae. Fruits d'outre-Mer* **24,** 365. (*Rev. Pl. Path.* **49,** 523, 1970.)

Frossard, P. (1970a) Précisions sur les propriétés du benomyl (benlate) vis-à-vis de l'anthracnose de blessures des Bananes. *Fruits d'outre-mer* **25,** 265. (*Rev. Pl. Path.* **50,** 2391, 1971.)

Frossard, P. (1970b) Désinfection des Ananas contre *Thielaviopsis paradoxa. Fruits d'outre-mer* **25,** 785. (*Rev. Pl. Path.* **50,** 3069, 1971.)

Frossard, P., Laville, E. and Motillon, J. (1973) Study of fungicide treatments applied postharvest to bananas. I. Effect of thiophanate and thiophanate-methyl. *Fruits d'outre-mer* **28,** 195.

Frost, A. J. P. B. and Pattisson, N. (1971) Some results obtained in the United Kingdom using dodemorph for the control of powdery mildew in roses and other ornamentals. *Proc. 6th Brit. Insectic. Fungic Conf.,* p. 349.

Frost, A. J. P., Jung, K. U. and Bedford, J. L. (1973) The timing of application of benodanil (BAS 3170 F) for the control of cereal rust diseases. *Proc. 7th Br. Insectic. Fungic. Conf.* **1,** 111.

Frost, A. J. P. and Brown, E. A. (1973) The effect of tridemorph alone and in mixtures with other fungicides on diseases other than *Erysiphe graminis. Proc. 7th Br. Insectic. Fungic. Conf.* **1,** 105.

Froyd, J. D. and Johnson, H. G. (1966) Sugar beets. *Cercospora* leaf spot. *Cercospora beticola. Fungicide–Nematicide tests, results of 1966,* **22,** 98. American Phytopathological Society.

Fuchs, A. (1964) Effect of treatment with anthraquinone derivatives on *Fusarium* infected tomato plants. *Host–parasite relations in plant pathology* (eds. Z. Király and G. Ubrizsy), p. 175. Budapest: Research Institute for Plant Protection.

Fuchs, A. and Bollen, G. J. (1975) Benomyl, after seven years. *Systemfungizide,* pp. 121–36. (eds H. Lyr and C. Polter) Berlin: Akademie-Verlag.

Fuchs, A., Doma, S. and Vörös, J. (1971) Laboratory and greenhouse evaluation of a new systemic fungicide, N,N'-bis-(1-formamido-2,2,2-trichloroethyl)piperazine (CELA W 524). *Neth. J. Pl. Path.* **77,** 42.

Fuchs, A. and Drandarevski, Ch. A. (1973) Wirkungsbreite und Wirkungsgrad von Triforine *in vitro* und *in vivo. Zeitschr. Pflanzenkrankheiten Pflanzenschutz* **80,** 403.

Fuchs, A. and Drandarevski, Ch. A. (1976) The likelihood of development of resistance to systemic fungicides which inhibit ergosterol biosynthesis. *Neth. J. Pl. Path.* **82,** 85.

Fuchs, A., Homans, A. L. and de Vries, F. W. (1970) Systemic activity of benomyl against *Fusarium* wilt of pea and tomato plants. *Phytopath. Z.* **69,** 330.

Fuchs, A. and Ost, W. (1976) Translocation, distribution and metabolism of triforine in plants. *Arch. Environm. Contam. Toxicol.* **4,** 30.

Fuchs, A. and Viets-Verweij, M. (1975) Permanent and transient resistance against triarimol and triforine in some pathogenic fungi. *Meded. Fakulteit Landbouwwetenschappen Gent* **40,** 699.

Fuchs, W. H. and Bauermeister, R. (1958) Uber die Wirkung von Thiosemicarbazid auf die Entwicklung von *Puccinia graminis tritici* auf Weizen. *Naturwissenschaften* **45,** 343.

Fukunaga, K., Misato, T. and Asakawa, M. (1955) Blasticidin, a new anti-phytopathogenic fungal substance. *Bull. Agr. Chem. Soc. Japan* **19,** 181.

Fulton, D., Cole, H., Goldberg, C. W. and Massie, L. B. (1971) *Fungicide and Nematicide Tests* **27,** 136.

Futrell, M. C. (1972) New concepts in chemical seed treatment of agronomic crops. *J. Environ. Qual.* **1,** 240.

Futreff, M. C. and Berry, J. W. (1964) The action of propionate on wheat stem rust and sorghum head smut. *Phytopathology* **54,** 893.

Gabrielson, R. L., Guildford, R. K. and Cochrane, D. R. (1971). Field control of white and gray mould of beans in Western Washington. *Pl. Dis. Reptr* **55,** 234.

Gaff, D. F., Chambers, T. C. and Markus, K. (1964) Studies of extrafascicular movement of water in the leaf. *Aust. J. Biol. Sci.* **17,** 581.

Gage, R. S. and Aronoff, S. (1960) Translocation. III. *Pl. Physiol. Lancaster* **35,** 53.

Gandy, D. G. (1971a) Experiments on the use of benomyl (Benlate) against *Verticillium. M.G.A. Bull.* **257,** 184.

Gandy, D. G. (1972) Observation on the development of *Verticillium malthousei* in mushroom crops and the role of cultural practices in its control. *Mushr. Sci.* **8,** 171.

Gandy, D. G. and Spencer, D. M. (1974) Fungicide tolerance and its implications. *Mushroom Journal,* **24,** 468.

Gardiner, J. A., Kirkland, J. J., Klöpping, H. L. and Sherman, H. (1974) Fate of benomyl in animals. *J. Agr. Food Chem.* **22,** 419.

Garibaldi, A. (1968) Sull 'efficacia del tiabendazolo contro la fialoforosi del garofano: Nota preliminaire *Inftore Fitopatol.* **18,** 52.

Gartel, W. (1974). Investigation on the effect of fungicides on spore discharge by fruiting bodies of the agent of dead arm disease (*Phomopsis viticola* Sacc.). *Weinberg und Keller* **21,** 267.

Gassner, G. and Hassebrauk, K. (1936) Untersuchungen zur Frage der Getreiderostbekämpfung mit chemischen Mitteln. *Phytopath. Z.* **9,** 427.

Gaunt, R. E. and Manners, J. G. (1971) Host–parasite relations in loose smut of wheat. II. *Ann. Bot.* **35,** 1141.

Gaze, R. H. and Fletcher, J. T. (1975) ADAS survey of mushroom diseases and fungicide usage 1974/75. *Mushroom Journal* **35,** 370.

Geary, T. F. and Kuntz, J. E. (1962) Effect of growth regulators on oak wilt development. *Phytopathology* **52,** 733.

Georgopoulos, S. G. (1962) Genetic nature of tolerance to *Hypomyces solani* f. *cucurbitae* to penta- and tetrachloronitrobenzene. *Nature, Lond.* **194,** 148.

Georgopoulos, S. G. (1963) Tolerance to chlorinated nitrobenzenes in *Hypomyces solani* f. *cucurbitae* and its mode of inheritance. *Phytopathology* **53,** 1086.

Georgopoulos, S. G. (1977) Development of fungal resistance to fungicides. In *Antifungal Compounds,* vol. 2 (eds Sisler, H. D. and Siegel, M. R.). Marcel Dekker, Inc. (in press).

Georgopoulos, S. G., Alexandri, E. and Chrysayi, M. (1972) Genetic evidence for the action of oxathiin and thiazole derivatives on the succinic dehydrogenase system of *Ustilago maydis* mitochondria. *J. Bacteriol.* **110,** 809.

Georgopoulos, S. G., Chrysayi, M. and White, G. A. (1975a) Carboxin resistance in the haploid, the heterozygous diploid, and the plant-parasitic dicaryotic phase of *Ustilago maydis. Pestic. Physiol.* **5,** 543.

Georgopoulos, S. G. and Dovas, C. (1973) A serious outbreak of strains of *Cercospora beticola* Sacc. resistant to benzimidazole fungicides in Northern Greece. *Plant Dis. Reptr* **57,** 321.

Georgopoulos, S. G., Geerligs, J. W. G. and Dekker, J. (1975b) Sensitivity of *Ustilago maydis* to pyrazophos and one of its conversion products, and failure to induce resistance with U.V.-treatment. *Neth. J. Plant Pathol.* **81,** 38.

Georgopoulos, S. G., Kappas, A. and Hastie, A. C. (1976) Induced sectoring in diploid *Aspergillus nidulans* as a criterion on fungitoxicity by interference with hereditary processes. *Phytopathology* **66,** 217.

Georgopoulos, S. G. and Panopoulos, N. J. (1966) The relative mutability of the cub loci in *Hypomyces. Can. J. Genetic. Cytol.* **8,** 347.

Georgopoulos, S. G. and Sisler, H. D. (1970) Gene mutation eliminating antimycin A-tolerant electron transport in *Ustilago maydis. J. Bacteriol.* **103,** 745.

Georgopoulos, S. G. and Zaracovitis, C. (1967) Tolerance of fungi to organic fungicides. *Ann. Rev. Phytopathol.* **5,** 109.

Gerecke, M., Kyburz, E., Planta, C. von and Brossi, A. (1962) Griseofulvin IV. Synthesis of griseofulvin analogues. *Helv. Chim. Acta.* **45,** 2241.

Germar, B. (1934) Über einige Wirkungen der Kieselsäure in Getreideptlanzen insbesondere auf deren Resistenz gegenüber Mehltau. *Z. Pflernähr. Düngung BodenKde A* **35,** 102.

Geuther, T. (1895) Über die Einwirkung von Formaldehydelösungen auf Getriedebrand. *Ber. dt. pharmacol. Ges.* **5,** 325.

Gibbs, J. N. and Clifford, D. R. (1974) Experiments with MBC derivatives for the control of Dutch elm disease. *Ann. appl. Biol.* **78,** 309.

Gibbs, J. N. and Dickinson, J. (1975) Fungicide injection for the control of Dutch elm disease. *Forestry* **48,** 165.

Gibson, I. A. S. and Howard, P. (1969) Graft failure in young *Cupressus lusitanica. East African Ag. and Forestry Jour.* **35,** 52.

Gigante, R. (1935) Ricerche sopra l'influenza del boro sulla resistenza della piante agli attachi parassitari. *Boll. Staz. Patol. veg. Roma* **15,** 471.

Gilchrist, A. J. and Cole, R. J. (1971) Control of diseases of soft fruit and other crops in the United Kingdom with thiophanate methyl. *Proc. 6th Br. Insectic. Fungic. Conf.* **2,** 332.

Gill, D. L. (1973) Fungicides and application frequency for arborvitae blight control. *Pl. Dis. Reptr* **57,** 739.

Gill, D. L. (1974) Control of Phomopsis blight of junipers. *Pl. Dis. Reptr* **58,** 1012.

Gilmour, J. (1971) Fungicidal control of mildew on spring barley in South-East Scotland. *Proc. 6th Br. Insectic. Fungic. Conf.,* **1,** p. 63.

Gilmour, J. and Fawcett, R. H. (1973) A risk analysis of the control of barley mildew with fungicides. *Proc. 7th Br. Insectic. Fungic. (1973) Conf.* **1,** 1.

Gilmour, J. W. and Vanner, A. L. (1971) Radiata pine needle blight (*Dothistroma pini*). *Fungicide and Nematicide Tests* **27,** 137.

Gilmour, J. W. and Vanner, A. L. (1972) Radiata pine needle blight (*Dothistroma pini*). *Fungicide and Nematicide Tests* **28,** 128.

Gilpatrick, J. D. (1969) Systemic activity of benzimidazoles as soil drenches against powdery mildew of apples and cherries. *Pl. Dis. Reptr* **53,** 721.

Gilpatrick, J. D. (1970) Apple: scab (*Venturia inaequalis*). *Fungicide and Nematicide Tests* **25,** 12.

Gilpatrick, J. D. (1973) Control of brown rot of stone fruits with thiophanate-methyl and a piperazine derivative fungicide. *Pl. Dis. Reptr* **57,** 457.

Gilpatrick, J. D. (1975) (ed.) Pome fruit disease reports. *Fungicide and Nematicide Tests* **30,** 6.

Gilpatrick, J. D. and Blowers, D. R. (1974) Ascospore tolerance to dodine in relation to orchard control of apple scab. *Phytopathology* **64,** 649.

Gilpatrick, J. D. and Bourke, J. (1973) Systemic activity of the fungicide triforine on apple seedlings. *Med. Facult. Landbouwwettensch. Riks Univ. Gent* **38,** 1537.

Gilpatrick, J. D., Fridrich, J., Smith, C. A. and Lienk, S. E. (1972) Orchard control of diseases and mites in apple with a new piperazine derivative fungicide. *Pl. Dis. Reptr* **56**, 451.

Gilpatrick, J. D. and Szkolnik, M. (1970) Unusual after-infection control of apple scab with α(2,4-dichlorophenyl)α-phenyl-5-pyrimidine methanol against *Venturia inaequalis*. *Phytopathology* **60**, 574.

Gilpatrick, J. D. and Szkolnik, M. (1972) Control of brown rot of plum with a new piperazine derivative fungicide. *Pl. Dis. Reptr* **56**, 456.

Goldberg, C. W. and Cole, H. (1972) *Fungicide and Nematicide Tests* **28**, 120.

Goldberg, C. W., Cole, H. and Burpee, L. (1973) *Fungicide and Nematicide Tests* **29**, 118.

Goodman, R. N. (1959) The influence of antibiotics on plants and plant disease control. *Antibiotics – Their Chemistry and Non-Medical Uses* (ed. H. S. Goldberg), p. 322. Princeton: van Nostrand.

Goodman, R. N. (1962a) The impact of antibiotics upon plant disease control. *Advances in Pest Control Research Vol. V* (ed. R. L. Metcalf), p. 1. New York: Wiley Interscience Publishers Inc.

Goodman, R. N. (1962b) Systemic effects of antibiotics. *Antibiotics in Agriculture* (ed. M. Woodbine), p. 165. London: Butterworths.

Goodman, R. N. (1967) Uses of antibiotics in plant pathology and production. *Antimicrob. Ag. Chemother. 1966*, 747.

Goodman, R. N. and Dowler, W. M. (1958) The absorption of streptomycin as influenced by growth regulators and humectants. *Pl. Dis. Reptr* **42**, 122.

Gorbach, S., Thier, W., Kellner, H. M., Schulze, E. F., Kuenzler, K. and Fisher, H. (1975) Environmental impact of pyrazophos. 1. Degradation in plants and in the rat. *Pesticides*, p. 840 (eds F. Coulston and F. Korte). Stuttgart: Georg Thieme Verlag.

Gottlieb, D. (1957) The effect of metabolites on antimicrobial agents. *Phytopathology* **47**, 59.

Gottlieb, D. and Kumar, K. (1970) The effect of thiabendazole on spore germination. *Phytopathology* **60**, 1451.

Gottlieb, D. and Shaw, P. D. (1967) *Antibiotics*. Vol. I. Mechanism of action. New York: Springer.

Gould, C. J. and Miller, V. L. (1970) Effectiveness of benzimidazole fungicides in controlling *Fusarium* basal rot of narcissus. *Pl. Dis. Reptr* **54**, 377.

Gould, C. J. and Miller, V. L. (1971) Improved control of *Fusarium* basal rot with benomyl and thiabendazole. *Pl. Dis. Reptr* **55**, 425.

Graham-Bryce, I. J. and Coutts, J. (1971) Interactions of pyrimidine fungicides with soil and their influence on uptake by plants. *Proc. 6th Br. Insectic. Fungic. Conf.* **2**, p. 419.

Gramlich, J. V., Schwer, J. F. and Brown, I. F. Jr. (1969) Characteristics and field performance of a new broad-spectrum fungicide. *Proc. 5th Br. Insectic. Fungic. Conf.* **2**, 576.

Grapov, A. F. and Mel'nikov, N. N. (1973) Organophosphorus fungicides. *Russ. Chem. Revs.* **42**, 772.

Gray, L. E. and Sinclair, J. B. (1971) Systemic uptake of ^{14}C-labelled 2-(4′-thiazolyl) benzimidazole in soybean. *Phytopathology* **61**, 523.

Gray, R. A. (1956) Increasing the absorption of streptomycin by leaves and flowers with glycerol. *Phytopathology* **46**, 105.

Gray, R. A. (1958) The downward translocation of antibiotics in plants. *Phytopathology* **48**, 71.

Greenaway, W. (1971) Relationship between mercury resistance and pigment production in *Pyrenophora avenae*. *Trans. Brit. mycol. Soc.* **56**, 37.

Greenaway, W. (1972) (private communication).

Greenaway, W. and Cowan, J. W. (1970) The stability of mercury resistance in *Pyrenophora avenae*. *Trans. Brit. mycol. Soc.* **54**, 127.

Greenaway, W. and Whatley, F. R. (1975) Resistance to systemic fungicides – causes and cures. *Current Advances in Plant Science* **20**, 335–47.

Gregory, G. F. and Jones, T. W. (1974) Protection of sand-grown red oak seedlings from oak wilt disease by drenching with benomyl. *Pl. Dis. Reptr* **58**, 65.

Gregory, G. F. and Jones, T. W. (1975) An improved apparatus for pressure-injecting fluid into trees. *US Forest Service Res. Note* NE-214. 6 pp.

Gregory, G. F., Jones, T. W. and McWain, P. (1971) Injection of benomyl into elm, oak and maple. N.E. Forest Expt. Sta., Upper Darby, Pa. *USDA Forest Serv. Res. Paper* NE-232. 7 pp.

Gregory, G. F., McWain, G. and Jones, T. W. (1972) Effectiveness of benomyl as a protectant and therapeutant for oak wilt disease in red oak seedlings. *Phytopathology* **62,** 761 (abs.).

Gregory, G. F., Jones, T. W. and McWain, P. (1973) Pressure injection of methyl-2-benzimidazole carbamate hydrochloride solution as a control for Dutch elm disease. *USDA Forest Serv. Res. Paper* NE-176. 9 pp.

Grewe, F. and Büchel, K. H. (1973) Ein neues Mehltanfungizid aus der Klasse der Trityltriazole. *Mitteil. Biol. Bundesanstalt Land u. Forstwiss.* **151,** 208.

Griffee, P. J. (1973) Resistance to benomyl and related fungicides in *Colletotrichum musae. Trans. Brit. mycol. Soc.* **60,** 433.

Griffee, P. J. and Burden, O. J. (1973) Banana diseases in the Windward Islands. *Proc. 7th Br. Insectic. Fungic. Conf.* **3,** 887.

Griffee, P. J. and Burden, O. J. (1974) Incidence and control of *Colletotrichum musae* on bananas in the Windward Islands. *Ann. appl. Biol.* **77,** 11.

Griffee, P. J. and Pinegar, J. A. (1974) Fungicides for control of the banana crown rot complex: *in vivo* and *in vitro* studies. *Tropical Science* **16,** 107.

Grosjean, J. (1951) Onderzoekingen over de mogelijkheid van een bestrijding van de loodglansziekte volgens de boorgat-methode. *Tijdschr. PlZiekt.* **57,** 103.

Grosse-Brauckmann, E. (1957) Über den Einfluss der Kieselsäure auf den Mehltaubefall von Getreide bei unterschiedlicher Stickstoffdüngung. *Phytopath. Z.* **30,** 112.

Grosse-Brauckmann, E. (1958) Massnahmen zur Mehltaubekämpfung bei verschiedener Gerstensorten, ihre Erfolgsaussichten und ihr Einfluss auf den Mineralstoffgehalt. *Z. PflKrankh. PflPath. PflSchutz.* **65,** 689.

Grossmann, F. (1957) Untersuchungen über die innertherapeutische Wirkung organischer Fungizide. I. Thiocarbamate und Thiurame. *Z. PflKrankh. PflPath. PflSchutz.* **64,** 718.

Grossmann, F. (1958a) Untersuchungen über die innertherapeutische Wirkung organischer Fungizide. II. Chlornitrobenzole. *Z. PflKrankh. PflPath. PflSchutz.* **65,** 594.

Grossmann, F. (1958b) Über die Hemmung pektolytischer Enzyme von *Fusarium oxysporum* f. *lycopersici* durch Gerbstoffe. *Naturwiss.* **45,** 113.

Grossmann, F. (1961) Die systemische Wirkung von Sulfonamiden gegen Pflanzenkrankheiten, hervorgerufen durch nicht obligate Erreger. *Z. PflKrankh. PflPath. PflSchutz.* **68,** 633.

Grossmann, F. (1962a) Untersuchungen über die Hemmung pektolytischer Enzyme von *Fusarium oxysporum* f. *lycopersici.* I. Hemmung durch definierte Substanzen *in vitro. Phytopath. Z.* **44,** 361.

Grossmann, F. (1962b) Therapeutische Wirkung von Pektinase-Hemmstoffen gegen die *Fusarium*-Welke der Tomate. *Naturwissenschaften* **49,** 138.

Grossmann, F. (1962c) Untersuchungen über die Hemmung pektolytischer Enzyme von *Fusarium oxysporum* f. *lycopersici.* II. Hemmung durch Pflanzenextrakte *in vitro. Phytopath. Z.* **45,** 1.

Grossmann, F. (1962d) Untersuchungen über die Hemmung pektolytischer Enzyme von *Fusarium oxysporum* f. *lycopersici.* III. Wirkung einiger Hemmstoffe *in vivo. Phytopath. Z.* **45,** 139.

Grossmann, F. (1963) Fungitoxische, enzymehemmende und therapeutische Wirkungen der Rufiansäure (1,4-Dioxy-Anthrachinon-Sulfonsäure-(2)). *Meded. LandbHoogesch. OpzoekStns. Gent* **28,** 604.

Grossmann, F. (1968a) Studies on the therapeutic effects of pectolytic enzyme inhibitors. *Neth. J. Pl. Path.* **74, Suppl. 1,** 91.

Grossmann, F. (1968b) Conferred resistance in the host. *Wld Rev. Pest Control* **7,** 176.

Grossmann, F. and Buchenauer, H. (1975) Triazinobenzimidazoles as systemic fungicides. *Proc. 8th Int. Plant Prot. Cong.* **2,** 275.

Grouet, D. and Allaire, E. (1973) Efficacité de l'oxycarboxine et de la triforine contre la rouille blanche de chrysanthème (*Puccinia horiana* P. Henn.). *Phyt.-Phytopharm.* **22,** 177.

Grove, J. F. (1963) Griseofulvin. *Quart. Rev.* **17,** 1.

Grümmer, G. and Beyer, H. (1960) The influence exerted by species of *Camelina* on flax by means of toxic substances. *Biology of Weeds* (ed. J. L. Harper), p. 153. Oxford: Blackwell.

Gunning, B. E. S. and Pate, J. S. (1969) 'Transfer cells': plant cells with wall ingrowths, specialised in relation to short distance transport of solutes. Their occurrence, structure and development. *Protoplasma* **68,** 107.

Gutter, Y. (1969a) Comparative effectiveness of benomyl, thiabendazole and other antifungal compounds for post-harvest control of Penicillium decay in Shamouti and Valencia oranges. *Pl. Dis. Reptr* **53,** 474.

Gutter, Y. (1969b) Effectiveness of pre-inoculation and post-inoculation treatments with sodium orthophenylphenate, thiabendazole and benomyl for green mold control in artificially inoculated Eureka lemons. *Pl. Dis. Reptr* **53,** 479.

Gutter, Y. (1970) Influence of application time on effectiveness of fungicides for green mold control in artificially inoculated oranges. *Pl. Dis. Reptr* **54,** 325.

Gutter, Y., Yanko, U., Davidson, M. and Rahat, M. (1974) Relationship between mode of application of thiabendazole and its effectiveness for control of green mold and inhibiting fungus sporulation on oranges. *Phytopathology* **64,** 1477.

Hacker, R. G. and Vaughn, J. R. (1957a) Cycloheximide analogues cause pre-infection resistance to *Puccinia graminis* var. *tritici* in spring wheat. *Phytopathology* **47,** 14.

Hacker, R. G. and Vaughn, J. R. (1957b) Chemically induced resistance to stem rust of wheat by derivatives of Acti-dione. *Pl. Dis. Reptr* **41,** 442.

Hagborg, W. A. F. (1970) Carboxanilido systemic chemical in the control of leaf and stem rusts of wheat. *Can. J. Pl. Sci.* **50,** 631.

Hagborg, W. A. F. (1971) 'Plantavax' emulsifiable concentrate in the control of leaf and stem rusts in wheat. *Can. J. Pl. Sci.* **51,** 239.

Hagborg, W. A. F., Spencer, J. F. T. and Chelack, W. S. (1961) Antibiotic P-9 in the control of cereal rusts. *Can. J. Bot.* **39,** 1725.

ten Haken, P. and Dunn, C. L. (1971) Structure-activity relationships in a group of carboxanilides systemically active against broad bean rust (*Uromyces fabae*) and wheat rust (*Puccinia recondita*). *Proc. 6th Br. Insectic. Fungic. Conf.* **2,** 453.

Hall, D. W. (1971) Control of mildew in winter barley with ethirimol. *Proc. 6th Br. Insectic. Fungic. Conf.*, p. 26.

Hallam, N. D. and Juniper, B. E. (1971) The anatomy of the leaf surface. *Ecology of leaf surface micro-organisms* (ed. T. F. Preece and C. H. Dickinson), p. 3. Academic Press.

Hamilton, J. M. (1931) Studies of the fungicidal action of certain dusts and sprays in the control of apple scab. *Phytopathology* **21,** 445.

Hamilton, J. M. (1948) Evaluation of organic fungicides for the control of *Venturia inaequalis* (Cke.) Wint. *Phytopathology* **38,** 313.

Hamilton, J. M. and Szkolnik, M. (1953) Factors involved in the performance of cycloheximide (acti-dione) against *Coccomyces hiemalis*. *Phytopathology* **43,** 109.

Hamilton, J. M., Szkolnik, M. and Sondheimer, E. (1956) Systemic control of cherry leaf spot fungus by foliar sprays of Acti-dione derivatives. *Science, N.Y.* **123,** 1175.

Hamilton-Miller, J. M. T. (1974) Fungal sterols and the mode of action of the polyene antibiotics. *Adv. Appl. Microbiol.* **17,** 109.

Hammerschlag, R. S. and Sisler, H. D. (1972) Differential action of benomyl and methyl-2-benzimidazole carbamate (MBC) in *Saccharomyces pastorianus*. *Pestic. Biochem. Physiol.* **2,** 123.

Hammerschlag, R. S. and Sisler, H. D. (1973) Benomyl and methyl-2-benzimidazole carbamate (MBC). Biochemical, cytological and chemical aspects of toxicity to *Ustilago maydis* and *Saccharomyces cerevisiae*. *Pestic. Biochem. Physiol.* **3,** 42.

Hammett, K. R. W. (1968) Root application of a systemic fungicide for control of powdery mildews. *Pl. Dis. Reptr* **52,** 754.

Hampel, M. and Locher, F. (1973) Control of cereal diseases with carbendazim. *Proc. 7th Br. Insectic. Fungic. Conf.* **1,** 127.

Hampel, M. and Pommer, E. H. (1971) Results of trials with tridemorph on the control of *Mycosphaerella musicola* on bananas. *Proc. 6th Br. Insectic. Fungic. Conf.* **3,** 842.

Hansing, E. D. (1967) Systemic oxathiin fungicide for control of loose smut (*Ustilago tritici*) of winter wheat. *Phytopathology* **57,** 814 (abs.).

Hardenburg, R. E. (1974) Benomyl and thiabendazole treatments reduce decay of apples in polyethylene consumer bags. *J. Amer. hort. Soc.* **99,** 236.

Harder, H. H. and Luepschen, N. S. (1975) Synergistic effect of SADH and benomyl against peach *Cytospora* canker *HortScience* **10,** 77.

Harding, P. R. (1964) Assaying for biphenyl resistance in *Penicillium digitatum* in California lemon packing houses. *Plant Dis. Reptr* **48,** 43.

Harding, P. R. (1967) Wax emulsion additives for control of storage decay in lemons. *Pl. Dis. Reptr* **51,** 781.

Harding, P. R. (1968) Comparison of Fungicide 1991, thiabendazole and sodium orthophenylphenate for control of Penicillium molds of postharvest citrus fruit. *Pl. Dis. Reptr* **52,** 623.

Harding, P. R. (1972) Differential sensitivity to thiabendazole by strains of *Penicillium italicum. Plant Dis. Reptr* **56,** 256.

Hardison, J. R. (1967) Chemotherapeutic control of stripe smut (*Ustilago striiformis*) in grasses by two derivatives of 1,4-oxathiin. *Phytopathology* **57,** 242.

Hardison, J. R. (1971) Chemotherapy of smut and rust pathogens in *Poa pratensis* by thiazole compounds. *Phytopathology* **61,** 1396.

Hardison, J. R. (1972a) Control of *Ustilago striiformis* and *Urocystis agropyri* in *Poa pratensis* by thiophanate fungicides. *Phytopathology* **61,** 1462.

Hardison, J. R. (1972b) Control of *Ustilago striiformis* and *Urocystis agropyri* in *Poa pratensis* by 2,5-dimethyl-3-furylanilide. *Pl. Dis. Reptr* **56,** 252.

Hardison, J. R. (1975) Control of *Puccinia striiformis* by two new fungicides BAY MEB 6447 and BAS 31708. *Pl. Dis. Reptr* **59,** 652.

Hare, R. (1973) Soil applications of systemics for prevention and eradication of fusiform rust. *Pl. Dis. Reptr* **57,** 776.

Harrison, M. B. (1970) *Fungicide and Nematicide Tests* **26,** 122.

Harper, F. R. (1968) Control of root rot in garden peas with a soil fungicide. *Pl. Dis. Reptr* **52,** 565.

Harrow, K. M. (1970) Onion smut: an examination of a potential hazard to commercial growers in New Zealand. *N.Z. Comml Grow. for March 1970,* p. 15.

Hart, H. and Allison, J. L. (1939) Toluene compounds to control plant disease. *Phytopathology* **29,** 978.

Hart, J. H. (1972) Control of Dutch elm disease with foliar applications of benomyl. *Pl. Dis. Reptr* **56,** 685.

Härtel, K. and Röchling, H. (1973). Triazinobenzimidazoles, a new systemic group of fungicidal active substances. *2nd. Int. Cong. Plant Pathology,* 0118 (abs.).

Harvey, A. E. and Helton, A. W. (1962) *In vitro* effects of seventeen chemical agents on *Cytospora cincta. Pl. Dis. Reptr* **46,** 593.

Hassebrauk, K. (1938) Weitere Untersuchungen über Getreiderostbekämpfung mit chemischen Mitteln. *Phytopath. Z.* **11,** 14.

Hassebrauk, K. (1940) Abschliessende Untersuchungen über die feldmässige Verwendungsmöglichkeit von *p*-Toluolsulfonamid als innertherapeutisch wirkendes Getreiderostbekämpfungsmittel. *Phytopath. Z.* **12,** 509.

Hassebrauk, K. (1951) Untersuchungen über die Einwirkung von Sulfonamiden und Sulfonen auf Getreideroste. I. Die Beeinflussing des Fruktifikationsvermögens. *Phytopath. Z.* **17,** 384.

Hassebrauk, K. (1952) Untersuchungen über die Einwirkung von Sulfonamiden und Sulfonen auf Getreideroste. II. Weitere Untersuchungen über die rosthemmende Wirkung. *Phytopath. Z.* **18,** 453.

Hastie, A. C. (1970) Benlate-induced instability of *Aspergillus* diploids. *Nature, Lond.* **226,** 771.

Hastie, A. C. (1971) Benomyl-induced instability of *Aspergillus* diploids. *Proc. 6th Br. Insectic. Fungic. Conf. at Brighton, 1971* **1,** 283.

Hastie, A. C. and Georgopoulos, S. G. (1971) Mutational resistance to fungitoxic benzimidazole derivatives in *Aspergillus nidulans. J. Gen. Microb.* **67**, 371.

Hawkins, D. D., Down, W. H., Chasseaud, L. F. and Lewis, J. D. (1974) The metabolism of tridemorph in rats. *Pestic. Science* **5**, 535.

Hay, S. J. B. (1971) The control of apple mildew with HOE2873. *Proc. 6th Br. Insectic. Fungic. Conf.* **1**, 134.

Hay, J. R. and Thimann, K. V. (1956) The fate of 2,4-dichlorophenoxyacetic acid in bean seedlings. II. *Pl. Physiol. Lancaster* **31**, 446.

Headford, D. W. R. and Douglas, D. (1967) Tuber necrosis following the desiccation of potato foliage with diquat. *Weed Res.* **7**, 131.

Headley, J. C. (1973) Environmental quality and the economics of agricultural pest control. E.M.P.P.O. Conference of Plant Protection Economy: Brussels, May 1973.

Hearn, C. J. and Childs, J. F. L. (1969) A systemic fungicide effective against sour orange scab disease. *Pl. Dis. Reptr* **53**, 203.

Hearn, C. J., Childs, J. F. L. and Fenton, R. (1971) Comparison of benomyl and copper sprays for control of sour orange scab of citrus. *Pl. Dis. Reptr* **55**, 241.

Hearn, C. J. and Fenton, R. (1970) Benomyl sprays for control of twig dieback of 'Robinson' tangerine. *Pl. Dis. Reptr* **54**, 869.

Heffernan, T. (1968) Advances in tree implantation. *Agr. Chemical West.* Jan. and Feb. 1968.

Heggeness, H. G. (1942) Effect of borax applications on the incidence of rust on flax. *Pl. Physiol.* **17**, 143.

Helser, T. L., Davies, J. E. and Dahlberg, J. E. (1971) Change in methylation of 16 S ribosomal RNA associated with mutation to kasugamycin resistance in *E. coli. Nature, New Biology, Lond.* **233**, 12.

Helton, A. W. and Kochan, W. J. (1967) Chemotherapeutic effects of cycloheximide thiosemicarbazone, phytoactin and 8-quinolinol benzoate, with and without dimethyl sulfoxide and pruning on the Cytospora disease of prune trees. *Pl. Dis. Reptr* **51**, 340.

Helton, A. W. and Rohrbach, K. G. (1967) Chemotherapy of Cytospora canker disease in peach trees. *Phytopathology* **57**, 442.

Helton, A. W. and Williams, R. E. (1970) Systemic distribution of three compounds in sprayed 'Jonathan' apple trees. *Pl. Dis. Reptr* **54**, 996.

Helweg, A. (1973) Persistence of benomyl in different soil types and microbial breakdown of the fungicide in soil and agar culture. *Tidsskr. Planteavl.* **77**, 232.

Heyns, A. J., Carter, G. A., Rothwell, K. and Wain, R. L. (1966) Investigations on fungicides. XIII. The systemic fungicidal activity of certain N-carboxymethyldithiocarbamic acid derivatives. *Ann. appl. Biol.* **57**, 33.

Hickey, K. D. (1971a) (ed.) Pome fruit disease reports. *Fungicide and Nematicide Tests* **26**, 4.

Hickey, K. D. (1971b) (ed.) Stone fruit disease reports. *Fungicide and Nematicide Tests* **26**, 42.

Hide, G. A. and Hirst, J. M. (1971) Chemical control of tuber diseases. *Rep. Rothamsted exp. Stn for 1970* **1**, 140.

Hide, G. A., Hirst, J. M. and Griffith, R. L. (1969) Control of potato tuber diseases with systemic fungicides. *Proc. 5th Br. Insectic. Fungic. Conf.*, p. 310.

Hignett, R. C. and Kirkham, D. S. (1967) The role of extracellular melanoproteins of *V. inaequalis* in host susceptibility. *J. gen. Microbiol.* **48**, 269.

Hilborn, M. T. (1953) Effect of various chemicals on infection by *Rhizoctonia solani* and *Verticillium albo-atrum. Phytopathology* **43**, 475.

Hilborn, M. T. (1969) Blueberry: powdery mildew (*Microsphaera penicillata* var. *vaccinii*). *Fungicide and Nematicide Tests* **24**, 34.

Hill, I. R. and Arnold, D. J. (1973) Degradation of ethirimol in soil. *Proc. 7th Br. Insectic. Fungic. Conf.* **1**, 47.

Hill-Cottingham, D. G. and Lloyd-Jones, C. P. (1968) Relative mobility of some nitrogenous compounds in the xylem of apple shoots. *Nature, Lond.* **220**, 389.

Hiltner, L. (1915) Über die Erfolge der im Herbst 1914 in Bayern durch geführten Beizung des Saatgutes von Winterroggen und Winterweizen mit sublimathaltigen Mittleln. *Prakt. Bl. Pflbau. PflSchutz.* **18,** 65.

Himelick, E. B. (1972) High pressure injection of chemicals into trees. *Arborists News* **37,** 97.

Himelick, E. B. and Neely, D. (1972) Sycamore anthracnose (*Gnomonia platani*). *Fungicide and Nematicide Tests* **28,** 130.

Himelick, E. B. and Ricketts, S. R. (1974) Sycamore anthracnose (*Gnomonia platani*). *Fungicide and Nematicide Tests* **30,** 129.

Hinds, T. E. and Peterson, R. S. (1966) Antibiotic tests on western gall rust and aspen cankers. *Pl. Dis. Reptr* **50,** 74.

Hine, R. B., Johnson, D. L. and Wenger, C. J. (1969) The persistence of two benzimidazole fungicides in soil and their fungistatic activity against *Phymatotrichum omnivorum. Phytopathology* **59,** 798.

Hinosan Information Bulletin, 1971.

Hirst, J. M. (1971) Mechanical dusting of tubers with fungicide. *Rep. Rothamsted exp. Stn for 1970* **1,** 141.

Hislop, E. C. (1967) Observations on the vapour phase activity of some foliage fungicides. *Ann. appl. Biol.* **60,** 265.

Hock, W. K. (1971) Sycamore anthracnose (*Gnomonia platani*). *Fungicide and Nematicide Tests* **27,** 138.

Hock, W. K. and Berry, F. H. (1975) Sycamore anthracnose in forest nursery diseases in the United States (eds G. W. Peterson and R. S. Smith). *USDA Forest Service Agr. Handbook* **470,** 88.

Hock, W. K. and Schreiber, L. R. (1971) Evaluation of benomyl for the control of Dutch elm disease. *Pl. Dis. Reptr* **55,** 58.

Hock, W. K. and Schreiber, L. R. (1971) Effect of adjuvants on the uptake of benomyl from planting media by American elm seedlings. *Pl. Dis. Reptr* **55,** 971.

Hock, W. K., Schreiber, L. R. and Roberts, B. R. (1970a) Suppression of Dutch elm disease in American elm seedlings by benomyl. *Phytopathology* **60,** 391.

Hock, W. K., Schreiber, L. R. and Roberts, B. R. (1970b) Factors influencing uptake, concentration and persistence of benomyl in American elm seedlings. *Phytopathology* **60,** 1619.

Hock, W. K. and Sisler, H. D. (1969a) Specificity and mechanism of antifungal action of chloroneb. *Phytopathology* **59,** 627.

Hock, W. K. and Sisler, H. D. (1969b) Metabolism of chloroneb by *Rhizoctonia solani* and other fungi. *J. Agric. Food Chem.* **17,** 123.

Hocking, D. and White, P. J. (1965) Fungicides for arabica coffee. III. Determination of translaminar, curative activity against leaf rust. *PANS* **B11,** 273.

Hodges, C. S. (1975) Cercospora blight of juniper in forest nursery diseases in the United States (eds G. W. Peterson and R. S. Smith) *USDA Forest Service Agr. Handbook* **470,** 80.

Hoffmann, J. A. (1969) *Fungicide and Nematicide Tests* **25,** 129.

Hoffmann, J. A. (1970) *Fungicide and Nematicide Tests* **26,** 146.

Hoffmann, J. A. (1971a) Control of common and dwarf bunt of wheat with systemic fungicides. *Pl. Dis. Reptr* **55,** 1132.

Hoffmann, J. A. (1971b) Control of common and dwarf bunt of wheat by seed treatment with thiabendazole. *Phytopathology* **61,** 1071.

Hoffmann, P. F. (1951) Screening chemotherapeutants for control of oak wilt. *Proc. Iowa Acad. Sci.* **58,** 139.

Hoffmann, P. F. (1952) Early trials in oak wilt chemotherapy. *Phytopathology* **42,** 11.

Hoffmann, P. F. (1953) Chemicals for therapy of oak wilt. *Phytopathology* **43,** 475.

Hoitink, A. A. J. and Schmitthenner, A. F. (1975) Comparative efficacy of 2-chloro-6-methoxy-4(trichloromethyl)pyridine and ethazole for control of *Phytophthora* root rot of rhododendron and soybean. *Phytopathology* **65,** 69.

Holloway, P. J. (1971) The chemical and physical characteristics of leaf surfaces. *Ecology of leaf surface micro-organisms* (eds T. F. Preece and C. H. Dickinson), p. 39. London: Academic Press.

Holly, K. (1956) Penetration of chlorinated phenoxyacetic acids into leaves. *Ann. appl. Biol.* **44,** 195.

Holmes, F. W. (1974) Dutch elm disease research progress report. *Mass. Coop. Rept. to Regional Project* NE-25 (Wilt Diseases).

Holmes, F. W. (1975) Elm rhizosphere inoculation of 'Benlate' benomyl suspensions: effects upon artificial and natural infection by *Ceratocystis ulmi. Proc. Amer. Phytopathological Soc.* **2,** 98.

Holowczak, J., Kuć, J. and Williams, E. B. (1962) Metabolism of DL- and L-phenylalanine in *Malus* related to susceptibility and resistance to *Venturia inaequalis. Phytopathology* **52,** 699.

Hood, I. A. (1973) Douglas fir Swiss needle cast (*Phaeocryptopus gaeumannii*). *Fungicide and Nematicide Tests* **29,** 122.

Hope, A. B. and Stevens, P. G. (1952) Electrical potential differences in bean roots and their relation to salt uptake. *Aust. J. sci. Res. Ser. B* **5,** 445.

Hopkins, D. L. (1974) Fungicidal control of bunch grape diseases in Florida. *Proc. Fla St. hort. Soc.* **86,** 329.

Hori, M., Eguchi, J., Kakiki, K. and Misato, T. (1974) Studies on the mode of action of polyoxins. VI. Effect of polyoxin B on chitin synthesis in polyoxin sensitive and resistant strains of *Alternaria kikuchiana. J. Antibiotics* (*Japan*) **27,** 260.

Hori, M., Kakiki, K., Suzuki, S. and Misato, T. (1971) Studies on the mode of action of polyoxins. III. Relation of polyoxin structure to chitin synthetase inhibition. *Agr. Biol. Chem.* **35,** 1280.

Hori, M., Kakiki, K. and Misato, T. (1974) Interaction between polyoxin and active center of chitin synthetase. *Agr. Biol. Chem.* **38,** 699.

Horn, N. L., Burnside, K. R. and Carver, R. B. (1972) Control of the crown rot phase of strawberry anthracnose through sanitation, breeding for resistance and benomyl. *Pl. Dis. Reptr* **56,** 515.

Horner, C. E. (1963) Chemotherapeutic effects of streptomycin on establishment and progression of systemic downy mildew infection in hops. *Phytopathology* **53,** 472.

Horner, C. E. and Maier, C. R. (1957) Antibiotics eliminate systemic downy mildew from hops. *Phytopathology* **47,** 525.

Horsfall, J. G. (1945) *Fungicides and their Action.* Waltham, Mass.: Chronica Botanica.

Horsfall, J. G. (1956) *Principles of Fungicidal Action.* Waltham, Mass.: Chronica Botanica.

Horsfall, J. G. (1961) Chemotherapy of plant disease. *Scientia, Sixième Série, Aug. 1961,* 1.

Horsfall, J. G. and Dimond, A. E. (1951a) Plant chemotherapy. *Trans. N.Y. Acad. Sci.* **13,** 338.

Horsfall, J. G. and Dimond, A. E. (1951b) Plant chemotherapy. *A. Rev. Microbiol.* **5,** 209.

Horsfall, J. G. and Dimond, A. E. (1957) Interaction of tissue sugar, growth substances and disease susceptibility. *Z. PflKrankh. PflPath. PflSchutz.* **64,** 415.

Horsfall, J. G., Dimond, A. E., Stoddard, E. M. and Chapman, R. A. (1951) Chemotherapy for controlling plant diseases. *Proc. 2nd int. Congr. Crop Prot. London 1949,* 1.

Horsfall, J. G. and Lukens, R. J. (1968) Aldehyde traps as antisporulants for fungi. *Conn. Agr. Exp. Station Bull.* **694,** 28 pp. New Haven.

Horsfall, J. G. and Zentmyer, G. A. (1942) Antidoting the toxins of plant diseases. *Phytopathology* **32,** 23.

Horst, R. K. and Hoitink, H. A. J. (1968) Occurrence of Cylindrocladium blights on nursery crops and control with fungicide 1991 on azalea. *Pl. Dis. Reptr* **52,** 616.

Hosford, R. M. Jr. and Busch, R. H. (1974) Losses in wheat caused by *Pyrenophora trichostoma* and *Leptosphaeria avenaria* f.sp. *triticea. Phytopathology* **64,** 184.

Hotson, H. H. (1952) The inhibition of wheat stem rust by sulfadiazine and the reversal of this effect by para-aminobenzoic acid and folic acid. *Phytopathology* **42,** 11.

Hotson, H. H. (1953) Some chemotherapeutic agents for wheat stem rust. *Phytopathology* **43,** 659.

Houseworth, L. D., Brunton, E. W. and Tweedy, B. G. (1971) The effect of triarimolon oxidation of glucose and acetate to carbon dioxide and on spore germination of *Aspergillus niger* and *Cladosporium cucumerinum. Phytopathology* **61,** 896 (abs.).

Howard, C. M. (1971) Control of strawberry anthracnose with benomyl. *Pl. Dis. Reptr* **55,** 139.

Howard, C. M. and Albregts, E. E. (1972) Control of leaf blight in Florida strawberry nurseries. *Pl. Dis. Reptr* **56,** 23.

Howard, F. L. (1941) Antidoting of toxins of *Phytophthora cactorum* as a means of plant disease control. *Science, N.Y.* **94,** 345.

Howard, F. L. and Caroselli, N. E. (1941) The role of tree injection in the control of bleeding canker of hardwoods. *Phytopathology* **31,** 12.

Howard, F. L. and Horsfall, J. G. (1959) Therapy. *Plant Pathology Vol. I* (eds J. G. Horsfall and A. E. Dimond), p. 563. New York and London: Academic Press.

Howard, F. L. and Machardy, W. E. (1969) The basis of elm twig wilt by *Ceratocystis ulmi* and its bearing on chemotherapy. *XI Internat. Botan. Congr. Abstr.*, p. 95. Seattle.

Howard, F. L. and Sorrell, M. B. (1943) Cationic phenyl mercury compounds as specific apple-scab eradicants on foliage. *Phytopathology* **33,** 1114.

Huang, K. T., Misato, T. and Asuyama, H. (1964) Selective toxicity of blasticidin S to *Piricularia oryzae* and *Pellicularia sasakii*. *J. Antibiotics* (Tokyo) A, **17,** 71.

Huber, D. M. and Mulanax, M. W. (1972) Benomyl rates and application time for wheat foot rot control. *Pl. Dis. Reptr* **56,** 342.

Hudson, J. C. and Jaffer, A. A. (1970) Effect of systemic fungicides on bean rust (*Uromyces appendiculatus* (Pers.) Lev.). *Misc. Rep. trop. Pestic.* no. 726, 7 pp.

Hunter, T. and Whiteley, L. A. (1976) Diseases of vine. *Rep. Long Ashton Res. Stn for 1975*. p. 122.

Iamomoto, T., Campacci, C. A., Suzuki, U. and Oi, A. (1971) Control of rust of string bean (*Phaseolus vulgaris*) caused by the fungus *Uromyces phaseoli* var. *typica*. *Biologico* **37,** 266.

Ibrahim, I. A. (1951) Effect of 2,4-D on stem rust development in oats. *Phytopathology* **41,** 951.

Iida, W. (1975) On the tolerance of plant pathogenic fungi and bacteria to fungicides in Japan. *Japan Pestic. Inform.* **23,** 13.

Il'ina, M. N., Maklakova, N. S. and Afanas'eva, T. N. (1974) Efficacy of systemic fungicides against clubroot in cabbage. *Khimiya v Sel'skom Khozyaistve* **12,** 36.

Ingram, J. (1971) Black currant: American gooseberry mildew/leaf spot 1970. *Rep. Luddington exp. Hort. Stn for 1970*, 40.

Ishii, K. (1970) New fungicides, thiophanate and its derivatives. *Summaries of Papers, VIIth Int. Cong. Plant Prot., Paris*, 200.

Ishiyama, T., Okamoto, H., Sato, K., Nakamura, K. and Nakamura, M. (1965a) Translocation of kasugamycin in the leaf tissue. *Ann. phytopath. Soc. Japan* **30,** 111 (abs.).

Ishiyama, T., Okamoto, H., Sato, K. and Nakamura, M. (1965b) Control of rice blast by application of kasugamycin through roots. *Ann. Phytopath. Soc. Japan* **30,** 111 (abs.).

Ishiyama, T., Sato, K., Nakamura, T., Takecuhi, T. and Umezawa, H. (1967) Absorption and translocation by rice plants of ^{14}C-kasugamycin. *J. Antibiotics Ser. B.* **20,** 357.

Isono, K., Nagatsu, J., Kobinata, K., Sasaki, K. and Suzuki, S. (1967) Studies on polyoxins, antifungal antibiotics. II. Isolation and characterisation of polyoxins C, D, E, F, G, H and I. *Agr. Biol. Chem.* **31,** 190.

Iwan, J. and Goller, D. (1974) Behaviour of SN 41703 (prothiocarb) in soils and plants. *3rd Int. Congr. Pesticide Chemistry*. p. 135 (abs.).

Iwan, J. and Goller, D. (1975) Behaviour of SN 41703 (prothiocarb) in soils and plants. *Pesticides* (eds F. Coulston and F. Korte). pp. 319–322. Stuttgart: Georg Thieme Verlag.

Iwata, K., Yamaguchi, H. and Hiratani, T. (1973) Mode of action of clotrimazole. *Sabouraudia* **11,** 158.

Jackson, N. and Fenstermacher, J. M. (1974) Fungicide control of stripe smut and melting out with consequent maintenance of sward density in 'Merion' Kentucky blue grass turf. *Pl. Dis. Reptr* **58,** 573.

Jacob, T. A., Carlin, J. R., Walker, R. W., Wolf, F. J. and Vandenheuvel, W. J. A. (1975) Photolysis of thiabendazole. *J. Agr. Food Chem.* **23,** 704.

Jacobs, C. J., Brodrick, H. T., Swarts, H. D. and Mulder, N. J. (1973) Control of post-harvest decay of mango fruit in South Africa. *Pl. Dis. Reptr* **57,** 173.

Jacobs, H. L. (1928) Injection of shade trees for the control of insects and diseases. *18th A. Conv. Davey Tree Expert Co.,* Kent, Ohio.

Jacobsen, B. J. and Williams, P. H. (1970) Control of cabbage clubroot using benomyl fungicide. *Pl. Dis. Reptr* **54,** 456.

Jacobsen, B. J. and Williams, P. H. (1971) Histology and control of *Brassica oleracea* seed infection by *Phoma lingam. Pl. Dis. Reptr* **55,** 934.

Jailloux, F. and Froidefond, G. (1974) Comparative effectiveness of different fungicides against *Fusicoccum amygdali* Del. agent of peach canker. *Phytiat.-Phytopharm.* **23,** 211. (*Rev. Pl. Path.* **55,** 322, 1975).

Jain, S. S. (1956) A preliminary note on the inactivation of *Fusarium oxysporum* f. *psidii* in guava plants by chemotherapeutic treatment. *Indian. J. Hort.* **13,** 102.

James, C. S. (1971) Private communication.

James, C. S., Gramlich, J. V. and Darge, G. D. (1971) Control of barley powdery mildew (*Erysiphe graminis*) with triarimol. British Association of Grain, Seed, Feed and Agricultural Merchants Ltd. Unpublished Report.

James, W. C. (1974) Assessment of plant diseases and losses. *Annual Rev. Phytopathology* **12,** 27–49.

Jank, B. and Grossmann, F. (1971) 2-Methyl-5,6-dihydro-4-H-pyran-3-carboxylic acid anilide: a new systemic fungicide against smut diseases. *Pestic. Sci.* **2,** 43.

Janutolo, D. B. and Stipes, R. J. (1975) Toxic effects of methyl-2-benzimidazole carbamate (MBC), MBC-HCl and MBC-H_3PO_4 on conidia of *Ceratocystis ulmi. Proc. Amer. Phytopath. Soc.* **2,** 32.

Jarvis, J. R. and Hargreaves, A. J. (1973) Tolerance to benomyl in *Botrytis cinerea* and *Penicillium corymbiferum. Plant Pathology* **22,** 139.

Jarvis, P. and Thaine, R. (1971) Strands in sections of sieve elements cut in a cryostat. *Nature, Lond.* **232,** 236.

Jaworski, E. G. and Hoffman, P. F. (1963) Chemotherapeutic control of wheat leaf rust with phenylhydrazones. *Phytopathology* **53,** 639.

Jaynes, R. A. and Van Alfen, N. K. (1974) Control of American chestnut blight by trunk injection with methyl-2-benzimidazole carbamate (MBC). *Phytopathology* **64,** 1479.

Jenkins, J. E. E. (1973) An evaluation of the control of cereal leaf diseases by fungicides in England and Wales. *Proc. 7th Br. Insectic. Fungic. Conf.* **3,** 781.

Jenkins, J. E. E. and Morgan, W. (1969) The effect of septoria diseases on the yield of winter wheat. *Pl. Path.* **18,** 152.

Jenkyn, J. F. (1974) Effects of mildew on the growth and yield of spring barley: 1969–72. *Ann. appl. Biol.* **78,** 281.

Jhooty, J. S. and Behar, D. S. (1970) Evaluation of different benomyl treatments for control of *Rhizoctonia* damping-off of peas. *Pl. Dis. Reptr* **54,** 1049.

Johnson, H. W. (1972) Control of powdery mildew of wheat by systemic seed treatments. *Can. Plant Dis. Surv.* **52,** 82.

Johnson, M. P. and Bonner, J. (1956) The uptake of auxin by plant tissue. *Physiologia Pl.* **9,** 102.

Johnson, R. P. C. (1973) Filaments, but no membranous transcellular strands in sieve pores in freeze-etched, translocating phloem. *Nature, Lond.* **244,** 464.

Johnston, H. and Tomita, M. S. (1966) Certain ethers of (trichloromethyl) pyridine compounds. US Patent 3,244,722.

Johnston, H. W. (1970) Control of powdery mildew of wheat by soil-applied benomyl. *Pl. Dis. Reptr* **54,** (2) 91.

Jones, A. L. (1975) Control of brown rot of cherry with a new hydantoin fungicide and with selected fungicide mixtures. *Pl. Dis. Reptr* **59,** 127.

Jones, A. L. and Burton, C. L. (1973) Heat and fungicide treatments to control post harvest brown rot of stone fruits. *Pl. Dis. Reptr* **57,** 62.

Jones, A. L., Burton, C. L. and Tennes, B. R. (1973) Postharvest fungicide and heat treatments for brown rot control on stone fruits. *Michigan State Univ. agric. Exp. Stn, East Lansing, Res. Rep.* No. 209 (10 pp.).

Jones, B. M. and Swartwout, H. G. (1961a) Systemic control of powdery mildew, *Sphaerotheca pannosa*, of roses with the semicarbazone derivative of actidione (cycloheximide). *Pl. Dis. Reptr* **45**, 366.

Jones, B. M. and Swartwout, H. G. (1961b) Systemic control of powdery mildew of roses, caused by the fungus *Sphaerotheca pannosa*, with derivatives of cycloheximide. *Pl. Dis. Reptr* **45**, 794.

Jones, J. P. and Barnett, R. D. (1968) Control of loose smut of wheat and barley in Arkansas with fungicides. US Dept. of Agr. CR42–68, 3.

Jones, J. P., Overman, A. J. and Geraldson, C. M. (1971) Fumigants for the control of *Verticillium* wilt of tomato. *Pl. Dis. Reptr* **55**, 26.

Jones, J. P. and Woltz, S. S. (1969) Effect of ethionine and methionine on the growth, sporulation and virulence of *Fusarium oxysporum* f. sp. *lycopersici*, race 2. *Phytopathology* **59**, 1464.

Jones, T. W. and Gregory, G. R. (1971) An apparatus for pressure injection of solutions into trees. N.E. For. Expt. Sta., Upper Darby, Pa. *USDA Forest Ser. Res. Paper* NE-233.

Jones, T. W., Gregory, G. R. and McWain, P. (1973) Pressure injections of solubilized benomyl for prevention and cure of oak wilt. *USDA Forest Serv. Res. Paper* NE-171. 4 pp.

Jooste, W. J. (1969) Control of ear and tassel smut of maize with 1,4-oxathiin derivatives. *Phytophylactica* **1**, 235.

Jordan, V. W. L. (1971) Strawberry: mildew (*Sphaerotheca macularis*). *Fungicide and Nematicide Tests* **26**, 61.

Jordan, V. W. L. (1972a) Evaluation of fungicides for the control of *Verticillium* wilt (*V. dahliae* Kleb.) of strawberry. *Ann. appl. Biol.* **70**, 163.

Jordan, V. W. L. (1972b) Diseases of strawberries: *Verticillium* wilt. *Rep. agric. hort. Res. Stn Bristol, 1971*, p. 140.

Jordan, V. W. L. (1973a) The effects of prophylactic spray programmes on the control of pre- and post-harvest diseases of strawberry. *Pl. Path.* **22**, 67.

Jordan, V. W. L. (1973b) The modes of action of two benzimidazoles and thiophanate-methyl used for the control of Verticillium wilt of strawberry. *Ann. appl. Biol.* **75**, 41.

Jordan, V. W. L. (1975) Control of Verticillium wilt transmission in strawberry runners with systemic benzimidazole-type fungicides. *Ann. appl. Biol.* **81**, 43.

Jordan, V. W. L. and Tarr, H. S. (1975) Diseases of plum. *Rep. Long Ashton Res. Stn for 1974*, p. 126.

Jovanovic, M. (1974) Trials on the possibility of curative control of *Uncinula necator* (Schw.) Burr. of grapevine by systemic fungicides. *Agrohenija*, **3/4**, 97.

Judd, R. W. and Walton, G. S. (1973) Benomyl dip for the control of *Ascochyta* decay of chrysanthemum cuttings. *Pl. Dis. Reptr* **57**, 624.

Judge, F. D., Jackson, D. and Hall, D. W. (1975) Control of yellow rust (*Puccinia striiformis*) in wheat with oxycarboxin. *Proc. 8th Br. Insectic. Fungic. Conf.* **3**, 967.

Jung, K. U. and Bedford, J. L. (1971) The evaluation of application time of tridemorph for the control of mildew in barley and oats in the United Kingdom. *Proc. 6th Br. Insectic. Fungic. Conf.* **1**, p. 75.

Kaars Sijpesteijn, A. (1961a) Chemotherapy and synthetic systemic fungicides. *Proc. 9th int. Bot. Congr. Montreal, 1959* **1**, 457.

Kaars Sijpesteijn, A. (1961b) New developments in the systemic combat of fungal diseases of plants. *Tijdschr. PlZiekt.* **67**, 11.

Kaars Sijpesteijn, A. (1969) Enzymic transformations of fungicides in the host plant. *Wld Rev. Pest Control* **8**, 138.

Kaars Sijpesteijn, A. (1970) Biochemical modes of action of agricultural fungicides. *Wld Rev. Pest Control* **9**, 85.

Kaars Sijpesteijn, A. (1977) Chapter 7 in *Systemic Fungicides* (ed. R. W. Marsh). London: Longman.

Kaars Sijpesteijn, A., Dekhuijzen, H. M. and Vonk, J. W. (1976) Biological conversions of fungicides in plants and microorganisms. *Antifungal compounds*, Vol. II, Chapter 3 (eds M. R. Siegel and H. D. Sisler). New York: Marcel Dekker.

Kaars Sijpesteijn, A. and van der Kerk, G. J. M. (1965) Fate of fungicides in plants. *A. Rev. Phytopath.* **3**, 127.

Kaars Sijpesteijn, A., Kerkenaar, A. and Overeem, J. C. (1974) Observations on selectivity and mode of action of prothiocarb (SN41703). *Meded. Faculteit Landbouwwetenschappen Gent* **39**, 1027

Kaars Sijpesteijn, A. and Pluijgers, C. W. (1962) On the action of phenylthioureas as systemic compounds against fungal diseases of plants. *Meded. LandbHoogesch. OpzoekStns. Gent* **27**, 1199.

Kaars Sijpesteijn, A., Pluijgers, C. W., Verloop, A. and Tempel, A. (1968) Phenylazothioformamide, the active principle in the fungitoxic action of 1-phenylthiosemicarbazide. *Ann. appl. Biol.* **61**, 473.

Kaars Sijpesteijn, A., Rombouts, J. E., van Andel, O. M. and Dekker, J. (1958) Investigations on the activity of pyridine-2-thiol-N-oxide as a systemic fungicide. *Meded. LandbHoogesch. OpzoekStns. Gent.* **23**, 824.

Kaars Sijpesteijn, A. and Sisler, H. D. (1968) Studies on the mode of action of phenylthiourea, a chemotherapeutant for cucumber scab. *Neth. J. Pl. Path.* **74**, 121.

Kable, P. F. (1970) Eradicant action of fungicides applied to dormant peach trees for control of brown rot (*Monilinia fructicola*). *J. hort. Sci.* **45**, 143.

Kable, P. F. (1971a) Better control measures for brown rot of stone fruits. *Agric. Gaz. N.S.W.* **82**, 194.

Kable, P. F. (1971b) Significance of short-term latent infections in the control of brown rot in peach fruits. *Phytopath. Z.* **70**, 173.

Kable, P. F. (1975) The frequency and timing of benomyl sprays for control of brown rot *Monilinia fructicola* in canning peaches. *J. hort. Sci.* **50**, 143.

Kado, M., Maeda, T., Yoshinaga, E., Iwakura, T. and Uchida, T. (1965) Studies on the rice blast control. I. Chemical structure and fungicidal activities of benzyl phosphoric esters. *Ann. Phytopathol. Soc. Japan* **30**, 109 (abs.).

Kado, M. and Yoshinaga, E. (1969) Fungicidal action of organophosphorus compounds. *Residue Rev.* **25**, 133.

Kais, A. G. (1975) Fungicidal control of *Scirrhia acicola* on longleaf pine seedlings. *Pl. Dis. Reptr* **59**, 686.

Kaiser, W. J., Okhovat, M. and Mossahebi, G. H. (1973) Effect of seed treatment fungicides on control of *Ascochyta rabiei* in chick pea seed infected with the pathogen. *Pl. Dis. Reptr* **57**, 742.

Kakiki, K., Maeda, T., Abe, T. and Misato, T. (1969) Studies on the mode of action of organophosphorus fungicide Kitazin. Part I: Effect on respiration, protein synthesis, nucleic acid synthesis, cell wall synthesis and leakage of cellular substances from mycelia of *Pyricularia oryzae. Nippon Nogeikagaku Kaishi* **43**, 37.

Kantzes, J. G. and Weaver, L. O. (1971) Pear: leaf spot (*Entomosporium maculatum*). *Fungicide and Nematicide Tests* **26**, 40.

Kappas, A. and Georgopoulos, S. G. (1970) Genetic analysis of dodine resistance to *Nectria haematococca* (syn. *Hypomyces solani*). *Genetics* **66**, 617.

Kappas, A. and Georgopoulos, S. G. (1971) Independent inheritance of avirulence and dodine resistance in *Nectria haematococca* var. *cucurbitae. Phytopathology* **61**, 1093.

Kappas, A., Georgopoulos, S. G. and Hastie, A. C. (1974) On the genetic activity of benzimidazole and thiophanate fungicides on diploid *Aspergillus nidulans. Mutation Res.* **26**, 17.

Kaspers, H., Grewe, F., Brandes, W., Scheinflug, H. and Büchel, K.-H. (1975) 1,2,4-Triazole derivatives, a new class of protective and systemic fungicides. *Proc. 8th Int. Plant Prot. Congr.* **3**, 398.

Kato, T., Tanaka, S., Ueda, M. and Kawase, Y. (1974) Effects of the fungicide S-1358 on general metabolism and lipid biosynthesis in *Monilinia fructigena. Agr. Biol. Chem.* **38**, 2377.

Kato, T., Tanaka, S., Ueda, M. and Kawase, Y. (1975a) Inhibition of sterol biosynthesis in *Monilinia fructigena* by the fungicide S-1358. *Agr. Biol. Chem.* **39**, 169.

Kato, T., Tanaka, S., Yamamoto, S., Kawase, Y. and Ueda, M. (1975b) Fungitoxic properties of a N-3-pyridylimidodithiocarbonate derivative. *Phytopathol. Soc. Japan* **41**, 1.

Kato, T. (1977) Mechanism of action of Denmert, an inhibitor of sterol biosynthesis. *Neth. J. Plant Pathol.* **83,** Suppl. 1.

Kavanagh, T. (1970) Black spot (*Diplocarpon rosae*) of rose. *Res. Rep. An Foras Taluntais, Hort. Div. 1970,* p. 87.

Kavanagh, T. (1972) Control of powdery mildew (*Sphaerotheca macularis*) of greenhouse strawberries with systemic fungicides. *Pl. Dis. Reptr* **56,** 235.

Kavanagh, T. and O'Callaghan, T. F. (1969) A comparison of fungicides for the control of Botrytis of outdoor and protected strawberries. *Proc. 5th Br. Insectic. Fungic. Conf.* **1,** 139.

Kavanagh, T. and O'Kennedy, N. D. (1971) Evaluation of apple scab (*Venturia inaequalis*) and canker (*Nectria galligena*) control programmes in Ireland 1961–1971. *Proc. 6th Br. Insectic. Fungic. Conf.* **1,** 110.

Kear, R. W. and Symons, J. P. (1973) Post-harvest deterioration of stored Dutch White Cabbage. *Meded. Fac. Landbouwwetensch. Rijksuniv. Gent* **38,** 1549.

Keen, N. T., Long, M. and Malca, I. (1970) Induction and repression of β-galactosidase synthesis by *Verticillium albo-atrum. Physiolog. Plantarum* **23,** 691.

Keen, N. T., Wang, M. C., Long, M. and Erwin, D. C. (1971) Dimorphism in *Verticillium albo-atrum* as affected by initial spore concentration and antisporulant chemicals. *Phytopathology* **61,** 1266.

Keiding, J. (1967) Persistence of resistant populations after the relaxation of the selection pressure. *Wld Rev. Pest Control* **6,** 115.

Keil, H. L. and Frohlich, H. P. (1961) US Patent 2971880.

Keil, H. L., Frohlich, H. P. and van Hook, J. O. (1958) Chemical control of cereal rusts. I. Protective and eradicative control of rye leaf rust in the greenhouse with various chemical compounds. *Phytopathology* **48,** 652.

Kennedy, J. (1961) Fumigation of agricultural products. XVII. Control of *Ascochyta* blight of peas by fumigation. *J. Sci. Fd. Agric.* **12,** 96.

Kent, N. L. (1941a) Absorption, translocation and ultimate fate of lithium in the wheat plant. *New Phytol.* **40,** 291.

Kent, N. L. (1941b) The influence of lithium salts on certain cultivated plants and their parasitic diseases. *Ann. appl. Biol.* **28,** 189.

van der Kerk, G. J. M. (1956) The present state of fungicidal research. *Meded. Landb-Hoogesch. OpzoekStns. Gent* **21,** 305.

van der Kerk, G. J. M. (1961) New developments in organic fungicides. *Soc. chem. Ind. Monograph* **15,** 67.

van der Kerk, G. J. M. (1971a) Differentiation between transportable fungicides and systemic infection-inhibiting agents, and their relative perspectives. *Acta phytopath. hung.* **6,** 311.

van der Kerk, G. J. M. (1971b) Systemic fungicides. New solutions and new problems. *Proc. 6th Br. Insectic. Fungic. Conf.* **3,** 791.

van der Kerk, G. J. M. (1973) Chemical and biochemical aspects of systemic fungicides. *EPPO Bull. No. 10,* p. 5.

Kerkenaar, A. and Kaars Sijpesteijn, A. (1977) On the mode of action of prothiocarb. *Neth. J. Plant. Pathol.* **83,** Suppl. 1.

Kerridge, D. (1958) The effect of actidione and other antifungal agents on nucleic acid and protein synthesis in *Saccharomyces carlsbergensis. J. Gen. Microbiol.* **19,** 497.

von Keyserlingk, N. and Wihrheim, S. E. (1971) Thiabendazole for the control of powdery mildew and scab in apples. *Proc. 6th Br. Insectic. Fungic. Conf.* **3,** 657.

Keyworth, W. G. and Dimond, A. E. (1952) Root injury as a factor in the assessment of chemotherapeutants. *Phytopathology* **42,** 311.

King, E. J. and Campana, R. J. (1974) Prophylactic and therapeutic effect of methyl-2-benzimidazole carbamate hydrochloride (MBC-HCl) against Dutch elm disease in nursery elms. *Proc. Amer. Phytopathological Soc.* **1,** 31 (abs.).

King, J. E. (1973) Cereal foliar disease surveys. *Proc. 7th Br. Insectic. Fungic. Conf.* **3,** 771.

King, J. M. (1971) The control of *Botrytis* pod rot in dwarf beans. *Proc. 6th Br. Insectic. Fungic. Conf.* **1,** p. 173.

Kingsland, G. (1970) Barley leaf stripe control by 'Vitavax'. *Phytopathology* **60,** 584.

Király, Z., Bócsa, E. and Vörös, J. (1962) Effect of treatment of wheat seed with gibberellic acid on bunt infection. *Phytopathology* **52,** 171.

Kirby, A. H. M. (1971) Phytotoxic effects of some new fungicides for apple mildew control. *Meded. Fac. Landbouw. RijksUniversiteit Gent* **36,** 355.

Kirby, A. H. M. (1972) Progress towards systemic fungicides. *PANS* **18,** 1.

Kirby, A. H. M., Butt, D. J. and Williamson, C. J. (1970) Trials in 1968 of new fungicides on apple and pear. *Rep. E. Malling Res. Stn, 1969*, p. 171.

Kirch, K. (1972) On the effect of du Pont benomyl against rotbrenner (*Pseudopeziza tracheiphila*). *Weinberg und Keller* **19,** 157.

Kirkham, D. S. (1957) The significance of polyphenolic metabolites of apple and pear on the host relations of *Venturia inaequalis* and *Venturia pirina. J. gen. Microbiol.* **17,** 491.

Kirkham, D. S. and Flood, A. E. (1963) Some effects of respiration inhibitors and *o*-coumaric acid on the inhibition of sporulation in *Venturia inaequalis. J. gen. Microbiol.* **32,** 123.

Kirkham, D. S. and Hunter, L. D. (1964) Systemic antifungal activity of *iso*butyl *o*-coumarate in apple. *Nature, Lond.* **201,** 638.

Kirkham, D. S. and Hunter, L. D. (1965) Studies of the *in vivo* activity of esters of *o*-coumaric and cinnamic acids against apple scab. *Ann. appl. Biol.* **55,** 359.

Kishi, K. (1971) History of chemical control for the main diseases and pests of Japanese pear (Nijusseiki). *Japan Pesticide Inf.* **6,** 11. (*Rev. Pl. Path.* **50,** 3965, 1971.)

Kline, D. M. (1972) Fungicides for the control of Helminthosporium stripe of barley. *Pl. Dis. Reptr* **56,** 183.

Klöpping, H. L. (1960) US patent 2933502; US patent 2933504.

Klos, E. J. and Fronek, F. R. (1964) Chemical eradication of cherry leaf spot fungus. *Quart. Bull. Mich. State Coll. agric. Exp. Sta.* **47,** 65.

Klos, E. J., Robb, J. and Miller, R. (1968) Grape: black rot and powdery mildew; *Guignardia bidwellii* and *Uncinula necator. Fungicide and Nematicide Tests* **23,** 58.

Klos, E. J., Yoder, K., Nowacka, H. and Riggle, J. (1971) Apple: scab (*Venturia inaequalis*). *Fungicide and Nematicide Tests* **26,** 26.

Knauss, J. F. (1974) Pyroxychlor, a new and apparently systemic fungicide for control of *Phytophthora palmivora. Pl. Dis. Reptr* **58,** 1100.

Knight, B. C. (1971) Private communication.

Koch, W. (1971) Behaviour of commercial systemic fungicides in conventional (non-systemic) tests. *Pestic. Sci.* **2,** 207.

Koffman, W. and Kable, P. F. (1975) Improved control of brown rot in harvested sweet cherries by triforine dip treatments. *Pl. Dis. Reptr* **59,** 586.

Kohmoto, K., Miyake, H., Nishimura, S. and Udagawa, H. (1974) Distribution and chronological population shift of polyoxin-resistant strains of black spot fungi of Japanese pear, *Alternaria kikuchiana*, in field. *Ann. Phytopath. Soc. Japan* **40,** 220 (abs.).

Kondo, E. W. (1972) A method for introducing water-soluble chemicals into mature elms. *Can. For. Serv., Sault Ste. Marie, Ont. Inf. Rep.* O-X-171. 11 pp.

Kondo, E. S. and Huntley, G. D. (1973) Root-injection field trials of MBC-phosphate in 1972 for Dutch elm disease control. *Can. For. Serv., Sault Ste. Marie, Ont. Inf. Rep.* O-X-182. 17 pp.

Kondo, E. S., Roy, D. N. and Jorgensen, E. (1973) Salts of methyl-2-benzimidazole carbamate (MBC) and assessment of their potential in Dutch elm disease control. *Can. J. For. Res.* **3,** 548.

König, K. H., Pommer, E. H. and Saune, W. (1965) N-substituted tetrahydro-1,4-oxazines: a new class of fungicidal compounds. *Angew. Chem.* (*Int. Ed.*) **4,** 336.

Koopmans, M. J. (1960) Systemic fungicidal action of some 5-amino-1-(bis(dimethylamido)phosphoryl triazoles-1,2,4. *Meded. LandbHoogesch. OpzockStns. Gent* **25,** 1221.

Kost, D. and Gurfinkel, E. (1975) Isolation and structure determination of two diastereoisomers of N-tridecyl-2,6-dimethylmorpholine (tridemorph). *J. Chromatogr.* **108,** 207.

Kosuge, T. (1969) The role of phenolics in host response to infection. *Ann. Rev. Phytopathol.* **7,** 195.

Kozaka, T. (1969) Chemical control of rice blast in Japan. *Rev. Pl. Protec. Res.* **2,** 53.

Kradel, J., Effland, H. and Pommer, E. H. (1969) Erfchrungen bei der Bekämpfung des Getreidemehtfaus (*Erysiphe graminis*). *Gesund Pflanzen* **21,** 121.

Kradel, J. and Pommer, E. H. (1967) Some remarks and results on the control of powdery mildew in cereals. *Proc. 4th Br. Insectic. Fungic. Conf.* **1,** 170.

Kradel, J., Pommer, E. H. and Effland, H. (1969) Response of barley varieties to the control of powdery mildew with cyclomorph and tridemorph. *Proc. 5th Br. Insectic. Fungic. Conf.* **1,** p. 16.

Kramer, P. J. (1957) Outer space in plants. *Science, N.Y.* **125,** 633.

Krüger, W. (1959) Antibiotics as seed protectants. *S. Afr. J. agric., Sci.* **2,** 207.

Krüger, W. (1960) Antibiotics as seed protectants (1958/59). *S. Afr. J. agric. Sci.* **3,** 409.

Kuć, J. (1963) The role of phenolic compounds in disease resistance. *Conn. Agric. Exp. Sta. Bull.* no. 663, p. 20.

Kuć, J. (1968) Biochemical control of disease resistance in plants. *Wld Rev. Pest Control* **7,** 42.

Kuć, J., Williams, E. B. and Shay, J. R. (1957) Increase in resistance to apple scab following injection of host with phenylthiourea and D-phenylalanine. *Phytopathology* **47,** 21.

Kuiper, J. (1965) Failure of hexachlorobenzene to control common bunt of wheat. *Nature, Lond.* **206,** 1219.

Kuiper, J. (1968) Field evaluation of a derivative of 1,4-oxathiin for the control of cereal smuts. *Aust. J. exp. Agric. Anim. Husb.* **8,** 756.

Kundert, J. (1975) Five years' experience with systemic fungicides in fruit growing. *Schweiz. Z. Obstu. Weinb.* **111,** 161. (*Rev. Pl. Path.* **54,** 5470, 1975.)

Kuprashvili, O. G. and Chekheidze, D. V. (1973) Gray mould control on grapevine. *Prace Instytutu Sadownictwa,* E. 2 (3 pp.).

Kurtzmann, R. H. (1966) Xylem sap flow as affected by metabolic inhibitors and girdling. *Pl. Physiol. Lancaster* **41,** 641.

Kylin, A. and Hylmö, B. (1957) Uptake and transport of sulphate in wheat. Active and passive components. *Physiologia Pl.* **10,** 467.

Lacy, M. L. (1973) Control of *Septoria apiicola* leaf spot of celery with systemic and non-systemic fungicides. *Pl. Dis. Reptr* **57,** 425.

Lade, D. H. (1970) Apple: scab (*Venturia inaequalis*); frogeye leaf spot (*Physalospora obtusa*). *Fungicide and Nematicide Tests* **25,** 25.

Lafon, R. and Bugaret, Y. (1968) Les traitements des semences d'oignon contre la pourriture blanche. *Phytiat.-Phytopharm.* **18,** 23.

Lampen, J. O., Arnow, P. M., Borowski, Z. and Laskin, A. I. (1962) Location and role of sterol at nystatin binding sites. *J. Bacteriol.* **84,** 1152.

Land, B. G. van't, and Dekker, J. (1972) Effect of antimetabolites and fungicides on elongation of germination hyphae of powdery mildew *in vitro*. *Neth. J. Pl. Path.* **78,** 242.

Langcake, P. and Wickins, S. G. A. (1975a) Studies on the mode of action of dichlorocyclopropane fungicides: effects of 2,2-dichloro-3,3-dimethyl cyclopropane carboxylic acid on the growth of *Piricularia oryzae* Cav. *J. gen. Microbiol.* **88,** 295.

Langcake, P. and Wickins, S. G. A. (1975b) Studies on the action of the dichlorocyclopropanes on the host-parasite relationship in the rice blast disease. *Physiol. Plant Pathol.* **7,** 113.

Larsen, P. O., Safford, J. and Saladini, J. L. (1974) *Fungicide and Nematicide Tests* **30,** 117.

Latorre, B. A. (1972) Late season fungicide application to control apple scab. *Pl. Dis. Reptr* **56,** 1079.

Laville, E. (1974) Action du R 23979 (Imazalil) sur les pourritures à *Penicillium* des agrumes. *Meded. Faculteit Landbouwwetenschappen Gent* **39,** 1121.

Leach, S. S. (1970) Evaluation of post-harvest-prestorage fungicidal treatments for the control of *Fusarium* tuber rot of potatoes. *Phytopathology* **60,** 1299.

Leaphart, C. D. and Wicher, E. F. (1968) The ineffectiveness of cycloheximide and phytoactin as chemical controls of the blister rust disease. *Pl. Dis. Reptr* **52,** 6.

Leath, K. T. and Berg, C. C. (1972) Long term control of powdery mildew of orchard grass with benomyl. *Pl. Dis. Reptr* **56,** 75.

Leben, C. and Arny, D. C. (1952) Seed treatment experiments with helixin B. *Phytopathology* **42,** 469.

Leben, C., Arny, D. C. and Keitt, G. W. (1953) Small grain seed treatment with the antibiotic, Helixin B. *Phytopathology* **43,** 391.

Leben, C., Arny, D. C. and Keitt, G. W. (1954) Effectiveness of certain antibiotics for the control of seed-borne diseases of small grains. *Phytopathology* **44,** 704.

Leben, C., Boone, D. M. and Keitt, G. W. (1955) *Venturia inaequalis* (Cke) Wint. IX. Search for mutants resistant to fungicides. *Phytopathology* **45,** 467.

Leben, C. and Keitt, G. W. (1952) Studies on helixin in relation to plant disease control. *Phytopathology* **42,** 168.

Leben, C. and Keitt, G. W. (1954) Antibiotics and plant disease. Effects of antibiotics in control of plant diseases. *J. agric. Fd Chem.* **2,** 234.

Lee, B. K. and Wilkie, D. (1965) Sensitivity and resistance of yeast strains to actidione and actidione derivatives. *Nature, Lond.* **206,** 90.

Lee, D. R., Arnold, D. C. and Fensom, D. S. (1971) Some microscopical observations of functioning sieve tubes of *Heracleum* using Nomarski optics. *J. exptl. Bot.* **22,** 25.

Lemaire, J. M., Doussinault, G. and Jouan, B. (1970) Nouvelle perspective de lutte contre le piétin-verse des céréales par traitement des semences. *C.R. Acad. Agri. France* **56,** 643.

Lemin, A. J. and Ford, J. H. (1960) Isocycloheximide. *J. Org. Chem.* **25,** 244.

Lemin, A. J., Klomparens, W. and Moss, V. D. (1960) The translocation of cycloheximide in white pines. *For. Science* **6,** 306.

Lemin, A. J. and Magee, W. E. (1957) Degradation of cycloheximide derivatives in plants. *Pl. Dis. Reptr* **41,** 447.

Lemin, A. J. and Thomas, R. C. (1961) The translocation and persistence of tritium-labeled cycloheximide in eastern white pine seedlings. *J. Agr. Food Chem.* **9,** 254.

Lester, E. (1971) Where do we stand with mildew. *Arable Farmer,* March 1971, p. 32.

Levitt, J. (1957) The significance of 'apparent free space' in ion absorption. *Physiologia Pl.* **10,** 882.

Lewis, D. (1963) Structural gene for the methionine activating enzyme and its mutation as a cause of resistance to ethionine. *Nature, Lond.* **200,** 151.

Lewis, F. H. (1968) Apple: scab, mildew, mites. *Venturia inaequalis, Podosphaera leucotricha, Panonychus ulmi* and *Tetranychus telarius. Fungicide and Nematicide Tests* **23,** 43.

Lindberg, G. D. (1969) *Fungicide and Nematicide Tests* **25,** 92.

Lindberg, G. D. (1974) Chemical control of blast in Louisiana. *66th Ann. Prog. Rept. Rice Expt. Stn, Louisiana,* p. 184.

Linder, P. J., Craig, J. C., Cooper, F. E. and Mitchell, J. W. (1958) Movement of 2,3,6-trichlorobenzoic acid from one plant to another through their root systems. *Ag. Food Chem.* **6,** 356.

Line, R. F. (1970) *Fungicide and Nematicide Tests* **26,** 147.

Line, R. F., Walder, J. T. and Hewitt, B. V. (1974a) *Fungicide and Nematicide Tests* **30,** 105.

Line, R. F., Walder, J. T. and Hewitt, B. V. (1974b) *Fungicide and Nematicide Tests* **30,** 106.

Little, R. and Doodson, J. K. (1971) The comparison of yields of some spring barley varieties in the presence of mildew and when treated with a fungicide. *Proc. 6th Br. Insectic. Fungic. Conf.* **1,** p. 91.

Littrell, R. H. (1974) Tolerance in *Cercospora arachidicola* to benomyl and related fungicides. *Phytopathology* **64,** 1377.

Livingston, J. E. (1953) The control of leaf and stem rust of wheat with chemotherapeutants, *Phytopathology* **43,** 496.

Livingston, J. E. and Hilborn, M. T. (1959) The control of plant diseases by chemotherapy. *Econ. Bot.* **13,** 3.

Locher, F. and Hampel, M. (1973) Neue fungizide zur Bekämpfung von Getreidekrankheiten. *BASF-Mitteilungen fur den haundbau.* **3/73,** 146.

Locke, S. B. (1969) Botran tolerance of *Sclerotium cepivorum* isolates from fields with different Botran treatment histories. *Phytopathology* **59,** 13.

Lockhart, C. L. (1971) Blueberry (low bush): red leaf (*Exobasidium vaccinii*). *Fungicide and Nematicide Tests* **26,** 55.

Lockhart, C. L. and Delbridge, R. W. (1974) Control of storage diseases of carrots with post-harvest fungicide treatments. *Can. Pl. Dis. Surv.* **54,** 52.

Lockhart, C. L., MacNab, A. A. and Bolwyn, B. (1969) A systemic fungicide for control of *Verticillium* wilt in strawberries. *Can. Pl. Dis. Surv.* **49,** 46.

Lockwood, J. L., Leben, C. and Keitt, G. W. (1954) Production and properties of antimycin A from a new *Streptomyces* isolate. *Phytopathology* **44,** 438.

Logan, C. (1974) The effect on gangrene incidence of dusting seed potatoes with benomyl at time of planting. *Ann. appl. Biol.* **78,** 261.

Long, P. G. (1970) Commercial control of stem-end rot disease (*Gloeosporium musarum*) of banana fruit. *Pl. Dis. Reptr* **54,** 93.

Long, P. G. (1971) Control of black leaf streak disease of bananas with benomyl. *Pl. Dis. Reptr* **55,** 50.

Lopatin, V. M., Chaban, V. S. and Antonyuk, V. S. (1973) Effectiveness of benomyl and topsin against clubroot of cabbage on peat soils. *Zakhyst Roslyn* **18,** 92.

Loux, H. M. (1961) Process for manufacture of certain alkyl esters of benzimidazole carbamic acids. US patent 3010968.

Lovrekovich, L. and Farkas, G. L. (1963) Kinetin as an antagonist of the toxic effect of *Pseudomonas tabaci. Nature, Lond.* **198,** 710.

Ludwig, R. A. (1960) Toxins. *Plant Pathology Vol. II* (eds J. G. Horsfall and A. E. Dimond), p. 315. New York and London: Academic Press.

Lukens, R. J. (1970) Melting-out of Kentucky bluegrass, a low-sugar disease. *Phytopathology* **60,** 1276.

Lukens, R. J. (1974) Bioassay for foliar protectant action of fungicides. *Proc. Am. Phytopath. Soc.* **1,** 139.

Lyda, S. D. and Burnett, E. (1970) Influence of benzimidazole fungicides on *Phymatotrichum omnivorum* and root rot of cotton. *Phytopathology* **60,** 726.

Lyr, H., Laussmann, B. and Casperson, G. (1975a) Wirkungsmechanismus von Terrazole. *Zeitschr. Allg. Mikrobiol.* **15,** 345.

Lyr, H., Ritter, G. and Banasiak, L. (1974) Detoxication of carboxin. *Zeitschr. Allg. Mikrobiol.* **14,** 313.

Lyr, H., Ritter, G. and Casperson, G. (1972) Wirkungsmechanismus des systemischen Fungicids Carboxin, *Zeitschr. Allg. Mikrobiol.* **12,** 271.

Lyr, H., Schewe, T., Müller, W. and Zanke, D. (1975b) Zum Problem der Selektivität sowie der Struktur-Rezeptor-Beziehungen von Carboxin und seinen Analogen. *Systemfungizide* (eds H. Lyr and C. Polter), p. 153. Intern. Symp. Reinhardsbrunn. Berlin: Akademie Verlag.

Maas, J. L. (1970) Fungicidal control of Botrytis fruit rot and powdery mildew on leaves of strawberries. *Pl. Dis. Reptr* **54,** 883.

Maas, J. L. (1971) Timing of benomyl application in relation to control of Botrytis strawberry rot. *Pl. Dis. Reptr* **55,** 883.

Maas, J. L. and MacSwan, I. C. (1970) Post-harvest fungicide treatments for reduction of Penicillium decay of Anjou pear. *Pl. Dis. Reptr* **54,** 887.

Maas, J. L. and Scott, D. H. (1971) Control of storage decay of strawberry plants with benomyl preharvest treatment. *Pl. Dis. Reptr* **55,** 1009.

Maas, J. L. and Scott, D. H. (1973) Preharvest benomyl application to increase survival and quality of cold-stored strawberry nursery plants. *HortScience* **8,** 121.

McCallan, S. E. A. and Wellman, R. H. (1943) A greenhouse method of evaluating fungicides by means of tomato foliage diseases. *Contr. Boyce Thompson Inst. Pl. Res.* **13,** 93.

McClintock, J. A. (1931) The relation of canker treatment to fireblight control. *Phytopathology* **21,** 901.

McClure, T. T. and Cation, D. (1951) Comparison of Acti-dione with some other spray chemicals for control of cherry leaf spot in Michigan. *Pl. Dis. Reptr* **35,** 393.

McCornack, A. A. and Brown, G. E. (1970) Benlate, an experimental postharvest citrus fungicide. *Proc. Fla. St. hort. Soc.* **82,** 235. (*Rev. Pl. Path.* **50,** 2278, 1971.)

McDonnell, P. F. (1970) Control of Nectria decay of apple fruits by benomyl. *Pl. Dis. Reptr* **54,** 83.

McDonnell, P. F. (1971) Control of Nectria decay of apple fruits by post-harvest chemical dip treatments. *Pl. Dis. Reptr* **55,** 771.

Mace, M. E. and Solit, E. (1966) Interactions of 3-indoleacetic acid and 3-hydroxytyramine in *Fusarium* wilt of banana. *Phytopathology* **56,** 245.

Macer, R. C. F., Cook, I. K. and Dash, C. R. (1969) Laboratory and field trials in the United Kingdom with the systemic fungicide carboxin. *Proc. 5th Br. Insectic. Fungic. Conf.* **1,** p. 55.

McGee, D. C. (1972) The effect of benomyl on *Gloeotinia temulenta* under laboratory and field conditions. *Aust. J. Exp. Agric. Anim. Husb.* **11,** 693.

McIntosh, D. L. (1969) A low-volume post-harvest spray of benomyl prevents ascospore production in apple leaves infected by *Venturia inaequalis. Pl. Dis. Reptr* **53,** 816.

McIntyre, J. L., Kuć, J. and Williams, E. B. (1974) Protection of Bartlett pear against fire blight with deoxyribonucleic acid from virulent and avirulent *Erwinia amylovora. Proc. Am. Phytopath. Soc.* **1,** 89.

MacLennan, D. H., Kuć, J. and Williams, E. B. (1963) Chemotherapy of the apple scab disease with butyric acid derivatives. *Phytopathology* **53,** 1261.

McMillan, R. T. (1970) Post-harvest control of *Sclerotinia sclerotiorum* of pole beans. *Proc. Fla. St. hort. Soc.* **82,** 139.

McMillan, R. T. (1973) Control of anthracnose and powdery mildew of mango with systemic and non-systemic fungicides. *Trop. Agric.* **50,** 245.

MacNab, A. A. (1975) (ed.) Stone fruit disease reports. *Fungicide and Nematicide Tests* **30,** 39.

McNabb, H. S., Heybroek, H. M. and MacDonald, W. L. (1970) Anatomical factors in resistance to Dutch elm disease. *Neth. J. Pl. Pathol.* **76,** 196.

McNall, E. C. (1960) Metabolic studies on griseofulvin and its mechanism of action. *Antibiot. Ann.*, p. 674.

McNew, G. L. and Sundholm, N. K. (1949) The fungicidal activity of substituted pyrazoles and related compounds. *Phytopathology* **39,** 721.

McOnie, K. C., Kellerman, C. and Kruger, D. J. (1969) Benlate, a highly promising new fungicide for the control of black spot. *S. Afr. Citrus J., 1969*, no. 423: p. 7. (*Rev. Pl. Path.* **49,** 2075, 1970.)

Macris, B. and Georgopoulos, S. G. (1973) Apparent effect of pentachloronitrobenzene on the hyphal wall polysaccharides of two strains of *Neurospora crassa. Zeitschr. Allg. Mikrobiol.* **13,** 415.

MacSwan, I. C. (1971a) Prune (Early Italian): brown rot fruit rot (*Monilinia fructicola*). *Fungicide and Nematicide Tests* **26,** 54.

MacSwan, I. C. (1971b) Blackberry (evergreen): storage rots (*Botrytis* sp., *Rhizopus* sp., *Penicillium* sp. and *Alternaria* sp.). *Fungicide and Nematicide Tests* **26,** 55.

McWain, P. and Gregory, G. (1971) Solubilization of benomyl for injections into trees for disease control. N.E. For. Expt. Sta., Upper Darby, Pa. *USDA Forest Serv. Res. Paper NE-234.* 6 pp.

McWain, P. and Gregory, G. F. (1973) A benomyl-derived fungitoxicant for tree wilt disease control. *USDA Forest Serv. Res. Note NE-162.* 3 pp.

Maeda, T., Abe, H., Kakiki, K. and Misato, T. (1970) Studies on the mode of action of organophosphorus fungicide, Kitazin. *Agr. Biol. Chem.* **34,** 700.

Magendans, J. F. C. and Dekker, J. (1966) A microscopic study of powdery mildew on barley after application of the systemic compound wepsyn. *Neth. J. Pl. Path.* **72,** 274.

Magie, R. O. (1970) Control methods for post-harvest diseases of flowers and bulbs. *Rep. 1st International Symp. on Flower Bulbs, Lisse, Netherlands (1970) Sect. 2*, p. 4.

Magie, R. O. and Wilfret, G. J. (1974) Tolerance of *Fusarium oxysporum* f. sp. *gladioli* to benzimidazole fungicides. *Plant Dis. Reptr* **58,** 256.

Maier, C. R. (1960) Streptomycin absorption, translocation and retention in hops. *Phytopathology* **50,** 351.

Mailman, R., Hodgson, E. and Huisingh, D. (1971) Effect of thiols in reversing the inhibition by methyl-1(butylcarbamoyl)-2-benzimidazole carbamate on *Saccharomyces cerevisiae. Pestic. Biochem. Physiol.* **1,** 401.

Malenin, I. and Kobokhidze, D. M. (1973) Testing Benlate for control of some grapevine diseases. *Zashch. Rast. Mosk.* **21,** 19.

Maloy, O. C. and Machtmes, R. (1974) Control of white rot by furrow and root dip applications of fungicides. *Pl. Dis. Reptr* **58,** 6.

Malz, H. *et al.* (1968) Belg. Pat. 701322.

Manking, C. J. and Wood, L. S. (1969) *Fungicide and Nematicide Tests* **25,** 106.

Manking, C. J. and Wood, L. S. (1970) *Fungicide and Nematicide Tests* **26,** 128.

Manners, K. B. and Corke, A. T. K. (1971) Bitter rot (caused by *Gloeosporium* spp.). *Rep. agric. hort. Res. Stn Bristol, 1970,* p. 116.

Manning, W. J. and Papia, P. M. (1972) Benomyl soil treatments and natural occurrence of Alternaria leaf spot on carnation *Pl. Dis. Reptr* **56,** 9–11.

Mariouw Smit, F. (1969) Diaethyl-methylethoxycarbonyl-pyrazolo-pyrimidino-yl fosforothiolaat, een nieuw systemisch werkzaam meeldauw bestrijdingsmiddel. *Meded. Fac. Landbouwwetensch. Rijksuniv. Gent.* **34,** 763.

Martin, J. T. and Juniper, B. E. (1970) *The cuticles of plants.* London: Edward Arnold Ltd.

Mason, T. G. and Maskell, E. J. (1928a) Studies on the transport of carbohydrates in the cotton plant. I. *Ann. Bot. Lond.* **42,** 189.

Mason, T. G. and Maskell, E. J. (1928b) Studies on the transport of carbohydrates in the cotton plant. II. *Ann. Bot. Lond.* **42,** 571.

Mason, T. G. and Phillis, E. (1933) On the simultaneous movement of solutes in opposite directions through the phloem. *Ann. Bot. Lond.* **50,** 161.

Massee, G. (1903) On a method of rendering cucumber and tomato plants immune against fungus parasites. *J. R. hort. Soc.* **28,** 142.

Masuda, T. and Kanazawa, J. (1973) Residue of kitazin P metabolite in rice plants and soil. *Agr. Biol. Chem.* **37,** 2931.

Mathre, D. E. (1968) Uptake and binding of oxathiin systemic fungicides by resistant and sensitive fungi. *Phytopathology* **58,** 1464.

Mathre, D. E. (1970) Mode of action of oxathiin systemic fungicides. I. Effect of carboxin and oxycarboxin on the general metabolism of several Basidiomycetes. *Phytopathology* **60,** 671.

Mathre, D. E. (1971a) Mode of action of oxathiin systemic fungicides. III. Effect on mitochondrial activities. *Pest. Biochem. Physiol.* **1,** 216.

Mathre, D. E. (1971b) Mode of action of oxathiin systemic fungicides, structure-activity relationships. *J. Agr. Fd Chem.* **19,** 872.

Matta, A. (1963) Attività polifenolossidasica resistenza indotta alla fusariosi nel Pomodoro. *Riv. Pat. veg., Pavia, Ser. 3,* **3,** 99.

Matta, A. and Garibaldi, A. (1970) Attività di fungicidi sistemici contro le Verticillosi del Pomodoro e della melanzana. *Agricoltore Ital.* **70, (25 NS),** 331.

Matta, A. and Gentile, I. A. (1971) Activation of the thiophanate systemic fungicides by plant tissues. *Meded. Fac. Landbouwwetensch. Rijksuniv. Gent.* **36,** 1151.

Matta, A., Gentile, I. A. and Giai, I. (1967) Variazioni del contenuto in feloli solubili indotte del *Fusarium oxysporum* f. *lycopersici* in piante di pomodoro suscettibili e resistenti. *Ann. Fac. Sci. Agr. della Univ. degli Studi di Torino* **4,** 17.

Maude, R. B. (1966) Pea seed infection by *Mycosphaerella pinodes* and *Ascochyta pisi* and its control by seed soaks in thiram and captan suspensions. *Ann. appl. Biol.* **57,** 193.

Maude, R. B. and Kyle, A. M. (1970a) Seed treatments with benomyl and other fungicides for the control of *Ascochyta pisi* of peas. *Ann. appl. Biol.* **66,** 37.

Maude, R. B. and Kyle, A. M. (1970b) Laboratory screening tests of systemic compounds as potential dry seed dressings. *Rep. natn. Veg. Res. Stn for 1969,* p. 105.

Maude, R. B. and Kyle, A. M. (1971) Seed dressings with systemic fungicides for the control of seed-borne fungal pathogens. *Rep. natn. Veg. Res. Stn for 1970*, p. 106.

Maude, R. B. and Moule, C. G. (1971a) Leaf spot (*Septoria apiicola*) of celery. *Rep. natn. Veg. Res. Stn for 1970*, p. 107.

Maude, R. B. and Moule, C. G. (1971b) Dark leaf spot (*Alternaria brassicicola*) of cabbage. *Rep. natn. Veg. Res. Stn for 1970*, p. 107.

Maude, R. B. and Presly, A. H. (1973) Onion neck rot and its control. *Proc. 7th Br. Insectic. Fungic. Conf.* **2,** p. 609.

Maude, R. B. and Shuring, C. G. (1969) Seed treatments with 'Vitavax' for the control of loose smut of wheat and barley. *Ann. appl. Biol.* **64,** 259.

Maude, R. B. and Shuring, C. G. (1970) *Septoria* blight of celery. *Rep. natn. Veg. Res. Stn for 1969*, p. 103.

Maude, R. B., Vizor, A. S. and Shuring, C. G. (1969) The control of fungal seed-borne diseases by means of a thiram seed soak. *Ann. appl. Biol.* **64,** 245.

May, H. C., Frost, A. J. P. and Woodcock, P. M. (1975) The use of benodanil and tridemorph plus 'Polyram' in the U.K. for the control of *Puccinia striiformis* in spring barley. *Proc. 8th Br. Insectic. Fungic. Conf.* **1,** 331.

Melin, P. (1970) Nouvelles perspectives de lutte contre la cercosporiose du Bananier – premières indications sur les résultats obtenus avec le Benlate dans les bananeraies du Cameroun. *Fruits d'outre-mer* **25,** 141. (*Rev. Pl. Path.* **50,** 2389, 1971.)

Melin, P. (1973) A study on new fungicides against *Cercospora* infection of banana. *Fruits d'outre-mer.* **28,** 429.

Mel'nikov, N. N. (1971) Chemistry of pesticides. *Residue Rev.* **36,** 1.

Melville, S. C. and Jemmett, J. L. (1971) The effect of glume blotch on the yield of winter wheat. *Pl. Path.* **20,** 14.

Mercer, R. T. (1971) Some studies on the systemic activity of the thiophanate fungicides in plants. *Pestic. Sci.* **2,** 214.

Meyer, C. von, Greenfield, S. A. and Seidel, M. (1970) Wheat leaf rust: control by 4-*n*-butyl-1,2,4-triazole a systemic fungicide. *Science N.Y.* **169,** 997.

Meyer, H. (1950) Über den Einfluss von Cadmium auf die Krankheitsbereitschaft der Weizens für *Erysiphe graminis tritici* Marchal. *Phytopath. Z.* **17,** 63.

Meyer, R. W. and Parmeter, J. R. (1968) Changes in chemical tolerance associated with heterocaryosis in *Thanatephorus cucumeris. Phytopathology* **58,** 472.

Meynhardt, J. T. and Malan, A. H. (1963) Translocation of sugars in double stem grape vines. *S. Afr. J. agr. Sci.* **6,** 337.

Miller, J. W. and McRitchie, J. J. (1973) *Cylindrocladium scoparium*, causal agent of a new disease of silver buttonwood and its control. *Pl. Dis. Reptr* **57,** 500.

Miller, M. W. and Fletcher, J. T. (1974) Benomyl tolerance in *Botrytis cinerea* isolates from glasshouse crops. *Trans. Brit. Mycol. Soc.* **62,** 99.

Miller, P. M. (1970) Reducing discharge of ascospores of *Venturia inaequalis* with a spring application of benomyl, thiabendazole or urea. *Pl. Dis. Reptr* **54,** 27.

Milholland, R. D. and Jones, R. K. (1975) Blueberry (*Vaccinium corymbosum*) leaf spot. *Fungicide and Nematicide Tests* **30,** 50.

Milthorpe, F. L. and Moorby, J. (1969) Vascular transport and its significance in plant growth. *A. Rev. Pl. Physiol.* **20,** 117.

Minabe, M. (1951) On the relation between the disease susceptibility of naked barley and the silicification of epidermal cells of leaves. *Proc. Crop. Sci. Soc. Japan* **19,** 255.

Misato, T. (1967) Blasticidin S. *Antibiotics.* Vol. I. *Mechanisms of action* (eds D. Gottlieb and P. D. Shaw) p. 434. New York: Springer.

Misato, T. (1969) Mode of action of agricultural antibiotics developed in Japan. *Residue Rev.* **25,** 93.

Misato, T. (1975) The development of agricultural antibiotics in Japan, *Proc. First Intersect. Congr. of Intern. Assoc. Microbiol. Soc. 1–7 September 1974, Tokyo* **3,** 589.

Misato, T., Ishii, M., Asakawa, M., Okimoto, Y. and Fukunaga, K. (1959) Antibiotics as protectant fungicides against rice blast. II. The therapeutic action of blasticidin-S. *Ann. phytopath. Soc. Japan* **24,** 302.

Misato, T., Yamaguchi, I. and Homma, Y. (1972) A new approach in the development of biodegradable pesticides in *Environmental Toxicology of Pesticides* (eds F. Matsumura, G. M. Boush and T. Misato). p. 587. New York: Academic Press.

Mitchell, J. E., Hotson, H. H. and Bell, F. H. (1950) The use of sulfanilamide derivatives as eradicant fungicides for wheat stem rust. *Phytopathology* **40**, 873.

Mitchell, J. W., Smale, B. C., Daly, E. J., Preston, W. H. Jr., Pridham, T. G. and Sharpe, E. S. (1959) Absorption and translocation of the F-17 complex by bean plants and subsequent effect on the rust fungus, *Uromyces phaseoli typica*. *Pl. Dis. Reptr* **43**, 431.

Mitchell, J. W., Smale, B. C. and Metcalf, J. W. (1960) Absorption and translocation of regulators and compounds used to control plant diseases and insects. *Adv. Pest Control Res.* **3**, 359.

Mitchell, J. W., Zaumeyer, W. J. and Preston, W. H. (1954) Absorption and translocation of streptomycin by bean plants and its effect on the halo and common blight organisms. *Phytopathology* **44**, 25.

Miura, H., Ito, H., Kimura, K. and Takahashi, S. (1973) Resistance of *Pyricularia oryzae* to kasugamycin. 1. Decline in effect of kasumin in Shonai district. *Ann. Phytop. Soc. Japan* **39**, 239.

Miura, H., Katagiri, M., Yamaguchi, T., Uesugi, Y. and Ito, H. (1976) Mode of occurrence of kasugamycin-resistant rice blast fungus. *Ann. Phytop. Soc. Japan* **42**, 117.

Moje, W., Kendrick, J. B. Jr. and Zentmyer, G. A. (1963) Systemic and fungicidal activity of D-, L- and DL-ethionine, S-alkyl-DL-homocysteine derivatives and methionine antagonists. *Phytopathology* **53**, 883.

Mokrzecki, S. (1903) Über die innere Therapie der Pflanzen. *Z. PflKrankh. PflPath. PflSchutz.* **13**, 257.

Moller, W. J. and Carter, M. V. (1969) A preliminary observation on apricot dieback prevention with chemicals. *Pl. Dis. Reptr* **53**, 828.

Moller, W. J. and Carter M. V. (1971) Field evaluation of benomyl for control of limb dieback (gummosis) in apricots. *Aust. J. exp. Agric. Anim. Husb.* **10**, 488.

Moller, W. J., Wicks, T. J. and Steed, J. N. (1969) Field evaluation of a new fungicide (benomyl) for apple scab. *Aust. J. exp. Agric. Anim. Husb.* **9**, 461. (*Rev. Pl. Path.* **49**, 198, 1970.)

Molzahm, S. W. and Woods, R. A. (1972) Polyene resistance and the isolation of sterol mutants in *Saccharomyces cerevisiae*. *J. Gen. Microbiol.* **72**, 339.

Moody, R. P. and Prasad, R. (1974) Impact of pesticides on forest ecosystems. Reduction of earthworm populations in forest soils following treatment of elm trees with benomyl. *Proc. Can. Fed. Biol. Societies* **17**, 100.

Moore, M. H. (1966) Glasshouse experiments on apple scab. II. Direct and translocated fungicidal activity. *Ann. appl. Biol.* **57**, 451.

Moore, M. H. (1967) Glasshouse experiments on apple scab. III. Fungicidal mixtures, curative translocation and the influence of mildew. *Ann. appl. Biol.* **59**, 239.

Moreshet, S. (1970) Effect of environmental factors on cuticular transpiration resistance. *Pl. Physiol. Lancaster* **46**, 815.

Morgan, N. G. (1952) Spray application problems. VI. A laboratory technique using *Botrytis fabae* on broad bean for the biological evaluation of fungicidal spray deposits. *Long Ashton 1952 Annual Report*, p. 99.

Morris, D. A. and Thomas, E. E. (1968) Distribution of ^{14}C-labelled sucrose in seedlings of *Pisum sativum* I. treated with indoleacetic acid and kinetin. *Planta* **83**, 276.

Morrison, L. S. (1972) Efficacy of benomyl (Benlate) and triarimal (EC-273) against juniper blight of *Juniperus sabina tamariscifolia*. *Phytopathology* **62**, 806 (abs.).

Moseman, J. G. (1968) Fungicidal control of smut diseases of cereals. US Dept. of Agr. CR42–68.

Moss, V. D. (1957) Actidione treatment of blister rust trunk cankers on western white pine. *Pl. Dis. Reptr* **41**, 709.

Moss, V. D. (1958) Actidione-stove oil treatment of blister rust trunk cankers on reproduction and pole western white pine. *Pl. Dis. Reptr* **42**, 703.

Moss, V. D. (1961) Antibiotics for control of blister rust on western white pine. *For. Sci.* **7**, 380.

Moss, V. D., Viche, H. J. and Klomparens, W. (1960) Antibiotic treatment of western white pines infected with blister rust. *J. For.* **58,** 691.

Muirhead, I. F. (1974) Resistance to benzimidazole fungicides in blue mould of citrus in Queensland. *Aust. J. exp. Agric. Anim. Husb.* **14,** 698.

Mukopadhyay, A. N. and Thakar, R. P. (1972) Systemic activity of triphenyltin chloride in sugar beet seedlings. *Pl. Dis. Reptr* **56,** 776.

Müller, A. (1926) *Die innere Therapie der Pflanzen*, Monographien der angewandten Entomologie, N. 8. Berlin: Paul Parey.

Muller, G. J., Linn, M. B. and Sinclair, J. B. (1972) Translocation, disease remission and persistence of thiabendazole and 5-ethoxy-3-(trichloromethyl)-1,2,4-thiadazole in tomato seedlings. *Pl. Dis. Reptr* **56,** 1054.

Müller, H. W. K. (1970) Results with benomyl against strawberry grey mould (*Botrytis cinerea*). *NachrBl. dt. PflSchutzdienst., Stuttg.* **22,** 81. (*Rev. Pl. Path.* **50,** 812, 1970.)

Müller, K. O., Klinkowski, M. and Meyer, G. (1939) Physiologisch-genetische Untersuchungen über die Resistenz der Kartoffel gegenüber *Phytophora infestans. Naturwissenschaften* **27,** 765.

Müller, K. O., Mackay, J. H. E. and Friend, J. N. (1954) Effect of streptomycin on host–pathogen relationship of a fungal pathogen. *Nature, Lond.* **174,** 878.

Münch, E. (1927) Versuche über den Saftkreislauf. *Ber. Deutsch Bot. Ges.* **45,** 340.

Mundy, E. J. (1973) The effect of yellow rust and its control on the yield of Joss Cambier winter wheat. *Pl. Path.* **22,** 171.

Mundy, E. J. and Page, R. A. (1973) Use of fungicides to control powdery mildew of spring barley. *Exp. Husb.* **24,** 94.

Murdoch, C. L., Laemmlen, F. F. and Parvin, P. E. (1973) Fungicidal control of *Puccinia striiformis* on three cultivars of *Poa pratensis* in Hawaii. *Pl. Dis. Reptr* **57,** 217.

Muse, R. R. (1970) *Fungicide and Nematicide Tests* **26,** 124.

Mussell, H. E. and Green, R. J. Jr. (1968) Production of polygalacturonase by *Verticillium albo-atrum* and *Fusarium oxysporum* f. sp. *lycopersici*, *in vitro* and in vascular tissue of susceptible and resistant hosts. *Phytopathology* **58,** 1061.

Mussell, H. W. and Green, R. J., Jr. (1970) Host colonization and polygalacturonase production by two tracheomycotic fungi. *Phytopathology* **60,** 192.

Nair, V. M. G., Walter, K. E. and Kuntz, J. E. (1969) The inhibition of tylosis and oak wilt development by the cytokinin 6-benzyl amino purine. *Phytopathology* **59,** 1042.

Nagatsu, J., Anzai, K. and Suzuki, S. (1962) Pathocidin, a new antifungal antibiotic. II. Taxonomic studies on the pathocidin producing organism *Streptomyces albus* var. *pathocidicus. J. Antibiotics* (*Tokyo*) **A15,** 103.

Nakamura, H. and Sakurai, H. (1968) Tolerance of *Piricularia oryzae* Cavara to Blasticidin-S. *Bull. Agr. Chem. Inspect. Stn.* **21,** 25.

Nakanishi, T. and Oku, H. (1969) Mechanism of selective toxicity: absorption and detoxication of an antibiotic, ascochytine, by sensitive and insensitive fungi. *Phytopathology* **59,** 1563.

Nakanishi, T. and Oku, H. (1970) Mechanism of selective toxicity of fungicides: absorption, metabolism and accumulation of pentachloronitrobenzene by phytopathogenic fungi. *Ann. Phytopath. Soc. Japan* **36,** 67.

Napier, E. J., Rhodes, A., Turner, D. I., Toothill, J. P. R. and Dunn, A. (1957) Systemic action of captan against *Botrytis fabae* (chocolate spot of broad bean). *J. Sci. Fd Agric.* **8,** 467.

Napier, E. J., Turner, D. I. and Rhodes, A. (1956) The *in vitro* action of griseofulvin against pathogenic fungi of plants. *Ann. Bot. N.S.* **20,** 461.

Napier, E. J., Turner, D. I., Rhodes, A. and Toothill, J. P. R. (1956) The systemic action against *Pseudomonas medicaginis* var *phaseoli* of a streptomycin spray applied to dwarf beans. *Ann. appl. Biol.* **44,** 145.

Natti, J. J. (1967) Bean white mold control with foliar sprays. *Fm Res.* **33,** 10.

Natti, J. J. (1970) Bean seed orientation and performance of fungicide treatment of seed. *Phytopathology* **60,** 577.

Natti, J. J. and Crosier, D. C. (1971) Seed and soil treatments for control of bean root rots. *Pl. Dis. Reptr* **55,** 483.

Natti, J. J. and Szkolnik, M. (1954) Influence of growth regulators on resistance of beans to *Colletotrichum lindemuthianum*. *Phytopathology* **44,** 111.

Neely, D. (1975) Treatment of foliar diseases of woody ornamentals with soil injections of benomyl. *Pl. Dis. Reptr* **59,** 300.

Neely, D. (1974) Efficacy of soil injections and foliar sprays of benomyl for control of Dutch elm disease in large nursery elms. *Pl. Dis. Reptr* **58,** 261.

Neely, D., Carter, J. C., Himelick, E. B., Born, G. and Black, W. M. (1973) Dutch elm disease research report. *Illinois Natural Hist. Survey Coop. Rept. to Regional Proj.* NCR-32 (Wilt Diseases).

Netzer, D. and Dishon, I. (1970) Effect of mode of application of benomyl on control of *Sclerotinia sclerotiorum* on musk melons. *Pl. Dis. Reptr.* **54,** 909.

Netzer, D., Dishon, I. and Krikun, J. (1970) Control of some diseases on greenhouse grown vegetables with benomyl as related to studies of its movement. *Proc. 7th Int. Congr. Pl. Prot.* (*Paris*), p. 222.

Nevins, D. J., English, P. D. and Albersheim, P. (1968) Changes in cell wall polysaccharides associated with growth. *Pl. Physiol.* **43,** 914.

Newby, L. and Tweedy, B. G. (1970) Comparison of amino acid exudates from leaves of two bean varieties. *Phytopathology* **60,** 6.

Nicholls, T. H. (1973) Fungicidal control of *Lophodermium pinastri* on red pine nursery seedlings. *Pl. Dis. Reptr* **57,** 263.

Nicholson, J. F., Meyer, W. A., Sinclair, J. B. and Buller, J. D. (1971) Turf isolates of *Sclerotinia homoeocarpa* tolerant to dyrene. *Phyt. Zeitschr.* **72,** 169.

Nicholson, J. F., Sinclair, J. B., White, J. C. and Kirkpatrick, B. L. (1972) Upward and lateral translocation of benomyl in strawberry. *Phytopathology* **62,** 1183.

Nickell, L. G., Gordon, P. N. and Goenaga, A. (1961) 2-*n*-Alkylmercapto-1,4,5,6-tetrahydropyrimidines, chemotherapeutic agents for plant rusts. *Pl. Dis. Reptr* **45,** 756.

Niemann, G. J. (1964) Investigations on the cnemical control of powdery mildew. *Thesis, University of Utrecht.*

Niemann, G. J. and Dekker, J. (1966a) Evaluation of the activity of a number of procaine derivatives and analogous compounds against powdery mildew (*Sphaerotheca fuliginea*) on cucumber. *Ann. appl. Biol.* **57,** 53.

Niemann, G. J. and Dekker, J. (1966b) Activity against rust and powdery mildew of some *para*phenylenediamines and related compounds. *Neth. J. Pl. Path.* **72,** 213.

Nishimura, M., Kohmoto, K. and Udagawa, H. (1973) Field emergence of fungicide-tolerance strains in *Alternaria kikuchiana* Tanaka. *Rept. Tottori Mycol. Inst.* (*Japan*) **10,** 677.

Noble, M., Maggarvie, Q. D., Hams, A. F. and Leafe, L. L. (1966) Resistance to mercury of *Pyrenophora avenae* in Scottish seed oats. *Pl. Path.* **15,** 23.

Noguchi, T. (1972a) A new approach in the development of biodegradable pesticides. *Environmental Toxicology of Pesticides.* (eds T. Matsumura, G. M. Boush and T. Misato), p. 607. New York and London: Academic Press.

Noguchi, T. (1972b) Environmental evaluation of systemic fungicides. *Environmental toxicology of pesticides.* (eds F. Matsumura, G. M. Boush and T. Misato). New York: Academic Press.

Noguchi, T. and Ohkuma, K. (1971) Chemistry and metabolism of thiophanate fungicides. *Symp. Terminal Pesticide Residues, Tel Aviv*, p. 263.

Noguchi, T., Ohkuma, K. and Kosaka, S. (1971) Relation of chemical structure to biological activity in thiophanate. *Proc. 2nd Int. Congress of Pesticide Chem.* **5,** 257.

North, C. P. (1962) Technique of injecting substances into woody plants. Chapter from *A decade of synthetic chelating agents in inorganic plant nutrition*, p. 138. Edited and published by A. Wallace, 2278 Parnell Avenue, Los Angeles, USA.

Norton, J. B. S. (1916) Internal action of chemicals on resistance of tomatoes to leaf diseases. *Md. Agric. Exp. Sta. Bull.* **192,** 17.

Noveroske, R. L. (1975a) Dowco 269: a new systemic fungicide for control of *Phytophthora parasitica* of tobacco. *Phytopathology* **65,** 22.

Noveroske, R. L. (1975b) Pyroxychlor – a new systemic fungicide for control of plant root diseases. *Down to Earth* **30,** 3.

Noveroske, R. L., Kuć, J. and Williams, E. B. (1964a) Oxidation of phloridzin related to resistance of *Malus* to *Venturia inaequalis. Phytopathology* **54,** 92.

Noveroske, R. L., Williams, E. B. and Kuć, J. (1964b) Beta-glucosidase and phenoloxidase in apple leaves and their possible relation to resistance to *Venturia inaequalis. Phytopathology* **54,** 98.

O'Brien, T. P. and Carr, O. J. (1970) A suberised layer in the cell walls of the bundle sheath of grasses. *Aust. J. biol. Sci.* **23,** 275.

Oeser, H., Gorbach, S. and Boerner, E. (1975) Translocation and metabolism of pyracarbolid in wheat. *Pesticides,* (ed. F. Coulston and F. Korte). pp. 557–61. Stuttgart: Georg Thieme Verlag.

Ogawa, J. M., Manji, B. T. and Blair, E. (1968a) Efficacy of Fungicide 1991 in reducing fruit rot of stone fruits. *Pl. Dis. Reptr* **52,** 722.

Ogawa, J. M., Manji, B. T. and Schreader, W. R. (1975) *Monilinia* life cycle on sweet cherries and its control by overhead sprinkler fungicide applications. *Pl. Dis. Reptr* **59,** 876.

Ogawa, J. M., Su, H. J., Tsai, Y. P., Chen, S. S. and Liang, C. H. (1968b) Protective and therapeutic action of 1-(butylcarbamoyl)-2-benzimidazole carbamic acid, methyl ester (F 1991) against the banana crown rot pathogens. *Pl. Prot. Bull. Taiwan* **10,** 1. (*Rev. Pl. Path.* **49,** 1100, 1970.)

Ogawa, J. M., Su, H. J., Tsai, Y. P. and Lee, I. M. (1969) Postharvest decay development as affected by systemic activity of benomyl in bananas, oranges and pineapples. *Phytopathology* **59,** 1043 (abs.).

Ogawa, J. M., Yates, W. E., Manji, B. T. and Cowden, R. E. (1972) Ground and aircraft applications of thiophanate-methyl on control of stone fruit brown rot blossom blight. *Phytopathology* **62,** 781.

Ohkuma, K. (1961) Chemical structure of toyocamycin. *J. Antibiotics* (*Tokyo*) **A19,** 343.

Ohmori, K. (1967) Studies on characters of *Piricularia oryzae* made resistant to kasugamycin. *J. Antibiotics* (*Tokyo*), A **20,** 109.

Ohta, N., Kakiki, K. and Misato, T. (1970) Studies on the mode of action of polyoxin D. Part II. Effect of polyoxin D on the synthesis of fungal cell wall chitin. *Agr. Biol. Chem.* **34,** 1224.

Okamoto, H. (1968) Fungicidal properties of kasugamycin. *Noyaku-Shunju.* **23,** 16.

Okimoto, Y. and Misato, T. (1963a) Antibiotics as protectant bactericide against bacterial leaf blight of rice plant. 2. Effect of cellocidin on growth, respiration and glycolysis of *Xanthomonas oryzae. Ann. Phytopath. Soc. Japan* **28,** 209.

Okimoto, Y. and Misato, T. (1963b) Antibiotics as protectant bactericide against bacterial leaf blight of rice plant. 3. Effect of cellocidin on TCA cycle, electron transport system and metabolism of protein in *Xanthomonas oryzae. Ann. Phytopath. Soc. Japan* **28,** 250.

Okuda, T., Suzuki, M., Egawa, Y. and Askino, K. (1958) Studies on cycloheximide and its new stereoisomeric antibiotics. *Chem. Pharm. Bull.* (*Tokyo*) **6,** 328.

Okuyama, A., Machiyama, N., Kinoshita, T. and Tanaka, N. (1971) Inhibition by kasugamycin of initiation complex formation on 30 *S* ribosomes. *Biochem. Biophys. Res. Comm.* **43,** 196.

Olsen, C. (1939) The employment for water culture experiments of distilled water containing traces of copper. *C.r. Trav. Lab. Carlsberg, Ser. chim.* **23,** 37.

Oort, A. J. P. (1971) Development of systemic fungicides: a historical background. *Meded. Fak. Landb. Wett. Gent* **36,** 58.

Oort, A. J. P. and van Andel, O. M. (1960) Aspects of chemotherapy. *Meded. LandbHoogesch. OpzoekStns. Gent* **25,** 981.

Oort, A. J. P. and Dekker, J. (1960) Experiments with rimocidin and pimaricin, two fungicidal antibiotics with systemic action. *Proc. IVth Int. Congr. Crop Prot., Hamburg, 1957,* **2,** 1565.

Oort, A. J. P. and Dekker, J. (1964) Systemic activity of 6-azauracil against powdery mildew. *Proc. Symp. Host–Parasite Relations, Budapest 1964,* p. 171.

Ordish, G. (1962) Economics and pest control. *World Rev. Pest Control.* No. 4, p. 31.

Ó'Ríordáin, F. (1973) Effects of benzimidazole compounds and other fungicides on leaf spot of black currant. *Proc. 7th Br. Insectic. Fungic. Conf.* **2,** 477.

Ó'Ríordáin, F., Kavanagh, T. and O'Callaghan, T. F. (1971) Comparisons of systemic and protectant fungicides for disease control on four soft fruits. *Proc. 6th Br. Insectic. Fungic. Conf.* **2,** 324.

Ost, W., von Bruchhausen, V. and Drandarevski, C. (1971) Transport of the systemic fungicide Cela W 524 in barley plants (Part 1). *Pestic. Sci.* **2,** 219.

Ost, W., Thomas, K., Jerchel, D. and Appel, K. R. (1969) Ger. offen. 1901421.

Ostermann, J. (1974) Trials of post infection treatments against pear scab (*Venturia pirina* Aderh.). *Phytiat.-Phytopharm.* **23,** 123 (*Rev. Pl. Path.* **54,** 1802, 1974.).

Otake, N., Takeuchi, S., Endo, T. and Yonehara, H. (1965) The structure of blasticidin-S. *Tetrahedron Letters* No. 19, 1411.

Otta, J. (1973) Eastern red cedar Phomopsis seedling blight (*Phomopsis juniperovora*). *Fungicide and Nematicide Tests* **29,** 123.

Otta, J. D. (1974) Benomyl and thiophanate methyl control Phomopsis blight of Eastern red cedar in a nursery. *Pl. Dis. Reptr* **58,** 476.

Otto, S. and Drescher, N. (1973) The behaviour of tridemorph in soil. *Proc. 7th Br. Insectic. Fungic. Conf.* p. 57.

Overbeek, J. van (1956) Absorption and translocation of plant regulators *A. Rev. Pl. Physiol.* **7,** 355.

Oxford, A. E., Raistrick, H. and Simonart, P. (1939) XXIX. Studies in the biochemistry of micro-organisms. LX. Griseofulvin $C_{17}H_{17}O_6Cl$, a metabolic product of *Penicillium griseofulvum* Dierckx. *Biochem. J.* **33,** 240.

Paisley, R. B. and Heuberger, J. W. (1972) Apple scab. VI. Prevention of primary scab infection in an isolated apple orchard – 1970–71. *Pl. Dis. Reptr* **56,** 736.

Palmer, L. T. (1973) *Fungicide and Nematicide Tests* **29,** 118.

Palmiter, D. H. (1970) Apple: scab (*Venturia inaequalis*). *Fungicide and Nematicide Tests* **25,** 31.

Papavizas, G. C. and Davey, C. B. (1963) Effect of sulfur-containing amino compounds and related substances on Aphanomyces root rot of peas. *Phytopathology* **53,** 109.

Papavizas, G. C. and Lewis, J. A. (1970) Suppression of Aphanomyces root rot of peas by soil amendments, fungicides and fumigants. *Phytopathology* **60,** 1306.

Papavizas, G. C. and Lewis, J. A. (1971) Black root rot of bean and tobacco caused by *Thielaviopsis basicola* as affected by soil amendments and fungicides in the greenhouse. *Pl. Dis. Reptr* **55,** 352.

Paravizas, G. C. and Lewis, J. A. (1975) Effect of seed treatment with fungicides on bean root rots. *Pl. Dis. Reptr* **59,** 24.

Papavizas, G. C., Lewis, J. A. and Russell, H. (1972) Chemical control of black root rot of bean and tobacco. *Pl. Dis. Reptr* **56,** 15.

Parle, J. N. and Dodanis, D. (1973) Control of *Botrytis cinerea* in grapes. *N.Z. Jl exp. Agric.* **1,** 81.

Parr, W. J. (1971) Private communication.

Parr, W. J. and Binns, E. S. (1970) Acaricidal activity of benomyl. *A Rep. Glasshouse Crops Res. Inst. 1969,* p. 113.

Partridge, A. D. and Rich, A. E. (1962) Induced tolerance to fungicides in three species of fungi. *Phytopathology* **52,** 1000.

Patel, P. V. and Johnson, J. R. (1968) Dominant mutation for nystatin resistance in yeast. *Appl. Microbiol.* **16,** 164.

Pathak, K. D. and Joshi, L. M. (1970) Seed treatment by systemic fungicides for the control of wheat rusts. *Indian Phytopath.* **13,** 655.

Patil, S. S. and Dimond, A. E. (1967a) Inhibition of *Verticillium* polygalacturonase by oxidation products of polyphenols. *Phytopathology* **57,** 492.

Patil, S. S. and Dimond, A. E. (1967b) Induction and repression of polygalacturonase synthesis in *Verticillium albo-atrum*. *Phytopathology* **57,** 825.

Patil, S. S. and Dimond, A. E. (1967c) Effect of sugars and sugar alcohols on production of polygalacturonase by *Fusarium oxysporum* f. sp. *lycopersici*. *Phytopathology* **57,** 825.

Patil, S. S. and Dimond, A. E. (1968) Repression of polygalacturonase synthesis in *Fusarium oxysporum* f. sp. *lycopersici* by sugars, and its effect on symptom reduction in infected tomato plants. *Phytopathology* **58,** 676.

Paul, R. and Tchelitcheff, S. (1955) Constitution chimique de l'inactone. Synthèse partielle de trois isomères de l'actidione. *Bull. Soc. Chim. France* **55,** 1316.

Paulus, A. O., Nelson, J., and Shibuya, F. (1972) Fungicides and methods of application for the control of cantaloup powdery mildew. *Plant Dis. Reptr* **56,** 935.

Paulus, A. O. and Netzer, D. (1974) Resistance in *Septoria leucanthemi* to benzimidazole type fungicides. *Proc. Amer. Phyt. Soc.* **1,** 45.

Paulus, A. O., Shibuya, F., Holland, A. H. and Nelson, J. (1970a) Timing interval for control of *Septoria apiicola* leaf spot of celery (*Apium graveolens*). *Pl. Dis. Reptr* **54,** 531.

Paulus, A. O., Shibuya, F., Nelson, J. and Harvey, D. A. (1970b) Systemic fungicide interval for control of sugarbeet *Cercospora* leaf spot. *Phytopathology* **60,** 1539.

Paulus, A. and Starr, G. H. (1951) Control of loose smut with antibiotics. *Agron. J.* **43,** 617.

Paulus, A. O., Voth, V., Bringhurst, R. S. and Welch, N. (1973) Comparison of fungicides and methods of application for the control of Ramularia leafspot of strawberry. *Proc. 7th Br. Insectic. Fungic. Conf.* **2,** 471.

Paulus, A. O., Voth, V., Shibuya, F., Bowen, H. and Holland, A. H. (1969) Fungicidal control of *Botrytis* fruit rot of strawberry. *Calif. Agric.* **23,** 15. (*Rev. appl. Mycol.* **48,** 1858, 1969.).

Peel, A. J. (1970) Further evidence for the relative immobility of water in the sieve tubes of willow. *Physiologia Pl.* **23,** 667.

Peel, A. J. (1974) *Transport of nutrients in plants.* London: Butterworths.

Pegg, G. F. and Vessey, J. C. (1973) Chitinase activity in *Lycopersicon esculentum* and its relationship to the *in vivo* lysis of *Verticillium albo-atrum* mycelium. *Physiol. Pl. Path.* **3,** 207.

Pegg, K. G. (1972) Control of *Sclerotinia sclerotiorum* rot in beans with benomyl. *Aust. J. Exp. Agric. Anim. Husb.* **12,** 81.

Pellegrini, G., Bugiani, A. and Tenerini, I. (1965) Systemic properties of a new class of fungicides. *Phytopath. Z.* **52,** 37.

Pellissier, M., Lacasse, N. L. and Cole, H., Jr. (1971) Effect of benomyl on the response to ozone in Pinto bean. *Phytopathology* **61,** 131.

Pepin, H. S. and Ormrod, D. J. (1968) Control of mummy berry of highbush blueberry. *Can. Pl. Dis. Surv.* **48,** 132.

Pepin, H. S. and Ormrod, D. J. (1974) Control of mummy berry of highbush blueberry. *Pl. Dis. Reptr* **58,** 840.

Person, C., Samborski, D. J. and Forsyth, F. R. (1957) Effect of benzimidazole on detached wheat leaves. *Nature, Lond.* **180,** 1294.

Pessanha, B. M. R., Iamamoto, T. and Scaranari, H. J. (1970) Benlate and cercobin – new fungicides in the control of leaf spots of strawberry (*Fragaria vesca* L.) (in Spanish). *Biológico* **36,** 121. (*Rev. Pl. Path.* **50,** 804, 1971.)

Petersen, D. and Cation, D. (1950) Exploratory experiments on the use of Actidione for the control of peach brown rot and cherry leaf spot. *Pl. Dis. Reptr* **34,** 5.

Peterson, C. A. and Edgington, L. V. (1969) Translocation of the fungicide benomyl in bean plants. *Phytopathology* **59,** 1044.

Peterson, C. A. and Edgington, L. V. (1970) Transport of the systemic fungicide benomyl in bean plants. *Phytopathology* **60,** 475.

Peterson, C. A. and Edgington, L. V. (1971) Transport of benomyl into various plant organs. *Phytopathology* **61,** 91.

Peterson, G. W. (1975a) Dothistroma needle blight of pines in forest nursery diseases in the United States. (eds G. W. Peterson and R. S. Smith) *USDA Forest. Serv. agric. handbook* **470,** 72.

Peterson, G. W. (1975b) Phomopsis blight of junipers in forest nursery diseases in the United States. *USDA Forest Serv. Agr. Handbook* **470,** 76.

Peterson, J. L. and Davis, S. H. (1973) Juniper (*Juniperus scopulorum*) tip blight (*Phomopsis juniperovora*). *Fungicide and Nematicide Tests* **29,** 122.

Peturson, B., Forsyth, F. R. and Lyon, C. B. (1958) Chemical control of cereal rusts. II. Control of leaf rust of wheat with experimental chemicals under field conditions. *Phytopathology* **48,** 655.

Phelps, R. H. (1969) Control of two citrus diseases by low volume spraying. *Bull. Citrus Res., Univ. W. Indies no. 16,* 4 pp. (*Rev. Pl. Path.* **50,** 1234, 1971.)

Phelps, W. R. and Kuntz, J. E. (1965) Translocation and persistence of cycloheximide and oligomycin in Northern pin oaks. *For. Sci.* **11,** 354.

Phelps, W. R., Kuntz, J. E. and Riker, A. J. (1957) Antibiotics delay oak wilt symptoms on inoculated northern pin oaks in central Wisconsin. *Phytopathology* **47,** 27.

Phelps, W. R., Kuntz, J. E. and Ross, A. (1966) A field evaluation of antibiotics and chemicals for control of oak wilt in northern pin oaks. *Pl. Dis. Reptr* **50,** 736.

Phelps, W. R. and Weber, R. (1966) A preliminary evaluation of antibiotics for control of white pine blister rust in the Lake States. *Pl. Dis. Reptr* **50,** 224.

Phelps, W. R. and Weber, R. (1967) Translocation and persistence of cycloheximide and phytoactin in Eastern white pine. *For. Sci.* **13,** 90.

Phelps, W. R. and Weber, R. (1969) An evaluation of chemotherapeutants for control of blister rust cankers in Eastern white pine. *Pl. Dis. Reptr* **53,** 514.

Phelps, W. R. and Weber, R. (1970) An evaluation of chemotherapeutic treatment of blister rust cankers in Eastern white pine. *Pl. Dis. Reptr* **54,** 1031.

Phillips, D. J. (1975) Detection and translocation of benomyl in postharvest-treated peaches and nectarines. *Phytopathology* **65,** 255.

Phillips, W. H., Ellis, G. A. V. and Merryweather, J. (1973) Field evaluation of carbendazim (BAS 346F) as a fungicide on soft fruit. *Proc. 7th Br. Insectic. Fungic. Conf.* **2,** 491.

Phillips, W. H. and Frost, A. J. P. (1975) The use of carbendazim and tridemorph in the United Kingdom for the control of *Rhynchosporium secalis* on barley. *Proc. 8th Br. Insectic. Fungic. Conf.* **2,** 437.

Phipps, P. M. and Stipes, R. J. (1972) The curative effects of two systemic, fungitoxic benzimidazoles on mimosa seedlings exhibiting Fusarium wilt symptoms. *Va. Jour. Sci.* **23,** 122 (abs.).

Phipps, P. M. and Stipes, R. J. (1973) An evaluation of two systemic fungitoxic benzimidazoles as control agents of Fusarium wilt of the mimosa tree. *Proc. 2nd Intern'l Congress of Plant Pathology* (abs. 0201).

Phipps, P. M. and Stipes, R. J. (1975) Control of Fusarium wilt of mimosa with benomyl and thiabendazole. *Phytopathology* **65,** 504.

Pieczarka, D. J. and Lorbeer, J. W. (1974) Control of bottom rot of lettuce by ridging and fungicide application. *Pl. Dis. Reptr* **58,** 837.

Pieroh, E. A., Ahrens, C. H. and Schumacher, J. (1975) Prothiocarb, a new systemic fungicide against phycomycetous fungi. *Reports and Information. Section III. Chemical control. Intern. Plant Protection Congress,* Moscow. Part I, p. 575.

Pierson, C. F. (1966) Fungicides for the control of blue mold rot of apples. *Pl. Dis. Reptr* **50,** 913.

Pinkas, Y., Shabi, E., Solel, Z. and Cohen, A. (1973) Infiltration and translocation of thiabendazole in apple trees by means of a pressure injection technique. *Phytopathology* **63,** 1166.

Pitblado, R. E. and Edgington, L. V. (1971) Effect of various surfactants on the 'solubility' of benomyl. *Proc. Can. Phytopathological Soc. 37th Session,* p. 20.

Pitman, M. G., Lütge, U., Kramer, D. and Ball, E. (1974) Free space characteristics of barley leaf slices. *Aust. J. Pl. Physiol.* **1,** 65.

van der Plank (1963) *Plant diseases: epidemics and control.* New York and London: Academic Press.

Pleisch, P., Lauber, J. and Preiser, F. A. (1971) Post-harvest disease control of pome fruits with thiabendazole. *Proc. 6th Br. Insectic. Fungic. Conf.* **1,** 181.

Pluijgers, C. W. (1959) Direct and systemic antifungal action of dithiocarbamic acid derivatives. *Thesis, University of Utrecht.*

Pluijgers, C. W. and Kaars Sijpesteijn, A. (1966) The antifungal activity of thiosemicarbazides *in vitro* and their systemic activity against cucumber scab *Cladosporium cucumerinum. Ann. appl. Biol.* **57,** 465.

Pomerleau, R. (1970) Pathological anatomy of the Dutch elm disease. Distribution and development of *Ceratocystis ulmi* in elm tissue. *Can. J. Bot.* **48,** 2043.

Pomerleau, R. and Mehran, A. R. (1966) Distribution of spores of *Ceratocystis ulmi* labelled with phosphorus-32 in green shoots and leaves of *Ulmus americana. Naturaliste can.* **93,** 577.

Pommer, E. H. (1970) The systemic efficacy of a new fungicide of the Furane carbon acid anilide group (BAS 3191F). *VII int. Congr. Pl. Prot. 1970.* (abs.)

Pommer, E. H. (1971a) The systemic activity of a new fungicide of the furane carbonic acid anilide group (BAS 3191F). *Proc. 2nd int. Congr. Pestic. Chem.* **5,** 397.

Pommer, E.-H. (1971b) Probleme bei der Entwicklung systemischer Fungizider. *Z. PflKrankh. PflPath. PflSchutz.* **78,** 513.

Pommer, E.-H. (1975) Private communication.

Pommer, E. H., Hampel, M. and Löcher, F. (1971) Results with 'BAS 3270F'. *Proc. 6th Br. Insectic. Fungic. Conf.* **2,** 587.

Pommer, E. H. and Kradel, J. (1967) Substitutierte dimethylmorpholin-derivate als neue fungizide zur bekämpfung echter mehltaupilze. *Meded. Rijksfaculteit Landbouw. Wetenschappen, Gent* **32,** 735.

Pommer, E. H. and Kradel, J. (1969) Mebenil (BAS 3050F), a new compound with specific action against some Basidiomycetes. *Proc. 5th Br. Insectic. Fungic. Conf.* **2,** 563.

Pommer, L. H., Otto, S. and Kradel, J. (1969) Some results concerning the systemic action of tridemorph. *Proc. 5th Br. Insectic. Fungic. Conf.* **2,** 347.

Ponchet, J. (1959) La maladie du piétin-verse des céréales: *Cercosporella herpotrichoides*, importance agronomique biologie, epiphytologie. *Ann. Épiphyt.* **10,** 45

Porter, R. H. (1956) Seed treatment tests for control of barley loose smut. *Pl. Dis. Reptr* **40,** 106.

Potter, A. W. and Milburn, J. A. (1970) Sub-aqueous transpiration in new perspective. *New Phytol.* **69,** 961.

Potter, H. S. and Sneider, C. J. (1973) Aerial spraying to control sugar beet leaf spot disease. *Agric. Aviation* **15,** 10.

Powell, C. C. and Leben, C. (1972) Scotch pine needle cast (*Lophodermium pinastri*). *Fungicide and Nematicide Tests* **28,** 129.

Powelson, R. L. (1974) *Fungicide and Nematicide Tests* **30,** 108.

Powelson, R. I. and Beaver, R. G. (1970) *Fungicide and Nematicide Tests* **26,** 108.

Powelson, R. L. and Cook, G. E. (1969) *Fungicide and Nematicide Tests* **25,** 97.

Powelson, R. L. and Shaner, G. E. (1966) An effective chemical seed treatment for systemic control of seedling infection of wheat by stripe rust (*Puccinia striiformis*). *Pl. Dis. Reptr* **50,** 806.

Powers, H. R. and Stegall, W. A. (1965) An evaluation of cycloheximide for control of white pine blister rust in the southeast. *Pl. Dis. Reptr* **49,** 342.

Pramer, D. (1953) Observations on the uptake and translocation of five actinomycete antibiotics by cucumber seedlings. *Ann. appl. Biol.* **40,** 617.

Pramer, D. (1955) Absorption of antibiotics by plant cells. *Science N.Y.* **121,** 507.

Pramer, D. (1956) Absorption of antibiotics by plant cells. II. *Archs. Biochem. Biophys.* **62,** 265.

Prasad, R. (1974a) Translocation of benomyl in elm (*Ulmus americana* L.) VII. Application of the trunk injection method for suppression of the Dutch elm disease (*C. ulmi*) in landmark and historical trees, *Chem. Control. Res. Instit., Ottawa, Canada. Inf. Rept.* CC-X-72.

Prasad, R. (1974b) Translocation of benomyl in elm (*U. americana*). VIII. Prevention of the Dutch elm disease (*Ceratocystis ulmi* (Buisn) Moreau) in mature trees following pressurized trunk injections. *Chem. Control Res. Inst., Ottawa, Can. Inf. Rept.* CC-X-73.

Prasad, R. and Moody, R. P. (1974) Translocation of benomyl in elm (*Ulmus americana* L.) IX. Some ecological consequences of the treatments on population dynamics of earthworms (*Lumbricus terrestes* L.). *Chem. Control Res. Inst., Ottawa, Can. Inf. Rept.* CC-X-81.

Prasad, R. and Travnick, D. (1973a) Translocation of benomyl in elm (*Ulmus americana* L.). IV. Uptake and movement through the foliage under laboratory and field conditions. *Chem. Control Res. Inst., Ottawa, Can. Inf. Rept.* CL-X-52.

Prasad, R. and Travnick, D. (1973b) Translocation of benomyl in elm (*Ulmus americana* L.). V. Distribution patterns in mature trees following trunk-injection under high pressure. *Chem. Control Res. Inst., Ottawa, Can. Inf. Rept.* LL-X-53.

Prasad, R. and Travnick, D. (1974) Fate of pesticides in forest trees. Transport, accumulation and excretion of benomyl by *Ulmus americana* L. *Proc. Can. Fed. of Biol. Societies* 17, 100 (abs.).

Pratella, C. and Tonini, G. (1968) Prevention and cure in the control of *Gloeosporium* rot (in Italian). *Frutticoltura* **30,** 235. (*Rev. Pl. Path.* **49,** 1708, 1970.)

Preston, W. H., Jr., Daly, E. G., Smale, B. C. and Mitchell, J. W. (1956) Effects of absorbed and translocated F-17 culture filtrate antibiotic factors on the bean rust organism. *Phytopathology* **46,** 469.

Preston, W. H., Mitchell, J. W. and Reeve, W. R. (1954) Movement of the alphamethoxyphenylacetic acid from one plant to another through their root systems. *Science, N.Y.* **119,** 437.

Price, D. (1970) Tulip fire caused by *Botrytis tulipae. A. Rep. Glasshouse Crop Res. Inst. 1969*, p. 124.

Price, D. and Briggs, J. B. (1974). The control of *Botrytis tulipae* (Lib.) Lind., the cause of tulip fire, by fungicidal dipping. *Expl. Hort.* **26,** 36.

Pridham, T. G. (1961) Plant disease antibiotics. *Developments in industrial microbiology, Vol. 2*, p. 141. New York: Plenum Press.

Pridham, T. G., Lindenfelser, L. A., Shotwell, O. L., Stodola, F. H., Benedict, R. G., Foley, C., Jackson, R. W., Zaumeyer, W. J., Preston, W. H., Jr. and Mitchell, J. W. (1956) Antibiotics against plant disease. I. Laboratory and greenhouse survey. *Phytopathology* **46,** 568.

Pringle, R. B. and Scheffer, R. P. (1964) Host-specific plant toxins. *A. Rev. Phytopath.* **2,** 133.

Purdy, L. H. (1964) Inhibition of wheat stripe rust by the antibiotic phleomycin in the greenhouse. *Pl. Dis. Reptr* **48,** 159.

Raa, J. (1968) Natural resistance of apple plants to *Venturia inaequalis.* A biochemical study of its mechanism. *Thesis, Utrecht.* 100 pp.

Raa, J. and Kaars Sijpesteijn, A. (1968) A biochemical mechanism of natural resistance of apple to *Venturia inaequalis. Neth. J. Pl. Pathol.* **74,** 229.

Raabe, R. D. and Hurliman, J. H. (1971) Control of *Thielaviopsis basicola* root rot of Poinsettia with benomyl and thiabendazole. *Pl. Dis. Reptr* **55,** 238.

van Raalte, M. H. (1954) A test for the translocation of fungicides through plant tissues. *C.r. IIIe Congrès International de Phytopharmacie, Paris, 1952, Vol. 2*, p. 76.

van Raalte, M. H., Kaars Sijpesteijn, A., van der Kerk, G. J. M., Oort, A. J. P. and Pluijgers, C. W. (1955) Investigations on plant chemotherapy. *Meded. LandbHoogesch. OpzoekStns. Gent.* **20,** 543.

Rader, W. E., Monroe, C. M. and Whetstone, R. R. (1952) Tetrahydropyrimidine derivatives as potential foliage fungicides. *Science* **115,** 124.

Ragsdale, N. N. (1975) Specific effects of triarimol on sterol biosynthesis in *Ustilago maydis. Biochim. Biophys. Acta* **380,** 81.

Ragsdale, N. N. and Sisler, H. D. (1970) Metabolic effects related to fungitoxicity of carboxin. *Phytopathology* **60,** 1422.

Ragsdale, N. N. and Sisler, H. D. (1972) Inhibition of ergosterol biosynthesis in *Ustilago maydis* by the fungicide triarimol. *Biochem. Biophys. Res. Commun.* **46,** 2048.

Ragsdale, N. N. and Sisler, H. D. (1973) Mode of action of triarimol in *Ustilago maydis. Pestic. Biochem. and Physiol.* **3,** 20.

Ramm, C. von, Lucas, G. B. and Marshall, H. V. (1962) The effect of maleic hydrazide on brown spot incidence of flue-cured tobacco. *Tobacco Sci.* **6,** 100.

Ramsdell, D. C., Nelson, J. W. and Myers, R. L. (1975a) Aerial application of systemic and non-systemic fungicides for control of mummy berry disease of highbush blueberry. *Pl. Dis. Reptr* **59**, 873.

Ramsdell, D. C., Nelson, J. W. and Myers, R. L. (1975b) Mummy berry disease of highbush blueberry: epidemiology and control. *Phytopathology* **65**, 229.

Ramsdell, D. C. and Ogawa, J. M. (1973) Systemic activity of methyl 2-benzimidazole carbamate (MBC) in almond blossoms following pre-bloom sprays of benomyl and MBC. *Phytopathology* **63**, 959.

Rangaswami, G. (1956) A preliminary report on the use of mycothricin complex in plants. *Pl. Dis. Reptr* **40**, 483.

Ranney, C. D. (1971) Studies with benomyl and thiabendazole on control of cotton diseases. *Phytopathology* **61**, 783.

Rao, K. V. (1960) E73: an anti-tumour substance II. structure. *J. Am. Chem. Soc.* **82**, 1130.

Rao, K. V. and Cullen, W. P. (1960) E73: an anti-tumour substance. *J. Am. Chem. Soc.* **82**, 1127.

Rao, S. S. and Grollman, A. P. (1967) Cycloheximide resistance in yeast: a property of the 60*S* ribosomal unit. *Biochem. Biophys. Res. Comm.* **29**, 696.

Rapilly, F. (1970) La détermination des dates de traitements fongicides appliqués par voie aérienne sur céréales en végétation-cas du blé d'hiver. *Phytiat.-Phytopharm.* **19**, 185.

Ravetto, D. J. and Ogawa, J. M. (1972) Penetration of peach fruit by benomyl and 2,6-dichloro-4-nitroaniline fungicides. *Phytopathology* **62**, 784.

Rawlinson, C. J. and Davies, J. M. L. (1971) Effect of eyespot (*Cercosporella herpotrichoides*) and seed treatments of benomyl, chlormequat and organomercury on rye. *Plant Pathology* **20**, 131.

Ray, J. (1901) Les maladies cryptogamiques des végétaux. *Revue gén. Bot.* **13**, 145.

Rea, B. L. and Brown, P. A. (1971) The control of barley powdery mildew using triarimol. *Proc. 6th Br. Insectic.*, p. 52.

Reed, H. E., and Chambers, A. Y. (1973) *Fungicide and Nematicide Tests* **29**, 99.

Reinbergs, E. and Edgington, L. V. (1968) Field control of loose smut in barley with the systemic fungicides 'Vitavax' and 'Plantavax'. *Can. J. Pl. Sci.* **48**, 31.

Renaud, R. (1968) Essais de traitements contre les Monilioses du Prunier sur fleurs et sur fruits. *Rev. Zool. agric. appl.* **3**, 81. (*Rev. appl. Mycol.* **48**, 1839, 1969.)

Renaud, R. (1970) Experiments with fungicides on Sclerotinia of French prune (*S. laxa* and *S. fructigena*). *Summaries of Papers, VIIth Int. Cong. Plant. Prot., Paris*, p. 267.

Reyes, A. A., Davidson, T. R. and Marks, C. F. (1974) Races, pathogenicity and chemical control of *Plasmodiophora brassicae* in Ontario. *Phytopathology* **64**, 173.

Rhodes, A. (1962) Status of griseofulvin in crop protection. *Antibiotics in Agriculture* (ed. M. Woodbine), p. 101. London: Butterworths.

Rhodes, A., Crosse, R., McWilliam, R., Toothill, J. P. R. and Dunn, A. T. (1957) Small-plot trials of griseofulvin as a fungicide. *Ann. appl. Biol.* **45**, 215.

Rhodes, A., Fantes, K. H., Boothroyd, B., McGonagle, M. P. and Crosse, R. (1961) Venturicidin: a new antifungal antibiotic of potential use in agriculture. *Nature, Lond.* **192**, 952.

Rhodes, R. C. and Peace, H. L. (1971) Fate of chloroneb in animals. *J. Agric. Food Chem.* **19**, 750.

Rhodes, R. C., Pease, H. L. and Brantley, R. K. (1971) Fate of ^{14}C-labelled chloroneb in plants and soils. *J. Agric. Food Chem.* **19**, 745.

Rice, E. L. (1948) Absorption and translocation of ammonium 2,4-dichlorophenoxyacetic acid by bean plants. *Bot. Gaz.* **109**, 301.

Rich, S. (1956) Foliage fungicides plus glycerine for the chemotherapy of cucumber scab. *Pl. Dis. Reptr* **40**, 620.

Rich, S. (1963) The role of stomata in plant disease. *Stomata and water relations* (ed. I. Zelitch) p. 102. *Conn. Agric. Exp. Stn. Bull. No. 664*, New Haven.

Rich, S. (1964) Ozone damage to plants. *Ann. Rev. Phytopathology* **2**, 253.

Rich, S., Ames, R. and Zukel, J. W. (1974) 1,4-Oxathiin derivatives protect plants against ozone. *Pl. Dis. Reptr* **58**, 162.

Rich, S. and Tomlinson, H. (1970) Private communication.

Richard, G. and Vallier, J. P. (1969) Treatment of cereal seed by the combination of carboxin with copper oxyquinolate. *Proc. 5th Br. Insectic. Fungic. Conf.* **1,** p. 45.

Richardson, L. T. (1973) Synergism between chloroneb and thiram applied to peas to control seed rot and damping-off by *Pythium ultimum. Pl. Dis. Reptr* **57,** 3.

Richardson, M. (1975) *Translocation in plants.* (2nd edition). London: Edward Arnold Ltd.

Richmond, D. V. and Phillips, A. (1975) The effect of benomyl and carbendazim on mitosis in hyphae of *Botrytis cinerea* and roots of *Allium cepa. Pestic. Biochem. Physiol.* **5,** 367.

Riehm, E. (1914) Prüfung einiger Mittel zur Bekämpfung des Steinbrandes. *Zentbl. Bakt. ParasitKde* (*Abt. II*) **40,** 424.

Rippon, L. E., Glennie-Holmes, M. and Gilbert, S. (1970) Post-harvest treatment of bananas with thiabendazole. *Agric. Gaz. N.S.W.* **81,** 416. (*Rev. Pl. Path.* **50,** 816, 1970.)

Ristich, S. S. and Cohen, S. I. (1961) Control of disease systemically by root absorption of derivatives of 2-pyridinethiol-1-oxide. *Phytopathology* **51,** 578.

Ritchie, D. F., Klos, E. J. and Yoder, K. S. (1975). Epidemiology of black knot of 'Stanley' plums and its control with systemic fungicides. *Pl. Dis. Reptr* **59,** 499.

Roach, W. A. (1934) Injection for the diagnosis and cure of physiological diseases of fruit trees. *Ann. appl. Biol.* **21,** 319.

Roach, W. A. (1938) Plant injection for diagnostic and curative purposes. *Tech. Commun. Imp. Bur. Pl. Crops, no. 10.*

Roach, W. A. (1939) Plant injection as a physiological method. *Ann. Bot. N.S.* **3,** 155.

Roach, W. A. (1942) The use of plant injection in plant pathology. *Trans. Br. mycol. Soc.* **25,** 338.

Robards, A. W. (1968) A new interpretation of plasmodesmatal ultrastructure. *Planta* **82,** 200.

Robards, A. W. and Robb, M. E. (1972) Uptake and binding of uranyl ions by barley roots. *Science, N.Y.* **178,** 980.

Robinson, H. J., Phares, H. F. and Graessle, O. E. (1964) Antimycotic properties of thiabendazole. *J. Invest. Dermatol.* **42,** 479.

Rodigin, M. N. and Krasnova, T. M. (1959) Seed treatment with microelements against brown rust of wheat. *Abstr. in Rev. appl. Mycol.* **39,** 405.

Rohrbach, K. U. (1973) Post-harvest control of brown rot on stone fruit by triforine. *Proc. 7th Br. Insectic. Fungic. Conf.* **2,** 631.

Rohrbach, K. G. and Apt, W. J. (1971) Field control of *Ceratocystis paradoxa* on pineapple asexual propagative parts. *Phytopathology* **61,** 1323 (abs.).

Rombouts, J. E. and Kaars Sijpesteijn, A. (1958) The chemotherapeutic effect of pyridine-2-thiol-N-oxide and some of its derivatives on plant diseases. *Ann. appl. Biol.* **46,** 30.

Rooy, M. de (1969) Use of benzimidazole compounds for the control of some pathogenic fungi in flower bulb culture. *21st Int. Symp. Phytofarm., Phytiat., Ghent,* p. 931.

Rosen, H. R. (1959) Preliminary report on the systemic activity of captan when applied to the soil for the control of black spot of roses. *Pl. Dis. Reptr* **43,** 1176.

Ross, I. S. (1974) Non-protein thiols and mercury resistance of *Pyrenophora avenae. Trans. Brit. Mycol. Soc.* **63,** 77.

Ross, I. S. and Old, K. M. (1973a) Mercuric chloride resistance of *Pyrenophora avenae. Trans. Brit. Mycol. Soc.* **60,** 293.

Ross, I. S. and Old, K. M. (1973b) Thiol compounds and resistance of *Pyrenophora avenae* to mercury. *Trans. Brit. Mycol. Soc.* **60,** 301.

Ross, R. G. (1973) Suppression of perithecium formation in *Venturia inaequalis* by seasonal sprays of benomyl and thiophanate-methyl. *Can. J. Pl. Sci.* **53,** 601.

Ross, R. G. and Newbery, R. J. (1975) Effects of seasonal fungicide sprays on perithecium formation and ascospore production in *Venturia inaequalis. Can. J. Pl. Sci.* **55,** 737.

Ross, R. G. and Sanford, K. H. (1971) Apple: spray injury. *Fungicide and Nematicide Tests* **26,** 35.

Rosser, W. R. (1969) Spray timing and the control of mildew in spring barley. *Proc. 5th Br. Insectic. Fungic. Conf.,* **1,** p. 20.

Rothwell, K. and Wright, S. T. C. (1967) Phytokinin activity in some new 6-substituted purines. *Proc. R. Soc. B.* **167**, 202.

Rouchaud, J. P., Decallonne, J. R. and Meyer, J. A. (1974) Metabolic fate of methyl-2-benzimidazole carbamate in melon plants. *Phytopathology* **64**, 1513.

Rowan, S. J. (1972) Selected systemic fungicides provide little control of fusiform rust of loblolly pine in forest tree nurseries. *Pl. Dis. Reptr* **56**, 628.

Rowell, J. B. (1967) Control of leaf and stem rust of wheat by an 1,4-oxathiin derivative. *Pl. Dis. Reptr* **51**, 336.

Rowell, J. B. (1968a) Chemical control of the cereal rusts. *Ann. Rev. Phytopathol.* **6**, 243.

Rowell, J. B. (1968b) Control of leaf and stem rusts of wheat by combinations of soil application of an 1,4-oxathiin derivative with foliage sprays. *Pl. Dis. Reptr* **52**, 856.

Rowell, J. B. (1969) *Fungicide and Nematicide Tests* **25**, 98.

Rowell, J. B. (1970) *Fungicide and Nematicide Tests* **26**, 110.

Roy, M. K. (1975) Radiation, heat and chemical combines in the extension of shelf life of apples infected with blue mold rot (*Penicillium expansum*). *Pl. Dis. Reptr* **59**, 61.

Royle, D. J. and Thomas, G. G. (1971) The influence of stomatal opening on the infection of hop leaves by *Pseudoperonospora humuli. Physiol. Pl. Path.* **1**, 329.

Rudd Jones, D. (1956) The systemic action of sulphonamides against plant diseases. *Outlook on Agriculture* **1**, 111.

Rumbold, C. (1920a) Giving medicine to trees. *Am. For.* **26**, 359.

Rumbold, C. (1920b) The injection of chemicals into chestnut trees. *Am. J. Bot.* **7**, 1.

Ruppel, E. G. (1975) Biology of benomyl-tolerant strains of *Cercospora beticola* of sugar beet. *Phytopathology* **65**, 785.

Ruppel, E. G. and Scott, P. R. (1974) Strains of *Cercospora beticola* resistant to benomyl in the USA. *Pl. Dis. Reptr* **58**, 434.

Rush, M. C., Boudreaux, R. A. and Heiman, J. L. (1974) Chemical control of rice seedling diseases. *66th Ann. Prog. Rept. Expt. Stn. Louisiana*, p. 167.

Rush, M. C. and Lindberg, G. D. (1974) Chemical control of rice seedling diseases. *Ann. Prog. Rept. Rice Expt. Stn. Louisiana*, p. 181.

Ryan, E. W. and Kavanagh, T. (1971) Comparisons of fungicides for control of leaf spot (*Septoria apiicola*) of celery. *Ann. appl. Biol.* **67**, 121.

Ryan, E. W., Staunton, W. P. and Kavanagh, T. (1971) Control of diseases of five vegetable crops with systemic fungicides. *Proc. 6th Br. Insectic. Fungic. Conf.*, p. 355.

Sakurai, H. and Shimada, T. (1974) Studies on polyoxin-resistant strains of *Alternaria kikuchiana* Tanaka. Part. 3. Characters on media and pathogenicity. *Bull. Agric. Chem. Inspect. Stn.* (*Japan*) **14**, 54.

Salerno, M. and Edgington, L. V. (1963) Similarity of movement of N-alkyl-quaternary ammonium compounds in aqueous systems on paper chromatograms and in plant xylem. *Phytopathology* **53**, 605.

Salerno, M. and Somma, V. (1971) Observations on the systemic nature of benomyl in Sour Orange seedlings and results of trials against Citrus 'mal secco' (in Italian). *Phytopath. Mediterranea* **10**, 99. (*Rev. Pl. Path.* **50**, 3744, 1971.)

Samborski, D. J., Forsyth, F. R. and Person, C. (1958) Metabolic changes in detached wheat leaves floated on benzimidazole and the effect of these changes on rust reaction. *Can. J. Bot.* **36**, 591.

Samborski, D. J., Person, C. and Forsyth, F. R. (1960) Differential effects of maleic hydrazide on the growth of leaf and stem rusts of wheat. *Can. J. Bot.* **38**, 1.

Sampson, M. J. (1969) The mode of action of a new group of species-specific pyrimidine fungicides. *Proc. 5th Br. Insectic. Fungic. Conf.* **2**, 483.

Sands, D. C. and Zucker, M. L. (1971) Impaired amino acid biosynthesis in phytopathogenic pseudomonads. *Phytopathology* **61**, 909.

Sasaki, S., Ohta, N., Yamaguchi, I., Kureda, S. and Misato, T. (1968) Studies on polyoxin action. Effect on respiration and synthesis of protein, nucleic acids and cell wall of fungi. *J. Agric. Chem. Soc. Jap.* **42**, 633.

Scheffer, R. R. and Pringle, R. B. (1967) Pathogen-produced determinants of disease and their effects on host plants. *The dynamics of molecular constituents on plant-parasite* *interaction* (eds C. J. Mirocha and I. Uritani) p. 217. St. Paul: Amer. Phytopathol Soc.

Scherer, C. M. (1927) Tree injection for control of fungus diseases and insect pests. *Phytopathology* **17,** 51.

Schicke, P., Adlung, K. G. and Drandarevski, C. A. (1971) Results of several years of testing the systemic fungicide Cela W-524 against mildew and rusts of cereals. *Proc. 6th Br. Insectic. Fungic. Conf.* **1,** p. 82.

Schicke, P. and Arndt, S. (1971) Die Bekämpfung von Zierpflanzenkrankheiten mit triforine. *23e Int. Symp. over Phytofarmacie en Phytiatrie Gent,* **36,** 1165.

Schicke, P. and Veen, K. H. (1969) A new systemic, Cela-W 524 (N,N′-bis(1-formamide-2,2,2-trichloroethyl)piperazine) with action against powdery mildew, rust and apple scab. *Proc. 5th Br. Insectic. Fungic. Conf.* **2,** 569.

Schloesser, E. (1965) Sterols and the resistance of *Pythium* species to filipin. *Phytopathology* **55,** 1075.

Schlüter, K. and Weltzien, H. C. (1971) Ein Beitrag zur Wirkungsweise systemischer Fungizide auf *Erysiphe graminis. Meded. Fac. Landbouwwetensch. Rijksuniv. Gent* **36,** 1159.

Schmeling, B. von and Clark, M. (1970) Oxathiin-induced plant growth stimulation. *VII Int. Congr. Pl. Prot. 1970,* p. 227 (abs.).

Schmeling, B. von and Kulka, M. (1966) Systemic fungicidal activity of 1,4-oxathiin derivatives. *Science N.Y.* **152,** 659.

Schmeling, B. von, Kulka, M., Thiara, D. S. and Harrison, W. A. (1966) *Control of plant diseases.* US patent 3249499.

Scholten, G. (1970) New possibilities for controlling foot rot diseases in carnations. *Hort. Abstr.* **40,** 8733.

Schönherr, J. and Bukovac, M. J. (1970a) Preferential polar pathways in the cuticle and their relationship to ectodesmata. *Planta* **92,** 189.

Schönherr, J. and Bukovac, M. J. (1970b) The nature of precipitates formed in the outer cell wall following fixation of leaf tissue with Gilson solution. *Planta* **92,** 202.

Schreiber, L. R. (1970) An evolution of thiabendazole, 2-(4-thiazoly) benzimidazole for the control of Dutch elm disease. *Pl. Dis Reptr* **54,** 240.

Schreiber, L. R. and Townsend, A. M. (1976) Naturally occurring tolerance in isolates of *Ceratocystis ulmi* to methyl-2-benzimidazole carbamate hydrochloride. *Phytopathology* **66,** 225.

Schroeder, W. T. and Provvidenti, R. (1969) Resistance to benomyl in powdery mildew of cucurbits. *Pl. Dis. Reptr* **53,** 271.

Schuepp, H. (1974) Are the systemic fungicides still suitable for controlling grey mould in viticulture? *Schweiz. Z. Obst.-u. Weinb.* **110,** 330.

Schumann, G. von (1967) Stand und Entwicklung der Bekämpfung von Getreid Krankheiten durch Saatgutbehändlung. *Biol. Bund für Landungforst. 2. Pflanzenkrankh. Pfl. Path. Pflanzenschutz* **74,** 155.

Schwer, J. F. (1971) Private communication.

Schwinn, F. J. (1969) Azepines, a new class of fungicides. *Proc. 5th Br. Insectic. Fungic. Conf.* **2,** 584.

Scott, D. H., Draper, A. D. and Maas, J. L. (1970) Benomyl for control of powdery mildew on strawberry plants in the greenhouse. *Pl. Dis. Reptr* **54,** 362.

Scott, K. J. and Roberts, E. A. (1967) Control in bananas of black end rot caused by *Gloeosporium musarum. Aust. J. exp. Agric. Anim. Husb.* **7,** 283. (*Rev. appl. Mycol.* **46,** 3147, 1967.)

Scott, K. J. and Roberts, E. A. (1970) Thiabendazole to reduce rotting in Packham's Triumph pears during storage and marketing. *Aust. J. exp. Agric. Anim. Husb.* **10,** 235. (*Rev. Pl. Path.* **49,** 3381, 1970.)

Seberry, J. A. and Baldwin, R. A. (1968) Thiabendazole and 2-aminobutane as post-harvest fungicides for Citrus. *Aust. J. exp. Agric. Anim. Husb.* **8,** 440. (*Rev. appl. Mycol.* **48,** 471, 1969.)

Seiler, J. P. (1972) The mutagenicity of benzimidazole and benzimidazole derivatives. I. Forward and reverse mutations in *Salmonella typhimurium* caused by benzimidazole and some of its derivatives. *Mutation Res.* **15,** 273.

Self, R. L. and Suberatz, F. J. (1974a) Root rot control studies. *Proc. So. Nurserymens Assoc. Res. Conf. 19th Annual Rept.*, p. 74.

Self, R. L. and Suberatz, F. J. (1974b) Alternaria blight control. *Proc. So. Nurserymens Assoc. Res. Conf. 19th Annual Rept.*, p. 68.

Selling, H. A., Vonk, J. W. and Kaars Sijpesteijn, A. (1970) Transformation of the systemic fungicide methyl thiophanate into 2-benzimidazole carbamic acid methyl ester. *Chem. and Ind.*, p. 1625.

Sempio, C. (1936) Influenza di varie sostanze sul parassitamento: ruggine del Fagiola, ruggine e mal bianco del Frumento. *Riv. Patol. veg., Padova* **26,** 201.

Sempio, C. (1938) Influenza di alcune sostanze, date alle piantine per assorbimento, sullo svillupo della carie del Grano. (Nota preventiva). *Riv. Patol. veg., Padova* **28,** 399.

Seth, A. K. and Wareing, P. F. (1967) Hormone directed transport of metabolites and its possible role in plant senescence. *J. exp. Bot.* **18,** 65.

Shabi, E. (1975) Control of apple powdery mildew by bupirimate and the influence of added spreader-sticker on its performance. *Proc. 8th Br. Insectic. Fungic. Conf.* **2,** 711.

Shabi, E., Pinkas, Y. and Solel, Z. (1974) Distribution of benzimidazole fungicides following pressure injection of pear trees at several growth stages. *Phytopathology* **64,** 963.

Sharples, R. O. (1963) Therapeutic activity of a streptothricin-like antibiotic against apple mildew. *Nature, Lond.* **198,** 306.

Sharvelle, E. G. (1952) Systemic fungicides in 1951: British investigations. *Pl. Dis. Reptr* **36,** 35.

Shatla, M. N. and Sinclair, J. B. (1963) Tolerance to PCNB and pathogenicity correlated in natural occurring isolates of *Rhizoctonia solani. Phytopathology* **53,** 1047.

Shephard, M. C. (1971) Report on Systemic Fungicide Symposium. *Nature, Lond.* **229,** 230.

Shephard, M. C. (1973) Barriers to the uptake and translocation of chemicals in herbaceous and woody plants. *Proc. 7th Br. Insectic. Fungic. Conf.* **3,** 841.

Shephard, M. C., Bent, K. J., Woolner, M. and Cole, A. M. (1975) Sensitivity to ethirimol of powdery mildew from UK barley crops. *Proc. 8th Br. Insectic. Fungic. Conf.* **1,** p. 59.

Sherald, J. L. and Sisler, H. D. (1970) Antimycin A-resistant respiratory pathway in *Ustilago maydis* and *Neurospora sitophila. Plant Physiol.* **46,** 180.

Sherald, J. L. and Sisler, H. D. (1972) Selective inhibition of antimycin A-insensitive respiration in *Ustilago maydis and Ceratocystis ulmi. Plant and Cell Physiol.* **13,** 1039.

Sherald, J. L. and Sisler, H. D. (1975) Antifungal mode of action of triforine. *Pestic. Biochem. Physiol.* **5,** 477.

Sherald, J. L., Ragsdale, N. N. and Sisler, H. D. (1973) Similarities between the systemic fungicides triforine and triarimol. *Pestic Sci.* **4,** 719.

Shillingford, C. A. (1970) Banana fruit rot control in Jamaica. *PANS* **16,** 69.

Shimada, T., Sakurai, H. and Yoshida, K. (1973) Studies on polyoxin resistance of the strain Y-33 of *Alternaria kikuchiana Tanaka. Bull. Agr. Chem. Insp. Stn. (Japan)* **13,** 39.

Shipton, W. A., Boyd, W. R. J., Rosielle, A. A. and Shearer, B. I. (1971) The common Septoria diseases of wheat. *Bot. Rev.* **37,** 231.

Shipton, W. A. (1968) Effect of Septoria diseases on wheat. *Aust. J. expt. Agric. Anim. Husb.* **8,** 89.

Shlevin, E. and Katan, J. (1975) *Rhizoctonia solani* disease of carrot seedlings and its control. *Pl. Dis. Reptr* **59,** 29.

Shone, M. G. T., Bartlett, B. O. and Wood, A. V. (1974) A comparison of the uptake and translocation of some organic herbicides and a systemic fungicide by barley. II. *J. exp. Bot.* **25,** 401.

Shone, M. G. T. and Wood, A. V. (1974) A comparison of the uptake and translocation of some organic herbicides and a systemic fungicide by barley. I. *J. exp. Bot.* **25,** 390.

Siddiqui, M. Q. and Manners, J. G. (1971) Some effects of general yellow rust (*Puccinia* *striiformis*) infection on ^{14}C-assimilation, translocation and growth in a spring wheat. *J. exp. Bot.* **22,** 792.

Siegel, M. R. (1973) Distribution and metabolism of methyl 2-benzimidazolecarbamate, the fungitoxic derivative of benomyl, in strawberry plants. *Phytopathology* **63,** 890.

Siegel, M. R. (1975) Benomyl-soil microbial interactions. *Phytopathology* **65,** 219.

Siegel, M. R., Kerkenaar, A. and Kaars Sijpesteijn, A. (1977) Studies on the mode of action of imazalil. *Neth. J. Plant. Pathol.* **83,** Suppl. 1.

Siegel, M. R. and Sisler, H. D. (1963) Inhibition of protein synthesis *in vitro* by cycloheximide. *Nature, Lond.* **200,** 675.

Siegel, M. R. and Sisler, H. D. (1964) Site of action of cycloheximide in cells of *Saccharomyces pastorianus.* I. Effect of the antibiotic on cellular metabolism. *Biochim. Biophys. Acta.* **87,** 70.

Siegel, M. R. and Sisler, H. D. (1965) Site of action of cycloheximide in cells of *Saccharomyces pastorianus. Biochim. Biophys. Acta.* **103,** 558.

Siegel, M. R. and Sisler, H. D. (1977) *Antifungal compounds,* Vol. II. Chapters 8–12. New York: Marcel Dekker.

Silverman, W. B. and Hart, H. (1954) Antibiotics tested for control of wheat stem rust. *Phytopathology* **44,** 506.

Simpson, W. R. and Fenwick, H. S. (1971) Suppression of corn head smut with carboxin seed treatments. *Pl. Dis. Reptr* **55,** 501.

Sims, J. J., Mee, H. and Erwin, D. C. (1969) Methyl 2-benzimidazole carbamate, a fungitoxic compound isolated from cotton plants treated with methyl 1-(butyl carbamoyl)-2-benzimidazole carbamate (benomyl). *Phytopathology* **59,** 1775.

Sinclair, W. A. (1975) Verticillium wilt of maples. Research progress report. *Cornell Univ. Coop. report to regional project* NE-25 (wilt diseases).

Singh, R. A. and Sharma, V. V. (1973) Systemic fungicidal activity of triphenyl tin chloride in rice seedlings. *Indian J. Mycol. Plant Pathol.* **3,** 141.

Sinha, A. K. and Wood, R. K. S. (1964) Control of *Verticillium* wilt of tomato plants with cycocel ((2-chloroethyl) trimethyl ammonium chloride). *Nature, Lond.* **202,** 824.

Sinha, A. K. and Wood, R. K. S. (1967) The effect of growth substances on *Verticillium* wilt of tomato plants. *Ann. appl. Biol.* **60,** 117.

Sisler, H. D. (1969) Effect of fungicides on protein and nucleic acid synthesis. *Ann. Rev. Phytopath.* **7,** 311.

Sisler, H. D. (1971) Mode of action of benzimidazole fungicides. *Proc. 2nd Int. Congress of Pesticide Chemistry IUPAC, Tel-Aviv, 1971,* p. 323.

Sisler, H. D. and Siegel, M. R. (1967) Cycloheximide and other glutarimide antibiotics. *Antibiotics. Vol. I. Mechanisms of action* (eds D. Gottlieb and P. D. Shaw), p. 283. New York: Springer.

Skene, K. G. M. (1972) Cytokinin-like properties of the systemic fungicide benomyl. *J. hort. Sci.* **47,** 179.

Skorda, E. A. (1975) The effect of some fungicides on powdery mildew of barley and oats in Greece. *Proc. 8th Br. Insectic. Fungic. Conf.* **2,** p. 445.

Skilling, D. D. and Waddell, C. D. (1975) Control of Rhizosphaera needle cast in blue spruce Christmas tree plantations. *Pl. Dis Reptr* **59,** 841.

Slade, P., Cavell, B. D., Hemmingway, R. J. and Sampson, M. J. (1971) Metabolism and mode of action of dimethirimol and ethirimol. *Proc. 2nd Intern. Congr. Pestic. Chem., Tel-Aviv,* p. 295.

Smale, B. C., Montgillion, M. D. and Pridham, T. G. (1961) Phleomycin, an antibiotic markedly effective for control of bean rust. *Pl. Dis. Reptr* **45,** 244.

Smalley, E. B. (1962) Prevention of Dutch elm disease by treatments with 2,3,6-trichlorophenyl acetic acid. *Phytopathology* **52,** 1090.

Smalley, E. B. (1967) Present status of Dutch elm disease research in Plant Pathology at Wisconsin. *Proc. U.S.D.A. Regional Conference on the Control of Dutch Elm Disease. Delaware, Ohio, June 29–30).* p. 16.

Smalley, E. B. (1971) Prevention of Dutch elm disease in large nursery elms by soil treatment with benomyl. *Phytopathology* **61,** 1351.

Smalley, E. B. (1976a) New controls for Dutch elm disease. *Park Maintenance* **29** (2), 15.

Smalley, E. B. (1976b) Dutch elm disease control with systemic fungicides. *Proc. 30th Annual Wisconsin Pest Control Conference with Industry*. Jan. 27, 28, 29, 1976.

Smalley, E. B., Meyers, C. J., Johnson, R. N., Fluke, B. C. and Vieau, R. (1973) Benomyl for practical control of Dutch elm disease. *Phytopathology* **63,** 1239.

Smiley, R. W. (1974) *Fungicide and Nematicide Tests* **30,** 123.

Smiley, R. W. and Harrison, M. B. (1974) *Fungicide and Nematicide Tests* **30,** 123.

Smith, B. D. and Corke, A. T. K. (1966) Effect of (2-chloroethyl)trimethyl ammonium chloride on the eriophyid gall mite *Cecidophyopis ribis* Nal., and three fungus diseases of blackcurrant. *Nature, Lond.* **212,** 643.

Smith, D., Muscatine, L. and Lewis, D. (1969) Carbohydrate movement from autotrophs to heterotrophs in parasitc and mutualistic symbiosis. *Biol. Rev.* **44,** 17.

Smith, P. M. (1970) The integrated control of *Botrytis cinerea* on glasshouse tomatoes. *Ph.D. Thesis Reading University*.

Smith, P. M. (1971a) Primary evaluation of new fungicides. *A. Rep. Glasshouse Crops Res. Inst.* 1970, p. 131.

Smith, P. M. (1971b) Narcissus leaf scorch caused by *Stagonospora curtisii. A. Rep. Glasshouse Crops Res. Inst.* 1970, p. 141.

Smith, P. M. and Spencer, D. M. (1971) Evaluation of systemic fungicides for use in the glasshouse industry. *Pestic. Sci.* **2,** 201.

Smith, P. M. and Worthing, C. R. (1973) The effect of soil sterilisation on efficiency of soil-applied benomyl and carbendazim for the control of some tomato diseases. *Proc. 7th Brit. Insectic. Fungic. Conf.* **1,** p. 201.

Smith, P. M. and Worthing, C. R. (1975) Microbial degradation of soil-applied fungicides. *A. Rep. Glasshouse Crops Res. Inst.* (1974), p. 106.

Smith, R. S., McCain, A. H. and Srago, M. D. (1973) Control of Botrytis storage rot of giant sequoia seedlings. *Pl. Dis. Reptr* **57,** 67.

Smith, W. L. (1971) Control of brown rot and Rhizopus rot of inoculated peaches with hot water or hot chemical suspensions. *Pl. Dis. Reptr* **55,** 228.

Smoot, J. J. and Brown, G. E. (1974) Occurrence of benzimidazole-resistant strains of *Penicillium digitatum* in Florida citrus packinghouses. *Pl. Dis. Reptr* **58,** 933.

Smoot, J. J. and Melvin, C. F. (1970) A comparison of postharvest fungicides for decay control of Florida oranges. *Proc. Fla St. hort. Soc.* **82,** 243. (*Rev. Pl. Path.* **50,** 2279, 1971.)

Smyly, W. G. and Filer, T. H. (1973) Benomyl controls Phomopsis blight of Arizona cypress in a nursery. *Pl. Dis. Reptr* **57,** 59.

Snel, M. and Edgington, L. V. (1970) Uptake, translocation and decomposition of systemic oxathiin fungicides in bean. *Phytopathology* **60,** 1708.

Snel, M. B. and Fletcher, J. T. (1971) Benomyl and thiabendazole for the control of mushroom diseases. *Pl. Dis. Reptr* **55,** 120.

Snel, M., Schmeling, B. von and Edgington, L. V. (1970) Fungitoxicity and structure-activity relationships of some oxathiin and thiazole derivatives. *Phytopathology* **60,** 1164.

Snell, B. K. (1968) Pyrimidines: Part I. The acylation of 2-amino-4-hydroxypyrimidines. *J. chem. Soc.* (C) 2358.

Snow, G. A., Czabator, F. J. and Sorrels, S. S. (1964) Cycloheximide derivatives for controlling brown spot on long leaf pine. *Pl. Dis. Reptr* **48,** 551.

Soeda, Y., Kosaka, S. and Noguchi, T. (1972a) Identification of alkyl 2-benzimidazole carbamates as a major metabolite of thiophanates fungicide in/on the bean plant. *Agr. Biol. Chem.* **36,** 817.

Soeda, Y., Kosaka, S. and Noguchi, T. (1972b) The fate of thiophanatemethyl fungicide and its metabolites on plant leaves and glass plates. *Agr. Biol. Chem.* **36,** 931.

Solel, Z. (1970a) The systemic fungicidal effect of benzimidazole derivatives and thiophanate against *Cercospora* leaf spot of sugar beet. *Phytopathology* **60,** 1186.

Solel, Z. (1970b) The performance of benzimidazole fungicides in the control of *Cercospora* leaf spot of sugar beet. *J. Am. Soc. Sug. beet Technol.* **16,** 93.

Solel, Z. (1971a) Vapour phase action of some foliar fungicides. *Pestic. Sci.* **2,** 126.

Solel, Z. (1971b) Effective control of *Cercospora* leaf spot of sugar beet by aerial spraying with benomyl. *Pl. Dis. Reptr* **55,** 408.

Solel, Z. and Edgington, L. V. (1973) Transcuticular movement of fungicides. *Phytopathology* **63,** 505.

Solel, Z., Pinkas, J. and Loebenstein, G. (1972) Evaluation of systemic fungicides and mineral oil adjuvants for the control of mal secco disease of lemon plants. *Phytopathology* **62,** 1007.

Solel, Z., Schooley, J. M. and Edgington, L. V. (1973) Uptake and translocation of benomyl and carbendazim (methyl benzimidazol-2yl-carbamate) in the symplast. *Pestic. Sci.* **4,** 713.

Somers, E. and Richmond, D. V. (1962) Translocation of captan by broad bean plants. *Nature, Lond.* **194,** 1194.

Soper, D. (1975) Evaluation of thiophanate-methyl for disease control and yield response in cereals. *Proc. 8th Br. Insectic. Fungic. Conf.* **1,** p. 313.

Spalding, D. H. (1970) Postharvest use of benomyl and thiabendazole to control blue mold rot development in pears. *Pl. Dis. Reptr* **54,** 655.

Spalding, D. H. and Hardenburg, R. E. (1971) Postharvest chemical treatments for control of blue mold of apples in storage. *Phytopathology* **61,** 1308.

Spalding, D. H. and Reeder, W. F. (1974) Post-harvest control of *Sclerotinia* rot of snap bean pods with heated and unheated chemical dips. *Pl. Dis. Reptr* **58,** 59.

Spalding, D. H., Vaught, H. C., Day, R. H. and Brown, G. A. (1969) Control of blue mold rot development in apples treated with heated and unheated fungicides. *Pl. Dis. Reptr* **53,** 738.

Spanner, D. C. (1958) The translocation of sugar in sieve tubes. *J. exp. Bot.* **9,** 332.

Spanner, D. C. (1970) The electro-osmotic theory of phloem transport in the light of recent measurements on *Heracleum* phloem. *J. exp. Bot.* **21,** 325.

Spencer, D. M. (1957) Studies on direct and systemic fungicidal action. *Ph.D. Thesis, University of London.*

Spencer, D. M. (1974) Pelargonium rust. *A Rep. Glasshouse Crops Res. Inst.* 1973, p. 110.

Spencer, D. M., Topps, J. H. and Wain, R. L. (1957) An antifungal substance from the tissues of *Vicia faba. Nature, Lond.* **179,** 651.

Spiertz, J. H. J. (1973) Effects of successive application of maneb and benomyl on growth and yield of five wheat varieties of different heights. *Neth. J. Agric. Sci.* **21,** 282.

Spinks, G. T. (1913) Factors affecting susceptibility to disease in plants. *J. agric. Sci. Camb.* **5,** 231.

Spotts, R. A., Hall, F. R., Ferree, D. C. and Jones, B. M. (1975) Yield reductions in apple related to benomyl-oil spray applications. *Pl. Dis. Reptr* **59,** 541.

Spotts, R. A., Lukezic, F. L. and Hamilton, R. H. (1974) The effect of benzimidazole on some membrane properties of ozonated pinto beans. *Phytopathology* **65,** 39.

Springer, J. K. (1970) Peach: Valsa canker (*Valsa cincta*). *Fungicide and Nematicide Tests* **25,** 48.

Spurr, H. W., Jr. and Chancey, E. L. (1970) The chemical control of powdery mildew by fumigant redistribution. *Phytopathology* **60,** 1062.

Stalder, L. and Barben, E. (1970) Benomyl gegen Wurzelbräune der Cyclamen. *Gartenweld Hamb.* **70,** 289.

Staron, T. and Allard, C. (1964) Propiétés antifongiques du 2-(4′-thiazolyl) benzimidazole ou thiabendazole. *Phytiat.-Phytopharm.* **13,** 163.

Staron, T., Allard, C., Darpoux, H., Grabowski, H. and Kollman, A. (1966) Persistance du thiabendazole dans les plantes. Propriétés systémiques de ses sels et quelques données nouvelles sur son mode d'action. *Phytiat.-Phytopharm.* **15,** 129.

Stingl, H., Härtel, K. and Heubach, G. (1970) Hoe 2989, a new systemic fungicide to control various Basidiomycetes (rusts and smuts) as well as *Rhizoctonia. 7th Int. Congress Plant Protection,* p. 205 (abs.).

Stipes, R. J. (1969a) Chemotherapeutic patterns of methyl 1-(butylcarbamoyl)-2-benzimidazolecarbamate and thiabendazole in the control of Dutch elm disease. *Phytopathology* **59,** 1560.

Stipes, R. J. (1969b) Uptake and translocation of thiabendazole and benomyl applied to phloem and xylem tissues of *Albizzia julibussin* Durozy. *Va. J. Sci.* **20** (3) (abs.).

Stipes, R. J. (1972a) Dutch elm disease (*Ceratocystis ulmi*). *Fungicide and Nematicide Tests* **28,** 124.

Stipes, R. J. (1972b) Dutch elm disease (*Ceratocystis ulmi*). *Fungicide and Nematicide Tests* **28,** 125.

Stipes, R. J. (1973) Control of Dutch elm disease in artificially-inoculated American elms with soil injected benomyl, captan and thiabendazole. *Phytopathology* **63,** 735.

Stipes, R. J. and Oderwald, D. R. (1971) Dutch elm disease; control with soil injected fungicides and surfactants. *Phytopathology* **61,** 913 (abs.).

Stipes, R. J., Phipps, P. M. and Oderwald, C. R. (1971) Prophylactic and therapeutic responses of Fusarium wilt of Mimosa to benomyl and thiabendazole. *Va. J. Sci.* **22,** 108 (abs.).

Stipes, R. J. and Weinke, K. E. (1972) Dutch elm disease control with soil-amended fungicides. *Pl. Dis. Reptr* **56,** 604.

Stoddard, E. M. (1946) Soil applications of oxyquinolin benzoate for the control of foliage wilting in elms caused by *Graphium ulmi. Phytopathology* **36,** 682.

Stoddard, E. M. (1951a) Chemotherapeutic control of *Fusarium* wilt of carnations. *Phytopathology* **41,** 33.

Stoddard, E. M. (1951b) A chemotherapeutic control of strawberry red stele. *Phytopathology* **41,** 34.

Stoddard, E. M. (1952) Chemotherapeutic control of *Rhizoctonia* on greenhouse stock. *Phytopathology* **42,** 476.

Stoddard, E. M. (1954) Chemotherapeutic control of cucumber scab. *Phytopathology* **44,** 507.

Stoddard, E. M. (1957) A *Fusarium* rot of geraniums and its control. *Pl. Dis. Reptr* **41,** 536.

Stoddard, E. M. and Dimond, A. E. (1949) The chemotherapy of plant diseases. *Bot. Rev.* **15,** 345.

Stoddard, E. M. and Dimond, A. E. (1951) The chemotherapeutic control of *Fusarium* wilt of carnations. *Phytopathology* **41,** 337.

Stoddard, E. M. and Miller, P. M. (1962) Chemical control of water loss in growing plants. *Science* **137,** 224.

Stoker, G. L. and Dewey, W. G. (1970) Effect of loose smut infection and 'Vitavax' seed treatment on barley yields. *Utah Sci.* **31,** (2) 46.

Stott, K. and Jefferies, C. J. (1972) Fruit-set. *Rep. agric. hort. Res. Stn Bristol, 1971*, p. 149.

Stover, R. H. (1969) The effect of benomyl on *Mycosphaerella musicola. Pl. Dis. Reptr* **53,** 830.

Straib, W. (1941) Über die Wirkung organischer Verbindungen als Spritzmittel gegen Rostpilzinfektion. *Zentbl. Bakt. ParasitKde* (*Abt. II*), **103,** 74.

Stringer, A. and Lyons, C. H. (1974) The effect of benomyl and thiophanate-methyl on earthworm populations in apple orchards. *Pestic. Sci.* **5,** 189.

Stringer, A. and Wright, M. A. (1973) The effect of benomyl and some related compounds on *Lumbricus terrestris* and other earthworms. *Pestic. Sci.* **4,** 165.

Strugger, S. (1949) *Praktikum der Zell-und Gewebephysiologie der Pflanze.* Berlin: Springer-Verlag.

Strugger, S. and Peveling, E. (1961) Die Elektronenmikroskopische Analyse der Extrafaszikulären Komponente des Transpirationsstromes mit Hilfe von Edelmetallsuspensoiden adäquater Dispersität. *Ber. dt. bot. Ges.* **74,** 300.

Stubbs, J. (1952) The evaluation of systemic fungicides by means of *Alternaria solani* on tomato. *Ann. appl. Biol.* **39,** 439.

Suberatz, F. J. and Self, R. L. (1974) Pestalotia blight and stem rot of azalea. *Proc. S. Nurserymens Assoc. Res. Conf. 19th Annual Rept.*, p. 69.

Suhara, Y., Maeda, K., Umezawa, H. and Ohno, M. (1966) Chemical studies on Kasugamycin V. The structure of Kasugamycin. *Tetrahedron Letters* **12,** 1239.

Suit, R. F. and Ducharme, E. P. (1971) Cause and control of pink pitting on grapefruit. *Pl. Dis. Reptr* **55,** 923.

Sumi, H., Takahi, Y., Nakagami, K. and Kondo, Y. (1968) Controlling effect of pentachlorobenzyl alcohol on rice blast. *Ann. Phytopathol. Soc. Japan* **34,** 114.

Suzuki, S., Isono, K., Nagatsu, J., Mizutani, T., Kawashima, Y. and Mizuno, T. (1965) A new antibiotic, Polyoxin A. *J. Antibiotics (Tokyo) Ser. A.* **11,** 1.

Suzuki, S., Nakamura, G., Okuma, K. and Tomiyama, Y. (1958). Cellocidin, a new antibiotic. *J. Antibiotics (Tokyo) Ser. A.* **11,** 81.

Swarts, D. H., Jacobs, C. and Nel, F. G. (1969) The control of collar rot in bananas. *Fmg S. Afr.* **45,** 15, 17. (*Rev. appl. Mycol.* **48,** 3072, 1969.)

Swinburne, T. R., Cartwright, J., Flack, N. J. and Brown, A. E. (1975) The control of apple canker (*Nectria galligena*) in a young orchard with established infections. *Ann. appl. Biol.* **81,** 61.

Szkolnik, M. (1969) Apple: scab (*Venturia inaequalis*). Cherry (red tart): leaf spot (*Coccomyces hiemalis*). *Fungicide and Nematicide Tests* **24,** 27.

Szkolnik, M. (1974) Unique post-infection control of cedar-apple rust on apple with triforine. *Pl. Dis. Reptr* **58,** 587.

Szkolnik, M. and Gilpatrick, J. D. (1969) Apparent resistance of *Venturia inaequalis* to dodine in New York apple orchards. *Pl. Dis. Reptr* **53,** 861.

Sztejnberg, A., Byrde, R. J. W. and Woodcock, D. (1975). Antisporulant action of fungicides against *Podosphaera leucotricha* on apple seedlings. *Pestic. Sci.* **6,** 107.

ten Haken – see under H.

Tahori, A. S., Zeidler, G. and Halevy, A. H. (1965) Effect of some plant growth-retarding compounds on three fungal diseases and one viral disease. *Pl. Dis. Reptr* **49,** 775.

Talboys, P. W. (1962) Systemic movement of some vascular pathogens. *Trans. Brit. mycol. Soc.* **45,** 280.

Talboys, P. W., Frick, E. L. and Davies, M. K. (1975) Duration of effects on growth and wilt (*Verticillium dahliae*) development in the strawberry cv. Cambridge Vigour, in relation to time of inoculation. *Ann. appl. Biol.* **81,** 305.

Tamari, K., Ogasawara, N., Kaji, J. and Togashi, K. (1966) On the effect of a piricularin-detoxifying substance, ferulic acid, on tissue resistance of rice plants to blast fungal infection. *Ann. phytopathol. Soc. Japan* **32,** 186.

Tammes, P. M. L. and Die, J. van (1966) Studies on the phloem exudate from *Yucca flacida* Haw. IV. *Koninkl. Ned. Akad. Wetenschap. Proc.* C **69,** 656.

Tanaka, F., Tanaka, N., Youchara, H. and Umezawa, H. (1962) Studies on bacithracin, a new antibiotic from *B. megatherium*. I. Preparation and its properties. *J. Antibiotics (Tokyo)* **A 16,** 86.

Tanaka, N. (1975) Aminoglycoside antibiotics. In *Antibiotics*. Vol. III, p. 340 (Ed. J. W. Corcoran and F. E. Hahn). New York: Springer.

Tanaka, N., Yamaguchi, H. and Umezawa, H. (1966) Mechanism of kasugamycin action on polypeptide synthesis. *J. Biochem.* (Tokyo) **60,** 429.

Tanton, T. W. and Crowdy, S. H. (1972a) Water pathways in higher plants. II. *J. exp. Bot.* **23,** 600.

Tanton, T. W. and Crowdy, S. H. (1972b) Water pathways in higher plants. III. *J. exp. Bot.* **23,** 619.

Tapley, R. G., Woodroffe, R. and Atkins, P. S. (1969) Field trials with benomyl on soft fruit in 1969. *Proc. 5th Br. Insectic. Fungic. Conf.* **1,** 130.

Tarjan, A. C. and Howard, F. L. (1953) Comparison of benzothiazole 1-2-thioglycolic acid derivatives with other chemicals for Dutch elm disease therapy. *Phytopathology* **43,** 486.

Taub, D., Kuo, C. H. and Wendler, N. L. (1962) Synthesis in the griseofulvin series: 7-fluoro-7-dechlorogriseofulvin, an active analogue. *Chem. and Ind.* 557.

Taylor, G. S. (1953) Control of tobacco blue mould by root application of zineb and ferbam. *Phytopathology* **43,** 486.

Taylor, G. S. (1970) Tobacco protected against fleck by benomyl and other fungicides. *Phytopathology* **60,** 578.

Taylor, J. (1971) A necrotic leaf blotch and fruit rot of apple caused by a strain of *Glomerella cingulata. Phytopathology* **61,** 221.

Taylor, J. S. and Waterhouse, S. (1975) The use of carbendazim in the United Kingdom for the control of *Cercosporella herpatrichoides* in winter wheat. *Proc. 8th Br. Insectic. Fungic. Conf.* **2,** p. 389.

Teal, G. and Cavall, B. D. (1975) Degradation of bupirimate fungicide on apples and in water. *Proc. 8th Br. Insectic. Fungic. Conf.* **1,** p. 25.

Temkin-Gorodeiski, N. and Peled, A. (1971) Apical rots of banana fruits on the plant and their control. *Fruits d'outre-mer* **26,** 733.

Tempel, A. and Kaars Sijpesteijn, A. (1967) Pentobarbital sodium salt, a systemic agent for control of powdery mildew of cucumber. *Nature, Lond.* **213,** 215.

Tempel, A., Meltzer, J. and van den Bos, B. G. (1968) Systemic fungicidal and insecticidal activities of 1- and 2-[bis(dimethylamido)phosphoryl]-3-alkyl-5-anilino-1,2,4-triazoles. *Neth. J. Pl. Path.* **74,** 133.

Tezuka, N. and Kiso, A. (1975) Emergence of resistant strains of *Botrytis* sp. to thiophanate-methyl on egg plants. *Ann. Phytopath. Soc.* **43,** 303 (abs.).

Thaine, R. (1962) A translocation hypothesis based on the structure of plant cytoplasm. *J. exp. Bot.* **13,** 152.

Thaine, R. (1969) Movement of sugars through plants by cytoplasmic pumping. *Nature, Lond.* **222,** 873.

Thanassoulopoulos, C. C., Giannopolitis, C. N. and Kitsos, G. T. (1970) Control of *Fusarium* wilt of tomato and watermelon with benomyl. *Pl. Dis. Reptr* **54,** 561.

Thapliyal, P. N. (1970) Systemicity and fungicidal activity of three systemic fungicides in soybean. *Ph.D. Thesis, Univ. Illinois, Urbana,* 41 pp.

Thayer, P. L., Ford, D. H. and Hall, H. R. (1967) Bis-(*p*-chlorophenyl)-3-pyridine methanol (EL 241): a new fungicide for the control of powdery mildew. *Phytopathology* **57,** 833.

Thiollière, J. (1970) Prospects in the control of grey mould (*Botrytis cinerea*) in vineyards in France. *Summaries of Papers, VIIth Int. Cong. Plant Prot., Paris,* p. 256.

Thiollière, J., Horrière, D. and Rossi, M. (1972) Practical effectiveness of the association of benomyl and maneb in the control of aerial diseases of cereals. *Phytiatrie-Phytopharmacie,* No. Special, p. 51.

Thomas, T. H. (1974) Investigations into the cytokinin-like properties of benzimidazole-derived fungicides. *Ann. appl. Biol.* **76,** 237.

Thomson, W. W., Platt, K. A. and Campbell, N. (1973) The use of lanthanum to delineate the apoplastic continuum in plants. *Cytobios* **8,** 57.

Thorn, G. D. (1973) Uptake and metabolism of chloroneb by *Phaseolus vulgaris. Pestic. Biochem. Physiol.* **3,** 137.

Threlfall, R. J. (1968) The genetics and biochemistry of mutants of *Aspergillus nidulans* resistant to chlorinated nitrobenzenes. *J. Gen. Microb.* **52,** 35.

Ticknor, R. L. and Tukey, H. B. (1957) Evidence for the entry of mineral nutrients through the bark of fruit trees. *Proc. Am. Soc. hort. Sci.* **69,** 13.

van Tiel, N. (1975) Principles and problems of pesticide legislation. *Pestic Sci.* **6,** 189.

Tillman, R. W. and Sisler, H. D. (1971) A chloroneb-resistant mutant of *Ustilago maydis. Phytopathology* **61,** 914.

Tillman, R. W. and Sisler, H. D. (1973) Effect of chloroneb on the growth and metabolism of *Ustilago maydis. Phytopathology* **63,** 219.

Timberlake, W. E. and Griffin, D. H. (1974) Differential effects of cycloheximide and other inhibitors of protein synthesis on *in vivo* ribosomal RNA synthesis in *Achlya bisexualis. Biochim. Biophys. Acta* **353,** 248.

Timmer, L. W. (1973) Efficacy of low concentrations of benzimidazole fungicides for control of postharvest decays of citrus fruits. *Pl. Dis. Reptr* **57,** 899.

Timmer, L. W. (1974) Evaluation of fungicides and application times for control of melanose on Texas grapefruit. *Pl. Dis. Reptr* **58,** 504.

Tisdale, M. J. and Lord, K. A. (1973) Uptake and distribution of thiabendazole by seed potatoes. *Pestic. Sci.* **4,** 121.

Tocco, D. J., Rosenblum, C., Martin, C. M. and Robinson, H. J. (1966) Absorption, metabolism, and excretion of thiabendazole in man and laboratory animals. *Toxicol. Appl Pharmacol.* **9,** 31.

Tomizawa, C. and Uesugi, Y. (1972) Metabolism of S-benzyl-*O*, *O*-diisopropyl phosphorothioate (Kitazin P) by mycelial cells of *Pyricularia oryzae. Agr. Biol. Chem.* **36,** 294.

Tomlinson, H. and Rich, S. (1973) Anti-senescent compounds reduce injury and steroid changes in ozonated leaves and their chloroplasts. *Phytopathology* **63,** 903.

Tomlinson, H. and Rich, S. (1974) Ozone injury and senescence of bean leaves inhibited by phenylurea. *Phytopathology* **64,** 587.

Tomlinson, J. A. (1974) Chemotherapy of plant virus diseases. *Rep. natn. Veg. Res. Stn 1974,* p. 117.

Tomlinson, J. A. and Webb, M. J. W. (1958) Control of turnip and cabbage mildew (*Erysiphe polygoni* D.C.) by zinc. *Nature, Lond.* **181,** 1352.

Tomlinson, J. A. and Webb, M. J. W. (1959) Powdery mildew diseases. *Rep. Nat. Veg. Res. Stn. for 1958,* p. 38.

Tomlinson, J. A. and Webb, M. J. W. (1960) Powdery mildew diseases. *Rep. Nat. Veg. Res. Stn. for 1959,* p. 43.

Torgeson, D. C. (1967) (ed.) *Fungicides:* **1,** New York and London: Academic Press.

Tramier, R. and Antonini, C. (1971) Essai de lutte contre les trachiomycoses de l'oeillet Americain par le benomyl. *Annls. Phytopath.* **3,** 189.

Tsai, Y. P., Su, H. J., Ogawa, J. M. and Chen, S. S. (1968) Control of crown rot and finger spotting on banana hands with 1-(butylcarbamoyl)-2-benzimidazole carbamic acid, methyl ester (F 1991). *Pl. Prot. Bull. Taiwan* **10,** 35. (*Rev. Pl. Path.* **49,** 1101, 1970.)

Tsao, P. H., Leben, C. and Keitt, G. W. (1960) The partial purification and biological activity of an antifungal antibiotic produced by a strain of *Streptomyces griseus. Phytopathology* **50,** 169.

Tugwell, B. L. and Wicks, T. J. (1969) New chemicals for mould control in Citrus. *J. Agric. S. Aust.* **72,** 372. (*Rev. appl. Mycol.* **48,** 2979, 1969.)

Tukey, H. B., Jr. (1970) The leaching of substances from plants. *A. Rev. Pl. Physiol.* **21,** 305.

Tuyl, J. M. van (1975a) Acquired resistance to carboxin in *Aspergillus nidulans. Neth. J. Pl. Path.* **81,** 122.

Tuyl, J. M. van (1975b) Genetic aspects of acquired resistance to benomyl and thiabendazole in a number of fungi. *Meded. Fakulteit Landbouwwetenschappen Gent* **40,** 691.

Tuyl, J. M. van, Davidse, L. C. and Dekker, J. (1974) Lack of cross-resistance to benomyl and thiabendazole in some strains of *Aspergillus nidulans. Neth. J. Pl. Path.* **80,** 165.

Tyree, M. T. and Fensom, D. S. (1970) Some experimental and theoretical observations concerning mass flow in the vascular bundles of *Heracleum. J. exp. Bot.* **21,** 304.

Uesugi, Y. (1974) Development of fungicide resistance in rice blast fungus, *Pyricularia oryzae* Cavara. *Japan Agric. Res. Quarterly* **7,** 185.

Uesugi, Y., Katagiri, M. and Fukanaga, K. (1969) Resistance in *Pyricularia oryzae* to antibiotics and organophosphorus fungicides. *Bull. Nat. Instit. Agric. Sci.* **C 23,** 93.

Uesugi, Y., Katagiri, M. and Noda, O. (1973) Negatively correlated cross-resistance and synergism between phosphoramidates and phosphorothiolates in their fungicidal actions on rice blast fungi. *Agr. Biol. Chem.* **38,** 907.

Uesugi, Y. and Tomizawa, C. (1971a) Metabolism of *O*-ethyl-*S,S*-diphenylphosphorodithioate (Hinosan) by mycelial cells of *Pyricularia oryzae. Agr. Biol. Chem.* **35,** 941.

Uesugi, Y. and Tomizawa, C. (1971b) Metabolism of *S*-benzyl-*O*-ethyl phenylphosphonothioate (Inezin) by mycelial cells of *Pyricularia oryzae. Agr. biol. Chem.* **36,** 313.

Uesugi, Y., Tomizawa, C. and Murai, T. (1972) Degradation of organophosphorus fungicides. *Environmental toxicology of pesticides*, (eds F. Matsumura, G. M. Boush and T. Misato), pp. 327–39. New York: Academic Press.

Ueyama, I. and Takase, I. (1975) Metabolic behavior of *O*-ethyl-*S,S*-diphenylphosphorodithiolate (edifenphos) in female goat. *Agr. Biol. Chem.* **39**, 1719.

Ueyama, I., Uesugi, Y., Tomizawa, C. and Murai, T. (1973) Metabolic fate of *O*-ethyl-*S,S*-diphenyl phosphorodithiolate (Hinosan) in rice plant. *Agr. Biol. Chem.* **37**, 1543.

Ulrich, J. T. and Mathre, D. E. (1972) Mode of action of oxathiin systemic fungicides. V. Effect on electron transport system of *Ustilago maydis* and *Saccharomyces cerevisiae*. *J. Bacteriol.* **110**, 628.

Umezawa, H., Okami, Y., Hashimoto, T., Suhara, Y., Hamada, M. and Takeuchi, T. (1965) A new antibiotic, Kasugamycin. *J. Antibiotics (Tokyo) Ser. A.* **18**, 101.

Umgelter, H. (1969) Erfahrungen mit plantvax gegen Rostkrankheiten bei Zierpflanzen und Gemüsekulturen. *Gesunde Pfl.* **21**, 53.

van Andel – see under A.

van den Bos – see under B.

von Meyer – see under M.

etc.

Vaartaja, O. (1955) Chemotherapeutic action of dibenzothiophene against *Rhizoctonia solani. Bi-m. Progr. Rep. Div. For. Biol., Dep. Agric. Can.* **11**, 2.

Valent, B. and Albersheim, P. (1974) A protein from the fungal pathogen *Colletotrichum lindemuthianum* inhibits an enzyme of its host capable of attacking the pathogen. *Proc. Am. Phytopath. Soc.* **1**, 52.

Vandenheuvel, W. J. A., Carlin, J. R., Ellsworth, R. L., Wolf, F. J. and Walker, R. W. (1974) Source of carboxamide nitrogen atom in 2-carboxamido-5-isopropoxycarbonylaminobenzimidazole, a metabolite of cambendazole. *Biomed. Mass Spectrom.* **1**, 190

Vanner, A. L. and Hood, I. A. (1974) Douglas fir Swiss needle cast (*Phaeocrytopus gaeumanii*). *Fungicide and Nematicide Tests* **30**, 127.

Vargas, J. M. (1973) A benzimidazole-resistant strain of *Erysiphe graminis. Phytopathology* **63**, 1366.

Venkata Ram, D. S. (1969) Systemic control of *Exobasidium vexans* on tea with 1,4-oxathiin derivatives. *Phytopathology* **59**, 125.

Venn, K. I., Nair, V. M. G. and Kuntz, J. E. (1968) Effects of TCPA on oak sapwood formation and the incidence and development of oak wilt. *Phytopathology* **58**, 1071.

Ventura, E., Bourdin, J. and Berthier, G. (1970) Étude de l'efficacité de quelques fongicides systémiques nouveaux vis-à-vis des principaux parasites des semences de céréales. *VII Int. Congr. Pl. Prot. 1970* (abs.).

Ventura, E., Bourdin, J. and Piedallu, C. (1968) Activité fongicide de la carboxin vis-à-vis de principaux parasites des semences de céréales. *Phytiat.-Phytopharm.* **3**, 197.

Viala, P. (1893) *Les maladies de la vigne*. 3rd edn. Paris: Montpellier et Cie.

Vidali, A. and Cifferi, R. (1951) Esperienze di lotta contro l'oidio del Tabacco (*Erysiphe cichoracearum*) a mezzo di tiosulfato e sali di litio. *Tabacco* **55**, 95.

Visser, R., Shanmuganathan, H. and Savanayagam, J. V. (1962) The influence of sunshine and rain on tea blister blight, *Exobasidium vexans* Massee, in Ceylon. *The Tea Quart.* **33**, 34.

Vojvodić, D. (1970) Investigation on the influence of late autumn treatment on the formation of perithecia of *Venturia inaequalis* (Cke.) Wint. *Zašt. Bilja* **21**, 151. (*Rev. Pl. Path.* **50**, 2345, 1971.)

Volger, C. (1959) Über die Möglichkeit einer systemischen Wirkung von T.M.T.D. Präparaten. *Meded. LandbHoogesch. OpzoekStns. Gent* **24**, 837.

Vomvoyanni, V. (1974) Multigenic control of ribosomal properties associated with cycloheximide sensitivity in *Neurospora crassa. Nature, Lond.* **248**, 508.

Vomvoyanni, V. and Georgopoulos, S. G. (1966) Dosage-response relationships in diphenyl tolerance of *Hypomyces. Phytopathology* **56**, 1330.

Vomvoyanni, V., Georgopoulos, S. G. and Kappas, A. (1971) Complexity of genetic control of biochemical processes in fungi as evidenced by studies on resistance to toxicants.

Radiation and radioisotopes for industrial microorganisms, Intern. Atomic Energy Agency, Vienna, p. 233.

Vonk, J. W. and Kaars Sijpesteijn, A. (1971) Methyl benzimidazol-2-ylcarbamate, the fungitoxic principle of thiophanate-methyl. *Pest. Sci.* **2,** 160.

Vonk, J. W. and Mihanović, B. (1977) A biochemical mechanism for the conversion of thiophanate-methyl into methyl benzimidazolyl carbamate (MBC) in plant tissue. *Neth. J. Pl. Path.* **83,** Suppl. 1 (in press).

Waard, M. A. de (1971a) Germination of powdery mildew conidia *in vitro* on cellulose membranes. *Neth. J. Pl. Path.* **77,** 6.

Waard, M. A. de (1971b) Effect of systemic and non-systemic compounds on the *in vitro* germination of powdery mildew conidia. *Meded. Fac. Landbouwwetensch. Rijksuniv. Gent* **36,** 113.

Waard, M. A. de (1972) On the mode of action of the organophosphorus fungicide *Hinosan. Neth. J. Pl. Path.* **78,** 186.

Waard, M. A. de (1974) Mechanisms of action of the organophosphorus fungicide pyrazophos. *Meded. Landbouwhogeschool Wageningen*, Nederland 74–14, 1.

Waard, M. A. de (1975) Mode of action of the organophosphorus fungicide pyrazophos. *Systemfungizide* (p. 197). Intern. Symp. Reinhardsbrunn. (eds H. Lyr and C. Polter) Berlin: Akademie-Verlag.

Waffelaert, P. (1969) Nouvelles perspectives de lutte contre le Monilia et l'Oidium de l'abricotier. *Phytiat. Phytopharm.* **18,** 51. (*Rev. Pl. Path.* **49,** 1433, 1970.)

Waggoner, P. E. (1956) Chemotherapy of *Verticillium* wilt of potatoes in Connecticut 1955. *Am. Pot. J.* **33,** 223.

Waggoner, P. E. and Dimond, A. E. (1952) Effect of stunting agents, *Fusarium lycopersici*, and maleic hydrazide upon phosphorus distribution in tomato. *Phytopathology* **42,** 24.

Waggoner, P. E. and Dimond, A. E. (1955) Production and role of extracellular pectic enzymes of *Fusarium oxysporum* f. *lycopersici. Phytopathology* **45,** 79.

Waggoner, P. E. and Dimond, A. E. (1957) Altering disease resistance with ionizing radiation and growth substances. *Phytopathology* **47,** 125.

Wagner, F. (1940) Die Bedeutung der Kieselsäure für das Wachstum einiger Kulturpflanzen, ihren Nährstoffhaushalt und ihre Anfälligkeit gegen echte Mehltaupilze. *Phytopath. Z.* **12,** 427.

Wain, R. L. (1951) Plant growth-regulating and systemic fungicidal activity: the aryloxyalkylcarboxylic acids. *J. Sci. Fd Agric.* **2,** 101.

Wain, R. L. (1952) Systemic fungicides. *Ann. appl. Biol.* **39,** 429.

Wain, R. L. (1953) Systemic fungicides. *Meded. LandbHoogesch. OpzoekStns. Gent* **18,** 394.

Wain, R. L. (1959) Some chemical aspects of plant disease control. *R. Inst. Chem. Monograph No. 3.*

Wain, R. L. (1963) Some developments in research on fungicides. *Meded. LandbHoogesch. OpzoekStns. Gent* **28,** 516.

Wain, R. L. (1969) Naturally occurring fungicides. *Proc. Symp. Potentials in Crop Protection, New York, 1969*, p. 26.

Wain, R. L. and Carter, G. A. (1967) Uptake, translocation and transformations by higher plants. *Fungicides, Vol. 1* (ed. D. C. Torgeson), p. 561. New York: Academic Press.

Wain, R. L. and Carter, G. A. (1977) Chapter 2 in *Systemic Fungicides* (ed. R. W. Marsh) London: Longman.

Wain, R. L., Spencer, D. M. and Fawcett, C. H. (1961) Antifungal compounds in seedlings of *Vicia faba. S.C.I. Monograph No. 15*, p. 109.

Walker, A. and Featherstone, R. M. (1973) Absorption and translocation of atrazine and linuron by plants with implications concerning linuron selectivity. *J. exp. Bot.* **24,** 450.

Wallace, H. A. H. (1968) Co-operative seed treatment trials, 1967. *Can. Pl. Dis. Surv.* **48,** (3) 82.

Wallen, V. R. (1955) Control of stem-rust of wheat with antibiotics. I. Greenhouse and field tests. *Pl. Dis. Reptr* **39,** 124.

Wallen, V. R. (1958) Control of stem rust of wheat with antibiotics. II. Systemic activity and effectiveness of derivatives of cycloheximide. *Pl. Dis. Reptr* **42,** 363.

Wallen, V. R. and Millar, R. L. (1957) The systemic activity of cycloheximide in wheat seedlings. *Phytopathology* **47,** 291.

Wallen, V. R. and Skolko, A. J. (1950) Antibiotic XG as a seed treatment for the control of leaf and pod spot of peas caused by *Ascochyta pisi. Can. J. Res. Ser. C* **28,** 623.

Wallen, V. R. and Skolko, A. J. (1951) Activity of antibiotics against *Ascochyta pisi. Can. J. Bot.* **29,** 316.

Wallnöfer, P. R. (1969) Der mikrobielle Abbau des 1,4-Oxathiinderivats, 2,3-dihydro-5-carboxanilido-6-methyl-1,4-oxathiin (DCMO). *Arch. Mikrobiol.* **64,** 319.

Wallnöfer, P. R. and Engelhardt, G. (1971) Der Abbau von Phenylamiden durch *Bacillus phaericus. Arch Mikrobiol.* **80,** 315.

Wallnöfer, P. R., Königer, M., Safe, S. and Hutzinger, O. (1972) Metabolism of the systemic fungicide 2,5-dimethyl-3-furancarboxylic acid anilide (BAS 3191) by *Rhizopus japonicus* and related fungi. *J. Agr. Food Chem.* **20,** 20.

Wallnöfer, P. R., Safe, S. and Hutzinger, O. (1971) Metabolism of the systemic fungicides 2-methylbenzanilide and 2-chlorobenzanilide by *Rhizopus japonicus. Pestic. Biochem. Physiol.* **1,** 458.

Wang, M. C., Erwin, D. C., Sims, J. J., Keen, N. T. and Borum, D. E. (1971) Translocation of ^{14}C-labelled thiabendazole in cotton plants. *Pestic. Physiol. Biochem.* **1,** 188.

Wardlaw, C. W. (1961) *Banana Diseases including Plantains and Abaca.* London: Longman.

Wardlaw, I. F. (1968) The control and pattern of movement of carbohydrates in plants. *Bot. Rev.* **34,** 79.

Wargo, P. M. (1975) Lysis of cell wall of *Armillaria mellea* by enzymes of forest trees. *Physiol. Pl. Path.* **5,** 99.

Waring, R. H. (1974) Metabolism of Vitavax by rats and rabbits. *Xenobiotica* **3,** 65.

Waring, R. H. and Wolfe, M. S. (1975) Distribution of tridemorph and its metabolites in barley plants grown to maturity. *Pestic. Science* **6,** 169.

Wark, M. C. and Chambers, T. C. (1965) Fine structure of the phloem of *Pisum sativum.* I. *Aust. J. Bot.* **13,** 171.

Warren, C. G., Cole, H., Jr. and Duich, J. M. (1974a) *Fungicide and Nematicide Tests* **30,** 127.

Warren, C. G., Sanders, P. and Cole, H. (1974b) *Sclerotinia homoecarpa* tolerance to benzimidazole configuration fungicides. *Phytopathology* **64,** 1139.

Watson, A. G. and Koons, C. E. (1973) Increased tolerance to benomyl in greenhouse populations of *Botrytis cinerea. Phytopathology* **63,** 1218 (abs.).

Watts Padwick, G. (1956) *Losses caused by plant diseases in the colonies.* Phytopathological Papers No. 1, published by C.M.I., Kew, Surrey.

Webster, R. K., Ogawa, J. M. and Bose, E. (1970) Tolerance of *Botrytis cinerea* to 2,6-dichloro-4-nitroaniline. *Phytopathology* **60,** 1489.

Weeks, R. E. and Hedrick, H. G. (1971) Screening microorganisms from two soils for utilization of the systemic fungicide Benlate. *Bact. Proc. 1971,* 14.

Weigun Yang and Lung-chi Wu (1971) Tolerance to PCNB and simulated pathogenic strength of *Sclerotium rolfsii. Memoirs Coll. Agric. Nat. Taiwan Univ.* **12,** 191.

Weindlmayr, J. (1963) Untersuchungen über eine Beeinflussung der Phytophthora-Anfälligkeit von Kartoffelpflanzen durch Gibberellin Behandlungen. *Z. PflKrankh. PflPath. PflSchutz.* **70,** 599.

Weinke, K. E., Lauber, J. J., Greenwald, B. W. and Preiser, F. A. (1969) Thiabendazole, a new systemic fungicide. *Proc. 5th Br. Insectic. Fungic. Conf.* **2,** 340.

Weir, L. C. (1963) Control of needle cast of Douglas fir by antibiotics. *Forestry Chron.* **39,** 205.

Welch, A. W. (1968) Grape: black rot, bitter rot; *Guignardia bidwellii, Melanconium fuligineum. Fungicide and Nematicide Tests* **23,** 59.

Welch, N. C., Greathead, A. S., Hall, D. H. and Little, T. (1969) Brussels sprout ring spot control with fungicides. *Calif. Agric.* **23,** 17.

Wells, J. M. (1971) Heated wax emulsions with benomyl and 2,6-dichloro-4-nitroaniline for control of post-harvest decay of peaches. *Phytopathology* **61**, 916 (abs.).

Wells, J. M. (1972) Heated wax emulsions with benomyl and 2,6-dichloro-4-nitroanilene for control of post-harvest decay of peaches and nectarines. *Phytopathology* **62**, 129.

Wells, J. M. and Gerdts, M. H. (1971) Pre- and post-harvest benomyl treatments for control of brown rot of nectarines in California. *Pl. Dis. Reptr* **55**, 69.

Wells, J. M. and Merwarth, F. L. (1973) Fungicide dips for controlling decay of carrots in storage for processing. *Pl. Dis. Reptr* **57**, 697.

Wensley, R. N. and Huang, C. M. (1970) Control of *Fusarium* wilt of musk melon and other effects of benomyl soil drenches. *Can. J. Microbiol.* **16**, 615.

Western, J. H. (1971) *Diseases of crop plants.* London and Basingstoke: The MacMillan Press Ltd.

Wettstein, F. O., Noll, H. and Penman, S. (1964) Effect of cycloheximide on ribosomal aggregates engaged in protein synthesis *in vitro. Biochim. biophys. Acta* **87**, 525.

Whan, J. H. (1976) Tolerance of *Sclerotinia fructicola* to benomyl. *Pl. Dis. Reptr* **60**, 200.

Wheeler, H. and Luke, H. H. (1963) Microbial toxins in plant disease. *A. Rev. Microbiol.* **17**, 223.

Whiffen, A. J. (1950) The activity *in vitro* of cycloheximide (acti-dione) against fungi pathogenic to plants. *Mycologia* **42**, 253.

White, G. A. (1971) A potent effect of 1,4-oxathiin systemic fungicides on succinate oxidation by a particulate preparation from *Ustilago maydis. Biochem. Biophys. Res. Comm.* **44**, 1212.

White, G. A. and Thorn, G. D. (1975) Structure-activity relationships of carboxamide fungicides and the succinic dehydrogenase complex of *Cryptococcus laurentii* and *Ustilago maydis. Pestic. Biochem. Physiol.* **5**, 380.

Whiteside, J. O. (1970) Effect of fungicides applied to citrus trees on perithecial development by the greasy spot fungus in detached leaves. *Pl. Dis. Reptr* **54**, 865.

Whiteside, J. O. (1973) Evaluation of spray materials for the control of greasy spot of citrus. *Pl. Dis. Reptr* **57**, 691.

Wicks, T. (1973) Control of apple scab with benomyl-oil-water emulsions. *Pl. Dis. Reptr* **57**, 560.

Wicks, T. (1974) Tolerance of the apple scab fungus to benzimidazole fungicides. *Pl. Dis. Reptr* **58**, 886.

Wiese, M. V. and Vargas, J. M. (1973) Interconversion of chloroneb and 2,5-dichloro-4-metoxyphenol by soil microorganisms. *Pestic. Biochem. Physiol.* **3**, 214.

Wilkie, D. and Lee, B. K. (1965) Genetic analysis of actidione resistance in *Saccharomyces cerevisiae. Genet. Res.* **6**, 130.

Williams, R. J. and Ayanaba, A. (1975) Increased incidence of Pythium stem rot in cowpeas treated with benomyl and related fungicides. *Phytopathology* **65**, 217.

Williamson, C. J. and Burchill, R. T. (1974) The perennation and control of pear scab (*Venturia pirina* Aderh.). *Pl. Path.* **23**, 67.

Willis, W. G. (1971) Scotch pine needle disease control. *Research progress report. Kan. Coop. Rept. to Regional Project* NCR-43 (Diseases of Landscape Plants).

Willis, W. G. (1972) Scotch pine needle disease control. *Research progress report. Kan. Coop. Rept. to Regional Project* NCR-43 (Diseases of Landscape Plants).

Wilson, E. M. and Ark, P. A. (1958) The effect of chlorophyll on the phytotoxicity of acti-dione on bean. *Pl. Dis. Reptr* **42**, 1069.

Wilson, G. J. (1974) Control of celery leaf spot. *N.Z.J. agric.* **129**, 33.

Winner, C. (1966) Untersuchungen über parasitogene Schäden an Wurzeln der Zuckerrübe, insbesondere durch Aphanomyces, und über Möglichkeiten ihrer Verhütung. III. Versuche zur Hemmung eines Aphanomyces-Befalls durch spezifisch wirkende Chemotherapeutica. *Phytopathol. Z.* **57**, 310.

Winter, A. G. and Willeke, L. (1951) Über die Aufnahme von Antibiotika durch höhere Pflanzen und ihre Stabilität in natürlichen Böden. *Naturwissenschaften* **38**, 457.

Witchalls, J. R. and Close, R. (1971) Control of eyespot lodging in wheat by benomyl. *Pl. Dis. Reptr* **55**, 45.

Witcher, W., Arnett, J. D., Baxter, L. W. and Cocke, M. L. (1975) Control of needle cast of Scotch pine in South Carolina. *Pl. Dis. Reptr* **59,** 881.

Witcher, W., Baxter, L. W. and Marbut, S. A. (1973) Benomyl promising chemical for leaf and stem diseases of red cedar and Arizona cypress. *Pl. Dis. Reptr* **57,** 315.

von Witsch, H. and Kasperlik, H. (1953) Infecktionhemmende Wirkung der Dichlorophenoxyessigsäure. *Naturwissenschaften* **40,** 294.

Wolfe, M. S. (1971) Fungicides and the fungus population problem. *Proc. 6th Insectic. Fungic. Conf. at Brighton, 1971,* **3,** 724.

Wolfe, M. S. Pathogen response to fungicide use. *Proc. 8th Brit. Insectic. Fungic. Conf. at Brighton, 1975* (in press).

Wolfe, M. S. and Dinoor, A. (1973) The problem of fungicide tolerance in the field. *Proc. 7th Brit. Insectic. Fungic. Conf. 1973* **1,** p. 11.

Wolkoff, A. W., Onuska, F. I., Comba, M. E. and Larose, R. H. (1975) Determination of 5,6-dihydro-2-methyl-1,4-oxathiin-3-carboxanilide (Vitavax) and two of its photoproducts by high speed liquid chromatography. *Anal. Chem.* **47,** 754.

Wood, R. K. S. (1960) Pectic and cellulolytic enzymes in plant diseases. *A. Rev. Pl. Physiol.* **11,** 299.

Wood, R. K. S. (1967) *Physiological plant pathology.* Oxford and Edinburgh: Blackwell Scientific Publications.

Woodcock, D. (1971) Chemotherapy of plant disease – progress and problems. *Chem. Brit.* **7,** 415.

Woodward, P. J. (1971) Gooseberry: American gooseberry mildew and Botrytis control 1970. *Rep. Luddington exp. Hort. Stn for 1970,* p. 56.

Wormald, H. (1954) The brown rot diseases of fruit trees. *Min. Agric. Tech. Bull. no. 3,* London: HMSO.

Wortley, W. R. S. (1936) The effect of salts of lithium on the resistance of certain plants to disease. *J. R. agric. Soc.* **97,** 492.

Wright, M. A. and Stringer, A. (1973) The toxicity of thiabendazole, benomyl, methyl benzimidazol-2ylcarbamate and thiophanate-methyl to the earthworm, *Lumbricus terrestris. Pestic. Sci.* **4,** 431.

Wuest, P. J., Cole, H. and Sanders, P. L. (1974) Tolerance of *Verticillium malthousei* to benomyl. *Phytopathology* **64,** 331.

Yakovleva, R. S. (1974) Fungicides protect black currant *Zashch. Rast., Mosk.* **11,** 26. (*Rev. Pl. Path.* **54,** 543, 1975.)

Yamada, M. (1955) The sporulation-inhibitive effects of sulfadiazine on wheat leaf rust. I. Some experimental results with wheat seedlings. *Ann. Phytopath. Soc., Japan* **19,** 146.

Yamaguchi, T. (1974) 1974 evaluation of candidate pesticides (B–I: Fungicides: Rice. *Japan Pesticide Information,* **24,** 13.

Yamaguchi, S. and Crafts, A. S. (1957) Translocation of 2,4-D in *Zebrina pendula* is greatly affected by growth rate. *Pl. Physiol. Lancaster* **32,** Suppl. xlii.

Yamamoto, H., Tomizawa, C., Uesugi, Y. and Murai, T. (1973) Absorption, translocation and metabolism of *O,O*-diisopropyl-*S*-benzyl phosphorothiolate (Kitazin P) in rice plant. *Agr. Biol. Chem.* **37,** 1553.

Yarham, D. J., Bacon, E. T. G. and Hayward, C. F. (1971) The effect on mildew development of the widespread use of fungicide on winter barley. *Proc. 6th Br. Insectic. Fungic. Conf.* **1,** p. 15.

Yarwood, C. E. (1947) Therapeutic treatments for bean rust. *Phytopathology* **37,** 24.

Yarwood, C. E. (1948) Therapeutic treatments for rusts. *Phytopathology* **38,** 542.

Yarwood, C. E. (1950) Effect of temperature on the fungicidal action of sulphur. *Phytopathology* **40,** 173.

Yarwood, C. E. (1959) Predisposition, *Plant Pathology Vol. 1* (eds J. G. Horsfall and A. E. Dimond), p. 521. New York: Academic Press.

Yasuda, Y., Hashimoto, S., Soeda, Y. and Noguchi, T. (1973) Metabolism of thiophanate-methyl by pathogenic fungi and antifungal activity of its metabolites. *Ann. Phytopath. Soc. Japan* **39,** 49.

Yoshinaga, E. (1969) A systemic fungicide for rice blast control. *Proc. 5th Br. Insectic. Fungic. Conf.* **2,** p. 593.

Yoshinaga, E., Uchida, T. and Iwakura, T. (1965) Studies on the rice blast control. IV. Rice blast control of kitazin by root treatment. *Ann. Phytopathol. Soc., Japan* **30,** 307 (abs.).

Zaracovitis, C. (1965) Activity of phenobarbitone in controlling marrow powdery mildew. *Nature, Lond.* **206,** 954.

Zaumeyer, W. J. (1957) Comparative protection of bean leaves from fungus infection by antibiotic treatments of upper and lower surfaces. *Phytopathology* **47,** 539.

Zaumeyer, W. J. (1958) Antibiotics in the control of plant diseases. *A. Rev. Microbiol.* **12,** 415.

Zaumeyer, W. J. and Wester, R. E. (1956) Control of several fungus diseases of beans and lima beans with antibiotics. *Phytopathology* **46,** 470.

Zehr, E. I. (1971a) Apple: scab (*Venturia inaequalis*). *Fungicide and Nematicide Tests* **26,** 40.

Zehr, E. I. (1971b) Grape: black rot (*Guignardia bidwellii*). *Fungicide and Nematicide Tests* **26,** 59.

Zelitch, I. (1961) Biochemical control of stomatal opening in leaves. *Proc. Nat. Acad. Sci., US* **47,** 1423.

Zelitch, I. (1963) The control mechanisms of stomatal movement. *Stomata and water relations* (ed. I. Zelitch), p. 18. *Conn. Agric. Exp. Stn Bull.* no. 664, New Haven.

Zentmyer, G. A. (1942) Toxin formation and chemotherapy in relation to Dutch elm disease. *Phytopathology* **32,** 20.

Zentmyer, G. A. (1954) Chemotherapy for the control of *Phytophthora* root rot of avocado. *Phytopathology* **44,** 511.

Zentmyer, G. A. and Gilpatrick, J. D. (1960) Soil fungicides for prevention and therapy of *Phytophthora* root rot of avocado. *Phytopathology* **50,** 660.

Zentmyer, G. A. and Horsfall, J. G. (1943) Internal therapy with organic chemicals in treatment of vascular diseases. *Phytopathology* **33,** 16.

Zentmyer, G. A., Horsfall, J. G. and Wallace, P. P. (1946) Dutch elm disease and its chemotherapy. *Conn. Agric. Exp. Stn Bull.* no. 498, p. 1.

Zentmyer, G. A., Moje, W. and Mircetich, S. M. (1962) Ethionine as a chemotherapeutant. *Phytopathology* **52,** 34.

Zimmermann, M. H. (1957) Translocation of organic substances in trees. I. *Pl. Physiol. Lancaster* **32,** 288.

Zolel, Z. (1970) The systemic fungicidal effect of benzimidazole derivatives and thiophanate against Cercospora leaf spot of sugar beet. *Phytopathology* **60,** 1186.

Zolel, Z. (1971) The systemic fungicidal activity of triphenyl tin acetate against *Cercospora beticola* on sugar beet. *Phytopathology* **61,** 738.

Zweep, W. van der (1961) The movement of labelled 2,4-D in young barley plants. *Weed Res.* **1,** 258.

Alphabetical list of systemic fungicides in commercial use

(1) *Common names*

Common name	Chemical name	Trade names of product
benodanil	2-iodobenzanilide	Calirus
benomyl	methyl-(1-butylcarbamoyl)benzimidazol-2-ylcarbamate	Benlate, Tersan 1991
bupirimate	5-butyl-2-ethylamino-6-methylpyrimidin-4-yl dimethylsulphamate	Nimrod
carbendazim	methyl benzimidazol-2-ylcarbamate	Bavistin, Derosal
carboxin	2,3-dihydro-6-methyl-5-phenylcarbamoyl-1,4-oxathiin	Vitavax, Murganic
chloroneb	1,4-dichloro-2,5-dimethoxybenzene	Demosan
dimethirimol	5-n-butyl-2-dimethylamino-4-hydroxy-6-methylpyrimidine	Milcurb
dodemorph	4-cyclododecyl-2,6-dimethylmorpholine	BAS 238F
ethirimol	5-n-butyl-2-ethylamino-4-hydroxy-6-methylpyrimidine	Milstem, Milgo
kasugamycin	[5-amino-2-methyl-6-(2,3,4,5,6-pentahydroxycyclohexyloxy)pyran -3-yl]amino-α-iminoacetic acid	Kasumin
	S-benzyl-*OO*-di-isopropyl phosphorothioate	Kitazin P
oxycarboxin	2,3-dihydro-6-methyl-5-phenylcarbamoyl-1,4-oxathiin 4,4-dioxide	Plantvax
pyracarbolid	2*H*-3,4-dihydro-6 methylpyran-5-carboxanilide	HOE 2989
pyrazophos	*0*-6-ethoxycarbonyl-5-methylpyrazolo[1,5*a*]pyrimidin-2-yl *OO*-diethyl phosphorothioate	Afugan
thiabendazole	2-(thiazol-4-yl)benzimidazole	Mertect, Tecto
thiophanate	1,2,-di-(3-ethoxycarbonyl-2-thioureido)benzene	Topsin, Cercobin
thiophanate-methyl	1,2-di-(3-methoxycarbonyl-2-thioureido)benzene	Mildothane
triadimefon	1-(4-chlorophenoxy)-3,3-dimethyl-1-(1,2,4-triazol-1-yl)butan-2-one	Bayer 6707
triamiphos	5-amino-1-(bisdimethylaminophosphinyl)-3-phenyl-1,2,4-triazole	Wepsyn
tridemorph	2,6-dimethyl-4-tridecylmorpholine	Calixin
triforine	1,4-di-(2,2,2-trichloro-1-formamidoethyl)piperazine	Cela W 524, Saprol

(2) *Trade names*

Product	Active ingredient	Firm
Afugan	pyrazophos	Hoechst
Bas 238F	dodemorph	Badische Anilin Soda Fabrik
Bayer 6707	triadimefon	Bayer
Bavistin	carbendazim	BASF
Benlate	benomyl	Du Pont
Calirus	benodanil	BASF
Calixin	tridemorph	BASF: Cyanamid
Cela W 524	triforine	Merck
Cercobin	thiophanate	Nippon Soda Co
Demosan	chloroneb	Du Pont
Derosal	carbendazim	Hoechst
HOE 2989	pyracarbolid	Hoechst
Kasumin	kasugamycin	Hokko Chemical Industry
Kitazin P	See (1)	Kumiai Chemical Industry
Mertect	thiabendazole	Merck
Mildothane	thiophanate-methyl	May and Baker
Milcurb	dimethirimol	Plant Protection
Milgo	ethirimol	Plant Protection
Milstem	ethirimol	Plant Protection
Murganic RPH	carboxin	Murphy: Rothwell Plant Health: Nickersons
Nimrod	bupirimate	Plant Protection
Plantvax	oxycarboxin	Uniroyal
Saprol	triforine	Shell
Tecto	thiabendazole	Rothwell Plant Health
Tersan	benomyl	Du Pont
Topsin	thiophanate	Nippon Soda Co
Vitavax	carboxin	Uniroyal
Wepsyn	triamiphos	Philips-Duphar

Index

(*Bold type denotes pages showing structural formulae*)

acetate metabolism inhibited by oxathiins, 53
Actidione = cycloheximide, *q.v.*
almond, 282
Alternaria brassicicola, 269
Alternaria kikuchiana, 48, 185
Alternaria longipes causing brown spot of tobacco, 124
Alternaria solani, use in bioassays, 14
2-aminobenzimidazole, 162
3-amino-1,2,4-triazoles, **33**
antibiotics, 23, 39, 87. *See also* under individual compounds
antifungal compounds in plants, 25
antimycin A, *Venturia inaequalis* resistant to, 178
antisporulants, 125, 280, 290
Aphanomyces
  cochlioides, sugar-beet root rot, 124
  euteiches, pea root rot, 266
apoplastic movement, 99
apple
  canker, *Nectria galligena*, antisporulants, 279
  cork spot, *Alternaria mali*, 47
  disease control, 275–80
  fly-speck, *Mycothyriella rubi*, 278
  mildew, *Podosphaera leucotricha*, 276
    effect of phenyl alanine, 118
  russeting, 277
  scab, *Venturia inaequalis*, control by
    benomyl, 275–7
    thiophanates, 276
    triforine, 276
  storage rots, *Gloeosporium* spp., control, 278
  yield increase after benomyl sprays, 277
application methods
  in glasshouses, 252
  on field crops, 195, 206
apricot diseases, 282
Ascochyta
  fabae, controlled by benomyl, 260
  rabiei on chick pea, 260
  pisi, controlled by benomyl, 260
    controlled by rimocidin, 49
Aspergillus
  midulans, resistance to benomyl, 181
  niger, effect of thiabendazole, 61
assessment of systemic and therapeutic activity, 9
avocado root rot, 123
8-azaadenine, 73, **74**
8-azaguanine, 73, **74**
6-azauracil, 3,5-dioxo-2,3,4,5-tetrahydro-1,2,4-triazine, **73**
  Cladosporium cucumerinum resistant to, 186
  metabolic conversion in host tissues, 117
  mode of action, 117, 145
azepines, **75,** 90

bacimethrin, **63**
banana
  diseases, 291
  Penicillium rots, control by thiabendazole, 272
barley
  covered smut, *Ustilago hordei*, 229
  loose smut, *U. nuda*, 52, 228
  mildew, *Erysiphe graminis*, 213–22
BAS 3170 (benodanil), 221
BAS 3170, 2-iodo-benzanilide, controls cereal rusts, 225
BAS 3191, 2,5-dimethylfuran-3-carboxylic acid anilide, 231
  controls smuts, 229
BAS 3270, 2,5-dimethylfuran-3-carboxylic acid cyclohexamide
  controls smuts, 229
BAS 346 F, (MBC) methyl benzimidazol-2-yl carbamate
  mammalian toxicity, 88
bean
  halo blight, *Pseudomonas medicaginis* var. *phaseolicola*, control by streptomycin, 42
  rust, *Uromyces phaseoli*, control, 270
  seed-borne fungi controlled by benomyl, 260
  white mould, *Sclerotinia sclerotiorum*, controlled by benomyl, 269
beet powdery mildew, *Erysiphe polygoni*, 68
Benlate T, 267
benodanil, 2-iodobenzanilide, 55, 221
benomyl, methyl (1-butylcarbamoyl)-benzimidazol-2-yl-carbamate, **61**

benomyl—*contd.*
against Dutch elm disease, 309
antifungal spectrum, 138
controls carrot liquorice rot, 272
decomposition to MBC, 60, 161
effect on fine structure of *Botrytis fabae*, 143
effect on soil microflora, 257
mammalian toxicity, 88
mode of action, 141
non-fungicidal effects on hosts, 255
post-harvest applications, 209
resistance of fungi to, 180
seed treatment against bunt, 230
seed treatment against rice blast, 234
seed treatment against smuts, 230
sporulostatic effect, 292
translocation, 110
use against cereal, fruit and vegetable diseases—*see under* individual crops
uses in forestry, 295
benzanilides, 53
benzimidazoles, 60
antifungal spectrum, 137
mode of action, 140
non-fungicidal effects of, 255
6-benzylamino purine vs. oak wilt, 304
O-benzylphosphates, **37**
S-benzylphosphorodithioates, **37**
biocidal effects, 88, 159
biological control, 256
blackberry diseases, 285
black currant
diseases, 284
leaf spot, *Pseudopeziza ribis*, control, 281
blast, *see Pyricularia oryzae*
blasticidin S
inhibits protein synthesis, **45,** 136, 185
mammalian toxicity, 87
resistance of *Pyricularia oryzae* to, 185
blueberry diseases, 285
Botryosphaeria ribs, 278
Botrytis allii, neck rot of onions, 41, 260
Botrytis cinerea
grey mould and ghost spot of tomatoes, 242
grey mould of black currants, 284
grey mould of strawberries, 286
on French bean pods, control, 270
on lettuce, 246
pear storage rot, 279
resistance to benomyl, 179, 194
resistance to dicloran, 182
sensitivity to systemics, 133
storage rot of sequoia, 300
Botrytis fabae
effect of benomyl on fine structure, 143
effect of captan, 13
effect of rimocidin, 49
use in bioassays, 13
Botrytis tulipae, tulip fire, 40, 250
Bremia lactucae, lettuce downy mildew, 247
brown rust of barley, *Puccinia hordei*, 221
brown patch, 236
Brussels sprouts ring-spot, *Mycosphaerella brassicicola*, control, 269
bulb and corm diseases, 250
bunt, *Tilletia caries*
control by benomyl and thiabendazole, 230
bupirimate, 5-butyl-2-ethylamino-6-methylpyrimidin-4-yl dimethylsulphamate, **67,** 89, 173
4-*n*-butyl-1,2,4-triazole (= RH 124), **70,** 223
translocation in wheat, 111
n-butyranilide, 56

cadmium, tolerance to, 177
caffeic acid products inactivate polygalacturonase, 126
calcium nutrition effects on cell-wall pectins, 123
Calonectria nivalis, control by fuberidazole, 62
captan used against *Botrytis fabae*, 13
carbendazim (*see also* MBC), **61**
mode of action, 141
suppresses virus expression, 271
carbendazim-HCl
prevents oak wilt, 304
treatment for Dutch elm disease, 314, 318
carbinols, **68**
carboxin, 2,3-dihydro-6-methyl-5-phenylcarbomoyl-1,4-oxathiin, **51,** 87
barrier effect of bean seed testa, 204
controls
carnation rusts, 248
Helminthosporium gramineum, 231
Urocystis agropyri, 230
Ustilago nuda and *U. tritici*, 228
metabolism, 165
mode of action, 134
tested against *Sphacelotheca reiliana*, 235
translocation, 110
carboxylic acid anilides, 51
carnation rusts, *Uromyces caryophyllinus* and *Puccinia arenaria*, 248
carnation wilts, 247
Carolate used against Dutch elm disease, 26
carrot liquorice rot, *Centrosperma acerina*, control by benomyl, 272
Cela W 524, triforine, *N,N'*-bis-(1-formamido-2,2,2-trichloroethyl)piperazine, **75,** 148

celery blights, *Septoria apiicola* and *Cercospora apii*, control by benomyl, 271
celery crown rot, *Centrospora acerina*, control by benomyl, 272
cellocidin, **49**
cell wall changes, 122
Centrospora acerina, rots of carrot and celery, control by benomyl, 272
Cephalosporium
  diospyri, oak decline, 306
  gregatum, stem rot of soy-bean, 263
Ceratocystis fagacearum, cause of oak wilt, 120, 304
Ceratocystis fimbriata f. *platani*, sycamore canker stain, 303
Ceratocystis paradoxa, pineapple fruit rot, 291
Ceratocystis ulmi, cause of Dutch elm disease, 306
  effect of vessel size on invasion, 119
  strains resistant to carbendazim, 319
Cercospora
  apii, early blight of celery, 268
  beticola, sugar-beet leaf spot, 268
  fusimaculans, 239
  oryzae, 235
  sequoiae, 300
  sp., pink pitting of citrus, 288
Cercosporella herpotrichoides, eye-spot of cereals, control by benomyl, 232
cereal rusts, treatments for, historical survey, 17
cherry diseases, 282
chestnut blight, *Endothia parasitica*, 7, 303
chick pea, control of *Ascochyta rabiei*, 260
chitin, fungal, inhibition of synthesis, 48, 146
chitinase, 129
chloramphenicol, 108
chloraniformethan, 1-(3,4-dichloranilino)-1-formylamino-2,2,2-trichloroethane, **75**, 76
  spray against cereal mildew, 217
chlorogenic acid inactivates polygalacturonase, 126
2-chloro-6-methoxy-4-(trichloromethyl) pyridine (Dow 269, pyroxychlor), **70**
chloroneb, 1,4-dichloro-2,5-dimethoxybenzene, **82**, 90
  mode of action, 144, 170
chlorquinox, 5,6,7,8-tetrachloroquinoxaline spray against cereal mildew, 217
'Christmas trees', 297
chrysanthemum
  control of *Ascochyta chrysanthemi* on cuttings, 249
  control of *Puccinia horiana* by carboxin, 249
circumvention, 188
citrus rots, *Penicillium* spp., control, 272
Cladosporium carpophilum, peach scab, 282
Cladosporium cucumerinum, cucumber scab
  control by reducing sugar level, 128
  resistance to 6-azuracil, 178
  use in bioassays, 14
  reduced by growth retardants, 124
  sensitivity to systemic fungicides, 133
Cladosporium fulvum, tomato leaf mould, 243
Coccomyces hiemalis, cherry leaf spot, 44, 282
cohnen
  controls *Pyricularia oryzae*, 233
Colletotrichum
  fragariae, strawberry anthracnose, 286
  gloeosporoides, mango anthracnose, 291
  lindemuthianum, bean anthracnose, control by benomyl, 260
  musae, banana crown rot, 291
compensation, 188
'conventional' fungicides, 177
conversion, metabolic, of systemic fungicides, 117, 160, 186
costs of fungicide treatments, 199
cotton
  boll rots controlled by benomyl, 209
  Rhizoctonia solani control by carboxin, 51
  wilt, control by benomyl, 60
Cronartium
  fusiforme, loblolly pine rust, 301
  ribicola, black currant rust, 284
    white pine blister rust, 301
cross-resistance, 189
crotonanilide, **56**
cucumber
  black root rot, *Phomopsis sclerotioides*, 246
  powdery mildew, *Sphaerotheca fuliginea*, 245
  scab, *Cladosporium cucumerinum*, 124
  stem rot, *Mycosphaerella melonis*, 246
cuticle structure, 93
2-cyano-*n*-(ethylaminocarbamyl)-2-(ethoxyimino)-acetamide, 271
cyclofuramid, 2,5-dimethyl-1,3-(cyclohexylcarbamoyl) furan, 55
cyclamen
  black root rot, *Thielaviopsis basicola*, control by benzimidazoles, 250
  heart rot, *Botrytis cinerea*, benomyl treatment, 179, 194
cycloheximide, β-(2-(3,5-dimethyl-2-oxocyclothexyl)-2-hydroxyethyl)glutarimide, 23, **43**
  inhibition of protein synthesis, 44

cycloheximide *contd.*
mammalian toxicity, 87
mode of action, 136
trials against white pine blister rust, 301
wheat rust control, 19
Cylindrocladium scoparium and *C. floridarum*, 295
cycocel, (2-chloroethyl)trimethylammonium chloride induces tylosis in tomato stems, 122
cypendazole, **61**, 62
cytokinin-like effects of benzimidazole compounds, 116, 255
cytomicin, **46**
Cytospora cincta, peach canker, 44, 282; sycamore canker, 303

2,4-D translocation, 107
damping-off diseases, 60
Dendrophoma sp., strawberry leaf spot, 286
detoxication of quintozene, 186
Deuterophoma tracheiphila, 'mal secco' of citrus, 288
Dexon, sodium-*p*-dimethylaminobenzenediazo sulphonate, 11
dialkylphosphorothioates, 36
Diaporthe medusae, citrus melanose, 288
diazinon, **64**
Dibotryon morbosum, plum black knot, 283
2,2-dichloro-3,3-dimethyl cyclopropane carboxylic acid, **156**
dicloran, resistance of *Botrytis cinerea* to, 182
Didymella
applanata, raspberry spur blight, 285
ligulicola, chrysanthemum ray blight, 249
lycopersici, tomato stem rot, 243
diffusion of liquids
into leaves, 98
into roots, 99
dimethirimol, 5-*n*-butyl-2-dimethylamino-4-hydroxy-6-methyl pyrimidine, **65**, 66
resistance to, 178
selectivity, 151
dinocap, 36, 207
Diplocarpon rosae, rose black spot, 248
Diplodia natalensis, stem-end rot of citrus, 288
disruption of cell structure, 131
distribution of benomyl and dinocap on leaves, 207
dithiouracil, **74**
DNA synthesis inhibited
by chloroneb, 144
by MBC, 137
dodemorph, 4-cyclododecyl-2,6-dimethylmorpholine, **72**, 90
controls rose powdery mildew, 249
dodine, resistance to, 177
dollar spot, 235
Dothiostroma pini, 299
Dow 269 = pyroxychlor *q.v.*
Dutch elm disease, *Ceratocystis ulmi* = *Ceratostomella ulmi*, 11, 119, 306
effect of trichlorophenyl-acetic acid, 120, 311
natural recovery, 120

EBC, ethyl benzimidazole-2-yl-carbamate, 138
ectodesmata, 94
edifenphos (= Hinozan), *O*-ethyl)-*S*-*S*-diphenyl phosphorodithioate, 37, **38**, 149, 170
EL 222, 217
EL 279, (α-2-chlorophenyl)-α-cyclohexyl-5-pyrimidine methanol, 224
elm disease, see Dutch elm disease
Elsinoe
ampelina, grape anthracnose, 289
fawcettii, citrus scab, 288
veneta, raspberry cane spot, 285
Endocronartium harknessii, pine gall rust, 302
Endothia parasitica, chestnut blight, 7, 303
Entomosporium maculatum, pear leaf spot, 279
enzymes, wall-digesting, produced by pathogens, 128
eradicant effects on barley mildew, 217
Erwinia amylovora, 42
Erysiphe cichoracearum, 68, 73, 270
Erysiphe graminis, 74, 76, 213–222
ethazole, 5-ethoxy-3-trichloromethyl-1,2,4-triazole, **70**, **155**
Ethionine
checks powdery mildews and pea root rot, 124
effect on cell walls, 123
ethirimol, 5-*n*-butyl-2-ethylamino-4-hydroxy-6-methyl pyrimidine, **65**, 66
mammalian toxicity, 88
mode of action, 151, 171
resistance of *Erysiphe graminis* to, 180
soil and spray application, 214
translocation, 110
Eutypa armeniacae, apricot dieback, 282
Exobasidium vaccinii, blueberry red leaf, 285
eyespot, *see Cercosporella herpotrichoides*

F 427, 2,3-dihydro-6-methyl-5-(2-diphenyl)carbamoyl-1,4-oxathiin, **51** (carboxin)

fenarimol, 147
ferulic acid detoxifies pyricularin, 130
fireblight, 42
5-fluorodeoxyuridine, anti-sporulant, 125
p-fluorophenylalanine, anti-sporulant, 125
fluotrimazole, **71**
foliar fungicides, application techniques, 206
folic acid antagonises dimethirimol, 151
free space, definition and estimates of, 95–9
French bean, grey mould, *Botrytis cinerea*, of pods, control by benomyl, 270
fuberidazole, 2-(2′-furyl)benzimidazole, **62,** 163
  cross-resistance with MBC, 138
Fulvia fulva, 243
furcarbanil 2,5-dimethyl-3 furanilide, **53**
Fusarium
  culmorum and *graminearum*, control by thiabendazole, 231
  nivale, 231, 238
  oxysporum 247, 251
    f.sp.*cubense*, spore migration in banana xylem, 120
    f.sp.*lycopersici*
      production of polygalacturonase, 122, 127
      tomato response to, 118
      use in bioassays, 12
    f.sp. *perniciosum* incites mimosa wilt, 304
  roseum, potato dry rot control by benomyl, 272
    banana fruit rot, 291
  roseum f.sp. *cerealis*, 238
  tricinctum f.sp. *poae*, 238
Fusicoccum amygdali, 283
future strategy for chemotherapy, 211

β-galactosidase synthesis in *Verticillium albo-atrum*, 127
genetic changes affecting fungal cell resistance, 182
geodin, **41**
ghost-spotting of tomato, 242
glasshouse acreages, 240
Gloeocercospora sorghi, 238
Gloeodes pomigena, apple sooty blotch, 278
Gloeosporium
  fruit rots, control by benomyl, 209, 278
  inconspicua, blueberry leaf spot, 285
  minus, blueberry leaf spot, 285
  musarum, banana anthracnose, 291
Gloeotinia temulenta, 239
Glomerella cingulata, apple necrotic leaf blotch, 278
β-1,3-gluconase, 129
glucosamine in fungal cell wall synthesis, 39
glucose spraying to control 'low-sugar' diseases, 128
Gnomonia
  leptostyla, walnut anthracnose, 298
  platani, sycamore anthracnose, 300
gooseberry, American mildew, control by benomyl, 281
  diseases, 284
granules
  carboxin, 223
  ethirimol, 214
  Kitazin P, 205, 233
grape diseases, 289
grass and turf diseases, 235
grisan-3,4-dione, **40**
grisenone, **40**
griseofulvamine, 42, **41**
griseofulvin, 24, 39, **40,** 108
growth-regulant effects
  on hosts, 119
  on pathogens, 123
Guignardia
  aesculi, leaf blotch of Ohio buckeye, 301
  bidwellii, grape black rot, 289
  citricarpa, citrus black spot, 288
gums obstruct spore migration in xylem, 121
Gymnosporangium juniperi-virginianae, cedar-apple rust, 278

haustorium formation inhibited by 6-azuracil, 145
Helminthosporium
  gramineum, control by carboxin, 231
  maydis, trials with carboxin, 235
  solani, control, 261
  vagans, melting-out disease of turf, control, 128
'high-sugar' diseases, 128
Hinozan, O-ethyl-S,S-diphenyl phosphorodithiolate, 37, **38,** 170, 233
HOE 2873, *see* pyrazophos
hops
  downy mildew, *Pseudoperonospora humuli*, 43
  powdery mildew, *Sphaerotheca humuli*, 36
host tissue modification, 116
hydroxypyrimidines, **63**
hyperauxiny, 121
Hypoxylon pruinatum, 303

imazalil (R23979), **150**
imidazole, 63
inactone, **44**
3-indole-acetic acid induces tylosis, 122
inezin, *S*-benzyl-*O*-ethylphenyl phosphorothioate, 37, **38,** 149, 170, 233

inhibition of
acetate metabolism, 53
biosynthesis, 131
cell multiplication, 63, 137
DNA synthesis, 44, 136
fungal chitin synthesis, 48, 145
haustorium formation, 74, 145
oxidative phosphorylation, 131
polynucleotide synthesis, 79
polyphenol oxidase, 80
protein synthesis, 46, 136
pyrimidine biosynthesis, 75
respiration, 63, 134
RNA synthesis, 53
tricarboxylic acid cycle, 54
injections into woody tissue, 208, 311, 315
insensitivity (*see also* resistance), 292
Isariopsis clavispora, grape leaf spot, 289
iso-prothiolane, 234

Japan, use of systemic fungicides in, 45
kasugamycin, **48,** 137
controls *Pyricularia oryzae*, 233
mammalian toxicity, 87
mode of action, 137
resistance of *Pyricularia oryzae* to, 187
Kasumin, 234
kinetin, 6-furfurylaminopurine, tested against *Sphaerotheca fuliginea*, 20
Kitazin, *O,O*-diethyl-*S*-benzyl phosphorothioate, **37**
application in irrigation water, 205, 234
mode of action, 38
Kitazin P, *O,O*-isopropyl-*S*-benzyl phosphorothioate, **37,** 149, 170

Ladogal, sodium salt of *p*-amino-benzylsulphonxymethylamide-*N*-*d*-glucoside sulphonic acid, 18
leaf disc tests, 20
leaf intake, 98
Leonardo da Vinci, 6
lethal synthesis, 186
lettuce
beet western yellows virus: symptoms suppressed by carbendazim, 271
downy mildew, *Bremia lactucae*, 246
lithium salts
injected against *Endothia parasitica*, 7
used against *Erysiphe graminis*, 7
Lignasan, 319
Lophodermium pinastri, 297
'low-sugar' diseases, 128

MBC (=carbendazim *q.v.*), methyl benzimidazol-2-yl-carbamate, **61**
action of, 137
decomposition product of benomyl, 137, 162
decomposition product of thiophanate-methyl, 138
inhibitor of DNA synthesis, 142
translocation, 111
maize head smut, *Sphacelotheca reiliana*, tests with carboxin, 235
'mal secco' of orange seedlings, 288
maleic hydrazide
increases cucumber scab, 124
reduces *Alternaria longipes* in tobacco, 124
mammalian toxicities of systemic fungicides, 86–91
mango diseases, 291
maple
bleeding canker, *Phytophthora cactorum*, 11
verticillium wilt, 305
Marssonina populi, *M. juglandis*, 300
Mauget trunk-injection needles, 312
mebenil, 2-toluanilide, **53,** 55
mechanisms of fungicidal action, 131
Melampsora medusae, 300
Melanconium fuligineum, grape bitter rot, 289
melting-out disease of turf, *Helminthosporium vagans*, control, 128
metabolic conversion, 160
metiram (=Polyram), 224
L-methionine checks powdery mildew and root rots, 124
miconazole, 150
Microsphaera grossulariae, 284
Microsporium gypseum, 40
Microthyriella rubi, apply fly speck, 278
Milcurb = 10% w/v solution of dimethirimol formulated as hydrochloride
mildew, cereal, *see Erysiphe graminis*
Milgo, Milstem = ethirimol formulations
mimosa wilt control, 304
mitochondria, energy production affected by fungicides, 183
Monilinia fructicola, brown rot of stone fruit, 282, 283
Monilinia vaccinii-corymbsi, blueberry blossom blight, 285
Monochaetia unicornis on Arizona cypress, 303
morphactins, esters of 9-hydroxy-fluoene-9-carboxylic acid, 124
morpholine, 71
mushrooms, 251
benomyl treatment, 252
cobweb disease, *Dactylium dendroides*, 251
dry bubble disease, *Verticillium fungicola*, 251
wet bubble disease, *Mycogone perniciosa*, 251
mutations, effect on resistance, 190

mycomycin, **50**
Mycosphaerella
brassicicola, ringspot of Brussels sprouts, control, 269
fijiensis, banana black leaf streak, 291
melonis, cucumber stem rot, 245
musicola, banana Sigatoka leaf spot, 292
sp., citrus greasy spot, 288

NF 35, thiophanate, *q.v.*
NF 44, thiophanate methyl, *q.v.*
NF 48, 2-(3-methoxycarbonyl-2 thioureido) aniline, **79**
α-naphthalene acetamide induces tylosis in tomato, 122
naphthaleneacetic acid, effect on cell-wall pectins, 123
naramycin B, isocycloheximide, 43
narcissus diseases, 251
nectarine diseases, 282
Nectria galligena, apple canker, anti-sporulants, 279
negative forecasting, 210
Neurospora crassa, antifungal mechanisms, 61, 63
nickel compounds used against wheat rusts, 19
Nigrosphaera sphaerica, banana squirter disease, 291
nonane, 4-phenyl-4-(3-pyridyl)-3-oxatricyclo-, **69**
nucleic acid metabolism affected by fungicides, 183

oak wilt, *Ceratocystis fagacearum*
altered susceptibility to, 120
control, 304
oats
covered smut, *Ustilago hordei* (= *U. kolleri*), 229
loose smut, *Ustilago avenae*, 228
Oidium chrysanthemi, control by griseofulvin, 40
oleandomycin, 49
onion
downy mildew, *Peronospora destructor*, 45
neck rot, *Botrytis allii*, 260
smut, *Urocystis cepulae*, 263
white rot, *Sclerotium cepivorum*, 263
Oospora pustulans, skinspot of potato, control, 261
orange
'mal secco' of seedlings, 288
mould (*Penicillium* spp.) control on fruit, 272
organomercury compounds, 45
organophosphorus compounds, 32–7
mammalian toxicity, 86
oxathiins, **51**
mode of action, 135
resistance to, 180
selectivity, 134, 158
oxazines, substituted, 71
oxycarboxin, **52**
against cereal rusts, 223
against flag smut, 230
mode of action, 134
metabolism, 165
ozone damage affected by benomyl, 116

paddy blast, *Pyricularia oryzae*, 45
parinol, bis-(*p*-chlorophenyl)-3-pyridine methanol, **68**
pathocidin, 63
pea
Ascochyta diseases, 260
Erysiphe polygoni, 74
Fusarium culmorum, 62
peach diseases, 282
pear diseases, 279
pectate lysases, 123
pectolytic enzymes, inactivation of, 126
pelargonium rust, *Puccinia pelargonii-zonalis*, 250
Pellicularia sasakii tolerant to blastocidin, 137
Penicillium
atroveratum, 61
brevicompactum, benomyl-resistant strain, 194
digitatum on banana and citrus, 272, 290
expansum, apple blue mould, 278
griseofulvum, 39
italicum on banana and citrus, 272, 290
nigricans, 40
pentachlorobenzyl alcohol, 36
Pezicula (*Gloeosporium* spp), 278
Phaeocryptopus gaeumannii, 299
phenols and derivatives, 82
phenolic compounds inactivate pectolytic enzymes, 126
phenylalanine induces resistance to apple scab, 118
phenyl mercuric acetate used on rice, 46
4-phenyl-4-(3-pyridyl)-3-oxatricyclo-nonane, **69**
phenyl-thiourea, **80**
phloem
exudates, 103
functions of, 103
microfibrils, 106
phloridzin, phloretin, effects on resistance to apple scab, 118
Phomopsis
juniperova, 299
sclerotioides, cucumber black root rot, 246
viticola, grape dead arm, 289
phosphoric esters of pyrazopyrimidines, 35

phosphorothioates, dialkyl, **37**
phosphorothionates, 37
Phragmidium mucronatum, rose rust, 248
Phycomycetes, effect of systemics on
Physalospora
  laricina, larch leaf disease, 48
  obtusa, apple frog-eye leaf spot, 278
phytoalexins, 117
Phytophthora
  cactorum, maple canker, 11
  cinnamomi, root rot of avocado, 123
  infestans, potato blight, 43, 60, 267
pineapple diseases, 291
piperazine, *N*,*N'*-bis(1-formamido-2,2,2-trichloroethyl)-piperazine = triforine, **75**
Piricularia, *see Pyricularia*
Plasmodiophora brassicae, club root, control with benomyl, 265
Plasmopara viticola, grape downy mildew, 60, 289
plum diseases, 283
Podosphaera
  leucotricha, apple mildew, control with benomyl, 277
  oxycanthae, cherry mildew, 282
poinsettia black root rot, *Thielaviopsis basicola*, control, 250
polyenes, 49
polygalacturonases, 122
  inactivated by chlorogenic and caffeic acid products, 126
  produced by *Fusarium oxysporum* f.sp. *lycopersici*, 122
polyoxins, **47**, 145
  polyoxin D, mode of action, 145
polyphenol oxidase, 118
polyynes, 49
post-harvest treatments, 209, 272
potato tuber diseases control by benomyl, 261
powdery mildews, treatments for, historical survey, 15
procaine hydrochloride used against cucumber powdery mildew, 16
protectant fungicides, mode of action, 157
protein synthesis, inhibition of in
  in *Pyricularia oryzae*, 46, 136
  in *Saccharomyces cerevisiae*, 187
prothiocarb, ethyl-*N*-(3-dimethylaminopropyl) thiocarbamate hydrochloride, **77, 155,** 174, 271
protoplast membrane permeability, fungicide effect on, 183
Pseudomonas
  medicagensis var. *phaseolicola*, bean halo blight, 42
  spp. control by kasugamycin, 49
Pseudoperonospora
  cubensis, cucumber mildew, 40
  humuli, hop downy mildew, 42
  trachciphila, grape red fire, 289
Pseudopeziza ribis, black currant leaf spot, control by benomyl, 281
Puccinia
  arenaria, carnation rust, 248
  brachypodia var. *poae-nemoralis*, 237
  coronata, crown rust, 237
  graminis, black rust, 223, 237
  hordei, brown rust of barley, 221
  horiana, white rust of chrysanthemum, 249
  pelargonii-zonalis, pelargonium rust, 250
  recondita (=*triticina*), brown rust of wheat, 223
  rubigo-vera tritici, wheat leaf rust, 52
  striiformis (=*glumarum*), yellow rust, 222, 237
Pucciniastrum goeppertianium, 299
purines, 73
pyracarbolid, 2-methyl-5,6-dihydro-4H-pyran-3-carboxylic acid anilide, **53,** 87
  controls *Urocystis agropyri*, 230
  controls *Ustilago nuda*, 229
pyran, substituted, 53
pyrazolopyrmidines, phosphoric esters of, **35**
pyrazophos, **35,** 154, 168, 276, 284
Pyrenochaeta lycopersici, tomato brown root rot, 245
Pyricularia oryzae, rice blast, control, 37, 46, 233
pyricularin, toxin produced by *P. oryzae*, 130
pyridines, 68–71
pyrimidines, 63, 73,
pyrocarbolid, **53,** 136, 167
pyroxychlor, 2-chloro-5-methoxy-4-trichloro-methyl pyridine, **70,** 111, **156,** 271
Pythium
  aphanidermatum, 236
  ultimum, 236

8-quinolinol sulphate
  for therapy of vascular wilt pathogens, 22
  for treating root rot of *Matthiola* sp., 10
  used against *Stereum purpureum*, 7
quintozene = pentachloronitro-benzene, detoxication of, 186

raspberry diseases, 285
residues, 91
resistance, fungal (*see also* insensitivity) 176–97
  of *Botrytis cinerea*, 179
  of *Penicillium brevicompactum*, 194
  to
    benomyl, 179, 254, 319
      induced by u.v. radiation, 188

resistance to—*contd.*
blasticidin S, 192
cadmium, 177
chlorinated nitrobenzenes, 178
chloroneb, 178
dimethirimol, 178, 255
dodine, 177
ethirimol, 180
kasugamycin, 180
oxathiins, 180
quintozene, 177
thiophanates, 180
respiration: effect of thiabendazole, 63
reversal of fungitoxicity of hydroxypyrimidines, 67
RH 124, 4-*n*-butyl-1,2,4-triazole, 32
used against rusts, 223
Rhabdocline pseudotsugae, 299
Rhizoctonia solani, 51, 52, 236, 261
on peas, 262
on potatoes, 261
Rhizosphaera kalkhoffii, 299
Rhodia, l-methoxycarbonyl-3-dimethylaminoacetamidophenyl thiourea HCl, 319
Rhyncosporium secalis, 228
riboflavin antagonises dimethirimol, 151
ribosomes, protein synthesis affected by fungicides, 183
rice paddy blast, *Pyricularia oryzae*, 36, 46, 233
control by kasugamycin, 233
rimocidin, **49**
RNA synthesis inhibited by oxathiins, 53
root uptake, 99
rose
black spot, *Diplocarpon rosae*, 248
powdery mildew, *Sphaerotheca pannosa*, 248
rust, *Phragmidium mucronatum*, 248
rufianic acid, 1,4-dioxyanthroquinone sulphonic acid, inactivates pectolytic enzymes, 12, 126
russeting of fruit, 277
rusts, *see Puccinia* and cereal rusts

S 1358, **70, 148**
Saccharomyces
carlsbergensis, 44
cerevisiae, 52
pastorianus, 44
saké, 44
salicylanilide, **51**
scab
apple, *Venturia inaequalis*, 275
pear, *Venturia prinia*, 279
Scirrhia acicola, 299
Sclerotinia
fructigena, apple brown rot, 278
homeocarpa, 40, 235
laxa f. *mali*, apple blossom wilt, 278
sclerotiorum, white mould of bean, control by benomyl, 269
spp., stone-fruit rots, control by benomyl, 280
Sclerotium cepivorum, onion white rot, control by benomyl, 263
Sclerotium oryzae, 234
Screening techniques, 9
Seed dressing
techniques, 203, 260
with ethirimol, 214
with fuberidazole, 62
with triforine, 76
selection pressure, 195
selectivity, mechanisms of, 158, 258
Septoria
apiicola, late blight of celery, control, 268, 271
musiva, poplar leaf spot, 300
nodorum, glume blotch of wheat controlled by benomyl spray, 227
sheep blowfly control, 64
sites of fungicidal action, 135
smut, head, *Sphacelotheca reiliana*, trials with carboxin, 235
smut, flag, *Urocystis agropyri*, controlled by carboxin, 230
smuts, *Ustilago* spp.
control by
benomyl, 229
carboxin, 228
pyracarbolid, 229
Sodium 2,3,6-trichlorophenyl acetate (TCPA), 311
soil application of ethirimol, 214
soil treatment, 204
sorghum kernel smut, *Sphacelotheca sorghi*, 50
soybean, 263
Sphacelotheca reiliana, head smut of maize, trials with carboxin, 235
Sphaerotheca
fuliginea, cucumber powdery mildew 245
control, 246
resistance to benomyl, 178
resistance to dimethirimol, 178
humuli, hop powdery mildew, 36
macularis, strawberry powdery mildew, 286
mors-uvae, American gooseberry mildew, control by benomyl, 281
pannosa,
grape mildew, 289
peach mildew, 283
rose mildew, 40, 248
spore migration in plant tissue, 121
sporulastatic effects, *see* antisporulants
spray application on cereals
ethirimol, 217
other systemic fungicides, 217
Stereum purpureum, silver-leaf of stone fruits, 7

Stigmina carpophila on peach, 283
stomatal closure barring entry of pathogens, 118
stone-fruit diseases, 280
strawberry
  diseases, 286
  wilt, *Verticillium dahliae*, control by benomyl, 287
Streptomyces
  albulus, 44
  albulus var. *pathocidus*, 63
  cacaoi var. *ascensis*, 47
  chibaensis, 49
  griseochromogenis, 45
  griseus, 44
  spp., 43, 49
streptomycin, **42,** 87
streptothricin tested against powdery mildews, 24
streptovitacins, 44
succinic dehydrogenase inhibition, 57
sugar beet
  leaf spot, *Cercospora beticola*, control, 268
  root rot, *Aphanomyces cochlioides*, 124
sugar levels in host tissues affect susceptibility, 128
sulphonamides, therapeutic activity, 18, 22
sulphonanilide, 36
surfactant effects, 208
sycamore anthracnose, 300
symplastic movement, 101

Taphrina deformans, peach leaf curl, 280
TCPA, trichlorophenyl acetate, used against Dutch elm disease, 311
terrazole, *see* ethazole
thiabendazole, 2-(4′-thiazolyl)benzimidazole, controls, **61**
    bunt, 230
    Fusarium graminearum and *F. nivale*, 231
    Septoria nodorum, 225
  mode of action, 143, 163
  translocation, 110
thiabendazole hypophosphite, 318
Thielaviopsis basicola, black root rot of beans, 266
  of cyclamen, 250
thiophanate, 1,2-bis-(3-ethoxycarbonyl-2-thioureido)benzene, **79**
  antifungal spectrum, 138
  mode of action, 140
  resistance to, 180
thiophanate-ethyl, 236
thiophanate-methyl, 1,2-bis-(3-methoxycarbonyl-2-thioureido) benzene, **79,** 90
  conversion to MBC, 139, 165
  resistance to, 180
  spray against cereal mildew, 217
thiosemicarbazides, substituted, 80
thiourea-based compounds, 78
thiram, 259
L-threo-β-phenyl serine reduces cucumber scab, 124
Tilletia caries and *T. contraversa*, wheat bunts, 229
time of attack, effects of on cereals, 219

times of spray applications, 209, 218
tobacco
  brown spot, *Alternaria longipes*, 124
  mosaic symptoms suppressed by carbendazim, 271
tolerance (*see also* resistance), 254
2-toluanilide (mebenil), **53,** 55
tomato
  brown root rot, *Pyrenochaeta lycopersici*, 245
  ghost spot and grey mould, *Botrytis cinerea*, 242
  leaf mould, *Cladosporium fulvum*, (*Fulvia fulva*) 243
  stem rot, *Didymella lycopersici*, 243
  verticillium wilt, *Verticillium* spp., 244
  wilt, *Fusarium oxysporum* f. *lycopersici*, varietal reactions, 118
toxicological data, 85–91
toxin inactivation, 129
toyocamycin, 63
'training' of fungal mycelium, 177
transcuticular movement, 94
translaminar movement, 13, 268
translocation downwards, 71, 111
transpiration stream, 101
Tranzschelia
  discolor, peach rust, 283
  pruni-spinosae, plum rust, 283
triadimefon, **71,** 89, **154**
triamiphos = Wepsyn, 1-bis-(dimethylamido)phosphoryl-3-phenyl-5-aminotriazole-1,2,4, **33,** 34
  mode of action, 152
triarimol, α-(2,4-dichlorophenyl)-α-phenyl-5-pyrimidine methanol, **69,** 89, 147
  used against cereal mildew, 217
triazinobenzimidazoles, **62**
1,2,4-triazoles, substituted, 32, **70**
tricarboxylic acid cycle inhibited by carboxin, 52
2,3,6-trichlorophenoxyacetic acid
  effect on
    bean chocolate spot, 13
    Dutch elm disease, 311
    oak wilt, 120
Trichoseptoria fructigena, 278
trickle irrigation, 253
tricyclozole, 234
tridemorph, *N*-tridecyl-2,6-dimethyl morpholine, **72,** 153, 173
  spray against cereal mildew, 219

triforine = Cela W 524, *N*,*N*'-bis(1-formamido-2,2,2-trichloroethyl)piperazine, **75**, 89, 148, 173
used against
carnation rusts, 248
cereal mildew, 222
cereal rusts, 225
cucumber powdery mildew, 246
rose black spot, 249
triphenyl tin acetate, 78
truban (= ethazole)
tubercidin, 63
tulip fire, *Botrytis tulipae*, 250
turf diseases, 235
tyloses obstructing spore migration, 120, 121

ultraviolet radiation induces resistance to benomyl, 188
Uncinula necator, grape powdery mildew, 289
uptake
by leaves, 98
by roots, 99
of ethirimol, 215
of oxathiins, 134
Urocystis
agropyri, flag smut control, 230 238
cepulae, onion smut, control by benomyl, 263
Uromyces
dianthi, carnation rust, 248
fabae, broad bean rust, 55
phaseoli, bean rust, control, 270
Ustilago
avenae, loose smut of oats, 229
hordei, covered smut of barley and oats, 228
kolleri = *hordei q.v.*
maydis, maize smut, 61, 235
nuda, loose smut of wheat, control with BAS 3191, 229
carboxin, 228
pyrocarbolid, 229
striiformis, 238
tritici = *nuda*, *q.v.*

vapour phase redistribution, 69
vascular pathogens, see also *Fusarium oxysporum*, 11
Venturia
inaequalis, apple scab, 276
effect of phenylalanine, 118
resistance to antimycin A, 178
nashicola, 279
pirina, pear scab, 279
venturicidin, 24
Verticillium
albo-atrum, tomato wilt, 244
control by benomyl synthesis of β-galactosidase, 127
cinerescens, carnation wilt, 247
dahliae, potato wilt, 265
strawberry wilt, 287
sycamore canker, 303
tomato wilt, 244
fungicola, dry bubble of mushrooms, 251
vessel group size, 119
virus symptom expression prevented by carbendazim, 271

Wepsyn, see triamiphos
wheat
bunt, *Tilletia caries*, 229
eyespot, *Cercosporella herpotrichoides*, 232
flag smut, *Urocystis agropyri*, 230
mildew, *Erysiphe graminis*, 221
rust, *Puccinia graminis* and *P. striiformis*, 17, 222
white pine blister rust, 301
wilt diseases, see vascular pathogens
winter barley, control of mildew, 219
wyerone, factor in natural plant resistance, 25

Xanthomonas
juglandis, walnut blight, 42
oryzae, rice bacterial leaf blight, 49

zinc chloride used against *Erwinia amylovora*, 7